MODERN MATHEMATICS
FOR BUSINESS DECISION MAKING

MODERN MATHEMATICS FOR BUSINESS DECISION MAKING

Second Edition

DONALD R. WILLIAMS

North Texas State University

WADSWORTH PUBLISHING COMPANY, INC.

Belmont, California

Design and Production: Greg Hubit Bookworks
Technical Illustration: Ayxa Art

© 1978 by Wadsworth Publishing Company, Inc.

© 1971, 1974 by Wadsworth Publishing Company, Inc., Belmont, California 94002. All rights reserved. No part of this book may be reproduced, stored in a retrieval system, or transcribed, in any form or by any means, electronic, mechanical, photocopying, recording, or otherwise, without the prior written permission of the publisher.

Printed in the United States of America

2 3 4 5 6 7 8 9 10—82 81 80 79

Library of Congress Cataloging in Publication Data

Williams, Donald R.
 Modern mathematics for business decision making.

 Includes index.
 1. Mathematics—1961–   2. Business—Decision making. I. Title.
 QA37.2.W54   1978   510   77-25317
ISBN 0-534-00558-6

PREFACE

This second edition of *Modern Mathematics for Business Decision Making* is designed for a one- or two-semester freshman or sophomore course in mathematics for business students. It is assumed that the reader has a mathematical foundation consisting of two years of high school mathematics or the equivalent.

The text is intended to be very readable from the students' viewpoint. The discussions are both lucid and mathematically correct. Each new concept is introduced through a simple, yet practical, business example and then developed for the general case. The fundamental reasoning underlying the basic concepts of mathematical analysis is presented in addition to their applications. This is accomplished through clear, informal discussions that are followed by practical examples with detailed solutions. Exercises are then given at the end of each section. Each exercise set is both ample and varied in difficulty and type of problem. Numerous application problems are included in the exercises, a large number of which are new to this edition.

The topics included in this text are a direct result of an extensive survey of business and mathematics instructors that was conducted by Wadsworth Publishing Company and me. Some new topics in this second edition include Markov chains and game theory. Many new application-oriented examples and exercises have been added throughout the book. Additionally, each chapter now contains a set of approximately ten review problems, mostly application oriented. The arrangement of topics in each chapter is such that the individual instructor can select the desired material.

Chapter 1 contains a discussion of sets, real numbers, relations, and functions; Chapter 2 covers linear and quadratic functions; and Chapter 3 covers exponential and logarithmic functions. These chapters should be covered in sequence, or Chapter 3 may be omitted entirely (however, it is a prerequisite for some portions of Chapter 11). Chapter 4 covers systems of linear equations and matrix algebra. Chapters 5 and 6 cover linear inequalities, linear programming, the graphic solution, inspection of extreme points, the simplex method, the dual problem, and applications; these topics should be covered in the order presented. However, Chapter 6 may be omitted if a very rudimentary coverage of linear programming is desired.

Chapter 7 contains a discussion of the essentials of probability. Chapter 8 is concerned with special probability distributions for discrete variables, Markov chains, and game theory. For full coverage of game theory, Chapter 6 is a prerequisite. The topics are intended to be covered in sequence or omitted (however, Chapter 7 is prerequisite to Section 13.6). Chapter 9 discusses sequences and series and incorporates the series concepts into the coverage of the mathematics of finance. Chapter 9 may be omitted without loss of continuity. Chapters 10 through 12 discuss differential calculus. They should be covered in sequence (however, Chapter 11 can be omitted by making adjustments in Chapter 12 for those discussions and problems involving logarithmic and exponential functions). Chapters 13 and 14 discuss integral calculus.

The following two tables are intended only as guides for selecting chapters for two separate one-semester courses. Of course, they may be modified to the individual needs of a class and combined in a suitable manner to fashion a two-semester course with the desired emphasis. For example, alternatives D and E could provide a two-semester course with a greater emphasis on finite mathematics. Obviously, one may wish to select alternatives that are not listed in the tables in order to better satisfy the objectives of a particular class.

Finite Mathematics

Alternative	Pace	Chapter Coverage	Emphasis
A	Slow	1–5, 7*, 9*	Minimal coverage
B	Average	1–6, 7*, 9*	Linear programming
C	Average	1–5, 7*–9*	Probability
D	Fast	1–9*	Probability and linear programming

*Chapter 7 or 9 (or both) may be modified by deleting sections to suit the needs of a particular course for adjusting either pace or coverage.

PREFACE

		Calculus*	
Alternative	Pace	Chapter Coverage	Emphasis
E	Slow	10, 11 (Secs. 11.1 and 11.5 only), 12, 13	Differential calculus
F	Average	10, 11 (Secs. 11.1 and 11.5 only), 12–14	Balanced
G	Average	10–13	Differential calculus
H	Fast	10–14	Balanced

*Sections 11.3 and 11.4 require at least a minimal coverage of Chapter 3. The other sections in the last five chapters can be covered without covering Chapter 3 by omitting the examples and exercises that utilize logarithms.

The following table shows which chapters are prerequisites for others.

Chapters (and Sections)	Prerequisite Chapters
1	—
2	1
3	1, 2
4	1, 2
5	1, 2, 4
6	1, 2, 4, 5
7	1, 2
8	1, 2, 7
(Sec. 8.6)	6
9	1, 2
10	1, 2
11	1, 2, 10
(Sec. 11.3, 11.4)	3
12	1, 2, 10
13	1, 2, 10, 12
(Sec. 13.6)	7
14	1, 2, 10, 12, 13

I wish to express my appreciation to the following reviewers of the manuscript for their extremely helpful suggestions: Sabra S. Anderson, University of Minnesota, Duluth; Delvis Fernandez, Chabot College, California; Dean W. Hoover, Alfred University, New York; Melvin Mitchell, Clarion State College, Pennsylvania; and Gordon Shilling, University of Texas, Arlington. I am also greatly indebted to Margaret Coy and Jim Hughes for their assistance in solving the exercises. Finally, I offer my most humble thanks to Sharon Hoerth for her typing assistance. Of course, any possible errors or omissions are my sole responsibility.

Donald R. Williams

CONTENTS

1 SETS, REAL NUMBERS, FUNCTIONS, AND GRAPHS, 1

1.1 Introduction, 1
1.2 Sets, 2
1.3 Properties of Real Numbers, 14
1.4 Functions, 19
1.5 Functions and Graphs, 23

2 LINEAR AND QUADRATIC FUNCTIONS, 33

2.1 Introduction, 33
2.2 Linear Equations, 33
2.3 Quadratic Equations, 42
2.4 Polynomial Functions, 52

3 EXPONENTIAL AND LOGARITHMIC FUNCTIONS, 57

3.1 Introduction, 57
3.2 Exponents, 58
3.3 Exponential Functions, 62
3.4 Logarithms, 68
3.5 Natural Logarithms, 76
3.6 Logarithmic Functions, 79

4 SYSTEMS OF LINEAR EQUATIONS AND MATRIX ALGEBRA, 82

4.1 Systems of Linear Equations, 82
4.2 Methods of Solution, 86
4.3 Summation Notation, 99
4.4 Matrix Algebra, 105
4.5 The Inverse of a Matrix, 117

5 LINEAR INEQUALITIES AND LINEAR PROGRAMMING, 131

5.1 Introduction, 131
5.2 Systems of Linear Inequality, 135
5.3 Linear Programming: Graphic Solution, 140
5.4 Inspection of Extreme Points, 145

6 LINEAR PROGRAMMING: SOLUTIONS AND APPLICATIONS, 152

6.1 The Simplex Algorithm, 152
6.2 The Dual Problem, 169

7 PROBABILITY, 181

7.1 Introduction, 181
7.2 The Principles of Counting, 186
7.3 Elementary Probability Concepts, 194
7.4 Postulates of Probability, 198
7.5 Joint, Marginal, and Conditional Probabilities, 205
7.6 Multiplication Rate, 214
7.7 Addition Rule, 223
7.8 Probability Distributions, 230
7.9 Mathematical Expectation, 237
7.10 Continuous Probability Distributions, 243
7.11 Summary, 246

8 SPECIAL PROBABILITY DISTRIBUTIONS AND APPLICATIONS, 252

8.1 Introduction, 252
8.2 Binomial Distribution, 253
8.3 Poisson Distribution, 266
8.4 Hypergeometric Distribution, 273
8.5 Markov Chains, 280
8.6 Game Theory, 284

9 MATHEMATICS OF FINANCE, 296

9.1 Introduction, 296
9.2 Sequences, 296
9.3 Simple and Compound Interest, 305
9.4 Annuities, 313

10 DIFFERENTIAL CALCULUS: BASIC METHODOLOGY, 322

10.1 Introduction, 322
10.2 Limit of a Function, 323
10.3 Derivative of a Function, 331
10.4 Rules of Differentiation, 342
10.5 Composite Functions, 351
10.6 Higher-Order Derivatives, 358

11 DIFFERENTIAL CALCULUS: ADVANCED METHODOLOGY, 365

11.1 Introduction, 365
11.2 Inverse Functions, 365
11.3 Exponential and Logarithmic Functions, 371
11.4 Logarithmic Differentiation, 380
11.5 Partial Derivatives, 382

12 DIFFERENTIAL CALCULUS: APPLICATIONS, 392

12.1 Introduction, 392
12.2 Maxima and Minima, 392
12.3 Marginal Analysis, 405
12.4 Extrema for Bivariate Functions, 412
12.5 Constrained Optima, 418

13 INTEGRAL CALCULUS: BASIC METHODOLOGY, 428

13.1 Introduction, 428
13.2 Indefinite Integral, 428
13.3 Rules of Integration, 430
13.4 Marginal Analysis, 433
13.5 Definite Integral, 436
13.6 Probability Density Functions, 449

14 INTEGRAL CALCULUS: ADVANCED METHODOLOGY, 459

    14.1 Introduction, 459
    14.2 Change of Variable, 459
    14.3 Integration by Parts, 465
    14.4 Integral Tables, 469

ANSWERS TO SELECTED EXERCISES, 475

APPENDIX, 493

    A.1 Common Logarithms of Numbers, 494
    A.2 Cumulative Binomial Distribution, 496
    A.3 Cumulative Poisson Distribution, 518
    A.4 Binomial Coefficients, 523
    A.5 Single-Payment Compound Amount Factor, 524
    A.6 Single-Payment Present Value Factor, 526
    A.7 Uniform-Payments Compound Amount Factor, 528
    A.8 Uniform-Payments Present Value Factor, 530
    A.9 Natural, or Naperian, Logarithms, 532
    A.10 Exponential Functions, 535

GLOSSARY OF TERMS AND FORMULAS, 540

INDEX, 545

# SETS, REAL NUMBERS, FUNCTIONS, AND GRAPHS

1.1 INTRODUCTION

Until recently the business community's use of mathematics was limited primarily to arithmetic, statistics, and some probability. However, with the development of electronic computers and the growth of the amount of information available to corporations, men and women in business today need a more sophisticated mathematical understanding. They need this in order to communicate effectively, to make comparisons, to express relationships, to understand statistics, and to reach conclusions. Thus the language of mathematics is very important in many areas of business. Consequently a major objective of this book is to promote proficiency in the use of this language in relating concepts of quantity and order.

The increasing emphasis on mathematical analysis has created a demand for people who understand and appreciate the role of mathematics in business. The fundamental reasoning underlying the mathematical concepts is important for those who use them in day-to-day business situations; it also prepares them to pursue more advanced work in management science, operations research, statistics, electronic data processing, econometrics, mathematical programming, production management, and similar fields. A course in business mathematics is not considered modern unless it contains topics such as linear programming, optimization models, matrix theory, decision theory, stochastic processes, and operations research. Of course it isn't possible to cover these topics without an understanding of the basic rules of algebra. So this first chapter is intended to give a quick review of some of the most elementary topics in algebra. Later chapters attempt to help you better understand the theoretical topics by relating them to most fields of business. This type of study will also help you in retaining the concepts and later in applying them to your chosen area of advanced study.

1.2 SETS

One of the most important and most fundamental concepts in modern mathematics and logic is that of *set*. In everyday terms, set is synonymous with *collection* and *aggregate*; in mathematics, we think of a set simply as a collection of objects. The objects that make up a set are called *elements*. The elements may be either real or abstract, and, for our purposes, we put no restrictions on their nature. For example, business problems may involve sets of data, sets of orders, sets of shipments, sets of accounts, sets of employees, or sets of decisions. We may consider the set consisting of three accountants, Smith, Taylor, and Jones, or the set consisting of a paperclip, a staple, and an eraser. A set may be of any size; that is, it may contain any number of elements. On occasion, we may wish to consider a set that contains *no elements*, such as the set of all Fords made by General Motors or the set of all glass windows made of paper. Such a set is called the *null*, or *empty*, set and is denoted by the symbol $\emptyset$. On the other hand, we may wish to consider the set of positive integers that would contain an infinite number of elements.

Sets can be specified simply by listing the individual elements. However, so that the list may be recognized as a set rather than some other type of list, it is usually enclosed by braces. For example, the set of three accountants, Smith, Taylor, and Jones, is denoted by

$$\{\text{Smith, Taylor, Jones}\}$$

and the set containing a paper clip, a staple, and an eraser is denoted by

$$\{\text{paper clip, staple, eraser}\}$$

The null set (which contains nothing) is denoted by

$$\emptyset = \{\ \}$$

Similarly, the set of results possible when a coin is tossed is denoted by

$$\{H, T\}$$

where H represents heads and T represents tails.

When we wish to denote a set without listing the elements, we generally denote it by a capital letter. Also, when we wish to denote a general element of the set A, we denote it by the lowercase letter a. Furthermore, the sets

$$A = \{\text{Smith, Taylor, Jones}\} \quad \text{and} \quad B = \{\text{Jones, Smith, Taylor}\}$$

are equal because each element of A is an element of B and each element of B is an element of A. Note that the order in which the elements appear within the braces does not matter; thus, we write $A = B$. To take advantage of this compact notation, consider the following definition.

Definition 1.1 Two sets A and B are equal and we write $A = B$ if and only if every element of A is an element of B and every element of B is an element of A.

To denote that a particular element a is an element belonging to the set A, we use a formalized version of the Greek letter ϵ (epsilon) and write $a \in A$. This reads "a is an element of A," or "a belongs to A." For the above sets, we can write Smith $\in A$, which reads "Smith is an element of A," or "Smith belongs to A." To denote that an element is not an element of the set A, we use the symbol $\notin$ and write $a \notin A$. This is read "a is not an element of A," or "a does not belong to A." For example, since Williams is not a member of set A, we write Williams $\notin A$, which reads "Williams is not an element of A." Similarly, if Definition 1.1 is not satisfied for two sets A and B, the two sets are said to be *unequal* and we write $A \neq B$. Thus, if $A = \{$Smith, Williams, Jones$\}$ and $B = \{$Smith, Jones$\}$, then $A \neq B$ because Williams is an element of A but not of B. Likewise, if $C = \{x, y, z\}$ and $D = \{r, s, t\}$, $C \neq D$ because there is at least one element in C which is not in D.

Suppose the sets A and B are defined such that $A = \{a, b, c\}$ and $B = \{a, c\}$; then $A \neq B$. However, each element in B is also an element in A. In this case, we say that the set B is *contained in* the set A, or that B is a *subset* of A. A more specific definition and the associated notation are given in the following definition.

Definition 1.2 A set A is a subset of a set B if and only if every element of A is an element of B. We write $A \subset B$, which reads "A is a subset of B," or "A is included in B."

Again, if the definition is not satisfied, we place a vertical slash through the appropriate symbol to indicate the negation of the statement. For example, if $A = \{a, b, c\}$ and $B = \{a, c\}$, we have $B \subset A$, which reads "B is a subset of A." We note, however, that A is not a subset of B; and we write $A \not\subset B$, which reads "A is not a subset of B." When B is a subset of A but A is not a subset of B, we say that B is a *proper subset* of A. Notationally, this is stated: If $B \subset A$ and $A \not\subset B$, then B is a proper subset of A. On the other hand, if $A \subset B$ and $B \subset A$, then Definition 1.1 is satisfied and we may write $A = B$. The null set is a subset of every set and a proper subset of every set except itself. We should note that the inclusion relation is a transitive relation; that is, if $A \subset B$ and $B \subset C$, then $A \subset C$.

In the previous discussion, we considered the possibilities of sets where one is wholly contained in another and where two sets are equal. There are times when it is desirable to use two or more sets to form a new set, and there are various ways in which this can be done. For example, suppose a company has an executive position which they wish to fill and they have narrowed the candidates to eight finalists. In reviewing their experience, it is determined that those having sales experience constitute the set

$$S = \{\text{Thomas, Smith, Jones, Lee}\}$$

those having management experience constitute the set

$$M = \{\text{Smith, Wilson, Henry, Lee}\}$$

and those having production experience constitute the set

$$P = \{\text{Henry, Thomas, Wilson, Johnson, Caldwell}\}$$

Rather than repeatedly recalling that the only ones with whom we are concerned are those who are still under consideration for the position, we shall consider this to be understood. To make it clear that we are concerned with no other elements, we call the set that contains all elements with which we are concerned in a given situation the *universal set*, and we shall denote it by U. Thus, for the above example,

$$U = \{\text{Thomas, Smith, Jones, Lee, Wilson, Henry, Johnson, Caldwell}\}$$

The company may want to know which candidates do not have sales experience, and in such a situation the following operation on sets is useful.

Definition 1.3 The complement of a set A relative to a universe U is the set of all elements of U that are not elements of A. The complement of A is denoted by A'.

Symbolically, Definition 1.3 can be written

$$A' = \{x \mid x \notin A\}$$

which reads "the complement of A is the set of all elements x in the universal set U such that x does not belong to the set A." Using Definition 1.3, we can determine those candidates without sales experience simply by finding S', that is, all candidates in U that are not in S. Thus,

$$S' = \{\text{Wilson, Henry, Johnson, Caldwell}\}$$

represents the set of candidates without sales experience. Similarly, those candidates without management experience are represented by

$$M' = \{\text{Thomas, Jones, Johnson, Caldwell}\}$$

and those without production experience are represented by

$$P' = \{\text{Smith, Jones, Lee}\}$$

The Union of Sets In many situations, it is desirable to combine collectively two or more sets into one set. The following definition provides us with the operation that will accomplish this task.

Definition 1.4 The union of sets A and B, denoted $A \cup B$ and read "A union B," or "A or B," is the set of all elements that belong to A or B or to both A and B.

Symbolically, Definition 1.4 can be written

$$A \cup B = \{x \mid x \in A \text{ or } x \in B\}$$

which reads "A union B is the set of all elements x such that x belongs to A or x belongs to B." The *or* in the definition is understood to be an *inclusive* or; i.e., either one or the other or both statements are true. Even though an element belongs to *both* A and B, it is listed only once in the enumeration of $A \cup B$.

Returning to our example, the company may wish to consider those candidates who have either sales or management experience or both. This set of candidates is represented by the set

$$S \cup M = \{\text{Thomas, Smith, Jones, Lee}\} \cup \{\text{Smith, Wilson, Henry, Lee}\}$$
$$= \{\text{Thomas, Smith, Jones, Lee, Wilson, Henry}\}$$

Similarly, the set of candidates with sales or production experience is given by

$$S \cup P = \{\text{Thomas, Smith, Jones, Lee}\}$$
$$\cup \{\text{Henry, Thomas, Wilson, Johnson, Caldwell}\}$$
$$= \{\text{Thomas, Smith, Jones, Lee, Henry, Wilson, Johnson, Caldwell}\}$$

Likewise, the set of candidates with management or production experience is given by

$$M \cup P = \{\text{Smith, Wilson, Henry, Lee}\}$$
$$\cup \{\text{Henry, Thomas, Wilson, Johnson, Caldwell}\}$$
$$= \{\text{Smith, Wilson, Henry, Lee, Thomas, Johnson, Caldwell}\}$$

If the company wishes to consider the set of candidates with sales, management, or production experience, the union of S, M, and P would accomplish this. However, at first glance, Definition 1.4 appears to consider only the union of two sets. Upon further consideration, however, we note that the union of two sets is itself a set; thus, $S \cup M$ is a set which may be combined with P through use of the union operation. Performing the union operation, we obtain

$$(S \cup M) \cup P = \{\text{Thomas, Smith, Jones, Lee, Wilson, Henry}\}$$
$$\cup \{\text{Henry, Thomas, Wilson, Johsnon, Caldwell}\}$$
$$= \{\text{Thomas, Smith, Jones, Lee, Wilson, Henry, Johnson, Caldwell}\}$$

Further investigation reveals that

$$(S \cup M) \cup P = U$$

which is not surprising, since the universal set U consists only of those candidates who have sales, management, or production experience.

The Intersection of Sets

In many situations it is desirable to form a set representing the common elements of two or more sets. The following definition provides the definition governing this task.

Definition 1.5 The intersection of sets A and B, denoted $A \cap B$ and read "A intersection B," or "A and B," is the set of all elements that belong to both A and B.

Symbolically, Definition 1.5 can be written

$$A \cap B = \{x \mid x \in A \text{ and } x \in B\}$$

which reads "A intersection B is the set of all elements x such that x belongs to A and x belongs to B." Thus in order for an element to belong to the intersection of two sets, it must belong to both sets. When the intersection of two sets is the null set, we say the sets are *mutually exclusive*.

Returning again to our example, the company may wish to consider those candidates who have sales *and* management experience. This set of candidates is represented by

$$S \cap M = \{\text{Thomas, Smith, Jones, Lee}\} \cap \{\text{Smith, Wilson, Henry, Lee}\}$$
$$= \{\text{Smith, Lee}\}$$

Similarly, the set of candidates with sales *and* production experience is given by

$$S \cap P = \{\text{Thomas, Smith, Jones, Lee}\}$$
$$\cap \{\text{Henry, Thomas, Wilson, Johnson, Caldwell}\}$$
$$= \{\text{Thomas}\}$$

Likewise, the set of candidates with management *and* production experience is given by

$$M \cap P = \{\text{Smith, Wilson, Henry, Lee}\}$$
$$\cap \{\text{Henry, Thomas, Wilson, Johnson, Caldwell}\}$$
$$= \{\text{Wilson, Henry}\}$$

Thus from the sets obtained through use of the intersection operation, the company can determine that (1) Smith and Lee are the only candidates with sales *and* management experience, (2) Thomas is the only candidate with sales *and* production experience, and (3) Wilson and Henry are the only candidates with management *and* production experience.

If the company wishes to consider the set of candidates with experience in sales, management, and production, the intersection of S, M, and P would accomplish this. However, we have a situation that is similar to the one for the union of three sets. We can apply the same logic to use the intersection of two sets as a single set by determining their intersection and then taking the intersection of the resulting set with the remaining set. For our example, we have

$$(S \cap M) \cap P = \{\text{Smith, Lee}\} \cap \{\text{Henry, Thomas, Wilson, Johnson, Caldwell}\}$$
$$= \{\ \}$$
$$= \emptyset$$

Since the set of candidates with experience in all three areas is the null set, there are no candidates in the universal set who have experience in all three areas.

Venn Diagrams In the previous examples, we listed the elements of each set and performed the indicated operation by comparing lists. Often it is very useful to represent sets and subsets graphically by means of so-called *Venn diagrams*. These diagrams received their name from the British logician John Venn, who used them in 1876 in a paper entitled "Boole's Logical System." A Venn diagram consists of a rectangular region, which represents the universal set, and circles or parts of circles, which represent the sets and subsets within the universal set.

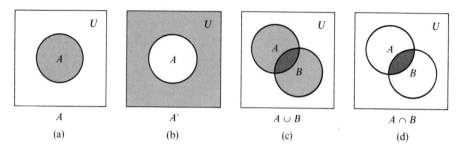

FIGURE 1.1

The shaded regions of Fig. 1.1a to d represent, respectively, the set A, the complement of A, the union of A and B, and the intersection of A and B. The sets A and B are represented by the circles nearest the corresponding labels. When we are considering three sets, we usually represent them by the corresponding circles, as shown in Figs. 1.2 and 1.3. The shaded area in Fig. 1.2 represents $A \cap (B \cap C)$, which contains each element that belongs to all

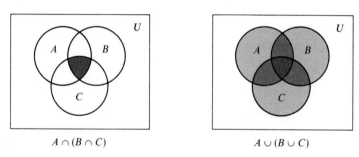

FIGURE 1.2 **FIGURE 1.3**

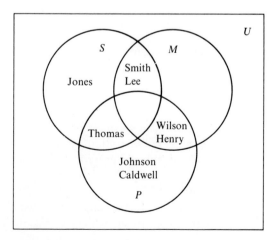

FIGURE 1.4

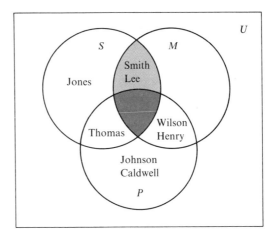

FIGURE 1.5

three sets. The shaded area in Fig. 1.3 represents $A \cup (B \cup C)$, which contains each element that belongs to any one or more of the three sets.

Venn diagrams are particularly helpful in representing the relationships among sets; their major advantage is that they give us a common-sense method for examining such relationships. For example, we can represent the three sets S, M, and P from our executive example graphically as in Fig. 1.4. In order to determine $S \cap M$, we simply shade the area common to the circles representing S and M, as shown in Fig. 1.5, and determine that

$$S \cap M = \{\text{Smith, Lee}\}$$

The other sets can be determined in a similar manner simply by shading the appropriate area and determining what elements are in the shaded area.

On many occasions it is beneficial to partition a group of sets into other sets that have no common elements.

Definition 1.6 Sets A and B are *disjoint sets* if $A \cap B = \emptyset$.

Let us consider an example illustrating the use of disjoint sets.

Example 1.1 Of 100 fast-food restaurants in Denver, 66 serve hamburgers, 25 serve pizza, 40 serve chicken, 15 serve *only* hamburger and chicken, 10 serve *only* hamburgers and pizza, 3 serve *only* pizza and chicken, and only 4 restaurants serve all three foods. Draw a Venn diagram to determine how many restaurants serve

(a) only hamburgers
(b) only pizza
(c) only chicken
(d) at least hamburgers and chicken

(e) at least pizza and chicken
(f) at least hamburgers and pizza
(g) none of the three foods

Solution Let H, P, and C represent the sets of restaurants serving hamburgers, pizza, and chicken, respectively. Since only 4 restaurants serve all these foods, we know that there are 4 restaurants belonging to the set $H \cap P \cap C$. Also, since 15 restaurants serve *only* hamburger and chicken, we know that the set $H \cap P' \cap C$ contains 15 restaurants. Likewise, $H \cap P \cap C'$ contains 10 restaurants and $H' \cap P \cap C$ contains 3 restaurants. Figure 1.6 shows the three sets and the number of restaurants in the sets that we have determined so far.

To obtain the number of restaurants in the remaining sets, we subtract the combined number found above from the total number given. That is, to get the number of restaurants in the set $H \cap P' \cap C'$, we note that H is made up of 4 disjoint sets $H \cap P' \cap C'$, $H \cap P \cap C'$, $H \cap P' \cap C$, and $H \cap P \cap C$. Since 66 restaurants serve hamburgers, we simply subtract the combined number of restaurants in sets $H \cap P \cap C'$, $H \cap P' \cap C$, and $H \cap P \cap C$ from 66 to get the number of restaurants in $H \cap P' \cap C'$. Thus we have $66-(10+15+4)=37$ restaurants in $H \cap P' \cap C'$. Similarly, we have $25-(10+3+4)=8$ restaurants in $H' \cap P \cap C'$ and $40-(15+3+4)=18$ restaurants in $H' \cap P' \cap C$. Finally we total all restaurants in the three sets of the Venn diagram in Fig. 1.7 and subtract from 100 to find the number of restaurants in the set $H' \cap P' \cap C'$ that represents those restaurants that do *not* serve any of the three foods. Thus we obtain $100-(37+10+8+4+3+15+18)=5$.

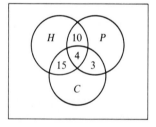

FIGURE 1.6   **FIGURE 1.7**

We can answer each part of this example by considering the appropriate disjoint sets as shown in Fig. 1.7.

(a) The restaurants serving only hamburgers are those in the set $H \cap P' \cap C'$, which contains 37 restaurants.
(b) The restaurants serving only pizza are those in the set $H' \cap P \cap C'$. Thus from Fig. 1.7 we see that there are 8 such restaurants.
(c) The restaurants selling only chicken are those in the sets $H' \cap P' \cap C$. Thus there are 18 restaurants serving only chicken.
(d) The restaurants that serve at least hamburger and chicken are those that serve both. Since this includes all restaurants serving both, whether or not they also serve pizza, the set representing these restaurants can be found by forming the union of the two disjoint

sets $H \cap P \cap C$ and $H \cap P' \cap C$; that is, $(H \cap P \cap C) \cup (H \cap P' \cap C) = H \cap C$. Since the two sets are disjoint sets, we can obtain the number of restaurants serving at least hamburgers and chicken by adding the number in the two disjoint sets. Thus the number of restaurants in $H \cap C$ is $4 + 15 = 19$ restaurants.

(e) Using logic similar to that in (d), we can obtain the number of restaurants serving at least pizza and chicken by forming the union of the disjoint sets $H \cap P \cap C$ and $H' \cap P \cap C$. Thus $(H \cap P \cap C) \cup (H' \cap P \cap C) = P \cap C$. Adding the number of restaurants in the disjoint sets, we have $4 + 3 = 7$ restaurants serving at least pizza and chicken.

(f) To determine the number of restaurants serving at least hamburger and pizza, we need to obtain the number in $H \cap P$. We get $H \cap P$ by obtaining the union of the disjoint sets $H \cap P \cap C$ and $H \cap P \cap C'$, which contains $4 + 10 = 14$ restaurants serving at least hamburgers and pizza.

(g) The set $H' \cap P' \cap C'$ contains only those restaurants that do not serve hamburgers, pizza, or chicken. From Table 1.1 we see that the number of restaurants in this set is 5.

From Table 1.1 and Fig. 1.7 we see the benefit of partitioning the sets H, P, and C into disjoint subsets. This particular practice is beneficial for solving many probability problems.

Table 1.1

Disjoint Sets	Number of Restaurants
$H \cap P' \cap C'$	37
$H \cap P \cap C'$	10
$H \cap P' \cap C$	15
$H \cap P \cap C$	4
$H' \cap P \cap C'$	8
$H' \cap P \cap C$	3
$H' \cap P' \cap C$	18
$H' \cap P' \cap C'$	5
	100

EXERCISES

1. Five brokerage firms in the Midwest are licensed to sell securities within certain states. The following sets represent the states in which the firms are licensed:

$A = \{\text{Illinois, Indiana, Missouri}\}$

$B = \{\text{Illinois, Iowa, Kansas, Nebraska}\}$

$C = \{\text{Illinois, Indiana, Missouri, Ohio}\}$

$D = \{\text{Iowa, Kansas, Missouri, Nebraska}\}$

$E = \{\text{Iowa, Kansas, Nebraska, Ohio}\}$

Determine whether the following statements are true or false:

(a) Indiana $\in C$
(b) Illinois $\notin D$
(c) Missouri $\in B$
(d) Nebraska $\in C$
(e) $A \subset C$
(f) $E \subset D$
(g) $C \not\subset A$
(h) $D \not\subset E$
(i) $A' = E$
(j) $C' = A$
(k) $A \cup C = C$
(l) $A \cup E = U$
(m) $A \cup D = U$
(n) $B \cup D = \{$Iowa, Kansas, Nebraska$\}$
(o) $A \cap B = \{$Illinois$\}$
(p) $A \cap C = A$
(q) $A \cap U = E'$
(r) $A \cap E' = \emptyset$
(s) $A \cap E = \emptyset$
(t) $B \cap D = B \cap E$

2. The following sets represent the products manufactured by four companies:

$A = \{$automobiles, radios, medicine, movies$\}$

$B = \{$copiers, radios, movies$\}$

$C = \{$detergents, medicine, movies$\}$

$D = \{$automobiles, copiers, movies$\}$

List the products contained in each of the following sets:

(a) A'
(b) B'
(c) C'
(d) $A \cup D$
(e) $A \cup C$
(f) $B \cap C$
(g) $A \cap D$
(h) $C \cap D$
(i) $A \cap B$
(j) $(A \cap B) \cap C$
(k) $A' \cap B$
(l) $A \cap U$

3. For two sets A and B, draw Venn diagrams for each of the following sets:

(a) $A \cup B$
(b) $A \cap B$
(c) A'
(d) B'
(e) $A \cup B'$
(f) $A' \cup B$
(g) $A' \cup B'$
(h) $A' \cap B$
(i) $A \cap B'$
(j) $A' \cap B'$
(k) $(A \cap B)'$
(l) $(A \cup B)'$

4. For the three sets A, B and C, draw Venn diagrams for each of the following sets:

(a) $(A \cap B) \cup C$
(b) $(A \cap B) \cup C'$
(c) $(A \cap B') \cap C$
(d) $(A \cap B') \cap C'$
(e) $(A \cap B)' \cap C$
(f) $(A \cap B)' \cap C'$
(g) $(A \cup B) \cap C$
(h) $(A \cup B) \cap C'$
(i) $(A \cup B') \cap C$
(j) $(A' \cup B') \cap C$

5. The following sets represent the supplies purchased from three companies:

$A = \{$typing paper, pens, pencils, paper clips$\}$
$B = \{$carbon paper, pens, letterhead, note pads$\}$
$C = \{$envelopes, ink pads, pencils, paper clips, pens$\}$

List the supplies contained in each of the following sets:

(a) C'
(b) $B \cap C$
(c) $A \cap B$
(d) $A \cup B$
(e) $A' \cap B$
(f) $A \cap C'$
(g) $(A \cup B)' \cap C$
(h) $(A \cup B) \cap C$
(i) $(A \cap B) \cup C$

Sec. 1.2 / SETS

6. Of 30 employees in a production department, 24 can work on the assembly line, 13 can package the product, 9 can inspect the product, 3 can perform all three jobs, 6 can work on the assembly line and package only, 3 can work on the assembly line and inspect only, and 1 can package and inspect only. Draw a Venn diagram to answer the following questions:

(a) How many employees can work only on the assembly line?
(b) How many can only package the product?
(c) How many can only inspect the product?

7. For three sets A, B, and C, draw Venn diagrams to represent the following:

(a) $A \subset B$ (b) $A \cap B = \{ \ \}$ (c) $A = B$
(d) $A \subset B \subset C$ (e) $A \cap B = C$

8. An automobile manufacturer makes 30 different types of major parts for the bodies of 3 automobiles: Chaffy, Gord, and Rumbler. Of these 30 types of parts, 4 will fit only on a Chaffy, 5 will fit only on a Gord, 2 will fit only on a Rumbler, 6 will fit on a Chaffy and on a Gord but not on a Rumbler, one will fit on a Chaffy and on a Rumbler but not on a Gord, 3 will fit on a Gord and on a Rumbler but not on a Chaffy, and 9 will fit on all three automobiles. Draw a Venn diagram for the sets of parts for the respective automobiles and determine how many different types of parts are needed for each of the following combinations:

(a) A Gord or a Chaffy
(b) A Gord or a Rumbler
(c) A Chaffy or a Rumbler
(d) A Gord and a Chaffy
(e) A Gord and a Rumbler
(f) A Chaffy and a Rumbler

9. Of the 500 managers of a large corporation, 353 received a raise in salary, 224 received a bonus, and 100 received both. How many received neither a raise nor a bonus?

10. In a group of 175 employees, 130 signed up for hospitalization insurance, 80 signed up for disability insurance, and 25 did not sign up for either insurance.

(a) How many employees signed up for both insurances?
(b) How many employees signed up for hospitalization insurance only?
(c) How many employees signed up for disability insurance only?

11. In a recent survey of 50 families, the numbers of subscribers to various sports magazines were found to be: *Sports Illustrated* (SI), 25; *Golf Digest* (GD), 18; and *Field and Stream* (FS), 5. Of those who subscribed to SI, nine also subscribed to GD but not to FS, two subscribed to FS but not to GD, and one subscribed to FS and GD. Of the remaining families who subscribed to GD, only one subscribed to FS but not to SI. How many families subscribed to

(a) *Sports Illustrated* only
(b) *Field and Stream* only
(c) *Golf Digest* only
(d) None of the magazines
(e) *Golf Digest* or *Sports Illustrated*
(f) *Golf Digest* or *Field and Stream*
(g) *Sports Illustrated* or *Field and Stream*
(h) *Sports Illustrated* or *Golf Digest* or *Field and Stream*

1.3 PROPERTIES OF REAL NUMBERS

Having considered some of the properties of sets, we will now consider some properties of real numbers. Our primary intent is to provide an introduction to or a review of some basic concepts of algebra of real numbers. Particularly we will consider the associative, distributive, and commutative laws and some basic rules associated with real numbers.

Addition and Multiplication

Suppose Mary Rice, a salesperson at Katz Department Store, works in the luggage department, and she sells 4 overnight cases on Friday and 7 cases on Saturday. If x represents the selling price of an overnight case and if the total dollar sales are $330, then we may write

$$4x + 7x = \$330 \quad \text{or} \quad 7x + 4x = \$330$$

In short, it doesn't matter whether we add Saturday's sales to Friday's sales or vice versa. We know from experience that it doesn't matter which way we add to get the total. However, behind each operation there is a law or postulate that assures us of the validity of certain operations. Keep in mind that we are dealing only with real numbers.

Postulate 1.1 $a + b = b + a$.   (Commutative Law of Addition)

Similarly, for multiplication we know that it doesn't matter whether we multiply $4 \cdot x$ or $x \cdot 4$. Using equivalent notation $a \cdot b = ab = a(b)$, we mean a times b. Thus an equivalent rule for multiplication is given by the following postulate.

Postulate 1.2 $ab = ba$.   (Commutative Law of Multiplication)

Both of these postulates can be extended to the sum or product, respectively, of three or more terms. For example, if Mary also sold three cases on Monday and her total sales were $360, then we would have

$$4x + 7x + 3x = \$360$$

Furthermore, it is immaterial whether we add Friday's and Saturday's sales before adding Monday's sales or whether we add Friday's sales to the total of Saturday's and Monday's sales. That is, we get the same results:

$$(4x + 7x) + 3x = 4x + (7x + 3x)$$

Thus the postulate for addition is as follows.

Postulate 1.3 $(a + b) + c = a + (b + c)$.   (Associative Law of Addition)

Sec. 1.3 / PROPERTIES OF REAL NUMBERS

Likewise, if for 2 days she sells 4 cases each day and her total sales for cases during that time amount to $240, it doesn't matter whether we figure the total sales as $2(4x) = \$240$ or as $(2 \cdot 4)x = \$240$.

Postulate 1.4 $a(bc) = (ab)c$. (Associative Law of Multiplication)

If Mary sells 3 cases each on the first 2 days of the week and 4 cases each on the next 2 days, she can determine the total number of cases sold by computing it either as $(2 \cdot 3) + (2 \cdot 4) = 14$ or as $2(3+4) = 14$. This is where the concept of "factoring out" a common multiplier is obtained. The expression $2(3+4)$ yields the equivalent expression $2 \cdot 3 + 2 \cdot 4$ when we *distribute* the multiplier 2 over the individual terms in the sum. This is known as the *distribution of multiplication over addition*.

Postulate 1.5 $a(b+c) = ab + ac$. (Distributive Law of Multiplication over Addition)

If we try to distribute addition over multiplication, we immediately see that it does *not* work. For example,

$$2 + (3 \cdot 4) \neq (2+3)(2+4)$$

Hence there is no law for distributing addition over multiplication.

If Mary sells 4 cases on Tuesday and none on Wednesday, the total number of cases sold is 4. That is, using the commutative law of addition in Postulate 1.1, we have $4 + 0 = 0 + 4$, which equals 4 cases. This is covered by the following law.

Postulate 1.6 $a + 0 = 0 + a = a$. (Identity Law of Addition)

Moreover, if Mary sells 4 cases on Tuesday and on Wednesday 4 cases are returned, the number of cases actually sold for the 2 days is 0. That is, $4 + (-4) = -4 + 4 = 0$. The law governing this situation is known as the *inverse law of addition*.

Postulate 1.7 $a + (-a) = -a + a = 0$, where $-a$ is called the *additive inverse of a*. (Inverse Law of Addition)

Two similar laws for multiplication also apply for real numbers.

Postulate 1.8 $a \cdot 1 = 1 \cdot a = a$. (Identity Law of Multiplication)

Postulate 1.9 $a \cdot a^{-1} = a^{-1} \cdot a = 1$, where $a \neq 0$ and a^{-1} is called the *multiplicative inverse of a*. (Inverse Law of Multiplication)

Subtraction and Division

We have just considered laws of addition and multiplication. The laws of subtraction and division are simply the laws of addition and multiplication applied to the respective inverses. That is, when we subtract 2 from 5, we write $5 - 2$ or $5 + (-2)$; thus we are simply adding 5 and the additive inverse of 2. Likewise, when we divide 6 by 3, we write $6 \div 3$ or $6 \cdot 3^{-1}$; thus we have the following two definitions concerning subtraction and multiplication, respectively.

Definition 1.7 $a - b = a + (-b)$.

Definition 1.8 $a \div b = \dfrac{a}{b} = ab^{-1}$, where $b \neq 0$.

We should note that for the special case where $a = 1$ we have

$$\frac{a}{b} = \frac{1}{b} = b^{-1} \qquad \text{where } b \neq 0$$

Thus for real numbers, the inverse of a number is simply its reciprocal (except for 0); that is

$$2^{-1} = \frac{1}{2}, \quad 4^{-1} = \frac{1}{4}, \quad 8 \cdot 2^{-1} = \frac{8}{2}, \quad \frac{9}{3^{-1}} = 9(3), \quad \left(\frac{2}{3}\right)^{-1} = \frac{3}{2}$$

The application of Definitions 1.7 and 1.8 is used in solving algebraic expressions. However, the following two rules make their application much simpler.

Rule 1.1 If two expressions are equal and we add the same number to each expression, the resulting two expressions are equal.

For example, if $2 + 3 = 5$ and we add 6 to the expression on each side of the equality, we have $2 + 3 + 6 = 5 + 6$. More generally, we may express the rule as: if $a = b$, then $a + c = b + c$.

Rule 1.2 If two expressions are equal and we multiply each expression by the same number, the resulting expressions are equal.

Sec. 1.3 / PROPERTIES OF REAL NUMBERS

For example, if $2+3=5$ and we multiply each expression by 6, we have $(2+3)6 = 5 \cdot 6$. More generally, we may express the rule as: if $a = b$, then $ac = bc$. Rules 1.1 and 1.2 can be extended to cover subtraction and division, respectively, since they are defined in terms of their respective additive and multiplicative inverses. However, when we are considering division, we must be certain that the divisor is not zero since the multiplicative inverse of zero is not defined. Therefore, Rules 1.1 and 1.2, in conjunction with their extensions to subtraction and division, may be stated simply that *when two expressions are equal we may add, subtract, multiply, or divide both expressions by the same number and the resulting expressions will be equal, with division by zero excepted*. Now that we have the basic rules and postulates, let us consider an example in which we apply them.

Example 1.2 Solve the following equations.

(a) $4x = 16$
(b) $5y = 20$
(c) $-2x = 6(-5)$
(d) $2(x+5) = 18$
(e) $2x + 1 + x = 4x + 5 - 2x$
(f) $3(x-2) - 2(x+4) = 6$

Solution

(a) Applying Rule 1.2 to $4x = 16$, we divide both expressions by 4 (i.e., multiplying by 4^{-1}) and obtain

$$\frac{4x}{4} = \frac{16}{4}$$

$$x = 4$$

(b) Dividing both expressions of $5y = 20$ by 5, we obtain

$$\frac{5y}{5} = \frac{20}{5}$$

$$y = 4$$

(c) Dividing both expressions of $-2x = 6(-5)$ by -2, we get

$$\frac{-2x}{-2} = \frac{6(-5)}{-2}$$

$$x = \frac{6(5)(-1)}{2(-1)}$$

$$x = 15$$

(d) Dividing both expressions of $2(x+5) = 18$ by 2, we get

$$\frac{2(x+5)}{2} = \frac{18}{2}$$

$$x + 5 = 9$$

Subtracting 5 from both expressions, we obtain

$$x + 5 - 5 = 9 - 5$$

$$x = 4$$

(e) For the equation $2x+1+x=4x+5-2x$, we first gather like terms to obtain

$$3x+1=2x+5$$
$$x+1=5$$

Next we subtract 1 from both expressions to get

$$x+1-1=5-1$$
$$x=4$$

(f) We first apply the distributive law of multiplication over addition for the two products on the left side of the equation and then collect like terms to obtain

$$3(x-2)-2(x+4)=6$$
$$3x-6-2x-8=6$$
$$x-14=6$$

Next we add 14 to both expressions to get

$$x-14+14=6+14$$
$$x=20$$

EXERCISES

1. Solve and check your answer.

   (a) $2x-4=x+3$
   (b) $3x-\frac{1}{2}=\frac{x}{3}$
   (c) $3x-4=2x+7$
   (d) $\frac{x+4}{2}=x-\frac{1}{2}$
   (e) $8x+4=x-11$
   (f) $\frac{2t+4}{5}=\frac{5-4t}{3}$
   (g) $2x=4(x+3)$
   (h) $8x=4x+24$
   (i) $2x-2(4-x)=3x-2$
   (j) $2(x+5)=4+3[x-(2+x)]$

2. A fisherman bought three flies at one store and five flies at another store. If the total cost to him was $4.00, determine the cost per fly.

3. Susan sold three books to one friend and two books to another friend. The total amount she received was $21. How much did she receive per book?

4. Tom sold 20 newspapers for $3.60. How much did he receive per newspaper?

5. Betty Adler, an advertising representative, drove her car on company business for which she received 15¢ per mile plus $25 for gas. She was given $75 to cover her traveling expenses, which included $5 for parking. How many miles did Betty drive?

6. Bill Johnson bought a new tire for his car. In addition to the retail price of the tire, he had to pay 5 percent sales tax and $2.25 federal excise tax. The total price Bill paid was $54.75. What was the retail price of the tire?

Sec. 1.4 / FUNCTIONS

7. Robin Hays bought a set of four steel-belted radial tires for her car. For each tire she had to pay the retail price of the tire plus a 5 percent sales tax and $2.60 federal excise tax. The total cost of the set of tires was $212. What was the retail price per tire?

8. Jennifer Anderson is planning a skiing trip to Colorado. She plans to spend several days skiing. She has estimated that her lodging will require one night less than the number of days she is skiing and that her share of the lodging will cost $15 per night. It will also cost her $10 per day for lift tickets and $12 per day for food. Also, she plans on buying food for one day more than the time actually spent skiing. The group that she is going with has chartered a bus that will cost her $20 for round-trip fare. Jennifer has saved $227 to cover the expense of the trip, including $25 allowance for miscellaneous items she may want to buy. How many days does Jennifer plan to spend skiing? How much does she plan to pay for her total lodging bill? Lift tickets? Food?

1.4 FUNCTIONS

Many everyday situations utilize the concept of *functions*. A common one is that of an address within a city. For example, if we were trying to locate 815 Elm Street, we would first locate Elm Street and then locate the street number of 815. This address constitutes an *ordered pair* in that we may consider the first member of the pair to be the street name and the second member to be the street number; thus (Elm, 815) is an ordered pair. The following is a more formal definition of an ordered pair.

> **Definition 1.9** An ordered pair is an element represented by (x,y), where x is called the first element and y is called the second element. The ordered pairs (x,y) and (s,t) are equal if and only if $x=s$ and $y=t$.

We see than an ordered pair is different from a set because the order of the elements in a set is immaterial. However, occasionally ordered pairs are constructed by combining the elements from two sets, and a certain operation permits us to combine the elements of two sets into a set called the *cartesian product*.

> **Definition 1.10** The cartesian product of the two sets X and Y, denoted by $X \times Y$, is the set of all ordered pairs (x,y) such that the first element is from the set X and the second element is from the set Y.

Definition 1.10 is represented symbolically by $X \times Y = \{(x,y) \mid x \in X$ and $y \in Y\}$, which is read "the cartesian product of X and Y is the set of all ordered pairs (x,y) such that $x \in X$ and $y \in Y$."

> **Example 1.3** Given in set notation, the number of employees who may be chosen from departments A and B to serve on a joint planning committee,

$X = \{1,2,3\}$ and $Y = \{4,5\}$, respectively determine both the cartesian product of the number of employees chosen from departments A and B and the cartesian product of the number chosen from departments B and A.

Solution The cartesian product of the number of employees chosen from departments A and B is

$$X \times Y = \{(1,4),(1,5),(2,4),(2,5),(3,4),(3,5)\}$$

The cartesian product of the number of employees chosen from departments B and A is

$$Y \times X = \{(4,1),(4,2),(4,3),(5,1),(5,2),(5,3)\}$$

Symbolically we can write

$$X \times Y = \{(x,y) \mid x \in \{1,2,3\} \text{ and } y \in \{4,5\}\}$$
$$Y \times X = \{(y,x) \mid y \in \{4,5\} \text{ and } x \in \{1,2,3\}\}$$

Furthermore, we note that the commutative law does not hold; i.e., $X \times Y \neq Y \times X$.

The basic concept of a *function* is based on the relationship, or correspondence, between the elements of two sets.

Definition 1.11 A function is a rule that assigns each element of a set D, called the domain, to one and only one element of a set R, called the range.

From Definition 1.11, we see that for each element in the *domain* there must correspond exactly one element in the *range*. In order to keep this assignment clear, we generally represent the relationship by use of ordered pairs. By convention, the first element of the ordered pair belongs to the domain and the second element is that one element from the range that is assigned by the first element.

Since a function is a *rule*, it may be expressed in nearly any form as long as the association between the domain and range elements is clearly defined. For example, let the domain consist of the officers of a particular organization such that

$$D = \{\text{president, vice-president, secretary, treasurer}\}$$

and suppose the ages of the officers are thirty-six, twenty-five, forty, and fifty-one years, respectively. Suppose the rule of interest is the age of the officer. Thus, the range is

$$R = \{25, 36, 40, 51\}$$

and verify the rule given by $f(x)$ is a function:

6. $f(x) = \sqrt{x}$ for $D = \{16, 25, 36\}$

7. $f(x) = -\sqrt{x}$ for $D = \{16, 25, 36\}$

8. $f(x) = x^3$ for $D = \{1, 2, 3\}$

9. $f(x) = 2x + 6$ for $D = \{1, 2, 3\}$

10. $f(x) = 2x + 6$ for $D = \{-1, -2, -3\}$

11. $f(x) = 1/x$ for $D = \{1, 2, 3, 4\}$

12. The positions of president, vice-president of marketing, and vice-president of production are to be filled by Clark, Lewis, and Smith. List the ordered pairs where the domain is the position and the range is the employee.

13. Given x is the cost of production per unit and y is the sales price per unit, determine the ordered pairs of (x, y) for

$$y = 1.5x + 3 \qquad \text{for } x \in \{10, 16, 23\}$$

14. A company makes three products that have production costs $y = 7.5x + 3$, where x is the direct labor hours used. Determine the ordered pairs of (x, y) for each product where $x \in \{2, 4, 6\}$.

1.5 FUNCTIONS AND GRAPHS

Suppose a salesman rents a car for which he must pay $15 per day and 12¢ per mile. The daily charge is a fixed cost and will not change regardless of whether the car is driven extensively or not at all; the total mileage charge is directly related to the number of miles driven. To obtain an equation that will yield the total cost for a given day, we must add the total variable cost and the fixed cost. The total variable cost can be obtained by multiplying the mileage cost of 12¢ by the number of miles driven, which varies and is represented by the variable x. Thus, the total cost in dollars, represented by y, for a day in which x miles are driven is given by

$$y = 0.12x + 15$$

In order to obtain the total cost, we determine the number of miles driven and substitute it into the above equation. There are many such practical uses of equations.

We previously noted that an ordered pair that related the element in the domain with its assigned element in the range was denoted by $(x, f(x))$, or by (x, y), where $y = f(x)$. Since $f(x)$ is itself a variable, we generally choose to represent it by a single symbol y. When we make such a representation, we refer to y as a *dependent variable* because its value depends on the value of x; the variable x is called an *independent variable* because its value does not depend on the value of any other variable.

For a function f, which has the rule

$$f(x) = x^2 + 3x + 4$$

the $f(x)$ portion is used to denote that the expression on the right is the rule for that particular function. In a strict sense, the function could be denoted by the expression

$$x^2 + 3x + 4$$

In most cases, however, we are interested in *equating* the value of the function with its associated rule. When we do this, we have what is generally called an *equation*, and the preceding equation is usually written to include the variable y rather than $f(x)$. Thus the equation with which we are concerned is

$$y = x^2 + 3x + 4$$

There are many different types of functions, and, of course, each has a particular name. A type of function of particular interest is one that has a rule given by the form

$$mx + b$$

where m and b are constants such that $m \neq 0$. Such a function is called a *linear function* of x and is usually written in equation form as

$$y = mx + b$$

This equation is referred to as a *linear equation in functional form.*

Often we want to represent the relationship between two variables geometrically. To do this, we construct a graph using two perpendicular lines called the *coordinate axes*. The point of intersection of the two axes is called the *origin*, or zero point. Each line is scaled as shown in Fig. 1.8. It is customary to label the horizontal line as the x axis and the vertical line as the y axis. Note in the figure that the y axis is scaled so that positive values lie above the origin and negative values are below it. On the x axis, the positive values are to the right and negative values to the left of the origin. These represent all possible values of the two variables whose relationship we wish to graph. However, the units of length need not be the same for both axes. The four regions formed by the axes are known as *quadrants*, numbered as indicated by the Roman numerals in Fig. 1.8. There we see that if x is *positive*, we go to the *right* on the x axis; and if it is *negative*, we go to the *left*. Similarly, if y is *positive*, we go *up* the y axis; and if it is *negative*, we go *down*.

From Fig. 1.8, we see that there are four directions in which we can go: up, down, left, or right. The ordered pairs that were introduced previously are used to indicate these movements. The ordered pair (x,y) tells us how many units to move and in which direction for x and y. The values assigned to x and y are called the x *coordinate* and y *coordinate*, respectively. For example,

Sec. 1.5 / FUNCTIONS AND GRAPHS

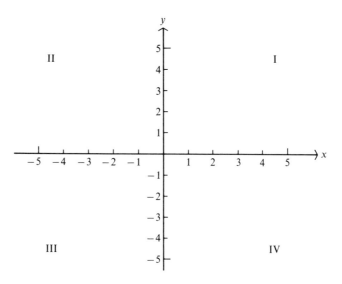

FIGURE 1.8

the ordered pair (2,3) tells us to move two units to the right of the origin on the x axis and three units above the origin on the y axis. In order that our movements will be coordinated (remember, they are called *coordinate* axes), we first move two units to the right and then move up three units parallel to the y axis. The ordered pair (2,3) represents the *point* at which our moves terminate; for the point (2,3), we say that its *abscissa*, or x *coordinate*, is 2 and its *ordinate*, or y *coordinate*, is 3. We note that the ordered pairs that could be represented in Fig. 1.8 are ordered pairs of real numbers. The coordinate system represented in Fig. 1.8 is the cartesian product of the set of real numbers with itself. Thus, the coordinates are called *cartesian coordinates*, or *rectangular coordinates*.

Figure 1.9 presents four different points with one in each quadrant. The point (2,3) is in quadrant I because both coordinates are positive. The point (−4,2) is in quadrant II because the first coordinate is negative, which means go left parallel to the x axis, and the second coordinate is positive, which means go up parallel to the y axis. The point (−3, −5) is in quadrant III because the coordinates indicate that we should move to the left and down. The point (4, −3) is in quadrant IV because the coordinates indicate that we should move to the right and down. The signs associated with the respective coordinates are given in each quadrant. Thus any point in quadrant I will be positive for both coordinates, i.e., (+, +). Similarly, any points with signs (−, +) will appear in quadrant II, those with signs (−, −) will appear in quadrant III, and those with signs (+, −) will appear in quadrant IV.

In the definition of a function, recall that for each element in the domain there exists one and only one element in the range. Thus for a function where the independent variable is x, the function value y will receive one and only

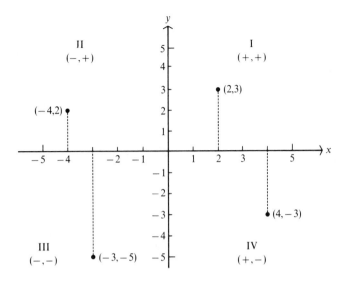

FIGURE 1.9

one value. In order to graph a function, we select a value for x and use the function rule to determine a value for y; next we determine the point (x,y) on the rectangular coordinates and plot the point. We continue this process for several x values until enough points are obtained to give the specific pattern of the function. In theory we should determine *all* possible (x,y) pairs and plot them in order to ensure that the graph of the function is correct. In practice, generally we determine enough points to sketch the graph by connecting the points by smooth curves or dashed lines.

Consider the function for the car rental example

$$y = 0.12x + 15$$

In order to graph this function, we need to compute the value of y for each of several different values of x. Since x represents the number of miles driven, it is not feasible for x to be negative, and so we shall consider only positive values of x. We compute the following values of y for the corresponding x values:

x	0	50	100	150	200
y	15	21	27	33	39

Plotting the values in the above table and connecting the points with a solid line, we obtain the graph given in Fig. 1.10. The solid line can be considered only as an estimate of where the function should be plotted because generally the domain consists of all real numbers of which there are an infinite number and the function value has not been determined for each of these values. When plotting the graph of a function, we never know how the function

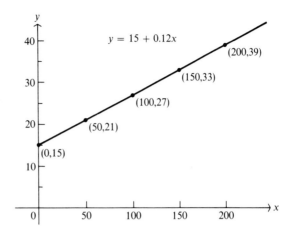

FIGURE 1.10

should look between two consecutive computed points unless we employ methods other than simply graphing the function. We shall consider some methods of this nature in later chapters.

For a better understanding of what can happen when plotting the graph of a function, consider the function

$$y = x^3 - 9x^2 + 24x - 10$$

First let us compute the values of y for values of x equal to 1, 3, and 5. We obtain the following values:

x	1	3	5
y	6	8	10

Plotting these values, we obtain three points, that appear to lie on a straight line. If we decide that this is a sufficient number of points to describe the function, we may connect the three points by solid lines and extend the lines beyond each of the outer points, as illustrated in Fig. 1.11.

However, if we include the values of x equal to 2 and 4, we see that the graph of the function is not a straight line. Evaluating the function for these additional x values, we obtain the following values:

x	2	4
y	10	6

When these values are incorporated and plotted with the initial three values and the points are connected with a solid smooth curve, we obtain the graph given in Fig. 1.12. This curve is considerably different from a straight line. Thus we should be very careful about using relatively few points when graphing functions because the function may be oscillating between adjacent plotted points.

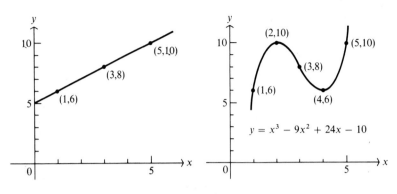

FIGURE 1.11   **FIGURE 1.12**

EXERCISES

1. If $f(x) = 3x + 10$ for any real number x, determine (a) $f(1)$, (b) $f(2)$, (c) $f(5)$, (d) $f(-1)$, and (e) $f(-5)$.

2. If $f(x) = x^2 + 2x + 3$ for any real number x, determine (a) $f(3)$, (b) $f(4)$, (c) $f(0)$, (d) $f(-3)$, and (e) $f(-10)$.

3. If $f(x) = x^3 + 3x^2 + 3x + 2$ for any real number x, determine (a) $f(5)$, (b) $f(3)$, (c) $f(0)$, (d) $f(-3)$, and (e) $f(-5)$.

4. Roger Caldwell, the sales manager of a small corporation, is paid a salary of $600 per month plus a commission of 3 percent of the gross sales for the month. Determine the function that expresses Mr. Caldwell's total monthly earnings as a function of the gross sales.

5. Ann Collins sells office copiers to local retailers. Each month the manufacturer pays her $50 times the *square* of the number of copiers she sells, plus a fixed amount of $200. However, she incurs a cost to herself of $100 per copier sold. Determine the function that expresses Ms. Collins' monthly net income as a function of the number of copiers sold.

6. If $f(x) = x^3 + 3x^2 + 3x + 2$ and the domain of the function is $D = \{-5, -3, 0, 3, 5\}$, determine the range of the function.

7. If $f(x) = 3x^2 + 5x + 2$ and $g(x) = 2x^2 - 3x + 1$ for any real number x, determine

   (a) $f(2) - f(3)$ (b) $f(3) - g(3)$ (c) $f(2) + g(4)$
   (d) $f(1) \cdot g(0)$ (e) $f(6)/g(3)$

8. For Fig. 1.13, determine the coordinates of (a) point A, (b) point B, (c) point C, (d) point D, (e) point E, (f) point F, (g) point G, and (h) point H.

9. Express each of the following linear equations in functional form:

   (a) $3y + 9x = 27$ (b) $20x = 50 - 10y$ (c) $7y - 7x = 35$

10. Graph each of the functions determined in Exercise 9 for the values of x equal to $-2, -1, 0, 1,$ and 2, and connect the plotted points by solid lines.

Sec. 1.5 / FUNCTIONS AND GRAPHS

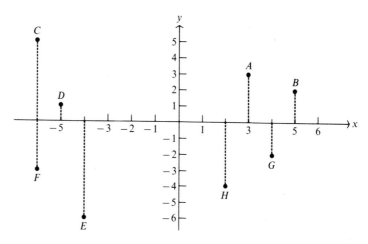

FIGURE 1.13

11. A plant has costs of $50,000 per month plus $12 per unit produced. Determine the function that expresses the monthly costs in terms of units produced.

12. A manufacturer has decided that the company's operating income is $25 per unit sold minus monthly fixed expenses of $2,000 and other expenses equal to 10¢ times the square of the number of units sold. Determine the function that expresses operating income in terms of units produced.

IMPORTANT TERMS AND CONCEPTS		
	Abscissa	Equation
	Associative law	Function
	of addition	Functional form
	of multiplication	Graph
	Axioms	Independent variable
	Cartesian coordinates	Intersection
	Cartesian product	Linear equation
	Commutative law	Linear function
	of addition	Null set
	of multiplication	Ordered pair
	Complement	Ordinate
	Coordinate axes	Origin
	Dependent variable	Postulates
	Distributive law of multiplication	Proper subset
	over addition	Quadrant
	Domain	Range
	Elements	Rectangular coordinates
	Empty set	Set
	Equal sets	Subset

Union
Universal set
Venn diagram
x axis

x coordinate
y axis
y coordinate

REVIEW PROBLEMS

1. In a survey of 125 people, the number of people who used various products were: product A, 57, product B, 58; product C, 42. Of those who use product A, 12 use B but not C, 16 use both B and C, and 3 use C but not B. Of those remaining people who use product C, 8 use B but not A. Draw a Venn diagram to determine how many people use

(a) Product A only
(b) Product B only
(c) Product C only
(d) None of the products
(e) Product A or B
(f) Product B or C
(g) Product A or C
(h) Product A or B or C
(i) Products B and C

2. The Gizmo Corporation supplies widgets and gadgets to 95 companies. They currently have orders to fill for widgets from 55 companies, for gadgets from 67 companies, and for widgets and gadgets from 30 companies.

(a) How many companies who order from Gizmo have no unfilled orders?
(b) How many companies have placed orders for widgets only?
(c) How many companies have placed orders for gadgets only?

3. Of the 75 people who work in the production department, 40 can work on the line for product A, 42 can work on the line for product B, and 29 can work on the line for product C, 6 can work on all three lines, 10 can work on the lines for products A and C but not for B, 2 can work on the lines for B and C but not for A, and 12 can work on the lines for A and B but not for C. Draw a Venn diagram for the sets of people for the respective lines and determine how many people can work on the lines for

(a) Product A only
(b) Product B only
(c) Product C only
(d) Product A or B
(e) Product B or C
(f) Product A or C
(g) Products A and B
(h) Products B and C
(i) Products A and C

4. A company has three vice-presidents (VP), Jones, Williams, and Todd; top financial personnel (F) are Jones, Taylor, and Davids; top production personnel (P) are Williams, Cox, and Smith; and top marketing personnel (M) are Todd, Reed, and Perry. List the names of those who are on the committees given by the following sets:

(a) $VP \cap M$ (b) $(VP \cap M) \cup (VP \cap P)$
(c) $(P \cup M) \cap (VP \cup M)$ (d) $(P \cup F) \cup (VP \cup P)$
(e) $(P \cup F)'$ (f) $(P \cup F)' \cap (P \cup M)'$

5. A high school has three substitute teachers for math, Rogers, Moore, and Ladd; two for government, Brown and Rogers; and four for English, Moore, Brown, Bradley, and Jackson. Show the logical symbols and list the members of the sets for

(a) Those who can teach math or government
(b) Those who can teach math or English
(c) Those who can teach math and government
(d) Those who can teach English but not math or government
(e) Those who can teach math but not English and government
(f) Those who can teach all three subjects
(g) Those who can teach exactly two subjects

6. A grocery store buys canned goods and frozen foods from supplier A; canned goods and bakery items from supplier B; bakery items, dairy products, and frozen foods from supplier C; nonfood items and confections from supplier D; and confections, bakery items, and canned goods from supplier E. Show the logical symbols and list the items purchased from

(a) Supplier A or B
(b) Supplier B or E
(c) Supplier C or D
(d) Suppliers A, B, and E
(e) Suppliers A and B or suppliers C and E
(f) Supplier C but not from supplier E
(g) Supplier C or D but not from supplier A or B
(h) Supplier B or C and from supplier D or E

7. The number of widgets made per hour is equal to $100X - 20$, where X is the number of machines being used. There are three machines that can produce widgets. Determine the range and the ordered pairs for the number of widgets that can be made in one hour.

8. The weekly variable cost of XYZ company is represented by the function $f(X) = 4X$, where X is the number of widgets manufactured weekly, production is 100, 140, or 160 widgets. Determine the range, the ordered pairs given by $f(X)$, and verify that $f(X)$ is a function.

9. A company has purchased three automobiles, a Chaffy, a Gord, and a Rumbler for three of its executives, Taylor, Smith, and Thompson. Determine the cartesian product of automobile (A) and executive (E). Also determine the cartesian product of executive and automobile.

10. A department store has weekly operating expenses of $2.25 per man-hour used and $750 fixed expenses. Determine the function that expresses the store's weekly operating expenses as a function of the man-hours used. Determine the weekly expenses if 500 man-hours are used in a week.

11. A company's daily profit is $3.50 per unit of sales less $25. Determine the function that expresses daily profits as a function of sales. Determine the profits if 10 units are sold, 30 units are sold, or 50 units are sold.

12. The time required to manufacture a batch of widgets is 5 hr times the square of the number of widgets produced in the batch less 6 hr times the number of widgets. Fixed set-up time required to manufacture a batch is 20 hr. Determine the function that expresses the total time required to manufacture a batch of widgets as a function of the number of widgets manufactured. Also determine the time required to manufacture a batch of 8 widgets, 10 widgets, and 15 widgets.

2 LINEAR AND QUADRATIC FUNCTIONS

2.1 INTRODUCTION

Linear and quadratic functions have many interesting applications to everyday problems. For example, we can use them to determine total monthly earnings for a given amount of gross sales, or total production cost for a specific number of units produced, or salary expense for a certain level of employment, or income for the number of hours worked, or income tax for a certain level of adjusted gross income, and so on. In addition to the numerous application possibilities, one of the properties of these functions is that the associated mathematics is fairly easy. Linear and quadratic functions are two very special cases of a more general type of function. However, they are of sufficient interest and importance to be considered separately. For each of these special functions, we shall consider ways of solving for the x values to obtain certain meaningful values of the functions.

2.2 LINEAR EQUATIONS

Many situations give rise to linear equations that are not in functional form. For example, if the ideal weight of a shipment of two chemicals is 12 tons and each unit of chemical 1 weighs 6 tons while a unit of chemical 2 weighs 4 tons, the number of units shipped of each chemical must satisfy the equation

$$6y + 4x = 12$$

where y and x represent the number of tons of chemicals 1 and 2, respectively. This equation is not in functional form but rather in what is commonly known as the *standard form* of a linear equation; i.e., $cx + dy = e$. However, the equation is converted to functional form simply by applying the rules of algebra, namely, by subtracting $4x$ from both sides of the equation and dividing both sides of the resulting equation by 6. The above linear equation is then converted to the functional-form linear equation

$$y = -\tfrac{2}{3}x + 2$$

In this section, we deal almost exclusively with linear equations that are already in functional form. As seen above, it is a simple matter to convert any linear equation into one in functional form.

In considering the linear equation $y = mx + b$ let us first consider the relationship of m and b to the graph of the function. The constants m and b are called the *coefficient* of x and the *constant term*, respectively. The constant term b does not create any *change* in y for different values of x; the coefficient of x indicates how much y increases as x increases by one unit. Thus, we note that y increases by m units for each unit increase in x. Consider, for example, the equation $y = 2x + 5$. Determining the values of y for several specific values of x, we obtain the following table:

x	0	1	2	3	4	5
y	5	7	9	11	13	15

In this table, we note that the y values change by two units for each one-unit change in the value of x. Plotting these points and connecting them with straight lines that extend beyond the points, we obtain the graph of the equation $y = 2x + 5$ (or, more precisely, the graph of the function $2x + 5$), as given in Fig. 2.1. We note without proof that a straight line that passes through any two computed points also passes through all other points that can be plotted for the linear equation $y = 2x + 5$. Thus the graph of a linear function is a straight *line*; this is why it is termed a *linear* function.

Slope-Intercept Formula   We see in Fig. 2.1 that changing x by one unit, from 3 to 4, causes y to change by two units; thus the *rate of change* of y per unit change in x is two units. This is checked for any two x values, say, x_1 and x_2, as indicated in Fig.

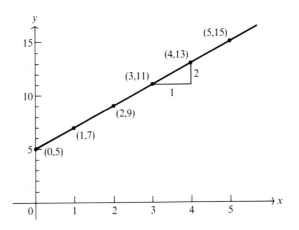

FIGURE 2.1

Sec. 2.2 / LINEAR EQUATIONS

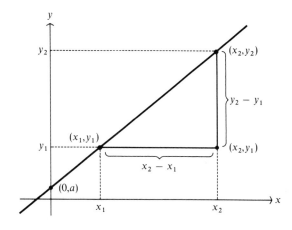

FIGURE 2.2

2.2. We take the ratio of the difference between the corresponding y values, say, y_1 and y_2, respectively, to the difference between the x values. Thus the rate of change for a linear function is given by

$$\text{slope} = \frac{\text{rise}}{\text{run}} = \frac{y_2 - y_1}{x_2 - x_1} \qquad x_2 \neq x_1 \tag{2.1}$$

This rate of change is called the *slope* of the straight line. In mathematics as in reading, we go from left to right. Thus, if the graph rises, the slope is positive; if it falls, the slope is negative; and if it is horizontal, then the slope is zero. In order to see how this slope is related to the constants m and b in the linear equation, let us determine the values of y_1 and y_2 as functions of x_1 and x_2, respectively. Letting y_1 and y_2 be the values of y when x equals x_1 and x_2, respectively, we have

$$y_1 = mx_1 + b$$

and

$$y_2 = mx_2 + b$$

Determining the appropriate differences for Eq. (2.1), we have

$$y_2 - y_1 = mx_2 + b - (mx_1 + b)$$
$$= mx_2 + b - mx_1 - b$$
$$= m(x_2 - x_1)$$

Substituting into Eq. (2.1), we obtain

$$\frac{y_2 - y_1}{x_2 - x_1} = \frac{m(x_2 - x_1)}{x_2 - x_1} = m$$

For specific values let us refer to the two points (3,11) and (4,13) in Fig. 2.1 and determine the slope from Eq. (2.1). That is, the slope m is given by

$$\frac{13-11}{4-3} = 2 = m$$

Hence the *slope* of the linear function (or, more precisely, the slope of the straight line that is the graph of the linear function) is given by the constant m, which is also the coefficient of x. We now know that the slope of the line segment connecting any two points on the graph of $y = mx + b$ will be constant.

Similarly, the constant b represents the value of the function when x is equal to zero. As seen in Fig. 2.1, for the equation $y = 2x + 5$, the value of b represents the value for which the graph of the function *intercepts* the y axis; thus the constant b is called the y *intercept*. With this knowledge concerning the two constants of the linear equation in functional form, we can determine the slope and y intercept for the graph of a linear function simply by observing the appropriate coefficients. Furthermore, since two points will geometrically determine a straight line, we need only two points to determine the graph of a linear function. In reality, we need only one more point when the linear equation is expressed in functional form because the y intercept automatically indicates one point, namely, $(0,b)$; the additional point can be computed from the linear equation for an arbitrarily selected x value. With this approach we can determine the equation of a straight line if we know its slope and intercept values; we can also determine the slope and intercept values of a straight line if we have the equation of the line expressed in functional form. For these reasons, a linear equation expressed in functional form is also said to be a *slope-intercept formula* for the equation of a straight line, given by

$$y = mx + b \tag{2.2}$$

Example 2.1 The Stemmons Metal Company produces metal fasteners. From available accounting information, they know that an initial cost of $525 will be incurred even if they make no fasteners. Furthermore, for all fasteners made, they will incur a material and labor cost of $3 per fastener. Determine the linear equation in functional form that will permit us to compute the total cost y of making x fasteners.

Solution The initial cost of $525 occurs even if Stemmons makes no fasteners. Thus the point $(0,525)$ must satisfy the linear equation and the intercept is 525. The rate of change of the cost for each additional fastener is $3, which is the slope m. Having determined the intercept value $b = 525$ and the slope $m = 3$, we obtain the linear equation through use of the slope-intercept formula:

$$y = 3x + 525$$

The cost of making 1,000 fasteners is $y = 3(1,000) + 525 = \$3,525$.

Point-Slope Formula In addition to the slope-intercept formula, two other fairly common methods are available for determining the equation of a straight line. The first, known as the *point-slope formula*, is slightly more general than the slope-intercept formula because it permits the use of any point on the line rather than the point where the line intercepts the y axis. To use the point-slope formula, we must know the coordinates (x_1, y_1) for a specific point P on the line and the slope of the straight line; that is, we must know the coordinates (x_1, y_1) of a point P and the slope m. Given this information, we can select an arbitrary point Q for which the coordinate values of x and y are unspecified and determine the equation of the straight line.

As illustrated graphically in Fig. 2.3, we know that the slope of the line segment connecting the points P and Q is equal to m. This slope must also be equal to the ratio of the change in y values to the change in x values for the two points because the slope of a straight line is constant for any two points on the line. Thus, from Fig. 2.3 and Eq. (2.1), we have

$$\frac{y - y_1}{x - x_1} = m \tag{2.3}$$

Using Eq. (2.3) and the given values for (x_1, y_1) and m, we can obtain the equation of the straight line having slope m and passing through the point (x_1, y_1). The formula in (2.3) is commonly known as the *point-slope formula*.

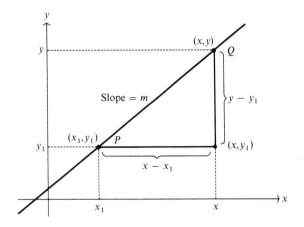

FIGURE 2.3

Example 2.2 The Spalding Kanine Kennel knows that two dogs in the same pen require 4 pounds (lb) of dog food per day. Each additional dog put in the pen requires another 2 lb of dog food per day. Obtain the equation in functional form that determines the required number of pounds of dog food for a given number of dogs in the same pen.

Solution Let x represent the number of dogs in the pen and y represent the number of pounds of dog food required. Since the required amount of dog food increases by 2 lb for each additional dog, the slope is $m=2$. Therefore, substituting this slope and $x_1=2$ and $y_1=4$ into Eq. (2.3), we have

$$\frac{y-4}{x-2}=2$$

Solving for the equation in functional form, we have $y=2x$.

Two-Point Formula

The third method for determining the equation of a straight line is known as the *two-point formula*. The name comes from the fact that it is a formula that requires that we know the coordinates for two points in order to determine the equation of the straight line passing through them. Suppose the points P and Q, as given in Fig. 2.4, are the points with known coordinates and the point R has unknown coordinates. Since a straight line has a constant slope for any two points on the line, we know that the slope of the line segment PQ in Fig. 2.4 must be equal to the slope of the line segment PR. Using the ratio of (2.1) for both line segments, we set the values for the two slopes as equal and obtain

$$\frac{y-y_1}{x-x_1}=\frac{y_2-y_1}{x_2-x_1} \tag{2.4}$$

The formula in (2.4) is the two-point formula for determining the equation of a straight line that passes through (x_1,y_1) and (x_2,y_2). Given the values of the coordinates of any two points, we can determine the equation of the straight line that passes through them.

Example 2.3 Jon Thompson hires college students to sell pennants at the Poindexter University football games. The students are each paid a salary plus a commission. Jim McDaniel and Bruce Woods worked for Jon at one

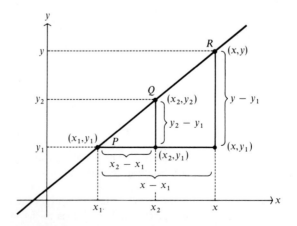

FIGURE 2.4

Sec. 2.2 / LINEAR EQUATIONS

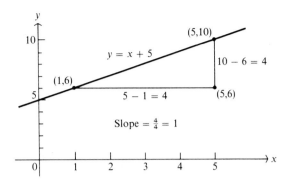

FIGURE 2.5

game where Jim sold one pennant and received $6, and Bruce sold five pennants and received $10. Determine the equation Jon uses to determine the amount he must pay as the result of the number of pennants sold.

Solution Letting x and y represent, respectively, the number of pennants sold and the amount paid to the student, we obtain for Jim and Bruce, respectively, the points (x_1, y_1) and (x_2, y_2), which are given by $(1,6)$ and $(5, 10)$. Substituting these values into (2.4), we have

$$\frac{y-6}{x-1} = \frac{10-6}{5-1}$$

which reduces to the equation

$$y = x + 5$$

Thus Jon pays each student a salary of $5 and a commission of $1 per pennant sold. The points and the appropriate differences are given in Fig. 2.5.

In summary, we have three methods for determining the equation of a straight line: the slope-intercept formula for the case where the slope and y intercept are known; the point-slope formula for the case where the slope and the coordinates of a point on the line are known; and the two-point formula for the case where the coordinates for two different points on the line are known.

In each of the linear equations that we have considered, the intercept and slope have been positive numbers. However, either or both of these values may be negative. For an understanding of what it means to have a negative intercept and/or slope, let us consider each briefly. The intercept is merely the value on the y axis where the straight line intercepts the axis. Thus if the value is positive, the point of intersection is above the x axis (remember from Sec. 1.4 that a positive value of y means it is above the x axis and a negative value of y means it is below the x axis); likewise, if the value is negative, the

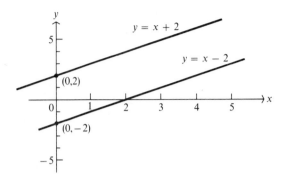

FIGURE 2.6

line intercepts the y axis below the x axis. This is demonstrated by the two equations given in Fig. 2.6, which differ only in the signs of the intercepts.

Since the value of the slope m represents the ratio of the change in y per unit change in x, the sign of the slope indicates simply whether the value of y increases or decreases as the value of x increases. For example, if the slope is positive, the value of y increases as the value of x increases; conversely, if the slope is negative, the value of y decreases as the value of x increases. Two equations that differ only in the signs of m are plotted in Fig. 2.7. The magnitude of the *rate* of change for the two lines is the same, but the *direction* of change is different because one is increasing and the other is decreasing.

Occasionally we need to determine the value of x that yields a specific value of y. To do this we substitute the specific value of y into the equation and solve for x by applying the rules of algebra. For example, suppose we need the value of x that yields a y value of 10 in the equation $y = 2x + 6$. We substitute 10 for y, obtaining

$$10 = 2x + 6$$

Subtracting 6 from both sides and dividing both sides of the resulting

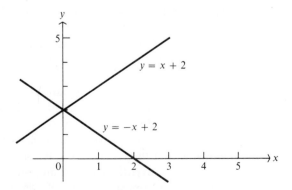

FIGURE 2.7

Sec. 2.2 / LINEAR EQUATIONS

equation by 2 yields

$$4 = 2x$$
$$2 = x$$

When the y value is set equal to zero, we are seeking the x value for which the straight line cuts the x axis; such a value is called the *root* of the equation. More precisely, a root of the function $f(x)$ is any value of x such that $f(x) = 0$. Linear equations have only one root; however, some equations have many roots.

EXERCISES

For each of the linear equations in Exercises 1 to 4, express the equation in functional form, graph the equation, and specify the values for the intercept and the slope.

1. $2y + 4x = 12$

2. $3y - 9x = 15$

3. $5y + 2x = 10$

4. $10y - x = 50$

5. Determine the equation in functional form for the straight line that has slope 4 and passes through the point (2, 6). Is the point (5, 18) on the line? Is the point (3, 12) on the line?

6. Determine the equation in functional form for the straight line that has slope -4 and passes through the point (0, 5). Is the point $(-2, 13)$ on the line? Is the point $(-1, 1)$ on the line?

7. Determine the equation in functional form for the straight line that has slope 3 and passes through the point (4, 10). Is the point (2, 6) on the line? Is the point (3, 9) on the line?

8. Determine the equation in functional form for the straight line that passes through the two points (1, 8) and (4, 23). Is the point (2, 13) on the line? Is the point (5, 28) on the line?

9. Determine the equation in functional form for the straight line that passes through $(-5, 10)$ and $(5, -10)$. Is the point (0, 0) on the line? Is the point (2, 4) on the line?

10. Determine the equation in functional form for the straight line that passes through $(-1, 1)$ and $(4, -14)$. Is the point $(5, -17)$ on the line? Is the point $(-5, 13)$ on the line?

Determine the root of each of the linear equations in Exercises 11 to 14.

11. $2y + 4x = 12$

12. $3y - 9x = 15$

13. $5y + 2x = 10$

14. $10y - x = 50$

For each of the equations in Exercises 15 to 17, determine the value of x that corresponds to the given y value.

15. $3y + 4x = 15$ when $y = 2$

16. $3y - 9x = 27$ when $y = 5$

17. $y = 5x + 15$ when $y = 10$

18. The profit y of a Gas-N-Go service station is a function of the gallons of gasoline x that it sells. The manager of the service station has determined that for every 5 gal of gasoline sold he makes 22¢ profit and the relationship between profit and amount sold is linear. Furthermore, he must sell 1,000 gal per day to break even. Determine the linear equation in functional form that gives the relationship between profit and sales on a daily basis. What is the profit if he sells no gasoline on a particular day? What is the profit if he sells 5,000 gal on a particular day?

19. A sales agent's reimbursement for expenses is a function of the miles (mi) traveled. One month she traveled 1,000 mi and was reimbursed $300. Another month, the reimbursement was $200 for 800 mi. Determine the linear equation that allows the agent to calculate her monthly reimbursement from the miles she travels.

20. For 1 week, a Fast-Chek store had expenses of $1500 and used 10 full-time employees. Each full-time employee averages $92 per week. Determine the linear equation that allows the manager to estimate weekly expenses based on the number of employees. What would be the expenses for the store if it were closed for 1 week?

21. The Davis County Coop sells corn for $4 per bushel (bu). It must sell 300 bu in order to break even. Determine the linear equation in functional form that gives the relationship between profit and bushels sold. How much will the Coop lose if only 200 bu of corn are sold? What is the profit if 600 bu are sold?

2.3 QUADRATIC EQUATIONS

Although there are many practical applications of linear functions, there are also numerous real-world situations that cannot be described by linear functions. Graphically this means that the graph of the function is a curve rather than a straight line. Perhaps the most elementary example of a function for which the graph is not a straight line is the *quadratic function*, which has the general form

$$f(x) = ax^2 + bx + c \qquad (2.5)$$

where a, b, and c are constant real numbers such that $a \neq 0$. Replacing the more cumbersome functional notation $f(x)$ by the single variable y, we obtain the *quadratic equation in functional form*

$$y = ax^2 + bx + c$$

This equation is also called the *standard form* of a quadratic equation.

Sec. 2.3 / QUADRATIC EQUATIONS

Let us first consider the graph of the quadratic equation given by

$$y = x^2 - 6x + 5$$

The points to be plotted are obtained by substituting the x values into the quadratic equation in order to determine the corresponding y value. For example, when $x = 2$, we obtaim

$$y = (2)^2 - 6(2) + 5 = -3$$

Similarly, we obtain the following values:

x	-1	0	1	2	3	4	5	6	7
y	12	5	0	-3	-4	-3	0	5	12

Plotting the points obtained in the above table and connecting the points by a smooth curve, we obtain the graph in Fig. 2.8.

To sketch the graph, we computed the coordinates for nine points. We might have been able to sketch the graph satisfactorily with fewer points. Nevertheless, the number of points required for graphing a quadratic equation is more than the two points required for graphing a linear equation. Upon examining Fig. 2.8, we note that the plotting of any number of points for which x is less than 3 would not give much of a clue as to the shape of the curve for those points having x values greater than 3; the converse is also true. Thus it is important to select carefully the values of x for the points that we wish to plot. Generally we need at least five well-chosen points to obtain an accurate sketch of a quadratic equation. Strictly speaking, three well-

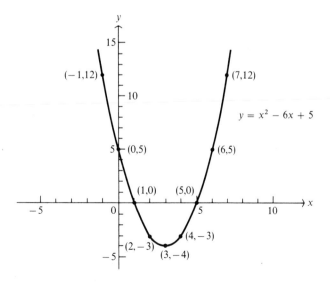

FIGURE 2.8

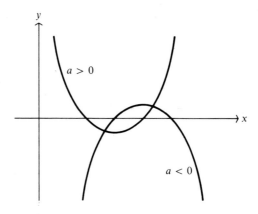

FIGURE 2.9

chosen points will suffice for a rough sketch; however, a better graph can be obtained from more points.

In order to determine the five best points for graphing a quadratic equation, let us first consider the general shape of the graph of a quadratic equation that is expressed in the functional form

$$y = ax^2 + bx + c \tag{2.6}$$

From the restriction on the constants given in Eq. (2.5), we know that $a \neq 0$ and hence it must be either positive or negative. The ax^2 term will eventually dominate the right side of the equation as x increases without regard to its sign, which means that the sign of the value of y will be the same as the sign of a for *extreme* values of x regardless of whether they are positive or negative values because the square of each will be positive. A general sketch of the graph of a quadratic equation is given in Fig. 2.9 for the two signs of a.

To consider the coordinates for the five best points, we examine the general graph of Eq. (2.6) for the case where a is positive, as given in Fig. 2.10. (Note that *identical* formulas will be obtained for the case where a is

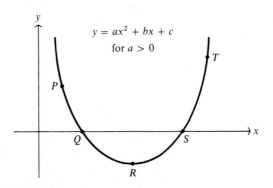

FIGURE 2.10

Sec. 2.3 / QUADRATIC EQUATIONS

negative; we are using the positive case here only for reference purposes.) In order to determine the five points, we need to determine the coordinates for the points P, Q, R, S, and T in Fig. 2.10. There, it is apparent that points Q, R, and S are significant in that Q and S constitute *roots* to the equation (i.e., values of x for which $y=0$) and R is the point for which the function is at its *minimum*. The points P and T are rather arbitrary except that the x value associated with P is less than the x value for Q and the x value for T is greater than the x value associated with S. Thus it appears to be extremely important to determine the coordinates for the roots and the minimum of the equation.

The Vertex To determine the coordinates for the minimum of the quadratic equation (2.6), we factor out the coefficient a and obtain

$$y = a\left(x^2 + \frac{b}{a}x + \frac{c}{a}\right)$$

Completing the square by adding and subtracting $b^2/4a^2$, we obtain

$$y = a\left[\left(x^2 + \frac{b}{a}x + \frac{b^2}{4a^2}\right) + \frac{c}{a} - \frac{b^2}{4a^2}\right]$$

By factoring and adding fractions, we obtain

$$y = a\left[\left(x + \frac{b}{2a}\right)^2 + \frac{4ac - b^2}{4a^2}\right]$$

Factoring -1 from the second term gives

$$y = a\left[\left(x + \frac{b}{2a}\right)^2 - \frac{b^2 - 4ac}{4a^2}\right] \tag{2.7}$$

To determine the value of x that yields the minimum value of y, we note that the first term within the brackets of Eq. (2.7) is the only term involving x and the second term is a constant that is being subtracted from the first term. Consequently, if the first term is as small as possible, the value of y will be the minimum. Since the first term is the square of a real number, it must be either positive or zero. When it is zero, the difference will be at its minimum. In order for the square term to be zero, the term being squared must be zero; that is,

$$x + \frac{b}{2a} = 0$$

which is only true when $x = -b/2a$. Thus the coordinates for the minimum point are

$$x = -\frac{b}{2a} \quad \text{and} \quad y = -\frac{b^2 - 4ac}{4a} \tag{2.8}$$

The graph of a quadratic function is called a *parabola*, and the coordinates for the minimum (or maximum if $a<0$) are given in (2.8); this point is called the *vertex* of the parabola.

The Quadratic Formula To determine the coordinates for roots of Eq. (2.6), we note that the terms within the brackets of Eq. (2.7) can be considered as the difference between two squares. By setting the right side of (2.7) equal to zero, we obtain

$$a\left[\left(x+\frac{b}{2a}\right)^2 - \frac{b^2-4ac}{4a^2}\right] = 0$$

which factors into

$$a\left[\left(x+\frac{b}{2a}\right) - \sqrt{\frac{b^2-4ac}{4a^2}}\right]\left[\left(x+\frac{b}{2a}\right) + \sqrt{\frac{b^2-4ac}{4a^2}}\right] = 0$$

Thus for a product to equal zero, one or more terms must equal zero. Since $a \neq 0$, we wish to determine all values of x for which

$$x + \frac{b}{2a} - \sqrt{\frac{b^2-4ac}{4a^2}} = 0$$

or

$$x + \frac{b}{2a} + \sqrt{\frac{b^2-4ac}{4a^2}} = 0$$

Taking the square root of the denominator under the square-root sign, solving for x in each of the above two equations, and adding fractions, we obtain

$$x = \frac{-b+\sqrt{b^2-4ac}}{2a}$$

and

$$x = \frac{-b-\sqrt{b^2-4ac}}{2a}$$

respectively. These equations are generally incorporated into one statement, called the *quadratic formula*, for determining the roots of a quadratic equation, namely,

$$x = \frac{-b \pm \sqrt{b^2-4ac}}{2a} \qquad (2.9)$$

Sec. 2.3 / QUADRATIC EQUATIONS

Hence the values of the x coordinates for the points Q and S in Fig. 2.10 are given by the two values obtained in Eq. (2.9), while the y coordinates are zero for both points.

Let us return to the example that is plotted in Fig. 2.8 and determine the values for the five points of interest. The quadratic equation is given by

$$y = x^2 - 6x + 5$$

Since the equation is already in functional form, we note that $a=1$, $b=-6$, and $c=5$ in order to conform to the general equation given by Eq. (2.6). The coordinates for the vertex of the associated parabola are obtained by substitution of these values into (2.8). Upon substitution, we obtain

$$x = -\frac{-6}{(2)(1)} = 3 \quad \text{and} \quad y = -\frac{(-6)^2 - 4(1)(5)}{(4)(1)} = -4$$

Thus the point R of Fig. 2.10 corresponds to the point $(3, -4)$ in Fig. 2.8. Similarly, the points corresponding to Q and S in Fig. 2.10 are determined by substituting the appropriate values into (2.9). Substituting and evaluating, we obtain

$$x = \frac{-(-6) \pm \sqrt{(-6)^2 - 4(1)(5)}}{(2)(1)} = \frac{6 \pm 4}{2}$$

or

$$x = 1 \quad \text{and} \quad x = 5$$

The points for Q and S are $(1,0)$ and $(5,0)$, respectively. Thus we have the three points associated with the roots and the vertex. In order to obtain points associated with P and T, we can choose a value of x that is greater than 5 and one less than 1 and determine their respective y values. If we choose $x=0$ for the left value and $x=6$ for the right value, we obtain the points $(0,5)$ and $(6,5)$, which correspond to P and T, respectively. Thus the five points that we plot for P, Q, R, S, and T in Fig. 2.10 are $(0,5)$, $(1,0)$, $(3,-4)$, $(5,0)$, and $(6,5)$, respectively.

We should note that the values obtained for the five points should always surround the vertex for the parabola. Obviously this is the most critical area of concern when graphing a quadratic function. The reader may wonder why we need to use the five points to determine the graph of the equation since we were able to do such a fine job in Fig. 2.8. The reason is that the points that were plotted in Fig. 2.8 were judiciously selected for illustration purposes, and not all quadratic equations have such simple values for the roots and vertex.

Now let us consider an example of a quadratic equation that does not have such simple values for the roots and vertex and that has a negative a value.

Example 2.4 Determine five points, which include the roots and the vertex, and sketch the graph of the quadratic equation $y = -9x^2 + 33x - 10$.

Solution The equation is already expressed in functional form; and to conform with the general equation in Eq. (2.6), we have $a = -9$, $b = 33$, and $c = -10$. To obtain the vertex, we substitute into (2.8) and obtain

$$x = -\frac{33}{(2)(-9)} = \frac{11}{6} = 1\frac{5}{6} \quad \text{and} \quad y = -\frac{(33)^2 - (4)(-9)(-10)}{(4)(-9)} = \frac{81}{4} = 20\frac{1}{4}$$

Thus the vertex point is given by $(1\frac{5}{6}, 20\frac{1}{4})$.

To obtain the roots, we substitute the constant values into Eq. (2.9) and obtain

$$x = \frac{-33 \pm \sqrt{(33)^2 - (4)(-9)(-10)}}{(2)(-9)} = \frac{-33 \pm 27}{-18}$$

or

$$x = \frac{1}{3} \quad \text{and} \quad x = \frac{10}{3} = 3\frac{1}{3}$$

Thus the points corresponding to the roots are $(\frac{1}{3}, 0)$ and $(3\frac{1}{3}, 0)$. To verify that the above two values of x are indeed the roots, we substitute them into the quadratic equation to determine that y is equal to zero for the respective x values.

Having already determined the roots, we now determine the points associated with P and T in Fig. 2.10 by choosing the first integer value for x that is less than $\frac{1}{3}$ (i.e., $x = 0$) and the first integer value of x that is greater than $3\frac{1}{3}$ (i.e., $x = 4$). We choose integer values only because the equation is easier to evaluate for integers. Any other two values would suffice as long as one is less than $\frac{1}{3}$ and the other is greater than $3\frac{1}{3}$; however, we generally do

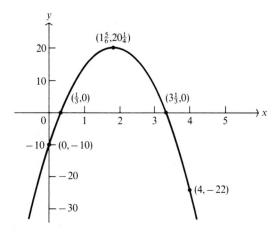

FIGURE 2.11

not look for x values that are either too near or too far away from the roots. Evaluating the quadratic equation for $x=0$, we obtain

$$y = -9(0)^2 + 33(0) - 10 = -10$$

The value of the equation for $x=4$ is

$$y = -9(4)^2 + 33(4) - 10 = -22$$

The two points corresponding to P and T in Fig. 2.10 are $(0, -10)$ and $(4, -22)$, respectively. The five points and a sketch of the graph are given in Fig. 2.11. If we wish to magnify the scale around the roots and the vertex, it is not necessary to plot the point $(4, -22)$ because it indicates what we want to know; that is, the curve is decreasing very rapidly between $3\frac{1}{3}$ and 4.

The Discriminant   In Fig. 2.9 it appears as though each quadratic equation will possess *two* real roots, but this is not necessarily true. The purpose of Fig. 2.9 was to demonstrate the general shape of a parabola, not to indicate the number of roots. By examining Eq. (2.9), we note that the term under the square root must be positive in order to have a real number to add to and subtract from the first term and obtain two distinct values of x. The term $b^2 - 4ac$ is called the *discriminant*, and in all cases considered to this point it has been positive, which results in two distinct real numbers for the roots of the quadratic equation. There do exist quadratic equations for which the discriminant is negative, and there are some for which it is equal to zero. We shall not consider those cases for which it is negative because the results are not real numbers.

Let us consider the case where the discriminant is zero. Examining Eqs. (2.8) and (2.9), we see that both involve the discriminant. Thus in Eq. (2.8), the y coordinate for the vertex is zero because $b^2 - 4ac = 0$. Similarly, the values for the two roots given by Eq. (2.9) will be equal because they differ only in that one is obtained by adding and the other by subtracting the square root of the discriminant, which is zero. The resulting x coordinates for the roots are both equal to $-b/2a$, which is the same as the x coordinate for the vertex. Likewise, the y coordinates for the roots, which are always zero, are equal to the y coordinate for the vertex. In summary, the coordinates for the two roots and the vertex are identical. When this is true, we say that the quadratic equation has a *double root*, which is also the vertex.

Thus the graph in Fig. 2.10 truly represents only the case for which the discriminant is positive. In order to determine the five points for the double-root case, we would determine the coordinates for two points having x values less than and two points having x values greater than the x value of the vertex. The points P, Q, R, S, and T would then appear as given in Fig. 2.12, where Q and S no longer represent the roots of the equation because the roots and the vertex are all the same and are represented by R. The parabola will be either all above or all below the x axis, except for point R, depending

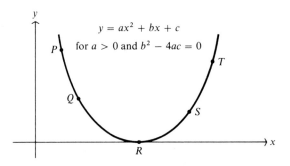

FIGURE 2.12

upon the sign of the coefficient *a*. A similar situation exists for the case where the discriminant is negative except the vertex is not in the *x* axis. Let us consider an example of a quadratic equation having a double root.

Example 2.5 Determine five points and sketch the graph for the quadratic equation

$$y = 4x^2 - 28x + 49$$

Solution The equation is in functional form with $a = 4$, $b = -28$, and $c = 49$. We determine the coordinates of the vertex to be

$$x = -\frac{b}{2a} = -\frac{-28}{(2)(4)} = \frac{7}{2} = 3\frac{1}{2}$$

and

$$y = -\frac{b^2 - 4ac}{4a} = -\frac{(-28)^2 - (4)(4)(49)}{(4)(4)} = 0$$

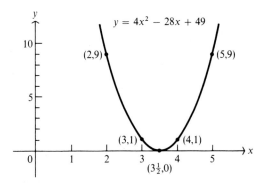

FIGURE 2.13

Sec. 2.3 / QUADRATIC EQUATIONS

The vertex point is $(3\frac{1}{2}, 0)$, which means the quadratic equation has a double root because the y coordinate of the vertex is zero.

In order to determine the remaining four points, we choose two x values to the left and two x values to the right of $x = 3\frac{1}{2}$. We shall select the integers 2, 3, 4, and 5. Upon substitution into the quadratic equation, we obtain the corresponding y coordinates and, consequently, the four points $(2,9)$, $(3,1)$, $(4,1)$, and $(5,9)$. These points are plotted and connected by a smooth curve in Fig. 2.13.

EXERCISES

Convert the quadratic equations in Exercises 1 to 4 to functional form.

1. $2y + 5x - 3x^2 = 6 + 13x^2 - 7x$

2. $3y + 9x^2 - 4x = 5y + 2x - 7x^2 + 14$

3. $31 + 7x - 4y + 6x^2 = -5x + 2y - 5$

4. $8y - 5x + 4x^2 + 6 = 6x^2 + 12 + 6y + 3x$

For each of the quadratic equations in Exercises 5 to 10, determine without graphing
    (a) Which way the parabola opens
    (b) The coordinates of the vertex
    (c) The coordinates associated with the roots of the equation (if the discriminant is not negative)

5. $y = 3x^2 - 15x + 12$

6. $y = 12x^2 - 2x - 4$

7. $y = -2x^2 + 11x - 12$

8. $y = -2x^2 - 32$

9. $y = x^2 - 16$

10. $y = x^2 + 8$

For each of the quadratic equations in Exercises 11 to 16, determine
    (a) Which way the parabola opens
    (b) Whether the parabola intersects the x axis
    (c) The coordinates of the vertex
    (d) The roots of the equation (if the discriminant is not negative)
    (e) The coordinates of five points (also plot the points and draw a smooth curve through them)

11. $y = x^2 - 9x + 18$

12. $y = 2x^2 - 12x + 18$

13. $y = -x^2 + 4x - 5$

14. $y = 2x^2 - 12x + 26$

15. $y = -3x^2 - 12x - 12$

16. $y = -3x^2 + 36x - 105$

17. The Wee Bakem Bakery bakes birthday cakes that cost them $3 each to make. They know the number of cakes they sell depends on the price per cake that they charge their customers. From past experience, they know they will be able to sell $600 - 50x$ cakes per week when they charge x dollars per cake.

   (a) Determine and graph the equation for computing the *total income* from the sale of birthday cakes.
   (b) From the equation in part (a), determine the maximum *total income* and the price per cake that they should charge in order to maximize the total income.
   (c) How many cakes will they sell at the price that maximizes the total income?
   (d) Determine and graph the equation for computing the *total profit* from the sale of birthday cakes.
   (e) From the equation in part (d), determine the maximum *total profit* and the price per cake that they should charge in order to maximize the total profit.
   (f) How many cakes will they sell at the price that maximizes the *total profit*?
   (g) How much profit will they lose if they charge the price per cake that maximizes total income instead of the price that maximizes total profit?

18. A factory has expenses of $x^2 + 15$ dollars, where x is the number of units produced. The product can be sold for $16 per unit.

   (a) Determine and graph the equation for computing the factory's profit based on the number of units produced and sold.
   (b) What is the maximum number of units that the factory should produce?
   (c) For what number of units produced will revenue equal expenses?

19. A shipping company has a shipping charge of $3x + 20$ dollars, where x is the number of pounds shipped. To prevent making shipments too large they have decided to increase the cost by $0.06x^2$. Determine and graph the equation for computing the shipping charge from the number of pounds shipped.

20. A manufacturer of two complementary products wants to determine how many units to manufacture. He knows that the demand of one is influenced by the demand of the other. The demand of product A is equal to one-half the square of the number of units of product B that are demanded less 250 units. Determine the equation that will compute the number of units to produce of product A based on the demand of product B. How many units of product B are demanded if the production of product A is zero?

2.4 POLYNOMIAL FUNCTIONS

As mentioned previously in this chapter, linear and quadratic functions are special cases of a more general type of function. This general type of function is called a *polynomial function*, and it is given by

$$f(x) = a_n x^n + a_{n-1} x^{n-1} + \cdots + a_1 x + a_0 \qquad (2.10)$$

Sec. 2.4 / POLYNOMIAL FUNCTIONS

where n is a nonnegative integer, $a_n \neq 0$, and $a_n, a_{n-1}, \ldots, a_1, a_0$ are real constants. The function of (2.10) is called a *polynomial function of degree n*. A linear function is a polynomial function of degree 1 because

$$f(x) = a_1 x^1 + a_0$$

is the equivalent of

$$f(x) = mx + b$$

where $a_1 = m$ and $a_0 = b$. Similarly, a quadratic function is a polynomial function of degree 2 because

$$f(x) = a_2 x^2 + a_1 x + a_0$$

is equivalent to

$$f(x) = ax^2 + bx + c$$

where $a_2 = a$, $a_1 = b$, and $a_0 = c$.

A polynomial function of degree 3, also called a *cubic function*, will have three roots, but like the quadratic function, these roots may be multiple roots and/or roots that are not real numbers. The graph of a cubic function was illustrated in Fig. 1.11. However, we do know that any polynomial function that has real constants for its coefficients and is of odd degree will have an odd number of real roots, and one of even degree will have an even number of real roots.

Polynomial functions of degrees greater than 2 can have graphs that have several peaks and valleys (or maxima and minima). The number of peaks and valleys is limited by the degree of the function. The graph of a polynomial function with even degree will have the extreme ends of the graph arriving and leaving the graph from the same direction. For a function of odd degree, the extemes arrive from one direction and depart to the opposite direction, as shown in Fig. 2.14.

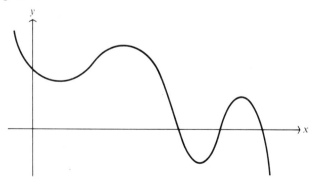

FIGURE 2.14

IMPORTANT TERMS AND CONCEPTS

Cubic function
Degree of a polynomial function
Double root
Functional form of a quadratic equation
Intercept
Parabola
Point-slope formula
Polynomial function
Quadratic equation
Quadratic function
Rate of change
Root
Slope
Slope-intercept formula
Standard form of a linear equation
Standard form of a quadratic equation
Two-point formula
Vertex of a parabola

REVIEW PROBLEMS

1. The yearly cost Y of an employee training program is a function of the number of employees X hired that year. The personnel manager has determined that the cost is $250 for every new employee, and the relationship between the cost and the number of employees hired is linear. Fixed costs for the training program are $2,500 per year. Determine the linear equation in functional form that gives the relationship between the number of employees hired and the cost of the training program. What is the cost if 10 employees are hired in a year? What is the cost if 16 employees are hired in a year?

2. The monthly salary Y of an insurance sales representative is a function of the total dollar premium X of the policies she sold that month. One month she sold $2,000 in premiums and her salary was $1,000. One month, when her salary was $1,200, she sold $3,000 in premiums. Determine the linear equation in functional form that gives the relationship between the total dollar value of the premiums of the policies sold and her monthly salary. What would her salary be if she sold $5,000 in premiums one month?

3. The profit Y of an adding machine manufacturer is a function of the number of adding machines X that are sold. For every two machines sold, the manufacturer makes a profit of $75. The relationship between profit and the number of machines sold is linear. Furthermore, the manufacturer must sell 100 adding machines per week to break even. Determine the linear equation in functional form that gives the relationship between profit and sales on a weekly basis. What is the profit if the manufacturer sells no machines one week? What is the profit if 300 machines are sold one week?

4. A manufacturer gives a 5¢ discount per dollar for cash purchases. Determine the linear equation in functional form that gives the relationship between the discount Y and the dollar amount of the cash purchase X. What is the discount if $1,000 of merchandise is purchased with cash?

5. The weekly cost Y of a delivery truck is a function of the number of miles X that it is driven. One week the truck was driven 75 mi and the cost was $46. Another week the truck was driven 33 mi and the cost was $34.24. Determine the linear equation in functional form that gives the relationship between the miles the truck is driven per week and the weekly cost of the truck. What is the cost if the truck is not driven one week? What is the cost if the truck is driven 88 mi in a week?

6. The cost Y of calling a repairman is linearly related to the time required for him to repair the machine. The repairman charges $7.50 per hour for working on the machine. One time, when the repairman charged $36.25, he worked on the machine for an hour and a half. Determine the linear equation in functional form that gives the relationship between the cost of calling a repairman and the time required for him to repair the machine. Determine the cost of calling the repairman if the machine does not need repairing. What is the cost if the repairman requires 3 hr to repair the machine?

7. The ABC Company has determined that the number of sales per week is approximately equal to $-0.5X^2 + 8X + 2$, where X is the number of calls made by their salesman per day.

   (a) Graph the equation for computing the number of sales from the number of calls.
   (b) From the graph in part (a), determine the maximum number of sales per week and the number of calls the salesman must make per day in order to maximize sales.

8. The Mini Manufacturing Company makes toy cars at a cost of $4 each. The number of toy cars sold depends on the sales price of the cars. From past experience, they know that they will be able to sell an average of $400 - 20X$ toy cars per week when the sales price is X dollars per car.

   (a) Determine and graph the equation for computing the total income from the sale of toy cars.
   (b) From the graph in part (a), determine the maximum total income and the price per toy car that they should charge in order to maximize the total income.
   (c) How many toy cars will they sell at the price that maximizes the total income?
   (d) Determine and graph the equation for computing the total profit from the sale of toy cars.
   (e) From the graph in part (d), determine the maximum total profit and the price per toy car that they should charge in order to maximize the total profit.
   (f) How many toy cars will they sell at the price that maximizes the total profit?
   (g) How much profit will they lose if they charge the price per toy car that maximizes total income instead of the price that maximizes total profit?

9. The number of defective batteries in a batch is given by the function $0.02X^2 - 5X + 325$, where X is the number of batteries produced in the batch.

   (a) Graph the function for computing the number of defective batteries in a batch.

(b) From the graph in part (a), determine the minimum number of defects per batch and the size of the batch that should be manufactured to minimize total defects.

10. The Sit-In Manufacturing Co. makes school desks at a variable cost of $8X^2 - 320X$, where X is the weekly production in hundreds. Weekly fixed costs are $12,500.

   (a) Determine and graph the equation for computing total costs from weekly production.
   (b) From the graph in part (a), determine the minimum total cost per week and the number of desks manufactured at that cost.
   (c) Determime and graph the total profit function from the sale of desks at $5.44 each.
   (d) From the graph in part (c), determine the maximum total profit and the number of desks sold to maximize profits.
   (e) How much profit will they lose if they manufacture to minimize total cost instead of to maximize total profit?

3 EXPONENTIAL AND LOGARITHMIC FUNCTIONS

3.1 INTRODUCTION

In numerous situations, people in business may be interested in the growth rate of certain phenomena such as sales, profit, costs, population, efficiency, or resource depletion. If the growth occurs in equal increments per unit of, say, time, it can be represented by a linear function. For example, if sales in units begin with 10,000 units today and increase, or grow, by 200 units per week, we can represent the sales s by the equation

$$s = 200x + 10{,}000$$

where x is in weeks and equals zero for the present week.

Frequently the growth of a particular phenomenon is not equal per unit of time; instead it may be growing by an equal *percentage* of the value for each successive period. If this is in fact the situation, a linear function will not adequately describe the growth relationship. A more complex function is necessary to represent growth. Thus in this chapter, we consider two very special types of functions that are of importance to business and economics as well as many other fields. First we consider *exponential functions*, which provide important models for describing economic and many other kinds of growth. Anyone studying exponential functions must first understand the basic *laws of exponents*; thus we review these laws prior to studying the functions. Second we consider *logarithmic functions*. Again we must first understand the *laws of logarithms* and the association between exponential functions and logarithms before we can effectively study logarithmic functions; thus we shall study the laws first and then the functions.

It is expected that the reader is already familiar with the *laws* of exponents and logarithms from elementary algebra; however, if not, he or she should be able to gain a basic understanding of and appreciation for them from this chapter. In short, the reader who has not been exposed to the laws of exponents and logarithms need not despair, while the reader who is somewhat familiar with them will find a good review of them in this chapter.

3.2 EXPONENTS

In general, mathematical notation is a kind of shorthand. It saves time and space and simplifies the representation of complex expressions. This is the primary purpose of exponential notation. For example, the term $bbbb$ is denoted by writing b^4, where the number b is called the base and 4 is called the *exponent*, or *power*, of b. In general, if n is a positive integer and b is any real number, we write

$$\underbrace{bbb \cdots b}_{n \text{ factors}} = b^n$$

for which the right side is read "b raised to the nth power." By convention, it is understood that the factor b by itself has an exponent of 1; in other words, the absence of an exponent is understood to mean that the exponent is 1. For example,

$$bbb = b^3$$
$$bb = b^2$$
$$b = b^1$$

By extending the above notational concept, we see that

$$b^m b^n = \underbrace{(bb \cdots b)}_{m \text{ factors}} \underbrace{(bb \cdots b)}_{n \text{ factors}} = \underbrace{(bb \cdots b)}_{m+n \text{ factors}} = b^{m+n}$$

Thus the product of two factors with the *same* base is simply the "base raised to the sum of the powers." Similarly, if we wish to raise to a power a factor that is already a base raised to a power, we have

$$(b^m)^n = \underbrace{b^m b^m \cdots b^m}_{n \text{ factors}} = b^{mn}$$

Thus when we raise to a power a base that is already raised to a power, we obtain the base raised to the product of the two exponents.

The above two relationships constitute two of the fundamental *laws of exponents*. Simply stated they are

$$b^m b^n = b^{m+n} \tag{3.1}$$

and

$$(b^m)^n = b^{mn} \tag{3.2}$$

Sec. 3.2 / EXPONENTS

In our previous discussion, the exponents m and n were considered as positive integers and the base b was considered as any real number. The laws actually apply for any *real* exponents m and n and for any real number b such that $b \neq 0$. However, when b is negative and/or the exponents are fractional, we run the risk of obtaining roots of negative numbers and/or roots of positive numbers for which there are more than one root. These situations generally require an extensive understanding of complex numbers. Therefore, we shall confine our consideration of b to the positive real numbers and our consideration of any factor involving a root to mean the positive real root.

Let us consider some of the special types of exponents, such as *negative* and *fractional* exponents. When a base raised to a negative exponent is encountered, it is merely the reciprocal of the base raised to the equal positive exponent; for example,

$$2^{-3} = \frac{1}{2^3}$$

Thus, in general, the meaning of negative exponents is explained by

$$b^{-m} = \frac{1}{b^m} \tag{3.3}$$

From Eq. (3.3), it is apparent that b cannot be zero because division by zero is undefined.

Applying the law of exponents given in (3.2), we could write

$$(\sqrt{3})^2 = (3^{1/2})^2 = 3^1 = 3$$

or

$$(\sqrt[3]{4})^3 = (4^{1/3})^3 = 4^1 = 4$$

In both cases, the term given by the base raised to a fractional exponent (that is, $3^{1/2}$ or $4^{1/3}$) represents the number that, when raised to a power equal to the reciprocal of its fractional exponent, is equal to the base. In general, we say that the number represented by $b^{1/m}$ is a number that will yield base b when raised to the mth power. It is also called the mth root of b, and it may be denoted by $\sqrt[m]{b}$.

As noted previously, the reason for stating that the mth root represents *a* number rather than *the* number is that b may have more than one real root. For example, for $b = 9$ and $m = 2$, we see that both 3 and -3 satisfy the above definition of $b^{1/m}$. However, in this text, we shall consider only the positive root of b when dealing with fractional exponents. In the preceding two examples, we considered raising the terms in parentheses to a power equal to the reciprocal of their respective fractional exponents. In Eq. (3.2), we are not limited to fractions and their reciprocals. As a matter of fact, it can be proved that we are limited only to real numbers m and n.

Using (3.1) and (3.3), we note that

$$b^m b^{-m} = b^{m-m} = b^0$$

which means that $b^0 = 1$ since

$$b^m b^{-m} = b^m \frac{1}{b^m} = 1$$

However, this is only true if $b \neq 0$, which is one of the restrictions on b. We can also see that

$$\frac{b^m}{b^n} = b^m b^{-n} = b^{m-n} \qquad (3.4)$$

Using the same approach that was employed for obtaining (3.1) and (3.2), we can show that

$$(ab)^m = a^m b^m \qquad (3.5)$$

and

$$\left(\frac{a}{b}\right)^m = \frac{a^m}{b^m} \qquad (3.6)$$

for any two real numbers a and b, such that $a \neq 0$ and $b \neq 0$.

Let us consider some examples using the laws of exponents as given in (3.1) to (3.6).

Example 3.1 Write each of the following terms using a single exponent:

(a) $\dfrac{3^3 3^5}{(3^4)^2}$

(b) $\dfrac{5^2 5^3}{\sqrt{5^4}}$

(c) $\dfrac{6^4 6^{1/2}}{\sqrt{6^3} \, 6^2}$

(d) $\dfrac{(2 \cdot 3^2)(4 \cdot \sqrt{3})}{(4 \cdot 3^{-5})(3^2)}$

Solution

(a) $\dfrac{3^3 3^5}{(3^4)^2} = \dfrac{3^{3+5}}{3^{4 \cdot 2}} = \dfrac{3^8}{3^8} = 3^{8-8} = 3^0 = 1$

(b) $\dfrac{5^2 5^3}{\sqrt{5^4}} = \dfrac{5^{2+3}}{(5^4)^{1/2}} = \dfrac{5^5}{5^2} = 5^{5-2} = 5^3$

(c) $\dfrac{6^4 \sqrt{6}}{\sqrt{6^3} \, 6^2} = \dfrac{6^{4+1/2}}{6^{3/2+2}} = \dfrac{6^{9/2}}{6^{7/2}} = 6^{9/2 - 7/2} = 6^1 = 6$

(d) $\dfrac{(2 \cdot 3^2)(4 \cdot \sqrt{3})}{(4 \cdot 3^{-5})(3^2)} = \dfrac{2 \cdot 3^{2+1/2}}{3^{-5+2}} = 2 \cdot \dfrac{3^{5/2}}{3^{-3}} = 2 \cdot 3^{5/2 + 3} = 2 \cdot 3^{11/2}$

Sec. 3.2 / EXPONENTS

Example 3.2 Using the laws of exponents, simplify the following expressions and state the answers in terms of positive exponents:

(a) $\dfrac{2x^2y(x^{-4})}{3xy^2(y^{-3})}$

(b) $\dfrac{4x^{-2}y^2(x^{1/2}y^{-1})}{2xy(x^{-1/2}y)}$

Solution

(a) $\dfrac{2x^2y(x^{-4})}{3xy^2(y^{-3})} = \dfrac{2x^{-2}y}{3xy^{-1}} = \dfrac{2}{3}x^{-3}y^2 = \dfrac{2y^2}{3x^3}$

(b) $\dfrac{4x^{-2}y^2(x^{1/2}y^{-1})}{2xy(x^{-1/2}y)} = \dfrac{4x^{-3/2}y}{2x^{1/2}y^2} = 2x^{-2}y^{-1} = \dfrac{2}{x^2y}$

Example 3.3 Using the laws of exponents, simplify the following expressions and determine the numerical answers:

(a) $(\tfrac{2}{3})^4(2\cdot 3)^{-4}\cdot 2^3\cdot 3^6$

(b) $(2\cdot 3)^2(4\cdot 5)^{-2}(\tfrac{5}{3})^3$

Solution

(a) $(\tfrac{2}{3})^4(2\cdot 3)^{-4}\cdot 2^3\cdot 3^6 = 2^4 3^{-4} 2^{-4} 3^{-4} 2^3 3^6$
$= 2^3 \cdot 3^{-2}$
$= \tfrac{8}{9}$

(b) $(2\cdot 3)^2(4\cdot 5)^{-2}(\tfrac{5}{3})^3 = 2^2\cdot 3^2\cdot (2^2)^{-2}\cdot 5^{-2}\cdot 5^3\cdot 3^{-3}$
$= 2^{-2} 3^{-1} 5$
$= \tfrac{5}{12}$

Example 3.4 Evaluate the following expressions:

(a) $(128)^{3/7}$ (b) $(243)^{3/5}$ (c) $(125)^{4/3}$ (d) $(64)^{2/3}$

Solution

(a) $(128)^{3/7} = (2^7)^{3/7} = 2^3 = 8$

(b) $(243)^{3/5} = (3^5)^{3/5} = 3^3 = 27$

(c) $(125)^{4/3} = (5^3)^{4/3} = 5^4 = 625$

(d) $(64)^{2/3} = (2^6)^{2/3} = 2^4 = 16$

EXERCISES

1. Use the laws of exponents to evaluate each of the following expressions:

(a) $\dfrac{2^3 3^2}{(3^6)^{1/2}}$

(b) $\dfrac{3^4 2^{-2}}{(\tfrac{3}{2})^3}$

(c) $\dfrac{(\tfrac{4}{3})^2(\tfrac{3}{2})^3}{(\tfrac{2}{3})^2}$

(d) $\dfrac{(\tfrac{3}{8})^2(\tfrac{2}{9})^3}{(8\cdot 9)^{-2}}$

2. Use the laws of exponents to simplify each of the following expressions:

(a) $\dfrac{x^2 y^3}{xy^2}$

(b) $\dfrac{(3xy^2)^2}{(x^2 y^2)^2}$

(c) $\dfrac{(x^2)^3 (y^2)^4}{(x^3)^3 (y^3)^2}$

(d) $\dfrac{x^2 y^3}{(y/x)^3 (x^3)^2}$

3. Simplify each expression and state the answers in terms of positive exponents only:

(a) $\dfrac{3x^2 y^3 z}{x^3 y z^2}$

(b) $\dfrac{(x^2 y)^3 (x/y^2)^4}{(x^3/y^2)^2}$

(c) $\dfrac{(2x+1)^3 (y+3)^2}{(2x+1)^2 (y+3)^{-4}}$

(d) $\dfrac{(x^3 y w^2)^2 (xy^2 w)^3}{(x^2 y/w)^3}$

4. Evaluate each expression:

(a) $2^3(1+2^{-4})$

(b) $\dfrac{3+5^{-2}}{2^2}$

(c) $\dfrac{2^{-3}+2^{-2}}{2^{-1}}$

(d) $2^{-2}(1+3^{-2})$

5. Simplify each expression:

(a) $\dfrac{(x^{1/2})^3 (y^{2/3})^2}{(x^{3/8})^4 (y^{2/3})^3}$

(b) $\dfrac{(x^{3/4})^2 (y^{1/2})^3}{(x^{1/2} y^{1/3})^6}$

(c) $\dfrac{(x^2 y)^{1/4} (y^3)^{1/2}}{(xy^2)^{1/3} (x^2)^{1/3}}$

(d) $\dfrac{(xy^{1/2})^{1/2} (x^{1/4} y^{1/2})^2}{(xy^2)^{1/4}}$

6. Evaluate each expression:

(a) $9^{3/2}$

(b) $(256)^{3/4}$

(c) $(16)^{-5/4}$

(d) $\left(\dfrac{9 \cdot 16}{25}\right)^{3/2}$

3.3 EXPONENTIAL FUNCTIONS

Any function $f(x)$ of the form

$$f(x) = a \cdot b^x \tag{3.7}$$

where a and b are nonzero real numbers such that b is positive and not equal to 1, is called an *exponential function* because the independent variable x

appears in the function as an *exponent*. Exponential functions are important in economics and business research because they can be used to describe growth and decline relationships among variables. The domain of the exponential function is the set of all real values x, and its range is either the set of all positive real numbers if a is positive or the set of all negative real numbers if a is negative. It can easily be seen that the sign of the function values is the same for all values of x; it is determined by the sign of a because b is required to be positive.

The graph of an exponential function can be determined essentially by the three key points $(-1, a/b)$, $(0, a)$, and $(1, ab)$. However, we may wish to plot more points in order to obtain a better idea of how the graph actually behaves in certain areas, but, with only the three key points, we can sketch the graph with reasonable accuracy. Exponential functions are similar to quadratic functions in that only three key points are required to sketch the graph of either function, although five points are *recommended* for quadratic functions.

Another characteristic of the exponential function is that its values change by a *constant percentage* with a constant change in the independent variable x. For example, when x increases from 1 to 2, 3, 4, and so forth, for the exponential function

$$f(x) = 3 \cdot 2^x$$

the function value $f(x)$ increases from 6 to 12, 24, 48, and so forth. Thus the percentage *change* in $f(x)$ for a unit change in x is a constant 100 percent regardless of which two x values we choose. That is, when x changes from 1 to 2, $f(x)$ changes from 6 to 12, and when x changes from 2 to 3, $f(x)$ changes from 12 to 24. Similarly, when x changes from 3 to 4, $f(x)$ changes from 24 to 48. Thus, in each case, a *one*-unit change in x produces a 100 percent change in $f(x)$. For exponential functions for which b is different from 2, the percentage change will differ from 100 percent.

Property 3.1 *If a is positive, the percentage change will consist of a $(b-1)100$ percent increase over the previous value for a one-unit change in x.*

The verification of Property 3.1 and the following statements is left as an exercise for the reader. If b is less than 1 and a is positive, the percentage increase will actually be negative, which indicates a decrease, i.e., a negative increase. If a is negative, the above increase simply becomes a decrease, and a negative decrease means a positive increase. In general, if a is negative, $(b-1)100$ will represent the percentage *decrease* over the previous value.

To better understand the characteristics of exponential functions, let us consider some examples.

Example 3.5 For the exponential function

$$f(x) = 3 \cdot 2^x$$

sketch the graph and determine the constant percentage increase of the function value by comparing the increases obtained for a one-unit increase in x for two such increases.

Solution To sketch the function, we determine the three key points $(-1, a/b)$, $(0, a)$, and $(1, ab)$, where $a = 3$ and $b = 2$. Thus, replacing $f(x)$ by the variable y and computing the values for the three key points, we obtain

$$y = 3 \cdot 2^x$$

and $(-1, \frac{3}{2})$, $(0, 3)$, and $(1, 6)$. Since we are not very familiar with the general shape of an exponential curve, we compute the values for several points on the graph in addition to the three key points, namely,

x	-3	-2	-1	0	1	2	3
y	$\frac{3}{8}$	$\frac{3}{4}$	$\frac{3}{2}$	3	6	12	24

Plotting these points and connecting them with a smooth curve, we obtain the graph given in Fig. 3.1.

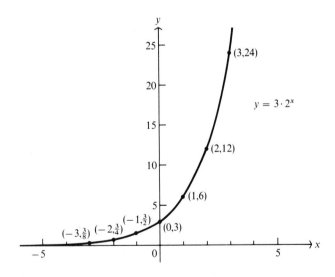

FIGURE 3.1

To determine the constant percentage increase, we may choose two sets of any two points for which the x values differ by one. Let us select $(-3, \frac{3}{8})$ and $(-2, \frac{3}{4})$ for one increase and $(2, 12)$ and $(3, 24)$ for the second increase. The *percentage increase* for a one-unit increase in x in the first set of points is determined by subtracting $\frac{3}{8}$ from $\frac{3}{4}$ to obtain the actual *increase* in the y value, dividing this difference by $\frac{3}{8}$, and multiplying the result by 100. The

Sec. 3.3 / EXPONENTIAL FUNCTIONS

percentage increase for the one-unit increase in x from -3 to -2 is

$$\frac{\frac{3}{8}}{\frac{3}{8}} = 1$$

or a 100 percent increase. For the second point, we obtain

$$\frac{12}{12} = 1$$

or a 100 percent increase. This verifies that the constant percentage increase is 100 percent. We could also compute this constant percentage increase by substituting the value of b into $(b-1)100$. If we do so, we obtain $(2-1)100 = 100$ percent.

In Fig. 3.1, we note that the graph of the exponential curve to the left of zero approaches the x axis as x *decreases*; however, it never actually touches the axis. When this happens, we say that the x axis is an *asymptote* of the curve. Intuitively we say that an asymptote of a curve is a line that the curve approaches more and more closely as x either increases or decreases, such that the curve may come as close as possible to the line but never actually touches the line. For an exponential function of the form given in (3.7), the asymptote will always be the x axis. A function that has an asymptote is said to *asymptotically* approach the line representing the asymptote.

Example 3.6 Sketch the graph and determine the constant percentage change of the growth rate of the sales of a car dealership, as represented by the exponential function

$$f(x) = (.8)^x$$

where x is the increase, in hundreds of dollars, in the price per car sold.

Solution To sketch the graph, we first determine the three key points $(-1, a/b)$, $(0, a)$, and $(1, ab)$. Since $a = 1$ and $b = .8$, we obtain the points

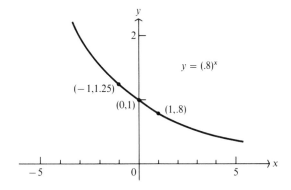

FIGURE 3.2

$(-1, 1.25)$, $(0, 1)$, and $(1, .8)$. Plotting these points and connecting them by a smooth curve, we obtain the graph given in Fig. 3.2.

To determine the constant percentage change, we evaluate $(b-1)100$ and obtain $(.8-1)100 = -20$ percent. This means that the growth rate of the dealership sales *decreases* by 20 percent for each one-unit *increase* in x.

Example 3.7 Sketch the graph and determine the constant percentage change for the change in the demand of rocking chairs. The change in demand is represented by the exponential function

$$f(x) = -2 \cdot 3^x$$

where x is the dollar increase in the price of a rocking chair.

Solution Since $a = -2$ and $b = 3$, we obtain the three key points $(-1, -\frac{2}{3})$, $(0, -2)$, and $(1, -6)$. Plotting these points and connecting them by a smooth curve, we obtain the graph in Fig. 3.3.

Since a is negative, the percentage decrease is given by $(b-1)100 = (3-1)100 = 200$ percent. This can also be determined by using the values for x equal to 0 and 1; that is,

$$\frac{-6 - (-2)}{-2} = 2$$

or 200 percent.

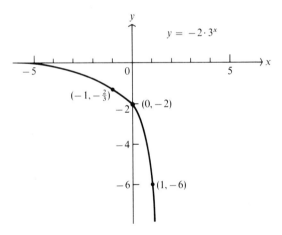

FIGURE 3.3

Example 3.8 Consider the exponential function that arises in determining the size of the market S for a product that is currently doing a sales volume of P dollars per year and growing at a rate of r per year. The formula for determining S is

$$S = P(1 + r)^n$$

Sec. 3.3 / EXPONENTIAL FUNCTIONS

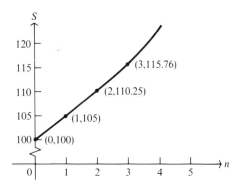

FIGURE 3.4

This formula fits the general exponential function form given in (3.7), where $a = P$, $b = 1 + r$, and $x = n$. Although n represents the number of years for which the sales are permitted to grow, it can actually assume any positive real value. It could also assume negative real values, but this gives a different meaning to the problem.

For the above exponential function with $P = 100$ and $r = .05$, sketch the graph of the function for nonnegative values of n and determine the constant percentage change.

Solution Substituting the appropriate values into the function, we obtain the exponential function

$$S = 100(1.05)^n$$

Computing and plotting the following values, we obtain the graph given in Fig. 3.4.

n	0	1	2	3
S	100	105	110.25	115.76

The graph does not appear to differ much from a straight line because b is *close* to 1; when b is 1, the function is a straight line. Hence this is why we do *not* permit $b = 1$ in the definition of an exponential function. The constant percentage change from one year to the next (i.e., a one-unit increase in x) is $(b - 1)100 = (1.05 - 1)100 = 5$ percent.

EXERCISES

Sketch the graph and determine the constant percentage change for each of the exponential functions in Exercises 1 to 10.

1. $f(x) = 4 \cdot 3^x$ **2.** $f(x) = 2 \cdot 10^x$ **3.** $f(x) = -2 \cdot 5^x$

4. The exponential function that arises in determining the amount of money for an investment of $50 at an annual interest rate of 30 percent

5. $f(x) = 3 \cdot (6.82)^x$ **6.** $f(x) = -(11.64)^x$ **7.** $f(x) = 4 \cdot 12^x$

8. $f(x) = -3 \cdot (6.5)^x$ **9.** $f(x) = 6 \cdot (.64)^x$ **10.** $f(x) = -2 \cdot (.8)^x$

11. For the past 5 years, a manufacturer has noticed that the rate of increase in sales of one of its products is decreasing. Sales records indicate that the rate of increase is two times 0.5^x, where x is the number of years since the product was introduced into the market. Determine and graph the equation to compute the rate of increase in sales each year if the current trend continues.

12. The operating time per week of a machine is equal to the number of hours per week the shop is open times 0.95 raised to a power equal to the number of jobs done on the machine that week. Determine and graph the equation to compute the operating time based on the number of jobs done if the shop is open 80 hr per week.

13. A chemical plant is facing the possibility of a strike and has asked you to help determine the possible cost if a strike does occur. The cost has been determined as $1,500 times 2 raised to a power equal to the number of days that the strike lasts. Determine and graph the equation to compute the cost of a strike based on the number of days the strike lasts.

3.4 LOGARITHMS

Logarithms are used in business primarily to simplify numerical calculations. Even though electronic calculators and computers are very widely used and will perform logarithmic computations for us, we must still understand the basic laws of logarithms to utilize calculators and computers. With logarithms we can reduce many complex operations of multiplication, exponentiation, and division to the simpler operations of addition and subtraction. Logarithms are related to exponents and, like them, conform to certain basic laws.

To examine the relationship between exponents and logarithms, let us return to the equation for the exponential functions considered in the last section:

$$y = a \cdot b^x \qquad (3.8)$$

To define a logarithm, we shall use the basic case $a = 1$; thus

$$y = b^x \qquad (3.9)$$

Definition 3.1 If b is a positive number and $y = b^x$, then the logarithm of y to the base b is a function that yields the value x and is written

$$\log_b y = x \qquad (3.10)$$

Functions defined by equations such as (3.10) are called *logarithmic functions*.

Sec. 3.4 / LOGARITHMS

In other words, the logarithm of a number y to the base b is the power to which b must be raised to obtain y. Thus we may write:

$$\log_b b^3 = 3$$
$$\log_b b^2 = 2$$
$$\log_b b^1 = 1$$
$$\log_b b^0 = 0$$
$$\log_b b^{-1} = -1$$
$$\log_b b^{-2} = -2$$
$$\log_b b^{-3} = -3$$

In the above relationships, we used only integers for the exponents, but the same is true for any real-number exponent of b. For example,

$$\log_b b^{1/2} = \tfrac{1}{2}$$
$$\log_b b^x = x$$
$$\log_b b^{x+y} = x + y$$
$$\log_b b^{\sqrt{3}} = \sqrt{3}$$

The most commonly used base for logarithmic computations is 10; thus the logarithms that have base 10 are called *common logarithms*. The first table of such logarithms was published early in the seventeenth century by the English mathematician Henry Briggs. One reason logarithms of base 10 are most commonly used is that the numbering system that is used almost exclusively in everyday situations is the decimal, or base 10, system.

Using Definition 3.1 and the decimal representation of 10^n for integer values of n, we can write the following:

$$\log_{10} 1000 = \log_{10} 10^3 = 3$$
$$\log_{10} 100 = \log_{10} 10^2 = 2$$
$$\log_{10} 10 = \log_{10} 10^1 = 1$$
$$\log_{10} 1 = \log_{10} 10^0 = 0$$
$$\log_{10} 0.1 = \log_{10} 10^{-1} = -1$$
$$\log_{10} 0.01 = \log_{10} 10^{-2} = -2$$
$$\log_{10} 0.001 = \log_{10} 10^{-3} = -3$$

As a matter of convention, the logarithm of a number to base 10 is generally written without the base 10; that is, whenever we encounter $\log N$, it is understood to mean $\log_{10} N$.

Basic Laws of Logarithms

Before we consider any computations with logarithms, let us consider the *basic laws of logarithms*. If we let

$$M = b^x$$
$$N = b^y$$

we have

$$MN = b^{x+y}$$

From the above three relations and the definition of a logarithm, we have

$$\log_b M = x$$
$$\log_b N = y$$
$$\log_b MN = x + y$$

Since the sum of the right side of the top two equations is equal to the right side of the third equation, the same must be true for the terms on the left side of the equations. Thus we have

$$\log_b MN = \log_b M + \log_b N \qquad (3.11)$$

This is one of the most important laws of logarithms: It states that *the logarithm of a product is simply the sum of the logarithms of the individual terms.* It can be extended to include the product of any number of terms for which the logarithm of the product will be the sum of the logarithms of the individual terms.

Another important law of logarithms is obtained using the notation $M = b^x$. By raising both terms to the Nth power, we obtain

$$M^N = (b^x)^N = b^{Nx}$$

Thus we have

$$\log_b M^N = Nx$$

However, since $x = \log_b M$, the above equation becomes

$$\log_b M^N = N \log_b M \qquad (3.12)$$

Hence we obtain another basic law concerning logarithms which states that *the logarithm of a number raised to a power is equal to the power times the logarithm of the number.*

We now have two basic laws of logarithms from which we can derive other laws. For example, we note that

$$\frac{M}{N} = MN^{-1}$$

Taking the logarithm of both sides to the base b, we obtain

$$\log_b \frac{M}{N} = \log_b MN^{-1}$$

Applying the rules given in (3.11) and (3.12) in that order, we obtain

$$\log_b \frac{M}{N} = \log_b MN^{-1} = \log_b M + \log_b N^{-1} = \log_b M - \log_b N$$

Thus we obtain a law pertaining to a quotient of two numbers that states *the logarithm of a quotient is the logarithm of the numerator minus the logarithm of the denominator.* In equation form, we have

$$\log_b \frac{M}{N} = \log_b M - \log_b N \qquad (3.13)$$

Now that we have at our disposal several laws concerning logarithms, let us consider the relationships between a number and its logarithm. Since logarithms to base 10 are most common, we shall consider the relationship between a number and its logarithm to base 10. Let us first consider the number that is the square root of 10: It can be written as $10^{1/2}$ or $10^{0.5}$, and by either longhand or trial and error, we can determine its approximate numerical value to be 3.1623. Using the exponential form of the number, we have the exponential form

$$3.1623 = 10^{0.5}$$

which, when converted to the logarithmic form, becomes

$$\log 3.1623 = 0.5000$$

Moreover, if we multiply both sides of the exponential equation by 10, 10^2, 10^3, and 10^4, we obtain

$$\begin{aligned} 31.623 &= 10^{1.5} \\ 316.23 &= 10^{2.5} \\ 3162.3 &= 10^{3.5} \\ 31623 &= 10^{4.5} \end{aligned} \qquad (3.14)$$

respectively. Converting the above exponential equations to their equivalent logarithmic equations, we have

$$\begin{aligned} \log 31.623 &= 1.5 \\ \log 316.23 &= 2.5 \\ \log 3162.3 &= 3.5 \\ \log 31623 &= 4.5 \end{aligned} \qquad (3.15)$$

respectively. We note that each time the exponent of 10 in (3.14) is increased by one unit the corresponding value of the logarithm in (3.15) increases by one unit. In other words, each time the decimal point is moved to the right one digit the corresponding value of the logarithm increases by 1.

The Mantissa and the Characteristic

Let us consider a more general exponent of 10 in order to more fully comprehend the relationship between exponents and logarithms. First we consider an exponent of 10 that is greater than or equal to 0 and less than 1. We shall represent this exponent by 0.MANTISSA, where the character to the left of the decimal point is zero and the term MANTISSA to the right of the decimal point represents any sequence of digits. Since the exponent 0.MANTISSA can be greater than or equal to 0, the number $10^{0.\underline{MANTISSA}}$ must be greater than or equal to 1. Similarly, since 0.MANTISSA is less than 1, the number $10^{0.\underline{MANTISSA}}$ must be less than 10. Hence the number $10^{0.\underline{MANTISSA}}$ will be greater than or equal to 1 and less than 10. Thus it will be equal to a number that will have exactly one digit to the left of the decimal point such that the digit will be nonzero. Let the decimal equivalent of $10^{0.\underline{MANTISSA}}$ be represented by N; the number N can equal 1, but it cannot equal 10. Expressing the above in equation form, we have

$$N = 10^{0.\underline{MANTISSA}} \tag{3.16}$$

Converting (3.16) to its logarithmic equivalent, we have

$$\log N = 0.\underline{MANTISSA} \tag{3.17}$$

Equation (3.17) tells us that the logarithm to base 10 of any number between 1 and 10 will be between 0 and 1. Moreover, if we were to multiply N by 10^K, we would obtain the exponential equation from Eq. (3.16)

$$N\,10^K = 10^{0.\underline{MANTISSA} + K} \tag{3.18}$$

The logarithmic equivalent of Eq. (3.18) is

$$\log(N\,10^K) = 0.\underline{MANTISSA} + K \tag{3.19}$$

Another way that this same relationship can be obtained is through use of Eqs. (3.11) and (3.12); that is,

$$\log(N\,10^K) = \log N + \log 10^K$$
$$= 0.\underline{MANTISSA} + K \log 10$$
$$= 0.\underline{MANTISSA} + K$$

Any real number can be expressed in the form given on the left side of Eq. (3.18) when K is permitted to assume *any integer* value. All that is

Sec. 3.4 / LOGARITHMS

required is that the decimal point be moved in either direction until there is only one digit to the left of the decimal point and this digit must be nonzero. The numerical value of K for the exponent of 10 is equal to the number of digits for which the decimal point is moved; and the sign of K is negative if the decimal point is moved to the right and positive if it is moved to the left. When a number is expressed in this form, it is said to be in *standard*, or *scientific*, form, and K is called the *characteristic* of the number. For example, the following numbers are expressed in decimal form and standard form:

$$265 = 2.65 \cdot 10^2$$
$$26.5 = 2.65 \cdot 10^1$$
$$2.65 = 2.65 \cdot 10^0$$
$$0.265 = 2.65 \cdot 10^{-1}$$
$$0.0265 = 2.65 \cdot 10^{-2}$$

Common Logarithms In order to determine the common logarithms of the above numbers, it is also necessary that we determine the MANTISSA portion of Eq. (3.17). This value is called the *mantissa* of N, where N is expressed in standard form and is determined from a *table of common logarithms*, for example, Table A.1 in the Appendix. The values in the body of the Table A.1 represent the first four digits for the *mantissa* of a number. (For more accurate computations, a table containing more than four digits is required.) To determine the mantissa of N, we must first locate N and then locate its mantissa. Let us select for N the number 2.65, which has a characteristic of 0. To find the mantissa of 2.65 in Table A.1, we go down the left column (which has N at the top) until we find the first two digits of 265, that is, 26. Next, we follow the row for 26 to the right until we intersect the column that has the third digit of 265 at the top, that is, 5. The value at the intersection of the 26 row and the 5 column is the mantissa of 2.65. Thus from the body of Table A.1, we read the mantissa of 2.65 to be 4232. Substituting into Eq. (3.17), we have

$$\log 2.65 = 0.4232$$

The common logarithms of the five numbers listed above are given below:

$$\log 265 = \log(2.65 \cdot 10^2) = 0.4232 + 2 = 2.4232$$
$$\log 26.5 = \log(2.65 \cdot 10^1) = 0.4232 + 1 = 1.4232$$
$$\log 2.65 = \log(2.65 \cdot 10^0) = 0.4232 + 0 = 0.4232$$
$$\log 0.265 = \log(2.65 \cdot 10^{-1}) = 0.4232 - 1 = -0.5768$$
$$\log 0.0265 = \log(2.65 \cdot 10^{-2}) = 0.4232 - 2 = -1.5768$$

As can be seen, the logarithms of numbers that have negative characteristics are also negative, and, furthermore, the digits to the right of the decimal point are not those of the manitssa. Consequently, since we shall also be using mantissas to look up numbers in the logarithm table, it is preferable to leave the logarithm of the number expressed in terms of the mantissa and the characteristic, such as in Eq. (3.19), *when the characteristic is negative.* Thus we would prefer to write

$$\log 0.265 = 0.4232 - 1$$

$$\log 0.0265 = 0.4232 - 2$$

However, for those situations where we need to convert a negative logarithm to the form of Eq. (3.19), we choose the positive integer that is one unit larger (regardless of how many digits) than the number to the left of the decimal point and add it to the negative number. We then write after this result minus the number that we added; for example, the logarithm with a value of -6.5072 would be expressed as

$$-6.5072 + 7 - 7 = 0.4928 - 7$$

We can readily determine the mantissa to be 4928 and the characteristic to be -7.

The Antilog   We have considered how to convert an exponential equation into logarithmic form, and now let us consider how to convert a logarithmic equation into an exponential equation. To do this, we must reverse the above process by converting an equation of the form of Eq. (3.19) into an equation of the form of (3.18). In particular, when we are seeking the logarithm of a number given by the left side of Eq. (3.18), we are seeking the value on the right side of Eq. (3.19). Conversely, when we are seeking the number whose logarithm is given by the right side of (3.19), we are seeking the value on the left side of (3.18). We shall refer to the process of converting a logarithm of the form (0.MANTISSA + K) into a standard number of the form $N \cdot 10^K$ as taking the *antilog* of (0.MANTISSA + K).

In order to determine the antilog of a number, we determine the characteristic and the number whose mantissa is given; this is done by determining the mantissa and finding its value in the body of Table A.1. The number associated with the mantissa is then formed by reading the first two digits from the far left column of the row in which the mantissa appears and the third digit from the top of the column in which it appears. For example, if the logarithm is 2.5966, that is, if the mantissa is 5966 and the characteristic is 2, we enter the body of Table A.1 and locate 5966. Locating 5966 in Table A.1, we read 39 from the left column of the row in which 5966 appears and 5 from the top of the column in which it appears. Thus 3.95 is the standard form of the number having 5966 as its mantissa. This number combined with the characteristic of 2 gives the antilog of 2.5966, and for this number we

Sec. 3.4 / LOGARITHMS

write

$$\text{antilog}(2.5966) = 3.95 \cdot 10^2 = 395$$

Similarly, if the logarithm is $0.8451-3$, we know that the mantissa is 8451 and the characteristic is -3. From Table A.1, we determine that 7.00 is the standard form of the number having 8451 as its mantissa. Thus

$$\text{antilog}(0.8451 - 3) = 7.00 \cdot 10^{-3} = 0.00700$$

Example 3.9 Determine the common logarithms of the following numbers. For those numbers with negative characteristics, express the logarithms in the form given by Eq. (3.19) without performing the subtraction.

(a) 731 (b) 298 (c) 1.27
(d) 0.00265 (e) 0.0000061 (f) 0.998

Solution Each number is first expressed in standard form. The characteristic and mantissa are then determined, and the logarithm is determined by combining the characteristic and the mantissa as given in Eq. (3.19):

(a) $731 = 7.31 \cdot 10^2$
$\log 731 = 0.8639 + 2 = 2.8639$

(b) $298 = 2.98 \cdot 10^2$
$\log 298 = 0.4742 + 2$
$ = 2.4742$

(c) $1.27 = 1.27 \cdot 10^0$
$\log 1.27 = 0.1038 + 0 = 0.1038$

(d) $0.00265 = 2.65 \cdot 10^{-3}$
$\log 0.00265 = 0.4232 - 3$

(e) $0.0000061 = 6.10 \cdot 10^{-6}$
$\log 0.0000061 = 0.7835 - 6$

(f) $0.998 = 9.98 \cdot 10^{-1}$
$\log 0.998 = 0.9991 - 1$

Example 3.10 Determine the antilog of each of the following numbers and express it in decimal form:

(a) 2.8319 (b) 1.6911 (c) 0.4609
(d) $0.4713 - 2$ (e) -1.3233 (f) -2.5129

Solution Each number is first expressed in the form of Eq. (3.19), and then the characteristic and mantissa are determined. The number corresponding to the mantissa is found from Table A.1, and the antilog is determined in standard form, which is then converted to decimal form:

(a) $2.8319 = 0.8319 + 2$
$\text{antilog}(2.8319) = 6.79 \cdot 10^2 = 679$
(b) $1.6911 = 0.6911 + 1$
$\text{antilog}(1.6911) = 4.91 \cdot 10^1 = 49.1$
(c) $0.4609 = 0.4609 + 0$
$\text{antilog}(0.4609) = 2.89 \cdot 10^0 = 2.89$
(d) $0.4713 - 2$
$\text{antilog}(0.4713 - 2) = 2.96 \cdot 10^{-2} = 0.0296$
(e) $-1.3233 = 0.6767 - 2$
$\text{antilog}(-1.3233) = 4.75 \cdot 10^{-2} = 0.0475$
(f) $-2.5129 = 0.4871 - 3$
$\text{antilog}(-2.5129) = 3.07 \cdot 10^{-3} = 0.00307$

EXERCISES

1. Does the logarithm to base 5 of 25 y equal 2?
2. Does the logarithm to base 2 of 9 y equal 3?
3. Does $\log 75 - \log 25 = \log 50$?
4. Does $\log 16 - \log 8 = 2$?
5. Does $\log 144 - \log 12 = \log 12$?
6. Does $\log 75 / \log 25 = \log 75 - \log 25$?
7. Does $(\log 15)(\log 5) = \log 15 + \log 5$?
8. Does $(\log 5)^2 = 2 \log 5$?
9. Does $(\log 7)^3 = \log 7^3$?
10. Does antilog($\log 15 + \log 5$) = antilog($\log 15$) + antilog($\log 5$)?
11. Determine the logarithm of each of the following numbers and express it such that the decimal portion is always positive:

    (a) 126 (b) 0.721 (c) 698 (d) 5.61
    (e) 72.3 (f) 0.00151 (g) 0.0230 (h) 9.96

12. Determine the antilog of each of the following numbers and express it in decimal form:

    (a) 1.8591 (b) −0.1421 (c) 2.1004 (d) 0.9983
    (e) 0.7490 (f) −2.8210 (g) 2.8439 (h) −1.6383

3.5 NATURAL LOGARITHMS

An exponential function of special interest is the one for which the base b is the *natural constant e*. The constant e is natural, just as the constant π (approximately equal to 3.14159) is natural. That is, the constant π is the ratio of the circumference of a circle to its diameter, and this ratio is constant for any circle. Similarly, the constant e is defined as the *limit* of the term $[1+(1/n)]^n$ as n increases without bounds and is approximately 2.718. More accurately, the value of e to 15 decimal places is 2.718281828459045. Furthermore, it is a never-ending decimal number that is also irrational, i.e., one that cannot be written as the ratio of two integers. (Recall that the rules for exponents and logarithms are valid for any positive base that does not equal 1. Thus we can utilize all previously developed rules for exponents and logarithms.)

The primary purpose for using a base of e is because of the relative simplicity of certain results in calculus when e is used as the base. This simplicity is of sufficient importance that tables of logarithms and exponents with base e are available. The logarithms are called *natural*, or *Naperian*, *logarithms* and are presented in Table A.9 in the Appendix. Table A.10 presents values of e^x for selected values of x. Natural logarithms are denoted by ln instead of the more cumbersome $\log_e$. As stated previously, the use of natural logarithms is mainly related to the simplicity of results obtained for

Sec. 3.5 / NATURAL LOGARITHMS

certain calculus problems. The use of *common* logarithms is usually the most convenient way to perform numerical computations, provided a choice exists.

We shall now consider an example for determining the values of $\ln x$ for various values of x. An explanation of how to determine the natural logarithm of a number is given at the beginning of Table A.9, and the values of $\ln x$ are given in the body of the table for x from 1.00 to 9.99. (The logic behind the explanation in Table A.9 is that of expressing a number in its standard form and taking the logarithm of a product.)

Example 3.20 Use Table A.9 to determine the following values:

(a) $\ln 6.45$ (b) $\ln 0.723$ (c) $\ln 19.6$
(d) $\ln 0.0644$ (e) $\ln 263$ (f) $\ln 0.0061$

Solution

(a) Directly from Table A.9, we obtain the value of $\ln 6.45$ by locating 6.4 in the column under N and moving across the row until we intersect the column for 5; thus $\ln 6.45 = 1.86408$.

(b) We express

$$\ln 0.723 = \ln (7.23 \cdot 10^{-1})$$
$$= \ln 7.23 - \ln 10$$

Determining the values from Table A.9, we have

$$\ln 0.723 = 1.97824 - 2.30259$$
$$= -0.32435$$

(c)
$$\ln 19.6 = \ln(1.96 \cdot 10^1)$$
$$= \ln 1.96 + \ln 10$$
$$= 0.67294 + 2.30259$$
$$= 2.97553$$

(d)
$$\ln 0.0644 = \ln(6.44 \cdot 10^{-2})$$
$$= \ln 6.44 - 2\ln 10$$
$$= 1.86253 - 4.60517$$
$$= -2.74264$$

(e)
$$\ln 263 = \ln(2.63 \cdot 10^2)$$
$$= \ln 2.63 + 2\ln 10$$
$$= 0.96698 + 4.60517$$
$$= 5.57215$$

(f)
$$\ln 0.0061 = \ln (6.1 \cdot 10^{-3})$$
$$= \ln 6.1 - 3\ln 10$$
$$= 1.80829 - 6.90776$$
$$= -5.09947$$

A property of interest is that every term of the form b^x can be expressed as a term of the form e^y. To establish the relationship, we set the two as equal and solve for y in terms of x. Thus we have $b^x = e^y$, and, taking the natural logarithm of both sides, we obtain

$$\ln b^x = \ln e^y$$

or

$$x \ln b = y \ln e$$

Thus since $\ln e = 1$, we have $y = x \ln b$. Therefore, we can always convert b^x to an exponent with base e; that is,

$$b^x = e^{x \ln b} \tag{3.20}$$

Example 3.21 Convert the following exponential functions to an exponential function with base e:

(a) 8^x (b) 6^x (c) 3.4^x

Solution

(a) Determining from Table A.9 that $\ln 8 = 2.07944$ and substituting into Eq. (3.20), we obtain

$$8^x = e^{2.07944x}$$

(b) $$6^x = e^{x \ln 6} = e^{1.79176x}$$

(c) $$3.4^x = e^{x \ln 3.4} = e^{1.22378x}$$

EXERCISES

Determine the natural logarithm for each of the numbers in Exercises 1 to 14.

1. 6.21
2. 5.36
3. 1.33
4. 9.87
5. 3.65
6. 10.9
7. 89.4
8. 0.614
9. 0.520
10. 0.0042
11. 153
12. 2,450
13. 505,000
14. 0.0000693

In Exercises 15 to 20 convert each of the exponential functions to an exponential function with base e.

15. 10^x **16.** 7^x

17. 5^x **18.** 2^x

19. 1.2^x **20.** 4.32^x

3.6 LOGARITHMIC FUNCTIONS

A function of the form

$$f(x) = \log_b x \qquad (3.21)$$

where x is a positive real number, is a *logarithmic* function. The graph of a logarithmic function is similar to the graph of an exponential function in that three points are generally sufficient to determine the graphic characteristics of the function. Points that can be used to sketch the graph of the function given in (3.21) can be obtained by using the property that $\log_b b = 1$. Three convenient points are $(1/b, -1)$, $(1,0)$, and $(b,1)$; these points are plotted in Fig. 3.5 for $b=2$ and are connected by a smooth curve. It can be seen that the function value decreases rapidly as x *decreases* toward zero *from* the positive direction. The logarithmic function differs from the exponential function in that the asymptote for the exponential function is the x axis and the asymptote for the logarithmic function is the y axis.

If we wish to plot more than three points, we could continue to use the property that $\log_b b = 1$ and obtain points for which the x value is of the form b^n for positive and negative values of n. For example, we could obtain the points $(1/b^2, -2)$, $(1/b^3, -3)$, $(1/b^4, -4)$, and so on, and $(b^2, 2)$, $(b^3, 3)$, $(b^4, 4)$, and so on. We can see that there is no shortage of easily computed values that can be plotted.

One of the great conveniences of working with logarithmic functions is that through the use of the basic laws of logarithms, the logarithm of a rather complex term usually can be reduced to a simple expression involving sums and differences of functions similar to that of (3.21).

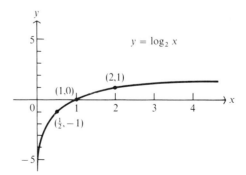

FIGURE 3.5

IMPORTANT TERMS AND CONCEPTS	Antilog Asymptote Asymptotically Base Characteristic Common logarithm Exponent Exponential function	Laws of exponents Laws of logarithms Logarithmic functions Logarithms Mantissa Natural logarithm Scientific form Standard form

REVIEW PROBLEMS

1. Simplify each of the following expressions and express the answers in terms of positive exponents only:

(a) $\dfrac{(3X)^2(Y^2Z^3)^3}{3Y(XZ^2)^4}$

(b) $\dfrac{X^2Y(Y^2Z)^2}{(XZ^2)^3(X^2Y)^2}$

(c) $\dfrac{(X^2Y^3)^2(X+1)^{-2}}{(X^3Y^2Z)^3(X+1)}$

(d) $\dfrac{(X^2Y^{-3})^2(XZ^{-3})^3}{Y^{-2}(X^2Y)^3}$

2. Simplify each of the following expressions:

(a) $\dfrac{(X^4Y^3)^{1/2}(XY)^{1/3}}{(X^4Y)^{1/3}(X^5Y)^{1/2}}$

(b) $\dfrac{(X-1)^{1/2}(XZ^2)^2}{X^2Y^3(X-1)^{3/2}}$

(c) $\dfrac{(X^2Y^{1/3})^2(Z^2)}{(X^{1/2}Y^2)^3(YZ)^{1/3}}$

(d) $\dfrac{(X^2Y^3)^{-1/2}(XY^2Z)^2}{(X^3Y)^{1/2}(Y^2Z^{-1})^3}$

Sketch the graph and determine the constant percentage change for each of the exponential functions in Exercises 3 to 6.

3. The exponential function that arises in determining the amount of money for a compound interest investment of $30 at an annual interest rate of 10 percent

4. The exponential function that arises in determining the amount to repay on a loan of $120 borrowed at an annual interest rate of 8 percent

5. The exponential function that arises in determining the amount of money for an investment of $75 at an annual interest rate of 25 percent

Chap. 3 / REVIEW PROBLEMS

6. The exponential function that arises in determining the collection amount if you loaned $10 at a monthly rate of 1 percent

Use logarithms to determine the values for the computations in Exercises 7 to 11.

7. $(65.4)^3$

8. $\dfrac{(2.4)(6.3)}{(10.2)}$

9. $\dfrac{(0.006)^{1/2}(0.98)}{4.01}$

10. $\dfrac{(22.4)^{1/2}(3.01)}{(21.4)} + \dfrac{(462)^{1/2}(20.5)^{1/2}}{(264)^{1/3}}$

11. Determine the amount to be repaid on a 7-month loan of $80 at a monthly interest rate of 1.2 percent.

12. Determine the natural logarithm for:

(a) 12.4
(b) 514
(c) 0.643
(d) 7.96

13. Convert each of the following exponential functions to an exponential function with base e:

(a) 4^x
(b) 7.7^x
(c) 3^x
(d) 3.68^x

SYSTEMS OF LINEAR EQUATIONS AND MATRIX ALGEBRA

4.1 SYSTEMS OF LINEAR EQUATIONS

On many occasions, the variables involved in a problem must satisfy more than one condition. For example, a manufacturer of tables and chairs must decide how many of each to produce in order to use all of the available time in the cutting and finishing departments. In the cutting department, a table requires 2 hr of cutting time, and a chair requires 4 hr. In the finishing department, a table requires 1 hr of finishing time, and a chair requires 5 hr. The cutting and finishing departments have available 160 hr and 170 hr, respectively, for the next week. Thus if x and y represent the number of tables and chairs produced, respectively, the times used in the cutting and finishing departments are given by the following equations:

$$2x + 4y = 160$$
$$x + 5y = 170$$

The values of x and y must satisfy both equations simultaneously. Consequently, the two equations are said to form a system of linear equations. In this chapter, we are concerned with obtaining solutions to such systems of linear equations. In Chap. 2, we considered linear equations and their graphs. The form $cx + dy = e$ was referred to as the *standard form* of the equation of a straight line. The values x and y are variables in that their values *vary*. However, the symbols c, d, and e do not vary but are constant for a specific linear equation. Such values are called *parameters*; and they may vary from one equation to the next, but for a given equation their values are fixed. For example, both $2x + 3y = 6$ and $3x + 4y = 10$ are linear equations, and x and y are the variables; however, the values of the parameters are different.

In considering the graph of an equation, we normally fix the value of the independent variable and determine the value of the dependent variable through use of the function rule. Previously we have considered equations involving only two variables. However, we could have equations such as

$$2x + 3y + 4z = 30$$
$$2p + 5q - 6r + s = 100$$

or

$$x_1 + 2x_2 + 4x_3 - 2x_4 - 3x_5 = 21$$

which are *linear equations* in *three, four,* and *five* variables, respectively. The graphs of linear equations in more than two variables are not straight lines, but are planes in three dimensions and hyperplanes in dimensions that cannot be visualized.

As can be seen from the above equations. it becomes increasingly more difficult to assign different letters for the variable names. In the linear equation in five variables, we used numerical *subscripts* for the variable x. The subscripts serve the purpose of identifying each variable, doing so by using only one letter for the variable name. When the number of variables in an equation is small, i.e., three or less, the variables are generally denoted by different letters. However, when more than three variables are involved, subscripted variables are generally used.

Generally the linear equations are expressed in standard form. For example, the following pair of linear equations constitutes a system of linear equations, or more explicitly, a system of *two* linear equations in *two* unknowns:

$$5x + 3y = 11$$
$$3x + 2y = 7$$

If all the equations are not initially represented in standard form, they should be changed to standard form.

When we encounter a system of linear equations, we generally wish to determine the *solution to the system of linear equations.* In order to understand what this means, let us consider a system of equations involving only two variables. The graph of each two-variable linear equation is a straight line, and there is one straight line for each equation. Thus, for a system of two linear equations in two unknowns (or variables), the graph of the equations in the system is two straight lines, as shown in Fig. 4.1*a*, *b*, and *c*. Thus when we speak of the *solution* to a system of linear equations, we mean *those values of the variables that satisfy each equation in the system.*

For a geometrical interpretation of the solution to a system of two equations in two unknowns, we refer again to Fig. 4.1*a*, *b*, and *c*. If the graph of the two equations is as shown in Fig. 4.1*a*, the solution to the system is given by the (x,y) coordinates of the point at which the two lines intersect because this represents the only values of x and y that satisfy both equations.

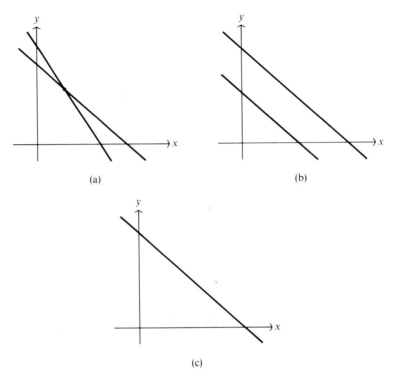

FIGURE 4.1

If the graph of the equations is as shown in Fig. 4.1b, the straight lines are parallel and consequently have no points in common. This means that there is no pair of values (x,y) that will satisfy both equations, and in this situation the system of linear equations has *no solution*. If the graph of the equations is as shown in Fig. 4.1c, the graph of the two equations is the same straight line. In this case, we have an unlimited number of solutions, and the two straight lines coincide and are called *coincident lines*.

The reader should verify that the following three systems of linear equations represent the three different possibilities shown in Fig. 4.1a, b, and c. A system that has a point of intersection, as shown in Fig. 4.1a, is given by

$$3x + 2y = 7$$
$$5x + 3y = 11$$

A system for which the graphs are similar to those shown in Fig. 4.1b is given by

$$2x + 3y = 10$$
$$4x + 6y = 24$$

Sec. 4.1 / SYSTEMS OF LINEAR EQUATIONS

A system that has coincident lines is given by

$$3x + 2y = 7$$
$$6x + 4y = 14$$

If a system of linear equations contains three equations in two unknowns, the graphs of the equations will consist of various combinations of the graphs shown in Fig. 4.1. The graph of the system will consist of three straight lines and will be one of the combinations given in Fig. 4.2. In Fig. 4.2a, a *unique solution* is indicated by the common point of intersection of all three lines. In Fig. 4.2b, c, and d, *no solution* is indicated because the three lines do not *all*

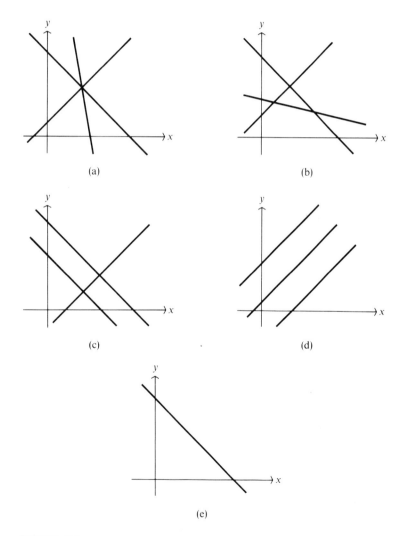

FIGURE 4.2

intersect at a common point; hence all three equations in the system cannot be satisfied by a single combination of x and y. Similarly, in Fig. 4.2e an *infinite number of solutions* is indicated because all three lines coincide and any point on one of the lines is also on the others.

This system of linear equations is one in which the number of equations exceeds the number of unknowns, and usually the graph for such a system will be similar to the one in Fig. 4.2b. When such is the case, we say that the system of equations is *overconstrained*. On the other hand, when there are fewer equations than unknowns, we usually have an *underconstrained* system for which there is an infinite number of solutions. A simple example of this is the system of one equation in two unknowns, i.e., a linear equation; the graph of the system constitutes one straight line and any point on that straight line satisfies each (i.e., one) equation in the system. In general, *the number of solutions to any system of linear equations is either zero, one, or an infinite number.*

4.2 METHODS OF SOLUTION

In the previous section, we considered the graphic representation of a system of linear equations. In this section, we shall consider several methods for determining the solution of a system of linear equations.

Method of Substitution

Suppose we wish to determine the solution of the following system of linear equations:

$$2x + 3y = 7$$
$$5x + 2y = 12 \tag{4.1}$$

One method of solving this system is to solve one of the equations for one variable in terms of the other variable, substitute this solution into the remaining equation, and solve for the value of the only remaining variable. For example, let us solve the first equation for x in terms of y. We obtain

$$x = \frac{7 - 3y}{2}$$

and, substituting this into the second equation in place of x, we have

$$5\left(\frac{7-3y}{2}\right) + 2y = 12$$

$$\frac{35 - 15y}{2} + 2y = 12$$

By expanding and multiplying both sides by 2, we obtain

$$35 - 15y + 4y = 24$$
$$-11y = -11$$

Sec. 4.2 / METHODS OF SOLUTION

Gathering terms and solving for y, we have

$$y = 1$$

We now have the two values

$$x = \frac{7 - 3y}{2}$$

and

$$y = 1$$

as the values for our solutions. However, we can substitute the numerical value of y into the term that expresses the value of x in terms of y and thus obtain a numerical value for x, that is

$$x = \frac{7 - 3(1)}{2} = 2$$

Therefore, the solution to the system of linear equations in (4.1) is $x = 2$ and $y = 1$. Relating this to the geometrical representation given in Fig. 4.1a, the solution of the system is the point (2,1), which is the point of intersection of the straight lines corresponding to the equations given in (4.1) and which is illustrated graphically in Fig. 4.3. This method of solution is called the *method of substitution*.

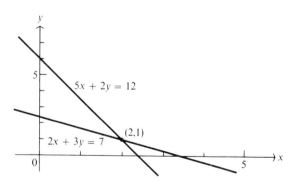

FIGURE 4.3

Example 4.1 The manager of a variety store has examined his sales receipts for 2 weeks. For the cosmetics department, he noted that the first week he sold 4 hair dryers and had 1 curling iron returned. The second week he sold 2 hair dryers and 1 curling iron. The profits for the first and second weeks were $14 and $4, respectively. Determine the system of linear equations and use the method of substitution to determine the profit for each hair dryer and curling iron.

Solution The system of linear equations is given by

$$4x - y = 14$$
$$2x + y = 4$$

where x and y represent the profit for each hair dryer and curling iron, respectively. Applying the method of *substitution*, we can solve the second equation for y in terms of x and obtain

$$y = 4 - 2x$$

Substituting the right side of this into the first equation in place of y, we have

$$4x - (4 - 2x) = 14$$

which yields $x = 3$. Substituting this into the value of y expressed in terms of x, we obtain

$$y = 4 - 2(3) = -2$$

Thus the solution to the above system of linear equations is $x = 3$ and $y = -2$. Therefore, the manager is making a profit of \$3 per hair dryer and a loss of \$2 per curling iron.

Method of Elimination by Addition or Subtraction

Another method for solution of a system of linear equations is known as the *method of elimination by addition or subtraction*. It is so-named because it utilizes the basic axioms of algebra and geometry that we may "add equals to equals," "subtract equals from equals," and "multiply equals by equals" and still have equivalent results. The purpose of this method is to *eliminate* one of the variables and solve for the value of the other variable.

Applying this method to the system given in (4.1), we may eliminate x from the two equations first by multiplying the first equation by 5 and the second equation by 2 to obtain an *equivalent* system of equations.

$$10x + 15y = 35$$
$$10x + 4y = 24$$

Next we subtract the second of the above equations from the first equation and obtain

$$11y = 11$$

which yields the value of $y = 1$. Substituting this value of y into either of the equations in (4.1), we obtain the associated x value. Choosing the first equation, we have $2x + 3(1) = 7$, which yields $x = 2$. Thus, by another method, the solution is again found to be $x = 2$ and $y = 1$.

Example 4.2 For the system of linear equations obtained in Example 4.1, use the method of elimination to determine the profit for each hair dryer and curling iron.

Sec. 4.2 / METHODS OF SOLUTION

Solution Applying the method of *elimination* to the system, we may add the second equation to the first and obtain

$$6x = 18$$

which yields $x = 3$. Substituting this value into either of the equations in the system will yield the value of y. Substituting into the second equation, we obtain

$$2(3) + y = 4$$

which yields $y = -2$. Therefore, the solution obtained through use of the method of elimination is $x = 3$ and $y = -2$, which is the same obtained previously by the method of substitution.

Gaussian Elimination Method The method of elimination requires that we determine respective numbers by which the first and second equations are multiplied in order to produce two new equations that have the same coefficient for the variable being eliminated. Rather than search for the two numbers, we could simply divide the first equation by its coefficient of the variable being eliminated. This would force the coefficient of the variable being eliminated to be equal to 1 for this equation. Thereafter it is a simple matter of subtracting a multiple of the first equation from the second equation and solving for the value of the remaining variable. For the system in (4.1), the elimination of x is accomplished by dividing the first equation by 2. We then obtain the following equivalent system:

$$x + \tfrac{3}{2} y = \tfrac{7}{2}$$
$$5x + 2y = 12$$

Subtracting five times the first equation from the second, we have the two equations

$$x + \tfrac{3}{2} y = \tfrac{7}{2}$$
$$(2 - \tfrac{15}{2}) y = 12 - \tfrac{35}{2}$$

or, by simplifying the terms and solving for y in the second equation,

$$x + \tfrac{3}{2} y = \tfrac{7}{2}$$
$$y = 1$$

With these equations, we can solve for the value of x by multiplying the second equation by $\tfrac{3}{2}$ and subtracting the result from the first equation. We then obtain the system

$$x = 2$$
$$y = 1$$

which is, in fact, the solution of the original system. This elimination procedure is known as the *gaussian elimination method* and is of paramount importance in solving linear programming problems.

Example 4.3 Use the gaussian elimination method to determine the profit for each hair dryer and curling iron for the system of linear equations obtained in Example 4.1.

Solution Recall that the system of equations is given by

$$4x - y = 14$$
$$2x + y = 4$$

Applying the method of *gaussian elimination*, we divide the first equation by 4 and obtain the system

$$x - \tfrac{1}{4}y = \tfrac{7}{2}$$
$$2x + y = 4$$

Subtracting two times the first equation from the second, we obtain the system

$$x - \tfrac{1}{4}y = \tfrac{7}{2}$$
$$\tfrac{3}{2}y = -3$$

Dividing the second equation of the resulting system by $\tfrac{3}{2}$ (that is, multiplying by $\tfrac{2}{3}$), we obtain the system

$$x - \tfrac{1}{4}y = \tfrac{7}{2}$$
$$y = -2$$

Multiplying the second equation by $\tfrac{1}{4}$ and adding the result to the first equation, we have

$$x = 3$$
$$y = -2$$

which is the solution obtained by the previous methods. The graphic representation of the original linear system is given in Fig. 4.4.

Let us use the gaussian elimination method to solve a general system of two linear equations in two unknowns. In so doing, we shall also determine a general symbolic solution for any system of two linear equations in two unknowns. We shall let the general system of two linear equations be represented by

$$a_{11}x_1 + a_{12}x_2 = b_1$$
$$a_{21}x_1 + a_{22}x_2 = b_2 \quad (4.2)$$

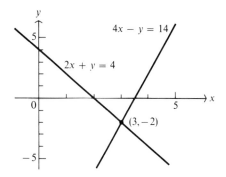

FIGURE 4.4

Even though there are only two variables, we have chosen to represent the variables by x_1 and x_2. The reasons for this notational change are that the system can easily be extended to include more variables (for example, x_3, $x_4, \ldots$) and more equations and that the subscripts on the parameters a_{ij} and b_i can be directly related to their location in the system. We note that x_1 and x_2 represent the first and second variables, respectively; thus x_i represents the ith variable in general. Also, we note that each coefficient of a variable has two subscripts; the first subscript indicates to which equation the coefficient belongs, and the second subscript indicates for which variable in that equation it is a coefficient. For example, a_{12} is the coefficient of the *second* variable in the *first* equation. In general, a_{ij} is the coefficient of the jth variable in the ith equation. The values on the right side of the equation have only one subscript, which denotes the equation to which they belong. For example, b_1 is the constant on the right side of the *first* equation, and b_2 is the constant on the right side of the *second* equation. In general, b_i is the constant on the right side of the ith equation.

The gaussian elimination method is well suited for machine computation, and generally such problems are indeed solved by computers. Now let us apply this method to determine the solution to the linear system in (4.2). For the sake of being definite, we assume that a_{11} and a_{22} are not equal to zero. If either or both of these quantities are zero, this does not prevent the use of the gaussian elimination method, but it means that we must choose a different equation for representing the solution value of the corresponding variable. In order to solve system (4.2) by use of the gaussian elimination method, we divide the first equation by a_{11} and obtain the system

$$x_1 + \frac{a_{12}}{a_{11}} x_2 = \frac{b_1}{a_{11}}$$

$$a_{21} x_1 + a_{22} x_2 = b_2$$

Multiplying the first equation of this system by a_{21} and subtracting this result from the second equation, we obtain the following system:

$$x_1 + \frac{a_{12}}{a_{11}}x_2 = \frac{b_1}{a_{11}}$$

$$\left(a_{22} - \frac{a_{21}a_{12}}{a_{11}}\right)x_2 = b_2 - \frac{a_{21}b_1}{a_{11}}$$

Dividing the second equation of this new system by its coefficient of x_2, provided it is nonzero, we obtain the following system, which contains the solution value for x_2:

$$x_1 + \frac{a_{12}}{a_{11}}x_2 = \frac{b_1}{a_{11}}$$

$$x_2 = \frac{a_{11}b_2 - a_{21}b_1}{a_{11}a_{22} - a_{21}a_{12}}$$

Multiplying the second equation by a_{12}/a_{11} and subtracting this result from the first equation, we have the system that gives the solution values for the original system in (4.2):

$$x_1 = \frac{a_{22}b_1 - a_{12}b_2}{a_{11}a_{22} - a_{21}a_{12}}$$

$$x_2 = \frac{a_{11}b_2 - a_{21}b_1}{a_{11}a_{22} - a_{21}a_{12}}$$

The purpose of the above exercise is not to have the reader memorize the relationships between the parameters that yield the solution values. Rather its purpose is to assist the reader in fully understanding the manipulations required to solve a system of linear equations by use of the gaussian elimination method so that eventually he or she will be able to extend this method to systems of linear equations that involve more than two variables and/or more than two equations.

Each of the three methods of solution can be applied to any system of linear equations. The methods of substitution and elimination become extremely cumbersome in large systems, but the method of gaussian elimination is not nearly as awkward and is generally the simplest and easiest method of solution. This does not mean that the gaussian elimination method requires little effort when used in large systems, but it requires less effort than the alternative methods.

Let us consider an example containing three equations in three unknowns.

Example 4.4 Ms. Jones received a commission of $8 by selling 3 pairs of socks, 4 ties, and 1 shirt. Ms. Smith received a commission of $10 by selling 5

Sec. 4.2 / METHODS OF SOLUTION

pairs of socks and 3 shirts and by having 2 ties returned. Mr. Black received a commission of $19 by selling 3 ties and 4 shirts and by having 1 pair of socks returned. Commissions are reduced for returned items. Use the methods of substitution and gaussian elimination to determine the commission for socks, ties, and shirts.

Solution The system of linear equations to compute the commission for each item is

$$3x + 4y + z = 8$$
$$5x - 2y + 3z = 10$$
$$-x + 3y + 4z = 19 \qquad (4.3)$$

where x, y, and z represent the commission for selling a pair of socks, a tie, and a shirt, respectively. To solve the system by substitution, we solve the first equation for z in terms of x and y and substitute this result into the second and third equations in place of z. Solving for z, we obtain

$$z = 8 - 3x - 4y \qquad (4.4)$$

Substituting into the second and third equations, we have

$$5x - 2y + 3(8 - 3x - 4y) = 10$$

and

$$-x + 3y + 4(8 - 3x - 4y) = 19$$

which simplifies to the following system:

$$-4x - 14y = -14$$
$$-13x - 13y = -13 \qquad (4.5)$$

We now use substitution on the system in (4.5). Solving the first equation of (4.5) for y in terms of x and substituting into the second equation, we have

$$y = \frac{14 - 4x}{14} \qquad (4.6)$$

and

$$-13x - 13\left(\frac{14 - 4x}{14}\right) = -13$$

Simplifying the last equation, we obtain $x = 0$; substituting $x = 0$ into Eq. (4.6), we determine that $y = 1$; substituting $x = 0$ and $y = 1$ into Eq. (4.4), we determine that $z = 4$. Thus the solution to the system of linear equations in (4.3) is $x = 0$, $y = 1$, and $z = 4$. That is, the salespeople receive no commission for selling socks and a commission of $1 for selling a tie and $4 for selling a shirt.

To solve the system by gaussian elimination, we divide the first equation of (4.3) by 3 to obtain the first equation of the system given below in (4.7).

We then use this equation to eliminate x from the remaining two equations of (4.3) by multiplying it by 5 and subtracting the resulting equation from the second equation of (4.3) to obtain the second equation of the system in (4.7). That is, we get the second equation as follows:

$$
\begin{array}{r}
5x - 2y + 3z = 10 \\
5x + \tfrac{20}{3} y + \tfrac{5}{3} z = \tfrac{40}{3} \\
\hline
-\tfrac{26}{3} y + \tfrac{4}{3} z = -\tfrac{10}{3}
\end{array}
$$

Adding the first equation of (4.7) to the third equation of (4.3), we obtain the third equation of (4.7) as follows:

$$
\begin{array}{r}
-x + 3y + 4z = 19 \\
x + \tfrac{4}{3} y + \tfrac{1}{3} z = \tfrac{8}{3} \\
\hline
\tfrac{13}{3} y + \tfrac{13}{3} z = \tfrac{65}{3}
\end{array}
$$

Thus we have the following system of equations:

$$
\begin{aligned}
x + \tfrac{4}{3} y + \tfrac{1}{3} z &= \tfrac{8}{3} \\
-\tfrac{26}{3} y + \tfrac{4}{3} z &= -\tfrac{10}{3} \\
\tfrac{13}{3} y + \tfrac{13}{3} z &= \tfrac{65}{3}
\end{aligned}
\qquad (4.7)
$$

Our next step is to eliminate y from the first and third equations of (4.7). We first divide the second equation by $-\tfrac{26}{3}$ to obtain the second equation given below in the system of (4.8). Next we multiply the second equation of (4.8) by $\tfrac{4}{3}$ and subtract the result from the first equation of (4.7), as illustrated below, which yields the first equation of (4.8).

$$
\begin{array}{r}
x + \tfrac{4}{3} y + \tfrac{1}{3} z = \tfrac{8}{3} \\
\tfrac{4}{3} y - \tfrac{8}{39} z = \tfrac{20}{39} \\
\hline
x \qquad\quad + \tfrac{7}{13} z = \tfrac{28}{13}
\end{array}
$$

Finally we multiply the second equation of (4.8) by $\tfrac{13}{3}$ and subtract the result from the third equation of (4.7), which yields the third equation of (4.8). That is, we get

$$
\begin{array}{r}
\tfrac{13}{3} y + \tfrac{13}{3} z = \tfrac{65}{3} \\
\tfrac{13}{3} y - \tfrac{2}{3} z = \tfrac{5}{3} \\
\hline
5z = 20
\end{array}
$$

Hence we have the following system of equations:

$$
\begin{aligned}
x \qquad\quad + \tfrac{7}{13} z &= \tfrac{28}{13} \\
y - \tfrac{2}{13} z &= \tfrac{5}{13} \\
5z &= 20
\end{aligned}
\qquad (4.8)
$$

Sec. 4.2 / METHODS OF SOLUTION

Our final step is to eliminate z from the first and second equations of (4.8). We first divide the third equation of (4.8) by 5 to obtain the third equation of (4.9). Next we multiply the third equation of (4.9) by $-\frac{2}{13}$ and subtract the result from the second equation of (4.8), which yields the second equation of (4.9) as given below:

$$\begin{array}{rl} y - \frac{2}{13}z = & \frac{5}{13} \\ -\frac{2}{13}z = & -\frac{8}{13} \\ \hline y \quad\quad = & 1 \end{array}$$

Finally we multiply the third equation of (4.9) by $\frac{7}{13}$ and subtract the result from the first equation of (4.8), which yields the first equation of (4.9) as follows:

$$\begin{array}{rl} x + \frac{7}{13}z = & \frac{28}{13} \\ \frac{7}{13}z = & \frac{28}{13} \\ \hline x \quad\quad = & 0 \end{array}$$

The equations in (4.9) are, in fact, solved for the three variables:

$$\begin{array}{rl} x & = 0 \\ y & = 1 \\ z & = 4 \end{array} \tag{4.9}$$

Thus the solution to the system of linear equations given in (4.3) is $x=0$, $y=1$, and $z=4$, and the salespeople receive no commission for selling socks, a commission of \$1 for selling a tie, and \$4 for selling a shirt.

In Sec. 4.1, we considered the graphic implications of having the number of equations equal the number of variables, and these implications were given in Fig. 4.1. In reviewing the figure, we note that we have considered only the case where a unique solution exists, as shown in Fig. 4.1a.

Let us consider what happens when the graphs of the two equations in two variables are parallel lines, as shown in Fig. 4.1b. The system of two linear equations in two unknowns that you were to verify as having a graph similar to the one shown in Fig. 4.1b is

$$2x + 3y = 10$$
$$4x + 6y = 24$$

Using the gaussian elimination method, we divide the first equation by 2, multiply the resulting equation by 4, and subtract this result from the second equation to obtain

$$x + \tfrac{3}{2} y = 5$$
$$0 = 4 \tag{4.10}$$

However, this last equation is *inconsistent* with our numbering system since 0

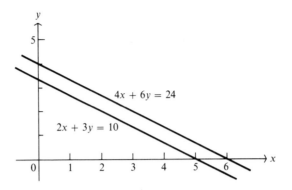

FIGURE 4.5

does not equal 4. The graphic representation of the original system is given in Fig. 4.5. Thus a system of linear equations that has a graph consisting of parallel lines is said to be *inconsistent*. In such a situation, no solution exists. If we were to use any of the other methods of solution, we would arrive at a similar inconsistency for the system.

A system from Sec. 4.1 that you were to verify as having coincident lines with a graph similar to Fig. 4.1c is

$$3x + 2y = 7$$
$$6x + 4y = 14$$

Using the gaussian elimination method, we divide the first equation by 3, multiply the resulting equation by 6, and subtract this result from the second equation to obtain

$$x + \tfrac{2}{3} y = \tfrac{7}{3}$$
$$0 = 0 \qquad (4.11)$$

The second equation of the resulting system is consistent with our numbering system; however, it does not contain any unknowns. This leaves us with one equation in two unknowns for which the graph is a straight line. Hence any point on the straight line satisfies the initial system, and the system has an infinite number of solutions. In this situation, the system is consistent but underconstrained, and any point satisfying one of the equations in the original system will also satisfy the other equation. This is illustrated in Fig. 4.6 for the above system.

So far, the only systems of linear equations that we have considered for the three methods of solution are those systems for which the number of equations equals the number of variables. For those situations where the number of equations is not equal to the number of variables, there will also be three possible types of results: no solution, a unique solution, or an infinite number of solutions. For the situation where no solution exists, the gaussian elimination method will yield at least one equation that is inconsistent with

Sec. 4.2 / METHODS OF SOLUTION

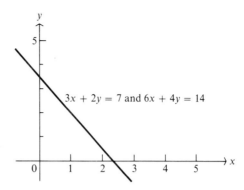

FIGURE 4.6

our numbering system, such as the last equation in system (4.10). When a unique solution exists, there will be one or more equations such as the last equation of (4.11). When an infinite number of solutions exist, there will be at least one remaining equation containing more than one variable, such as the first equation of (4.11).

EXERCISES

1. Use each of the three methods of solution to solve each of the following systems of linear equations:

(a) $2x + 4y = 8$
$\quad\ 3x - 2y = 4$

(b) $\quad 5x + 2y = 5$
$\quad -4x + 3y = -27$

(c) $3w + 6z = 9$
$\quad\ 4w - z = 3$

(d) $-2x + 4y + z = 25$
$\quad\ \ x + 2y - z = -3$
$\quad\ 3x - 3y + 2z = -11$

(e) $2x_1 + x_2 - 3x_3 = -7$
$\quad\ x_1 - 2x_2 + x_3 = 4$
$\quad 3x_1 + 4x_2 + x_3 = 2$

2. Extend the notation of system (4.2) to write

(a) Three linear equations in x_1, x_2, and x_3
(b) Four linear equations in x_1, x_2, x_3, and x_4

3. Determine whether the two lines represented by graphing the two linear equations in each of the following systems have the same slope. If they do not have the same slope, determine the solution to the system, i.e., the point of intersection; if they do have the same slope, determine whether they are parallel lines or coincident lines.

(a) $\quad 2x + 3y = 7$
$\quad\ \ 6x + 9y = 21$

(b) $10x + 4y = 10$
$\quad\ 5x + 2y = 7$

(c) $\quad 6x + 3y = 24$
$\quad -2x + 3y = 8$

(d) $10x + 5y = 15$
$\quad\ 6x + 3y = 6$

(e) $\quad 3x + 6y = 12$
$\quad\ \ 6x + 3y = 15$

(f) $\quad 4x + 2y = 6$
$\quad 10x + 5y = 15$

4. Use the gaussian elimination method to determine whether each of the following systems of linear equations has zero, one, or an infinite number of solutions. If only one solution exists, determine the values of the variables for the solution; if the system has an infinite number of solutions, give the values of the variables for at least one solution.

(a) $2x + y + 3z = 12$
$x - y + 4z = 8$
$3x + 2y - 2z = 6$

(b) $2x + 3y - z = 5$
$3x - 2y + 2z = 5$
$3x + 11y - 5z = 10$

(c) $3x + 2y + z = 14$
$2x + 3y - 2z = 10$
$5x + 5y - z = 15$

(d) $2x + 3y - z = 19$
$3x - 2y + 3z = 7$

(e) $2x + 3y = 14$
$5x + 2y = 13$
$3x - y = -1$

(f) $4x + 3y = 5$
$3x + 5y = 1$
$5x + y = 12$

5. A rubber company manufactures two types of tires. The first type requires 5 lb of rubber and 3 hr of labor, and it is produced at a cost of $13.50; the second type requires 4 lb of rubber and 5 hr of labor, and it is produced at a cost of $16.00. Determine the cost of rubber per pound and the cost of labor per hour.

6. Laura Henderson, a regional representative for a leading manufacturer of ladies' apparel, drove 1,050 mi during the past month. She plans her trips to travel by freeway as much as possible. However, Laura knows that she drove 600 mi on nonfreeway roads. How many miles did she drive on freeways?

7. Bill worked as a clerk for TG&E Stores and was responsible for checking out his cash register. He knows that he has a total of $19 in quarters and dimes. If he has 115 coins in quarters and dimes, how many quarters does he have? How many dimes?

8. Dr. Davis, an orthodontist, bought 8 drills and 4 handtools. A drill cost $4 more than a handtool. He spent a total of $212 for the order. How much does a handtool cost? A drill?

9. Wyatt Henderson has started playing trombone for a popular jazz group. He receives $150 for each concert and $100 for each dance for which the group plays. According to Wyatt's records, he has participated in 25 appearances and has received $3,000. How many concerts has Wyatt played? How many dances?

10. The V&I Sugar Company packages sugar in 5 lb and 10 lb bags. They make a profit of 4¢ per lb on the 5 lb bags and 3¢ per lb on the 10 lb bags. If they sell the Kruger Grocery Store 150 bags of sugar and made a profit of $37.50 from the sale, how many 5 lb bags did V&I sell to Kruger? How many 10 lb bags?

11. Dee Ann Williams and her brother Don operate a concession stand at the county fair. They sell hot dogs and hamburgers for 60¢ and 75¢, respectively. Their sales from hot dogs and hamburgers total $63 for the evening. From the count of the wrappers they used, they know they sold a total of 90 hot dogs and hamburgers. How many hot dogs did they sell? How many hamburgers?

12. A candy company produces three types of chocolate candy: peanut clusters, chocolate-coated caramels, and chocolate-coated almonds. A large box of chocolates contains 15 peanut clusters, 9 chocolate-coated caramels, and 3 chocolate-coated

almonds; a medium box of chocolates contains 6 peanut clusters, 9 chocolate-coated caramels, and 12 chocolate-coated almonds; a small box of chocolates contains 5 peanut clusters, 5 chocolate-coated caramels, and 6 chocolate-coated almonds. Determine how many boxes of chocolates of each size the company can package if they have available 88 peanut clusters, 73 chocolate-coated caramels, and 60 chocolate-coated almonds and must use all they have available.

13. A men's clothing store purchased 20 short-sleeve shirts and 15 long-sleeve shirts for $62 from one of its suppliers. Another time, they purchased 16 short-sleeve shirts and 25 long-sleeve shirts for $70.40 from the same supplier. Determine the cost of the long-sleeve and short-sleeve shirts from this supplier.

14. An electronics company makes two types of calculators that must be processed by two departments. The first type requires 3 hr in department A and 2 hr in department B. The second type requires $1\frac{1}{2}$ hr in department A and $2\frac{1}{2}$ hr in department B. The time available to process these two types of calculators is $145\frac{1}{2}$ hr in department A and $161\frac{1}{2}$ hr in department B. Determine how many of each type can be produced if all of the time available must be used.

15. A bedspread manufacturer has 1,000 square feet (sq ft) of quilted material and 1,450 sq ft of unquilted material of the same pattern. A regular size bedspread requires 35 sq ft of quilted material and 50 sq ft of unquilted material. A twin size bedspread requires 25 sq ft of quilted material and 40 sq ft of unquilted material. How many of each type of bedspread can the manufacturer make if all of the material is to be used?

16. A manufacturer of desks, tables, and chairs is trying to decide how many of each to produce in order to use all of the time available in each job shop. In the cutting department, a desk requires $8\frac{1}{2}$ hr, a chair requires 11 hr, and a table requires 4 hr. The time required in the polishing department is $4\frac{1}{2}$ hr for a desk, 6 hr for a chair, and 8 hr for a table. The time required in the finishing department is $9\frac{1}{2}$ hr for a desk, $7\frac{1}{2}$ hr for a chair, and 4 hr for a table. The time available per week is 756 hr for cutting, 480 hr for polishing, and 590 hr for finishing. Determine how many desks, chairs, and tables should be produced each week.

4.3 SUMMATION NOTATION

In dealing with mathematical statements, occasionally it is advantageous to express the statements in a compact form. When dealing with a sum of several numbers, we can use *summation notation* to express the sum in a very compact form. The Greek letter Σ (sigma) is the symbol used to represent a summation. The expression

$$\sum_{j=1}^{4} j$$

is read "the summation of j, j going from 1 to 4," and it means to substitute the consecutive values of j from 1 to 4 and sum the resulting numbers; that is,

$$\sum_{j=1}^{4} j = 1 + 2 + 3 + 4 = 10$$

In general, the expression that follows the summation symbol Σ is evaluated and summed for the values of the variable appearing below the summation symbol, starting with the integer appearing below the summation symbol and continuing for each consecutive integer value through the integer value appearing above the symbol. For example,

$$\sum_{j=1}^{4} 3j = 3(1) + 3(2) + 3(3) + 3(4) = 30$$

Similarly,

$$\sum_{i=2}^{4} 2i = 2(2) + 2(3) + 2(4) = 18$$

$$\sum_{x=4}^{6} x^2 = 4^2 + 5^2 + 6^2 = 77$$

and

$$\sum_{p=2}^{5} \left(\frac{p}{3} + 4\right) = \left(\frac{2}{3} + 4\right) + \left(\frac{3}{3} + 4\right) + \left(\frac{4}{3} + 4\right) + \left(\frac{5}{3} + 4\right) = 20\frac{2}{3}$$

Also,

$$\sum_{i=1}^{3} x_i = x_1 + x_2 + x_3$$

and

$$\sum_{j=1}^{4} (y+j) = (y+1) + (y+2) + (y+3) + (y+4) = 4y + 10$$

There are three important properties concerning summation notation. These properties are obtained by expanding, rearranging, and contracting summation notation to establish relationships between summations of different arrangements of terms. We shall state the properties as *rules* and verify each one.

Rule 4.1 If c is any constant, then

$$\sum_{i=1}^{n} cx_i = c \sum_{i=1}^{n} x_i$$

The verification of Rule 4.1 is as follows:

$$\sum_{i=1}^{n} cx_i = cx_1 + cx_2 + \cdots + cx_n$$
$$= c(x_1 + x_2 + \cdots + x_n)$$
$$= c \sum_{i=1}^{n} x_i$$

Rule 4.2

$$\sum_{i=1}^{n} (x_i + y_i) = \sum_{i=1}^{n} x_i + \sum_{i=1}^{n} y_i$$

The verification of Rule 4.2 is as follows:

$$\sum_{i=1}^{n} (x_i + y_i) = (x_1 + y_1) + (x_2 + y_2) + \cdots + (x_n + y_n)$$
$$= (x_1 + x_2 + \cdots + x_n) + (y_1 + y_2 + \cdots + y_n)$$
$$= \sum_{i=1}^{n} x_i + \sum_{i=1}^{n} y_i$$

Rule 4.3

$$\sum_{i=1}^{n} c = nc \quad \text{and} \quad \sum_{i=m}^{n} c = (n-m+1)c$$

The verification of Rule 4.3 is as follows:

$$\sum_{i=1}^{n} c = \underbrace{c + c + \cdots + c}_{n \text{ terms}}$$
$$= \underbrace{(1 + 1 + \cdots + 1)}_{n \text{ terms}} c$$
$$= nc$$

and

$$\sum_{i=m}^{n} c = \underbrace{c + c + \cdots + c}_{(n-m+1) \text{ terms}}$$
$$= \underbrace{(1 + 1 + \cdots + 1)}_{(n-m+1) \text{ terms}} c$$
$$= (n - m + 1)c$$

Example 4.5 Use the three summation rules (Rules 4.1 to 4.3) to determine the values of the following summations:

(a) $\sum_{i=1}^{6} 5x_i^2$

(b) $\sum_{i=1}^{3} (i^2 + 3i)$

(c) $\sum_{i=1}^{6} 4$

(d) $\sum_{i=5}^{8} 2$

Solution

(a) Using Rule 4.1, we have

$$\sum_{i=1}^{6} 5x_i^2 = 5 \sum_{i=1}^{6} x_i^2$$

(b) Using Rule 4.2, we obtain

$$\sum_{i=1}^{3} (i^2 + 3i) = \sum_{i=1}^{3} i^2 + \sum_{i=1}^{3} 3i$$

Evaluating the first term and applying Rule 4.1 to the second term and then evaluating it, we obtain

$$\sum_{i=1}^{3} (i^2 + 3i) = (1 + 4 + 9) + 3(1 + 2 + 3)$$

$$= 32$$

(c) Using Rule 4.3, we have

$$\sum_{i=1}^{6} 4 = 4 \cdot 6 = 24$$

(d) Using Rule 4.3, we have

$$\sum_{i=5}^{8} 2 = (8 - 5 + 1)2 = 8$$

One of the major advantages of summation notation is that it reduces large equations or systems of equations to simple forms. For example, we have

$$\sum_{i=1}^{n} x_i = x_1 + x_2 + \cdots + x_n$$

and

$$\sum_{i=1}^{n} a_i = a_1 + a_2 + \cdots + a_n$$

Sec. 4.3 / SUMMATION NOTATION

Also, we can write

$$\sum_{j=1}^{n} a_j x_j = a_1 x_1 + a_2 x_2 + \cdots + a_n x_n$$

The linear equation in two variables

$$a_1 x_1 + a_2 x_2 = b$$

can be expressed in summation notation as

$$\sum_{j=1}^{2} a_j x_j = b$$

Similarly, a linear equation in n variables

$$a_1 x_1 + a_2 x_2 + \cdots + a_n x_n = b$$

can be expressed in summation notation as

$$\sum_{j=1}^{n} a_j x_j = b$$

which requires the same space and effort as the summation-notation form for a linear equation in two variables.

If we were to take the second equation of the symbolic representation of the general system of two linear equations in two variables as given in (4.2), we would have

$$a_{21} x_1 + a_{22} x_2 = b_2$$

which in summation form becomes

$$\sum_{j=1}^{2} a_{2j} x_j = b_2$$

We could represent the ith equation of that symbolic representation by

$$\sum_{j=1}^{2} a_{ij} x_j = b_i$$

Since i can assume the values 1 and 2 for the two equations, the system of linear equations in two variables as given in (4.2) can be represented in summation notation as

$$\sum_{j=1}^{2} a_{ij} x_j = b_i \qquad i = 1, 2$$

Similarly, a system of m linear equations in n variables, which is symbolically represented by the general system

$$a_{11}x_1 + a_{12}x_2 + \cdots + a_{1n}x_n = b_1$$
$$a_{21}x_1 + a_{22}x_2 + \cdots + a_{2n}x_n = b_2$$
$$\cdots\cdots\cdots\cdots\cdots\cdots\cdots\cdots\cdots$$
$$a_{m1}x_1 + a_{m2}x_2 + \cdots + a_{mn}x_n = b_m \qquad (4.12)$$

can be simply represented by summation notation as

$$\sum_{j=1}^{n} a_{ij}x_j = b_i \qquad i = 1, 2, \ldots, m \qquad (4.13)$$

Thus a system of m equations in n unknowns can be represented in a much more compact form through the use of summation notation.

EXERCISES

1. Determine the numerical values of the following terms:

   (a) $\sum_{k=1}^{4} k$
   (b) $\sum_{j=3}^{7} 2j$
   (c) $\sum_{i=32}^{45} 6$
   (d) $\sum_{x=1}^{3} x^3$
   (e) $\sum_{s=11}^{13} s^2$
   (f) $\sum_{y=1}^{10} 5$
   (g) $\sum_{i=1}^{5} (i^2 + 3i + 2)$
   (h) $\sum_{i=2}^{4} (6i - 5)$

2. Write the following summations in expanded form:

   (a) $\sum_{j=2}^{4} 3x_j^2$
   (b) $\sum_{j=4}^{7} jx_j^3$
   (c) $\sum_{j=1}^{4} b_j x_j$
   (d) $\sum_{j=1}^{4} a_j x_j = 20$
   (e) $\sum_{j=1}^{3} a_{2j} x_j = 16$
   (f) $\sum_{j=1}^{3} a_{ij} x_j = 3i \qquad i = 1, 2$

3. Write the following expressions in compact form using summation notation:

   (a) $2 + 4 + 6 + 8$
   (b) $y_4 + y_5 + y_6$
   (c) $2x_2 + 3x_3 + 4x_4 + 5x_5$
   (d) $a_1 x_1 + a_2 x_2 + a_3 x_3$
   (e) $x_1 - y_1 + x_2 - y_2 + x_3 - y_3$
   (f) $2 + 4 + 8 + 16 + 32$

4. Write the following systems of linear equations in expanded form:

(a) $\sum_{j=1}^{2} a_{ij}x_j = b_i \quad i=1,2$

(b) $\sum_{j=1}^{3} a_{ij}x_j = b_i \quad i=1,2$

(c) $\sum_{j=1}^{2} a_{ij}x_j = b_i \quad i=1,2,3$

(d) $\sum_{j=1}^{3} a_{ij}x_j = b_i \quad i=1,2,3$

(e) $\sum_{j=1}^{4} a_{ij}x_j = b_i \quad i=1,2$

(f) $\sum_{j=1}^{2} a_{ij}x_j = b_i \quad i=1,2,3,4$

5. Write the following systems of linear equations in compact forms using summation notation:

(a) $a_{11}x_1 + a_{12}x_2 = b_1$
$a_{21}x_1 + a_{22}x_2 = b_2$
$a_{31}x_1 + a_{32}x_2 = b_3$

(b) $c_{11}y_1 + c_{12}y_2 = d_1$
$c_{21}y_1 + c_{22}y_2 = d_2$

(c) $c_{11}x_1 + c_{12}x_2 + c_{13}x_3 = b_1$
$c_{21}x_1 + c_{22}x_2 + c_{23}x_3 = b_2$

(d) $a_{11}x_1 + a_{12}x_2 + a_{13}x_3 + a_{14}x_4 = b_1$
$a_{21}x_1 + a_{22}x_2 + a_{23}x_3 + a_{24}x_4 = b_2$
$a_{31}x_1 + a_{32}x_2 + a_{33}x_3 + a_{34}x_4 = b_3$

4.4 MATRIX ALGEBRA

In Sec. 4.3 summation notation was introduced as a means of representing a summation or a series of summations as a single compact term. In this section, we shall consider a means of representing a group of numbers by a single term. For this reason, this and the following section may appear to be somewhat more difficult than the preceding sections; however, they are simply extensions of the previous material in conjunction with a few new concepts.

A single number is quite often sufficient to describe a particular item of interest. For example, the gross income of a business may be the only item of interest for a particular situation; however, in another situation, items of interest may include gross income, rental expense, labor expense, materials expense, and advertising expense. These values may be represented by a row of values such as

$$[4{,}724 \quad 800 \quad 1{,}732 \quad 752 \quad 175]$$

An array of numbers such as this is called a *row vector*. If the numbers are listed in column form such as

$$\begin{bmatrix} 4{,}724 \\ 800 \\ 1{,}732 \\ 752 \\ 175 \end{bmatrix}$$

the array is called a *column vector*. Generally such an array is simply called a *vector* with the row or column designation being understood for a particular situation.

In the above example, there may be need to consider simultaneously the values for several businesses. By representing the values for a particular business as a row vector, we could represent values for several businesses as a rectangular array. For example, such a rectangular array for three businesses could be given by

$$\begin{bmatrix} 4{,}724 & 800 & 1{,}732 & 752 & 175 \\ 3{,}156 & 400 & 1{,}318 & 630 & 280 \\ 8{,}639 & 1{,}500 & 2{,}537 & 1{,}841 & 923 \end{bmatrix}$$

Such a rectangular array is called a *matrix*. In this example, the rows correspond to different businesses and the columns correspond to different characteristics of the business.

In general, a *matrix* is a rectangular array of numbers; for our purposes, the numbers will be real numbers. The components of a matrix are called *elements*. A matrix is usually denoted by a bold-faced capital letter, and its size is indicated by the number of rows and columns that it contains. A matrix that contains m rows and n columns is said to be of *order* $m \times n$ (read "m by n"). Occasionally it is advantageous to refer to specific elements of a matrix $\mathbf{A}$. Thus the element in the ith row and jth column of $\mathbf{A}$ is denoted by a_{ij}. An alternative way of representing a matrix $\mathbf{A}$ is by reference to a general element of $\mathbf{A}$ in brackets. Thus a matrix can be represented by either $\mathbf{A}$ or $[a_{ij}]$ or the full rectangular array, with the preference being dictated by the purpose for which the notation is to be used. That is,

$$\mathbf{A} = [a_{ij}] = \begin{bmatrix} a_{11} & a_{12} & \cdots & a_{1n} \\ a_{21} & a_{22} & \cdots & a_{2n} \\ \cdots & \cdots & \cdots & \cdots \\ a_{m1} & a_{m2} & \cdots & a_{mn} \end{bmatrix}$$

Thus a 3×2 matrix $\mathbf{A}$ could be represented by

$$\mathbf{A} \quad \text{or} \quad [a_{ij}] \quad \text{or} \quad \begin{bmatrix} a_{11} & a_{12} \\ a_{21} & a_{22} \\ a_{31} & a_{32} \end{bmatrix}$$

Two matrices are *equal* if they are of the same order and the corresponding elements are equal. That is, $\mathbf{A} = \mathbf{B}$ if and only if $\mathbf{A}$ and $\mathbf{B}$ are both of order $m \times n$ and $a_{ij} = b_{ij}$ for all i and j such that $i = 1, 2, \ldots, m$ and $j = 1, 2, \ldots, n$. For example, the matrices

$$\mathbf{A} = \begin{bmatrix} 1 & 2 & -1 \\ 0 & 4 & 3 \end{bmatrix} \quad \text{and} \quad \mathbf{B} = \begin{bmatrix} \frac{2}{2} & \frac{4}{2} & -\frac{3}{3} \\ 0 & \frac{16}{4} & 3 \end{bmatrix}$$

are equal because they are both of order 2×3 and each element of $\mathbf{A}$ is equal to the corresponding element of $\mathbf{B}$. However, for the matrix

$$\mathbf{C} = \begin{bmatrix} 0 & 4 & 3 \\ 1 & 2 & -1 \end{bmatrix}$$

we note that **A** ≠ **C** even though the same numbers are used to form the two matrices; they are not equal because at least one element of **A** is not equal to the corresponding element of **C**. This is a situation where the representation of the matrices **A** and **B** by the notation $[a_{ij}]$ and $[b_{ij}]$ is advantageous because we write $[a_{ij}] = [b_{ij}]$ if and only if $a_{ij} = b_{ij}$ for all $i = 1, 2, \ldots, m$ and $j = 1, 2, \ldots, n$.

Matrix Addition In order to add two matrices **A** and **B**, we require that they be of the same order. If this is satisfied, the matrix that represents the sum of **A** and **B** is simply the matrix whose elements are the sum of the corresponding elements of **A** and **B**. Notationally

$$[a_{ij}] + [b_{ij}] = [a_{ij} + b_{ij}] \tag{4.14}$$

For example, if

$$\mathbf{A} = \begin{bmatrix} 2 & 1 \\ 3 & 1 \\ 0 & 4 \end{bmatrix} \quad \text{and} \quad \mathbf{B} = \begin{bmatrix} 4 & 6 \\ 3 & 5 \\ 1 & 0 \end{bmatrix}$$

the two matrices are both of order 3×2 and their sum is given by

$$\mathbf{A} + \mathbf{B} = \begin{bmatrix} 2 & 1 \\ 3 & 1 \\ 0 & 4 \end{bmatrix} + \begin{bmatrix} 4 & 6 \\ 3 & 5 \\ 1 & 0 \end{bmatrix} = \begin{bmatrix} 2+4 & 1+6 \\ 3+3 & 1+5 \\ 0+1 & 4+0 \end{bmatrix} = \begin{bmatrix} 6 & 7 \\ 6 & 6 \\ 1 & 4 \end{bmatrix}$$

We note the matrix that results from the addition of two matrices is of the same order as the matrices being summed. If an attempt is made to add two matrices of different order, the addition is *undefined* because there are elements of one matrix that do not have corresponding elements in the other matrix. The reader should verify that the commutative law of addition is satisfied for matrices; that is, $\mathbf{A} + \mathbf{B} = \mathbf{B} + \mathbf{A}$.

If we wish to multiply a matrix **A** by a single number α, we multiply each element of **A** by α; for example, if $\alpha = 2$ and

$$\mathbf{A} = \begin{bmatrix} 1 & 2 & 2 \\ 0 & 1 & 3 \end{bmatrix}$$

then

$$\alpha \mathbf{A} = 2 \begin{bmatrix} 1 & 2 & 2 \\ 0 & 1 & 3 \end{bmatrix} = \begin{bmatrix} 2 \cdot 1 & 2 \cdot 2 & 2 \cdot 2 \\ 2 \cdot 0 & 2 \cdot 1 & 2 \cdot 3 \end{bmatrix} = \begin{bmatrix} 2 & 4 & 4 \\ 0 & 2 & 6 \end{bmatrix}$$

Notationally we write

$$\alpha \mathbf{A} = \alpha [a_{ij}] = [\alpha a_{ij}] \tag{4.15}$$

The number α is called a *scalar*. There are no limitations on the order of a matrix for the multiplication by a scalar to be defined.

We can combine the definitions of addition and scalar multiplication of matrices in order to determine the difference between two matrices of the same order; that is, we can write

$$\mathbf{A} + (-1)\mathbf{B} = \mathbf{A} - \mathbf{B} = [a_{ij} - b_{ij}] \qquad (4.16)$$

From (4.16) we can see that if we subtract an $m \times n$ matrix $\mathbf{A}$ from itself, we obtain an $m \times n$ matrix $\mathbf{A} - \mathbf{A}$ whose elements are all zeroes. A matrix of all zero elements is called a *zero matrix*, or *null matrix*, and is denoted by $\mathbf{0}$. The null matrix serves the same role in matrix algebra as the number zero serves in common arithmetic; that is, $\mathbf{A} + \mathbf{0} = \mathbf{0} + \mathbf{A} = \mathbf{A}$, where $\mathbf{A}$ is any $m \times n$ matrix and $\mathbf{0}$ is an $m \times n$ matrix of zeroes. The same notation is used for any null matrix; however, just because we have two null matrices $\mathbf{0}$ and $\mathbf{0}$, they are not necessarily equal because they may not be of the same order. For example,

$$\mathbf{0} = \begin{bmatrix} 0 & 0 \\ 0 & 0 \end{bmatrix} \quad \text{and} \quad \mathbf{0} = \begin{bmatrix} 0 & 0 & 0 \\ 0 & 0 & 0 \end{bmatrix}$$

are both null matrices, but they are not equal. This should not create a problem for us because the order of the null matrix must be such that the matrix operation in which it is involved is defined.

Matrix Multiplication In the above discussion, we have considered addition and subtraction of matrices, where the matrices must be of the same order for the operations to be defined. Next we shall consider the multiplication of matrices. The general notational representation for the product $\mathbf{AB}$, where $\mathbf{A}$ is an $m \times p$ matrix and $\mathbf{B}$ is a $p \times n$ matrix, is the $m \times n$ matrix given by

$$\mathbf{AB} = [c_{ij}] = \left[\sum_{k=1}^{p} a_{ik} b_{kj} \right] \qquad (4.17)$$

where $c_{ij} = \sum_{k=1}^{p} a_{ik} b_{kj}$. Equation (4.17) tells us that the element in the ith row and jth column of the product matrix is determined by the summation

$$\sum_{k=1}^{p} a_{ik} b_{kj}$$

When we multiply two matrices $\mathbf{AB}$, we mean that the element in the ith row and jth column of the matrix that represents the result of the product is obtained by multiplying the first element in the ith row of $\mathbf{A}$ (that is, a_{i1}) by the first element in the jth column of $\mathbf{B}$ (that is, b_{1j}) and adding to this product each of the similar products for the second, third, etc., elements of the ith row of $\mathbf{A}$ and the jth column of $\mathbf{B}$. In order for this element-by-element multiplication to be consistent, it is necessary that the number of elements in the ith row of $\mathbf{A}$ be equal to the number of elements in the jth column of $\mathbf{B}$; in other words, the number of columns in $\mathbf{A}$ must equal the

Sec. 4.4 / MATRIX ALGEBRA

number of rows in **B**. Thus, if **A** is a 2×3 matrix, it is necessary that **B** have exactly three rows. Actually in the product **AB**, it does not matter how many rows **A** has or how many columns **B** has, as long as the number of columns in **A** is equal to the number of rows in **B**. Thus in order for the matrix multiplication **AB** to be defined, **A** must be an $m \times p$ matrix and **B** must be a $p \times n$ matrix; the matrix that results from the multiplication will be an $m \times n$ matrix.

Consider the three matrices

$$\mathbf{A} = \begin{bmatrix} 0 & 1 \\ 2 & 3 \end{bmatrix} \quad \mathbf{B} = \begin{bmatrix} 4 & 5 \\ 6 & 7 \\ 8 & 9 \end{bmatrix} \quad \mathbf{C} = \begin{bmatrix} 10 & 11 & 12 \\ 13 & 14 & 15 \end{bmatrix}$$

We note that **A** is a 2×2 matrix, **B** is a 3×2 matrix, and **C** is a 2×3 matrix. Let us consider each of the six possible products: **AB, BA, AC, CA, BC,** and **CB**. The product **AB** is not defined because **A** is a 2×2 matrix and **B** is a 3×2 matrix. The product **BA** is defined because **B** has the same number of columns as the number of rows of **A**; thus we have a 3×2 matrix times a 2×2 matrix, which is defined and will result in a 3×2 matrix for the product. Performing the multiplication, we obtain

$$\mathbf{BA} = \begin{bmatrix} 4 & 5 \\ 6 & 7 \\ 8 & 9 \end{bmatrix} \begin{bmatrix} 0 & 1 \\ 2 & 3 \end{bmatrix} = \begin{bmatrix} 4 \cdot 0 + 5 \cdot 2 & 4 \cdot 1 + 5 \cdot 3 \\ 6 \cdot 0 + 7 \cdot 2 & 6 \cdot 1 + 7 \cdot 3 \\ 8 \cdot 0 + 9 \cdot 2 & 8 \cdot 1 + 9 \cdot 3 \end{bmatrix} = \begin{bmatrix} 10 & 19 \\ 14 & 27 \\ 18 & 35 \end{bmatrix}$$

For the matrices **A** and **C**, the product **AC** indicates that we have a 2×2 matrix times a 2×3 matrix, which is defined and results in a 2×3 matrix for the product. Performing the multiplication, we obtain

$$\mathbf{AC} = \begin{bmatrix} 0 & 1 \\ 2 & 3 \end{bmatrix} \begin{bmatrix} 10 & 11 & 12 \\ 13 & 14 & 15 \end{bmatrix}$$

$$= \begin{bmatrix} 0 \cdot 10 + 1 \cdot 13 & 0 \cdot 11 + 1 \cdot 14 & 0 \cdot 12 + 1 \cdot 15 \\ 2 \cdot 10 + 3 \cdot 13 & 2 \cdot 11 + 3 \cdot 14 & 2 \cdot 12 + 3 \cdot 15 \end{bmatrix} = \begin{bmatrix} 13 & 14 & 15 \\ 59 & 64 & 69 \end{bmatrix}$$

The product **CA** indicates that we have a 2×3 matrix times a 2×2 matrix. This multiplication is not defined because the number of columns of **C** is not equal to the number of rows of **A**.

For the matrices **B** and **C**, the product **BC** indicates that we have a 3×2 matrix times a 2×3 matrix, which is defined and will result in a 3×3 matrix for the product. Performing the multiplication, we obtain

$$\mathbf{BC} = \begin{bmatrix} 4 & 5 \\ 6 & 7 \\ 8 & 9 \end{bmatrix} \begin{bmatrix} 10 & 11 & 12 \\ 13 & 14 & 15 \end{bmatrix} = \begin{bmatrix} 4 \cdot 10 + 5 \cdot 13 & 4 \cdot 11 + 5 \cdot 14 & 4 \cdot 12 + 5 \cdot 15 \\ 6 \cdot 10 + 7 \cdot 13 & 6 \cdot 11 + 7 \cdot 14 & 6 \cdot 12 + 7 \cdot 15 \\ 8 \cdot 10 + 9 \cdot 13 & 8 \cdot 11 + 9 \cdot 14 & 8 \cdot 12 + 9 \cdot 15 \end{bmatrix}$$

$$= \begin{bmatrix} 105 & 114 & 123 \\ 151 & 164 & 177 \\ 197 & 214 & 231 \end{bmatrix}$$

The product **CB** indicates that we have a 2×3 matrix times a 3×2 matrix, which is defined and results in a 2×2 matrix for the product. Performing the multiplication, we obtain

$$CB = \begin{bmatrix} 10 & 11 & 12 \\ 13 & 14 & 15 \end{bmatrix} \begin{bmatrix} 4 & 5 \\ 6 & 7 \\ 8 & 9 \end{bmatrix} = \begin{bmatrix} 10 \cdot 4 + 11 \cdot 6 + 12 \cdot 8 & 10 \cdot 5 + 11 \cdot 7 + 12 \cdot 9 \\ 13 \cdot 4 + 14 \cdot 6 + 15 \cdot 8 & 13 \cdot 5 + 14 \cdot 7 + 15 \cdot 9 \end{bmatrix}$$

$$= \begin{bmatrix} 202 & 235 \\ 256 & 298 \end{bmatrix}$$

From the above example, we can conclude that when dealing with matrices it is critical which matrix is on the left and which matrix is on the right in the multiplication. Thus we *cannot* say generally that **AB** is equal to **BA** for any two matrices **A** and **B**; that is, the commutative law is not satisfied for matrix multiplication. With this result, we have encountered something that is contrary to our rules for real numbers. However, we must realize that we are dealing with more than individual real numbers; that is, we are dealing with *rectangular arrays* of real numbers. When a matrix is an $n \times n$ matrix, it is called a *square* matrix. If **A** and **B** are square matrices of the same order, then both **AB** and **BA** are defined but are not necessarily equal.

Example 4.6 For the matrices

$$A = \begin{bmatrix} 1 \\ 2 \end{bmatrix} \quad B = \begin{bmatrix} 0 & -3 \\ 1 & 6 \end{bmatrix} \quad C = \begin{bmatrix} 1 & 2 & 1 \\ 2 & 1 & 2 \end{bmatrix}$$

$$D = \begin{bmatrix} 1 & 2 \\ 2 & 1 \\ 1 & 2 \end{bmatrix} \quad \text{and} \quad E = \begin{bmatrix} 1 & 1 & 1 \end{bmatrix}$$

determine the following products:

(a) **AE**   (b) **BA**   (c) **BC**   (d) **CD**
(e) **DA**   (f) **DB**   (g) **DC**   (h) **ED**
(i) **AB**   (j) **CE**

Solution

(a)

$$AE = \begin{bmatrix} 1 \\ 2 \end{bmatrix} \begin{bmatrix} 1 & 1 & 1 \end{bmatrix} = \begin{bmatrix} 1 \cdot 1 & 1 \cdot 1 & 1 \cdot 1 \\ 2 \cdot 1 & 2 \cdot 1 & 2 \cdot 1 \end{bmatrix} = \begin{bmatrix} 1 & 1 & 1 \\ 2 & 2 & 2 \end{bmatrix}$$

(b)

$$BA = \begin{bmatrix} 0 & -3 \\ 1 & 6 \end{bmatrix} \begin{bmatrix} 1 \\ 2 \end{bmatrix} = \begin{bmatrix} 0 \cdot 1 - 3 \cdot 2 \\ 1 \cdot 1 + 6 \cdot 2 \end{bmatrix} = \begin{bmatrix} -6 \\ 13 \end{bmatrix}$$

(c)

$$BC = \begin{bmatrix} 0 & -3 \\ 1 & 6 \end{bmatrix} \begin{bmatrix} 1 & 2 & 1 \\ 2 & 1 & 2 \end{bmatrix} = \begin{bmatrix} 0 \cdot 1 - 3 \cdot 2 & 0 \cdot 2 - 3 \cdot 1 & 0 \cdot 1 - 3 \cdot 2 \\ 1 \cdot 1 + 6 \cdot 2 & 1 \cdot 2 + 6 \cdot 1 & 1 \cdot 1 + 6 \cdot 2 \end{bmatrix}$$

$$= \begin{bmatrix} -6 & -3 & -6 \\ 13 & 8 & 13 \end{bmatrix}$$

(d)
$$CD = \begin{bmatrix} 1 & 2 & 1 \\ 2 & 1 & 2 \end{bmatrix} \begin{bmatrix} 1 & 2 \\ 2 & 1 \\ 1 & 2 \end{bmatrix} = \begin{bmatrix} 1\cdot 1+2\cdot 2+1\cdot 1 & 1\cdot 2+2\cdot 1+1\cdot 2 \\ 2\cdot 1+1\cdot 2+2\cdot 1 & 2\cdot 2+1\cdot 1+2\cdot 2 \end{bmatrix} = \begin{bmatrix} 6 & 6 \\ 6 & 9 \end{bmatrix}$$

(e)
$$DA = \begin{bmatrix} 1 & 2 \\ 2 & 1 \\ 1 & 2 \end{bmatrix} \begin{bmatrix} 1 \\ 2 \end{bmatrix} = \begin{bmatrix} 1\cdot 1+2\cdot 2 \\ 2\cdot 1+1\cdot 2 \\ 1\cdot 1+2\cdot 2 \end{bmatrix} = \begin{bmatrix} 5 \\ 4 \\ 5 \end{bmatrix}$$

(f)
$$DB = \begin{bmatrix} 1 & 2 \\ 2 & 1 \\ 1 & 2 \end{bmatrix} \begin{bmatrix} 0 & -3 \\ 1 & 6 \end{bmatrix} = \begin{bmatrix} 1\cdot 0+2\cdot 1 & 1\cdot(-3)+2\cdot 6 \\ 2\cdot 0+1\cdot 1 & 2\cdot(-3)+1\cdot 6 \\ 1\cdot 0+2\cdot 1 & 1\cdot(-3)+2\cdot 6 \end{bmatrix} = \begin{bmatrix} 2 & 9 \\ 1 & 0 \\ 2 & 9 \end{bmatrix}$$

(g)
$$DC = \begin{bmatrix} 1 & 2 \\ 2 & 1 \\ 1 & 2 \end{bmatrix} \begin{bmatrix} 1 & 2 & 1 \\ 2 & 1 & 2 \end{bmatrix} = \begin{bmatrix} 1\cdot 1+2\cdot 2 & 1\cdot 2+2\cdot 1 & 1\cdot 1+2\cdot 2 \\ 2\cdot 1+1\cdot 2 & 2\cdot 2+1\cdot 1 & 2\cdot 1+1\cdot 2 \\ 1\cdot 1+2\cdot 2 & 1\cdot 2+2\cdot 1 & 1\cdot 1+2\cdot 2 \end{bmatrix}$$
$$= \begin{bmatrix} 5 & 4 & 5 \\ 4 & 5 & 4 \\ 5 & 4 & 5 \end{bmatrix}$$

(h)
$$ED = \begin{bmatrix} 1 & 1 & 1 \end{bmatrix} \begin{bmatrix} 1 & 2 \\ 2 & 1 \\ 1 & 2 \end{bmatrix} = \begin{bmatrix} 1\cdot 1+1\cdot 2+1\cdot 1 & 1\cdot 2+1\cdot 1+1\cdot 2 \end{bmatrix} = \begin{bmatrix} 4 & 5 \end{bmatrix}$$

(i) **AB** is undefined since **A** is a 2×1 matrix and **B** is a 2×2 matrix.
(j) **CE** is undefined since **C** is a 2×3 matrix and **E** is a 1×3 matrix.

Systems of Linear Equations in Matrix Form

When we considered the general form for a system of two equations in two unknowns, we obtained the general system given in (4.2), namely

$$a_{11}x_1 + a_{12}x_2 = b_1$$
$$a_{21}x_1 + a_{22}x_2 = b_2 \tag{4.18}$$

Since one of the advantages of matrix algebra is its compact notation, let us represent the above system of linear equations in matrix form. Let us represent the coefficients, variables, and constants, respectively, by the three matrices

$$\mathbf{A} = \begin{bmatrix} a_{11} & a_{12} \\ a_{21} & a_{22} \end{bmatrix} \quad \mathbf{X} = \begin{bmatrix} x_1 \\ x_2 \end{bmatrix} \quad \text{and} \quad \mathbf{B} = \begin{bmatrix} b_1 \\ b_2 \end{bmatrix}$$

Calculating the product **AX**, we obtain

$$AX = \begin{bmatrix} a_{11} & a_{12} \\ a_{21} & a_{22} \end{bmatrix} \begin{bmatrix} x_1 \\ x_2 \end{bmatrix} = \begin{bmatrix} a_{11}x_1 + a_{12}x_2 \\ a_{21}x_1 + a_{22}x_2 \end{bmatrix}$$

The values in the resulting matrix are the values on the left side of (4.18); thus, using the right side of (4.18) and substituting, we obtain

$$AX = \begin{bmatrix} a_{11}x_1 + a_{12}x_2 \\ a_{21}x_1 + a_{22}x_2 \end{bmatrix} = \begin{bmatrix} b_1 \\ b_2 \end{bmatrix} = B$$

In matrix notation, the system of linear equations in (4.18) is represented by **AX = B**.

Similarly, the system of m linear equations in n unknowns as given in (4.12) can be represented in matrix form by letting

$$A = \begin{bmatrix} a_{11} & a_{12} & \cdots & a_{1n} \\ a_{21} & a_{22} & \cdots & a_{2n} \\ \vdots & & & \vdots \\ a_{m1} & a_{m2} & \cdots & a_{mn} \end{bmatrix} \quad X = \begin{bmatrix} x_1 \\ x_2 \\ \vdots \\ x_n \end{bmatrix} \quad \text{and} \quad B = \begin{bmatrix} b_1 \\ b_2 \\ \vdots \\ b_m \end{bmatrix}$$

Calculating the product **AX**, we obtain

$$AX = \begin{bmatrix} a_{11} & a_{12} & \cdots & a_{1n} \\ a_{21} & a_{22} & \cdots & a_{2n} \\ \vdots & & & \vdots \\ a_{m1} & a_{m2} & \cdots & a_{mn} \end{bmatrix} \begin{bmatrix} x_1 \\ x_2 \\ \vdots \\ x_n \end{bmatrix} = \begin{bmatrix} a_{11}x_1 + a_{12}x_2 + \cdots + a_{1n}x_n \\ a_{21}x_1 + a_{22}x_2 + \cdots + a_{2n}x_n \\ \vdots \\ a_{m1}x_1 + a_{m2}x_2 + \cdots + a_{mn}x_n \end{bmatrix}$$

The values in the resulting equation are the values on the left side of (4.12); thus using the right side of (4.12) and substituting, we obtain

$$AX = \begin{bmatrix} a_{11}x_1 + a_{12}x_2 + \cdots + a_{1n}x_n \\ a_{21}x_1 + a_{22}x_2 + \cdots + a_{2n}x_n \\ \vdots \\ a_{m1}x_1 + a_{m2}x_2 + \cdots + a_{mn}x_n \end{bmatrix} = \begin{bmatrix} b_1 \\ b_2 \\ \vdots \\ b_m \end{bmatrix} = B$$

which is simply

$$AX = B \tag{4.19}$$

where **A** is called the *coefficient* matrix and **B** is termed the *constant* matrix. We note that the order of the respective matrices is compatible since we multiply an $m \times n$ matrix by an $n \times 1$ matrix and obtain an $m \times 1$ matrix. The system in (4.12) is considerably less awkward to write in matrix form than in the equation form of (4.12), and the matrix form is even slightly easier to write than the summation form for the system of equations as given in (4.13). The matrix form does have a definite advantage over the summation form

Sec. 4.4 / MATRIX ALGEBRA

owing to the algebraic operations that are available for manipulating matrices.

Before considering the main advantages of the matrix form, let us consider the representation of some specific systems of linear equations in matrix form.

Example 4.7 Express the following systems in matrix form:

(a) $3x_1 + 4x_2 + x_3 = 12$
$x_1 - 3x_2 + 5x_3 = 18$

(b) $2x_1 + 3x_2 = 10$
$3x_1 - 4x_2 = 15$
$5x_1 - x_2 = 25$

(c) $3x_1 + 2x_2 + x_3 = 7$
$2x_1 + x_2 = 3$
$x_1 + 3x_3 = 4$

(d) $2x_1 - 11x_2 = 19$
$3x_1 + 15x_2 = 60$

Solution

(a) $\begin{bmatrix} 3 & 4 & 1 \\ 1 & -3 & 5 \end{bmatrix} \begin{bmatrix} x_1 \\ x_2 \\ x_3 \end{bmatrix} = \begin{bmatrix} 12 \\ 18 \end{bmatrix}$

(b) $\begin{bmatrix} 2 & 3 \\ 3 & -4 \\ 5 & -1 \end{bmatrix} \begin{bmatrix} x_1 \\ x_2 \end{bmatrix} = \begin{bmatrix} 10 \\ 15 \\ 25 \end{bmatrix}$

(c) $\begin{bmatrix} 3 & 2 & 1 \\ 2 & 1 & 0 \\ 1 & 0 & 3 \end{bmatrix} \begin{bmatrix} x_1 \\ x_2 \\ x_3 \end{bmatrix} = \begin{bmatrix} 7 \\ 3 \\ 4 \end{bmatrix}$

(d) $\begin{bmatrix} 2 & -11 \\ 3 & 15 \end{bmatrix} \begin{bmatrix} x_1 \\ x_2 \end{bmatrix} = \begin{bmatrix} 19 \\ 60 \end{bmatrix}$

EXERCISES

1. For the matrices

$$A = \begin{bmatrix} 1 & 3 & 2 \\ 6 & 0 & 4 \end{bmatrix} \quad B = \begin{bmatrix} 3 & 4 & 2 \\ 1 & 1 & 5 \end{bmatrix} \quad \text{and} \quad C = \begin{bmatrix} 6 & 4 & 3 \\ 2 & 1 & 8 \end{bmatrix}$$

determine the following:

(a) $A+B$
(b) $C+A$
(c) $5A$
(d) $3B+2A$
(e) $C-A$
(f) $3A-2B$
(g) $2B+C+3A$
(h) $4A-2B+3C$

2. For each of the following pairs of matrices, determine whether **AB**, **BA**, both, or neither are defined and, for those that are defined, specify the order of the resulting matrix:

(a) **A** is a 1×2 matrix and **B** is a 2×1 matrix.
(b) **A** is a 2×2 matrix and **B** is a 2×1 matrix.
(c) **A** is a 3×2 matrix and **B** is a 2×2 matrix.
(d) **A** is a 3×2 matrix and **B** is a 1×4 matrix.
(e) **A** is a 3×2 matrix and **B** is a 2×3 matrix.
(f) **A** is a 3×3 matrix and **B** is a 2×3 matrix.
(g) **A** is a 3×3 matrix and **B** is a 3×3 matrix.
(h) **A** is a 3×4 matrix and **B** is a 2×2 matrix.

3. For the matrices

$$A = \begin{bmatrix} 1 & 2 & 0 \\ 1 & 1 & 3 \\ 2 & 3 & 1 \end{bmatrix} \quad B = \begin{bmatrix} 2 & 3 & 1 \\ 1 & 2 & 3 \end{bmatrix} \quad \text{and} \quad C = \begin{bmatrix} 4 & 3 \\ 2 & 1 \\ 0 & -1 \end{bmatrix}$$

determine the following products that are defined, or, for those that are not defined, specify that they are undefined:

(a) **AB**   (b) **AC**   (c) **BA**
(d) **BC**   (e) **CA**   (f) **CB**

4. For the matrices

$$A = \begin{bmatrix} 2 & 1 \\ 3 & 2 \end{bmatrix} \quad \text{and} \quad B = \begin{bmatrix} 4 & 3 \\ 1 & 5 \end{bmatrix}$$

determine **AB** and **BA** and show that the commutative law, that is, **AB = BA**, is not valid for matrix multiplication.

5. The *transpose* of a matrix **A**, written **A**′, is a matrix that has as its first column the first row of **A**, as its second column the second row of **A**, and so on. For example, the transpose of

$$A = \begin{bmatrix} 1 & 3 \\ 2 & 0 \\ 1 & 2 \end{bmatrix}$$

is given by

$$A' = \begin{bmatrix} 1 & 2 & 1 \\ 3 & 0 & 2 \end{bmatrix}$$

For the matrix **A** given above and the matrix

$$B = \begin{bmatrix} 1 & 1 & 3 \\ 4 & -2 & 0 \end{bmatrix}$$

determine the following:

(a) The order of **B**′   (b) **(AB)**′
(c) **B**′**A**′   (d) **A**′**B**′
(e) **(BA)**′

6. A survey was made of the prices of bread, butter, milk, and eggs at four supermarkets in a large metropolitan city. The results of the survey are presented in the following price matrix, where the elements in the rows represent the prices of the four commodities for a specific supermarket and the elements in the columns represent the prices charged by the supermarkets for a specific commodity. The rows are labeled according to the supermarket that they represent, and the columns are labeled according to the commodities that they represent. The elements of the matrix are expressed in cents, and the prices for the commodities are expressed for the following units: 1 loaf of bread, 1 lb of butter, 1 gal of milk, and 1 doz eggs.

$$\text{Supermarket} \begin{array}{c} A \\ B \\ C \\ D \end{array} \begin{bmatrix} \text{Bread} & \text{Butter} & \text{Milk} & \text{Eggs} \\ 28 & 89 & 109 & 56 \\ 31 & 94 & 125 & 57 \\ 37 & 84 & 129 & 55 \\ 32 & 88 & 115 & 50 \end{bmatrix}$$

(a) Give the matrix that is required for determining the *average* price of one unit of each commodity when multiplying the above price matrix by it.

(b) Determine the matrix that represents the average prices for the four commodities.

(c) Give the matrix that, when multiplied by the above price matrix, is required for determining the total cost in cents of a basket containing one unit of each commodity at each supermarket.

(d) Determine the matrix that represents at each supermarket the total cost of a basket that contains one unit of each commodity.

(e) Give the matrix that, when multiplied by the above price matrix, will have in the first column of the product the total cost at each supermarket of a basket that contains 4 loaves of bread, 1 lb butter, 2 gal milk, and 3 doz eggs; and in the second column of the product the total cost at each supermarket of a basket that contains 2 loaves of bread, 2 lb butter, 3 gal milk, and 4 doz eggs.

(f) Multiply the matrix from part (e) by the price matrix to determine the cost of each of the three baskets at each of the four supermarkets.

7. A manufacturer utilizes the services of three different departments to remanufacture four different automobile parts. The number of hours of labor required from each department for each part is given in the following time-allocation matrix:

$$\text{Department} \begin{array}{c} 1 \\ 2 \\ 3 \end{array} \begin{bmatrix} \text{Part} \\ A & B & C & D \\ 1 & 2 & 2 & 3 \\ 3 & 1 & 2 & 1 \\ 4 & 5 & 3 & 2 \end{bmatrix}$$

(a) Give the matrix that, when the above time-allocation matrix is multiplied by it, will yield a matrix containing the *total* number of hours spent remanufacturing each product.

(b) Determime the matrix that represents the total number of hours spent remanufacturing each product.

(c) Give the matrix that, when multiplied by the above time-allocation matrix, will yield the total time spent by each department in remanufacturing 3 units of A, 2 units of B, 4 units of C, and 10 units of D.

(d) Determine the matrix that represents the total time spent by each department in remanufacturing the mixture of units given in part (c).

(e) Give the matrix that, when multiplied by the above time-allocation matrix, will yield in the first column of the product the total time spent by each department in remanufacturing a group of products containing 3 units of A, 2 units of B, 4 units of C, and 10 units of D; in the second column the total time spent in remanufacturing a group of 4 units of A, 6 units of B, 7 units of C, and 3 units of D; and in the third column the total time spent in remanufacturing a group of 9 units of A, 3 units of B, 8 units of C, and 1 unit of D.

(f) Determine the matrix that represents the total time spent by each department in remanufacturing each of the three groups of products given in part (e).

8. Express the following systems of linear equations in equation form:

(a) $\begin{bmatrix} 2 & 3 \\ 1 & -2 \end{bmatrix} \begin{bmatrix} x \\ y \end{bmatrix} = \begin{bmatrix} 9 \\ 4 \end{bmatrix}$ (b) $\begin{bmatrix} 1 & 4 & -1 \\ 3 & 2 & 2 \end{bmatrix} \begin{bmatrix} x_1 \\ x_2 \\ x_3 \end{bmatrix} = \begin{bmatrix} 12 \\ 21 \end{bmatrix}$

(c) $\begin{bmatrix} 1 & 4 \\ 3 & -2 \\ 2 & -1 \end{bmatrix} \begin{bmatrix} x \\ y \end{bmatrix} = \begin{bmatrix} 10 \\ 5 \\ 6 \end{bmatrix}$

9. Express the following systems of linear equations in matrix form:

(a) $3x + 4y = 5$
$2x + y = 10$
$4x - 8y = 1$
$6x + 4y = 20$

(b) $4x_1 + 5x_2 + 6x_3 = 17$
$8x_1 - 4x_2 + 11x_3 = 25$
$3x_1 + 7x_2 - 8x_3 = 24$

(c) $3x_1 + 2x_2 + 4x_3 - 6x_4 + 8x_5 = 15$
$9x_1 - 3x_2 - 8x_3 + 7x_4 - 9x_5 = 29$
$4x_1 + 7x_2 - 10x_3 + 11x_4 - 2x_5 = 57$

10. A service station sells regular, unleaded, and premium gas at self-service and full-service pumps. The price per gallon in cents of each for both types of service is given in the following matrix:

	Regular	Unleaded	Premium
Self-service	52	53	56
Full-service	56	57	60

(a) At the self-service pumps, the station sells an average of 700 gal of regular, 650 gal of unleaded, and 940 gal of premium on weekdays. The average sales for a weekday at the full-service pumps are 520 gal of regular, 480 gal of unleaded, and 360 gal of premium. Give the matrix that is required for determining the total sales at each type of pump for a weekday.
(b) Determine the total sales at each type of pump for a weekday.
(c) On the weekend, average sales per day at the self-service pumps are 810 gal of regular, 940 gal of unleaded, and 860 gal of premium. At the full-service pumps, average sales per day are 610 gal of regular, 460 gal of unleaded, and 320 gal of premium. Give the matrix that is required for determining the total sales at each type of pump for a day on the weekend.
(d) Determine the total sales at each type of pump for a day on the weekend.

11. A Pizza House puts some combination of five ingredients on its three most popular pizzas. These are shown in the following ingredient matrix where a one means the ingredient is used, and a zero means the ingredient is not used on the pizza:

	A	B	C
Cheese	1	1	1
Tomato sauce	1	1	1
Ground beef	1	1	0
Green pepper	1	0	0
Pepperoni	0	0	1

(a) Give the matrix that is required for determining the number of times each ingredient is used if one each of the pizzas is sold.
(b) Determimne the number of times each ingredient is used by multiplying the ingredient matrix and the matrix determined in part (a).
(c) On a typical Saturday night, 43 A's, 32 B's, and 64 C's are sold. Give the matrix that is required for determining the number of times each ingredient is used on a typical Saturday night.
(d) Determine the number of times each ingredient is used on a typical Saturday night by multiplying the ingredient matrix and the matrix obtained for part (c).

4.5 THE INVERSE OF A MATRIX

In Chap 3, we noted that when using exponents for real numbers, $a \cdot a^{-1} = 1$ if $a \neq 0$. This was the meaning attached to a^{-1}; that is, $a^{-1} = 1/a$ and is such that when multiplied by a the product is equal to 1. Thus the equation $ax = b$ could be solved when $a \neq 0$ by dividing both sides of the equation by a to obtain $x = b/a$; by dividing both sides of the equation by a, we are essentially multiplying both sides of the equation by a^{-1} or $1/a$. The concepts of addition, subtraction, and multiplication were readily extended to matrix operations with only a minimum amount of adaptation. The concept of the *reciprocal* of a real number, however, is not so readily adaptable to matrix operations. In order to extend this concept to matrices let us first consider the matrix equivalent to the real number 1.

The real number 1 is called the *identity number* because a real number multiplied by the identity number is equal to itself; that is, $a \cdot 1 = a$. The matrix counterpart to the identity number is the *identity matrix*, denoted by **I**. The identity matrix is a *square* matrix that has 1's on the diagonal and 0's elsewhere. For example, the 2×2 identity matrix is

$$\mathbf{I} = \begin{bmatrix} 1 & 0 \\ 0 & 1 \end{bmatrix}$$

the 3×3 identity matrix is

$$\mathbf{I} = \begin{bmatrix} 1 & 0 & 0 \\ 0 & 1 & 0 \\ 0 & 0 & 1 \end{bmatrix}$$

the 4×4 identity matrix is

$$\mathbf{I} = \begin{bmatrix} 1 & 0 & 0 & 0 \\ 0 & 1 & 0 & 0 \\ 0 & 0 & 1 & 0 \\ 0 & 0 & 0 & 1 \end{bmatrix}$$

and so forth. It is clear that any $k \times 2$ matrix multiplied on the *right* by the

2×2 identity matrix will yield the original $k\times 2$ matrix; for example, for the 4×2 matrix

$$A = \begin{bmatrix} 1 & 2 \\ 3 & 1 \\ 6 & 7 \\ 0 & 9 \end{bmatrix}$$

we determine

$$AI = \begin{bmatrix} 1 & 2 \\ 3 & 1 \\ 6 & 7 \\ 0 & 9 \end{bmatrix} \begin{bmatrix} 1 & 0 \\ 0 & 1 \end{bmatrix} = \begin{bmatrix} 1 & 2 \\ 3 & 1 \\ 6 & 7 \\ 0 & 9 \end{bmatrix}$$

Similarly, any $2\times k$ matrix that is multiplied on the *left* by the 2×2 identity matrix will yield the original $2\times k$ matrix; for example, for the 2×5 matrix

$$A = \begin{bmatrix} 1 & 3 & 6 & 9 & 4 \\ 5 & 8 & 3 & 2 & 1 \end{bmatrix}$$

we determine

$$IA = \begin{bmatrix} 1 & 0 \\ 0 & 1 \end{bmatrix} \begin{bmatrix} 1 & 3 & 6 & 9 & 4 \\ 5 & 8 & 3 & 2 & 1 \end{bmatrix} = \begin{bmatrix} 1 & 3 & 6 & 9 & 4 \\ 5 & 8 & 3 & 2 & 1 \end{bmatrix}$$

Thus, for any $m\times n$ matrix **A**, the relationship

$$AI = IA = A$$

is true when multiplying on the right by the $n\times n$ identity matrix **I** and on the left by the $m\times m$ identity matrix.

Let us now consider the following definition.

Definition 4.1 If **A** and **B** are square matrices such that $AB = I$ and $BA = I$, where **I** is the identity matrix of the same order as **A**, then **B** is called the inverse of **A** and is denoted by A^{-1}.

For the matrix

$$A = \begin{bmatrix} 2 & 1 \\ 3 & 2 \end{bmatrix}$$

its inverse is given by

$$A^{-1} = \begin{bmatrix} 2 & -1 \\ -3 & 2 \end{bmatrix}$$

Sec. 4.5 / THE INVERSE OF A MATRIX

This is easily verified by showing that $\mathbf{AA}^{-1}=\mathbf{I}$; that is,

$$\mathbf{AA}^{-1} = \begin{bmatrix} 2 & 1 \\ 3 & 2 \end{bmatrix} \begin{bmatrix} 2 & -1 \\ -3 & 2 \end{bmatrix} = \begin{bmatrix} 1 & 0 \\ 0 & 1 \end{bmatrix}$$

It should also be noted that $\mathbf{A}^{-1}\mathbf{A}=\mathbf{I}$; that is,

$$\mathbf{A}^{-1}\mathbf{A} = \begin{bmatrix} 2 & -1 \\ -3 & 2 \end{bmatrix} \begin{bmatrix} 2 & 1 \\ 3 & 2 \end{bmatrix} = \begin{bmatrix} 1 & 0 \\ 0 & 1 \end{bmatrix}$$

We shall see later that not every matrix has an inverse; however, if a matrix does have an inverse, it is unique. For example, suppose the matrix $\mathbf{A}$ has two different inverses $\mathbf{A}^{-1}$ and $\mathbf{B}$. According to the definition of an inverse, we have

$$\mathbf{AB} = \mathbf{I}$$

However, multiplying both sides of the equation by $\mathbf{A}^{-1}$, we obtain

$$\mathbf{A}^{-1}\mathbf{AB} = \mathbf{A}^{-1}\mathbf{I}$$

which reduces to

$$\mathbf{IB} = \mathbf{A}^{-1}$$

or simply

$$\mathbf{B} = \mathbf{A}^{-1}$$

Hence $\mathbf{A}^{-1}$ and $\mathbf{B}$ are not different inverses but are indeed the same, which means that the inverse of $\mathbf{A}$ is unique. Furthermore, if the matrix $\mathbf{A}$ has an inverse $\mathbf{A}^{-1}$, the inverse of $\mathbf{A}^{-1}$ is $\mathbf{A}$ itself.

Matrix Solution of a System of Linear Equations The inverse of a matrix can be used in determining the solution of a system of linear equations. Thus for a system of n equations in n unknowns, we may write the system in the matrix form

$$\mathbf{AX} = \mathbf{B} \qquad (4.20)$$

where $\mathbf{A}$ is an $n \times n$ matrix of *coefficients*, $\mathbf{X}$ is an $n \times 1$ matrix of *variables*, and $\mathbf{B}$ is an $n \times 1$ matrix of *constants*. If $\mathbf{A}$ has an inverse, then we may multiply both sides of (4.20) by $\mathbf{A}^{-1}$ and obtain

$$\mathbf{A}^{-1}\mathbf{AX} = \mathbf{A}^{-1}\mathbf{B}$$

which reduces to

$$\mathbf{IX} = \mathbf{A}^{-1}\mathbf{B}$$

or simply

$$X = A^{-1}B \qquad (4.21)$$

Therefore, as can be seen from (4.21), the solution of a system of n linear equations in n unknowns can be determined through use of the inverse of the *coefficient* matrix **A** if such an inverse exists. How to determine whether a matrix has an inverse and how to determine the inverse if it does exist have not yet been discussed. Essentially the procedure for solving both of these problems has been given in Sec. 4.2.

Let us use matrix notation as a shorthand method of denoting the values given in the system of (4.20). The shorthand notation will consist of *augmenting* the *coefficient* matrix **A** by the *constant* matrix **B**; that is, we represent the system in (4.20) by

$$[A | B] \qquad (4.22)$$

which represents an $n \times n$ matrix augmented by an $n \times 1$ matrix, which results in an $n \times (n+1)$ matrix. Further investigation reveals that the values that make up the elements in the ith row of (4.22) are the coefficients of the n variables and the right-side constant for the ith equation in the $n \times n$ system of equations.

If we perform operations on the rows of the matrix of (4.22) in the same manner as we did when applying the gaussian elimination method to the system of equations in Sec. 4.2, we shall obtain an identity matrix in the first n columns and the values of the variables in the $(n+1)$th column, i.e., an identity matrix to the left of the vertical dividing line and the solution values to the right of the line. For example, using the system of two linear equations in two unknowns given in Example 4.1, we may write the augmented matrix as

$$\begin{bmatrix} 4 & -1 & | & 14 \\ 2 & 1 & | & 4 \end{bmatrix}$$

Dividing the first row by 4, we obtain

$$\begin{bmatrix} 1 & -\frac{1}{4} & | & \frac{7}{2} \\ 2 & 1 & | & 4 \end{bmatrix}$$

Subtracting two times the first row from the second row, we have

$$\begin{bmatrix} 1 & -\frac{1}{4} & | & \frac{7}{2} \\ 0 & \frac{3}{2} & | & -3 \end{bmatrix}$$

Dividing the second row by $\frac{3}{2}$; that is, multiplying by $\frac{2}{3}$, we obtain

$$\begin{bmatrix} 1 & -\frac{1}{4} & | & \frac{7}{2} \\ 0 & 1 & | & -2 \end{bmatrix}$$

Multiplying the second row by $\frac{1}{4}$ and adding the result to the first row, we obtain

$$\begin{bmatrix} 1 & 0 & | & 3 \\ 0 & 1 & | & -2 \end{bmatrix}$$

which, upon replacing the variables, becomes

$$\begin{bmatrix} 1 & 0 \\ 0 & 1 \end{bmatrix} \begin{bmatrix} x \\ y \end{bmatrix} = \begin{bmatrix} 3 \\ -2 \end{bmatrix}$$

or

$$\begin{bmatrix} x \\ y \end{bmatrix} = \begin{bmatrix} 3 \\ -2 \end{bmatrix}$$

You should verify that these steps are identical to those for the gaussian elimination method in Example 4.3.

From the above operations, we see that we can perform row operations on the augmented matrix in order to determine the solution to the original system of equations. These row operations are called *elementary row operations*, and they consist of any combination of the following row operations:

1. Multiplying or dividing each element of a row by the same nonzero number
2. Adding or subtracting each element of a row to or from the corresponding elements of any other row
3. Interchanging two rows

The solution obtained above is the same as the one given in (4.21); that is,

$$\mathbf{X} = \mathbf{A}^{-1}\mathbf{B} = \begin{bmatrix} 3 \\ -2 \end{bmatrix}$$

This suggests that by following the gaussian elimination procedure (i.e., a special sequence of elementary row operations) for the augmented matrix, we shall obtain $[\mathbf{I} | \mathbf{A}^{-1}\mathbf{B}]$. Thus we could begin with the augmented matrix $[\mathbf{A} | \mathbf{B}]$ and perform the gaussian elimination procedure to obtain the matrix $[\mathbf{I} | \mathbf{A}^{-1}\mathbf{B}]$. If instead of beginning with $[\mathbf{A} | \mathbf{B}]$ we were to begin with $[\mathbf{A} | \mathbf{I}]$, where $\mathbf{I}$ is the $n \times n$ identity matrix, and apply the gaussian elimination method to the rows of the augmented matrix: we would obtain $[\mathbf{I} | \mathbf{A}^{-1}]$.

Using the system of equations from Example 4.1 and augmenting the coefficient matrix with the identity matrix, we have

$$\begin{bmatrix} 4 & -1 & | & 1 & 0 \\ 2 & 1 & | & 0 & 1 \end{bmatrix}$$

Performing the same sequance of row operations as above, we obtain $\mathbf{A}^{-1}$. Thus dividing the first row by 4, we obtain

$$\begin{bmatrix} 1 & -\frac{1}{4} & | & \frac{1}{4} & 0 \\ 2 & 1 & | & 0 & 1 \end{bmatrix}$$

Subtracting two times the first row from the second row, we have

$$\begin{bmatrix} 1 & -\frac{1}{4} & | & \frac{1}{4} & 0 \\ 0 & \frac{3}{2} & | & -\frac{1}{2} & 1 \end{bmatrix}$$

Dividing the second row by $\frac{3}{2}$, we obtain

$$\begin{bmatrix} 1 & -\frac{1}{4} & | & \frac{1}{4} & 0 \\ 0 & 1 & | & -\frac{1}{3} & \frac{2}{3} \end{bmatrix}$$

Multiplying the second row by $\frac{1}{4}$ and adding the result to the first row, we have

$$\begin{bmatrix} 1 & 0 & | & \frac{1}{6} & \frac{1}{6} \\ 0 & 1 & | & -\frac{1}{3} & \frac{2}{3} \end{bmatrix}$$

If we have performed all of the operations properly, the matrix to the right of the vertical line should represent $\mathbf{A}^{-1}$, where

$$\mathbf{A} = \begin{bmatrix} 4 & -1 \\ 2 & 1 \end{bmatrix}$$

To verify that it is indeed the inverse of $\mathbf{A}$, we multiply it by $\mathbf{A}$ and obtain

$$\begin{bmatrix} 4 & -1 \\ 2 & 1 \end{bmatrix} \begin{bmatrix} \frac{1}{6} & \frac{1}{6} \\ -\frac{1}{3} & \frac{2}{3} \end{bmatrix} = \begin{bmatrix} \frac{2}{3}+\frac{1}{3} & \frac{2}{3}-\frac{2}{3} \\ \frac{1}{3}-\frac{1}{3} & \frac{1}{3}+\frac{2}{3} \end{bmatrix} = \begin{bmatrix} 1 & 0 \\ 0 & 1 \end{bmatrix}$$

Thus the matrix is the inverse of $\mathbf{A}$, and we may write

$$\mathbf{A}^{-1} = \begin{bmatrix} \frac{1}{6} & \frac{1}{6} \\ -\frac{1}{3} & \frac{2}{3} \end{bmatrix}$$

Sec. 4.5 / THE INVERSE OF A MATRIX

Furthermore,

$$X = A^{-1}B = \begin{bmatrix} \frac{1}{6} & \frac{1}{6} \\ -\frac{1}{3} & \frac{2}{3} \end{bmatrix} \begin{bmatrix} 14 \\ 4 \end{bmatrix} = \begin{bmatrix} \frac{14}{6} + \frac{4}{6} \\ -\frac{14}{3} + \frac{8}{3} \end{bmatrix} = \begin{bmatrix} 3 \\ -2 \end{bmatrix}$$

If we wish to determine A^{-1} and $A^{-1}B$ simultaneously, we can augment A with the two matrices I and B to obtain $[A|I|B]$. By performing the operations for the gaussian elimination method, we obtain $[I|A^{-1}|A^{-1}B]$. Thus for the above example the reader may verify that we could begin with

$$\begin{bmatrix} 4 & -1 & | & 1 & 0 & | & 14 \\ 2 & 1 & | & 0 & 1 & | & 4 \end{bmatrix}$$

and, applying the gaussian elimination method, obtain

$$\begin{bmatrix} 1 & 0 & | & \frac{1}{6} & \frac{1}{6} & | & 3 \\ 0 & 1 & | & -\frac{1}{3} & \frac{2}{3} & | & -2 \end{bmatrix}$$

Since the gaussian elimination method is applicable for any number of equations, we can extend the above procedure to any system of n linear equations in n unknowns. The procedure will obtain the inverse of A *if it exists*; however, if A does not have an inverse, we obtain a row of zeroes to the left of the first vertical line, and the system of linear equations will not contain a *unique* solution.

Example 4.8 For the problem given in Example 4.4, determine the solution to the system by simultaneously finding the inverse of the coefficient matrix and the solution values of the system. Verify the inverse and the solution values by matrix multiplication.

Solution Recall that the obtained system of linear equations is

$$\begin{aligned} 3x + 4y + z &= 8 \\ 5x - 2y + 3z &= 10 \\ -x + 3y + 4z &= 19 \end{aligned}$$

Representing the system of equations in matrix form, we obtain

$$\begin{bmatrix} 3 & 4 & 1 \\ 5 & -2 & 3 \\ -1 & 3 & 4 \end{bmatrix} \begin{bmatrix} x \\ y \\ z \end{bmatrix} = \begin{bmatrix} 8 \\ 10 \\ 19 \end{bmatrix}$$

Augmenting the coefficient matrix with the identity matrix and the matrix of right-side constants, we have

$$\begin{bmatrix} 3 & 4 & 1 & | & 1 & 0 & 0 & | & 8 \\ 5 & -2 & 3 & | & 0 & 1 & 0 & | & 10 \\ -1 & 3 & 4 & | & 0 & 1 & 1 & | & 19 \end{bmatrix}$$

Applying the gaussian elimination method to the augmented matrix, we divide the first row by 3 and obtain

$$\begin{bmatrix} 1 & \frac{4}{3} & \frac{1}{3} & \frac{1}{3} & 0 & 0 & \frac{8}{3} \\ 5 & -2 & 3 & 0 & 1 & 0 & 10 \\ -1 & 3 & 4 & 0 & 0 & 1 & 19 \end{bmatrix}$$

In order to obtain the first column of the 3×3 identity matrix, we first multiply the first row by 5 and subtract the resulting row from the second row, and then multiply the first row by 1 and add the resulting row to the third row. Performing the operations, we obtain

$$\begin{bmatrix} 1 & \frac{4}{3} & \frac{1}{3} & \frac{1}{3} & 0 & 0 & \frac{8}{3} \\ 0 & -\frac{26}{3} & \frac{4}{3} & -\frac{5}{3} & 1 & 0 & -\frac{10}{3} \\ 0 & \frac{13}{3} & \frac{13}{3} & \frac{1}{3} & 0 & 1 & \frac{65}{3} \end{bmatrix}$$

In the next step, we divide the second row by $-\frac{26}{3}$ to obtain

$$\begin{bmatrix} 1 & \frac{4}{3} & \frac{1}{3} & \frac{1}{3} & 0 & 0 & \frac{8}{3} \\ 0 & 1 & -\frac{2}{13} & \frac{5}{26} & -\frac{3}{26} & 0 & \frac{5}{13} \\ 0 & \frac{13}{3} & \frac{13}{3} & \frac{1}{3} & 0 & 1 & \frac{65}{3} \end{bmatrix}$$

By first multiplying the second row of the resulting matrix by $\frac{4}{3}$ and subtracting the resulting row from the first row, and then multiplying the second row by $\frac{13}{3}$ and subtracting the resulting row from the third row, we obtain

$$\begin{bmatrix} 1 & 0 & \frac{7}{13} & \frac{1}{13} & \frac{2}{13} & 0 & \frac{28}{13} \\ 0 & 1 & -\frac{2}{13} & \frac{5}{26} & -\frac{3}{26} & 0 & \frac{5}{13} \\ 0 & 0 & 5 & -\frac{1}{2} & \frac{1}{2} & 1 & 20 \end{bmatrix}$$

For the final result, we perform the following three steps; first we divide the third row by 5 to obtain the third row of the matrix given below, second we multiply the resulting third row by $\frac{2}{13}$ and add the resulting row to the second row to obtain the second row of the matrix given below, and third we multiply the resulting third row by $\frac{7}{13}$ and subtract the resulting row from the first row to obtain the first row of the matrix

$$\begin{bmatrix} 1 & 0 & 0 & \frac{17}{130} & \frac{1}{10} & -\frac{7}{65} & 0 \\ 0 & 1 & 0 & \frac{23}{130} & -\frac{1}{10} & \frac{2}{65} & 1 \\ 0 & 0 & 1 & -\frac{1}{10} & \frac{1}{10} & \frac{1}{5} & 4 \end{bmatrix}$$

Sec. 4.5 / THE INVERSE OF A MATRIX

If we have not made an error in our computations, the solution values expressed in matrix form are

$$\begin{bmatrix} 1 & 0 & 0 \\ 0 & 1 & 0 \\ 0 & 0 & 1 \end{bmatrix} \begin{bmatrix} x \\ y \\ z \end{bmatrix} = \begin{bmatrix} 0 \\ 1 \\ 4 \end{bmatrix}$$

or simply

$$\begin{bmatrix} x \\ y \\ z \end{bmatrix} = \begin{bmatrix} 0 \\ 1 \\ 4 \end{bmatrix}$$

In order to verify by matrix multiplication that this solution is correct, we must first verify that the 3×3 matrix obtained between the two vertical lines in the final augmented matrix is indeed $\mathbf{A}^{-1}$. To check this, we determine whether the product $\mathbf{A}\mathbf{A}^{-1}$ is equal to $\mathbf{I}$. Multiplying, we obtain

$$\begin{bmatrix} 3 & 4 & 1 \\ 5 & -2 & 3 \\ -1 & 3 & 4 \end{bmatrix} \begin{bmatrix} \frac{17}{130} & \frac{1}{10} & -\frac{7}{65} \\ \frac{23}{130} & -\frac{1}{10} & \frac{2}{65} \\ -\frac{1}{10} & \frac{1}{10} & \frac{1}{5} \end{bmatrix} = \begin{bmatrix} 1 & 0 & 0 \\ 0 & 1 & 0 \\ 0 & 0 & 1 \end{bmatrix}$$

and therefore we may write

$$\mathbf{A}^{-1} = \begin{bmatrix} \frac{17}{130} & \frac{1}{10} & -\frac{7}{65} \\ \frac{23}{130} & -\frac{1}{10} & \frac{2}{65} \\ -\frac{1}{10} & \frac{1}{10} & \frac{1}{5} \end{bmatrix}$$

Now, to verify the solution values obtained in the final augmented matrix, we shall determine $\mathbf{A}^{-1}\mathbf{B}$; thus we have

$$\mathbf{A}^{-1}\mathbf{B} = \begin{bmatrix} \frac{17}{130} & \frac{1}{10} & -\frac{7}{65} \\ \frac{23}{130} & -\frac{1}{10} & \frac{2}{65} \\ -\frac{1}{10} & \frac{1}{10} & \frac{1}{5} \end{bmatrix} \begin{bmatrix} 8 \\ 10 \\ 19 \end{bmatrix} = \begin{bmatrix} 0 \\ 1 \\ 4 \end{bmatrix}$$

EXERCISES

1. For the matrix

$$\mathbf{A} = \begin{bmatrix} 1 & 3 & 1 & 4 \\ 2 & 5 & 3 & 1 \\ 1 & 4 & 2 & 6 \end{bmatrix}$$

verify that $\mathbf{AI} = \mathbf{A} = \mathbf{IA}$, where $\mathbf{I}$ is the identity matrix of the proper order in each case.

Chap. 4 / SYSTEMS OF LINEAR EQUATIONS AND MATRIX ALGEBRA

2. Determine AA^{-1} and $A^{-1}A$ to verify that the inverse of

$$A = \begin{bmatrix} 2 & 3 \\ 2 & -1 \end{bmatrix} \quad \text{is} \quad A^{-1} = \begin{bmatrix} \frac{1}{8} & \frac{3}{8} \\ \frac{1}{4} & -\frac{1}{4} \end{bmatrix}$$

Use A^{-1} to solve the following systems of linear equations by computing $A^{-1}B$. Verify each answer by substituting the solution values into the system of equations.

(a) $2x + 3y = 5$
$2x - y = 1$
(b) $2x + 3y = 12$
$2x - y = 4$
(c) $2x + 3y = -4$
$2x - y = 0$
(d) $2x + 3y = 31$
$2x - y = 11$

3. Determine AA^{-1} and $A^{-1}A$ to verify that the inverse of

$$A = \begin{bmatrix} 1 & 1 & 0 \\ 1 & 2 & 5 \\ 2 & 4 & 3 \end{bmatrix} \quad \text{is} \quad A^{-1} = \begin{bmatrix} 2 & \frac{3}{7} & -\frac{5}{7} \\ -1 & -\frac{3}{7} & \frac{5}{7} \\ 0 & \frac{2}{7} & -\frac{1}{7} \end{bmatrix}$$

Use A^{-1} to solve the following systems of linear equations by computing $A^{-1}B$. Verify each answer by substituting the solution values into the system of equations.

(a) $x + y = 5$
$x + 2y + 5z = 5$
$2x + 4y + 3z = 10$
(b) $x + y = 2$
$x + 2y + 5z = 21$
$2x + 4y + 3z = 21$
(c) $x + y = 5$
$x + 2y + 5z = 20$
$2x + 4y + 3z = 26$
(d) $x + y = 3$
$x + 2y + 5z = 15$
$2x + 4y + 3z = 9$

4. Determine the inverse of

$$A = \begin{bmatrix} 3 & 5 \\ 4 & 7 \end{bmatrix}$$

and use it to determine the solutions of the following systems of linear equations. Verify each answer by substituting the solution values into the systems of equations.

(a) $3x + 5y = 8$
$4x + 7y = 11$
(b) $3x + 5y = 1$
$4x + 7y = 1$
(c) $3x + 5y = 20$
$4x + 7y = 28$
(d) $3x + 5y = 5$
$4x + 7y = 6$

5. Determine the inverse of

$$A = \begin{bmatrix} 1 & 2 & 4 \\ 1 & 4 & 6 \\ 3 & 5 & 13 \end{bmatrix}$$

Sec. 4.5 / THE INVERSE OF A MATRIX

and use it to determine the solutions of the following systems of linear equations. Verify each answer by substituting the solution values into the system of equations.

(a) $x_1 + 2x_2 + 4x_3 = -1$
$x_1 + 4x_2 + 6x_3 = -1$
$3x_1 + 5x_2 + 13x_3 = -5$

(b) $x_1 + 2x_2 + 4x_3 = 4$
$x_1 + 4x_2 + 6x_3 = 6$
$3x_1 + 5x_2 + 13x_3 = 11$

(c) $x_1 + 2x_2 + 4x_3 = 3$
$x_1 + 4x_2 + 6x_3 = 5$
$3x_1 + 5x_2 + 13x_3 = 4$

(d) $x_1 + 2x_2 + 4x_3 = 8$
$x_1 + 4x_2 + 6x_3 = 12$
$3x_1 + 5x_2 + 13x_3 = 26$

6. Solve the following system of linear equations by performing the gaussian elimination method on the augmented matrix $[A|B]$. Verify the solution values by substituting the values into the original equations.

$$2x + 5y = 8$$
$$6x - 2y = 58$$

7. Solve the following system of linear equations by performing the gaussian elimination method on the augmented matrix $[A|B]$. Verify the solution values by substituting the values into the original equations.

$$2x_1 + 6x_2 + 4x_3 = 12$$
$$x_1 + 4x_2 + 6x_3 = 11$$
$$x_1 + 3x_2 + 3x_3 = 7$$

8. Determine the inverse of A and the solution values for the following system of linear equations by applying the gaussian elimination method to the augmented matrix $[A|I|B]$:

$$2x + 5y = 8$$
$$6x - 2y = 58$$

9. Determine the inverse of A and the solution values for the following system of linear equations by applying the gaussian elimination method to the augmented matrix $[A|I|B]$:

$$2x_1 + 6x_2 + 4x_3 = 12$$
$$x_1 + 4x_2 + 6x_3 = 11$$
$$x_1 + 3x_2 + 3x_3 = 7$$

10. A customer tells you she needs 77 sq ft of carpeting for one room and 120 sq ft for another. In addition, she wants 288 sq ft of wall paneling in the first room and none in the other. The most she is willing to pay is $2,206.15 for the first room and $1,194 for the second room. Since she does not wish to exceed her budget, determine by performing the gaussian elimination method on the augmented matrix $[A|B]$ the maximum price per square foot she can pay to stay within her budget.

11. During one quarter a company had storage costs of $31.50 with 3 units of product x, 5 units of product y, and 2 units of product z in inventory. During another quarter when average inventory was 6 units of x, 8 units of y, and 3 units of z, storage costs were $53.50. During a third quarter, average inventory was 2 units of x, 1 unit of y, and 4 units of z, and storage costs were $19.50. Determine by performing the gaussian elimination method on the augmented matrix $[A\,|\,I\,|\,B]$ the per unit storage cost of each product.

12. To make one unit of product x requires 3 hr in division A and 2 hr in division B. One unit of product y requires 2 hr in division A and 5 hr in division B. Division A has available 85 hr per week, and division B has available 130 hr per week. By applying the gaussian elimination method to the augmented matrix $[A\,|\,I\,|\,B]$ determine the inverse of A and the number of units of each product that can be made if all of the available time must be used.

13. A fruit stand sold 3 pints (pt) of strawberries, 5 lb of peaches, and 2 lb of plums to one customer for $5.54. Another customer purchased 1 pt of strawberries, 3 lb of peaches, and 3 lb of plums for $3.48. A third customer purchased 4 lb of peaches and 6 lb of plums for $4.48. By applying the gaussian elimination method to the augmented matrix $[A\,|\,I\,|\,B]$, determine the inverse of A and the price per pint of strawberries and per pound of peaches and plums.

IMPORTANT TERMS AND CONCEPTS

Augmented matrix
Coefficient matrix
Coincident lines
Column vector
Elementary row operations
Elements
Gaussian elimination method
Identity matrix
Inconsistent equations
Inverse of a matrix
Matrix
Matrix addition
Matrix multiplication
Method of elimination
Method of substitution
Null matrix
Overconstrained system
Parameters
Row vector
Scalar
Subscripts
Summation notation
Systems of linear equations
Underconstrained system
Zero matrix

REVIEW PROBLEMS

1. Use each of three methods of solution to solve each of the following systems of linear equations:

   (a) $4X - 6Y = 2$
   $5X + 3Y = 4$

   (b) $2X - Y + 3Z = 17$
   $4X + 2Y + Z = 10$
   $X + 3Y + 5Z = 19$

(c) $3X + Y + 5Z = 20$
$X + 3Y + 7Z = -4$
$5X - 2Y - 4Z = 28$

(e) $X_1 - 2X_2 + X_3 = 0$
$3X_1 + X_2 - 2X_3 = -1$
$5X_1 - 2X_2 + 4X_3 = 5$

(d) $X_1 + X_2 + X_3 = 4$
$2X_1 - X_2 + 3X_3 = 6$
$4X_1 - 8X_2 + 5X_3 = 10$

2. ABC Company sold 10 cases of product A, 15 cases of product B, and 12 cases of product C for \$495.50. Five cases of product A, 2 cases of product B, and 9 cases of product C sold for \$184.05. Two cases of product A, 6 cases of product B, and 12 cases of product C sold for \$243.40. Determine the cost of a case of product A, a case of product B, and a case of product C.

3. The molding department can mold 15 doors for compact cars and 12 doors for trucks in 8 machine hours. Another possible production schedule is 10 doors for compact cars and 16 doors for trucks in 8 hr. Determine the time required to mold a door for a compact car and to mold a door for a truck. (Hint: convert hours to minutes.)

4. A customer of the Wax Shop purchases 2 six-inch (in.) candles and 6 ten-in. candles for \$4.80. Ten six-in. candles and 4 ten-in. candles were purchased by another customer for \$7.10. Determine the sales price for six-in. candles and ten-in. candles.

5. A pet shop sold 4 guppies, 2 goldfish, and 1 box of fish food for \$6.55. A boy purchased 2 guppies and 1 box of fish food for \$2.27. Another customer purchased 3 guppies, 4 goldfish, and a box of fish food for \$8.16. Determine the sales price of a guppy, a goldfish, and a box of fish food.

6. Express the systems of linear equations in Problem 1 of this section in matrix form.

7. A manufacturer classifies its employees as unskilled, skilled, or supervisors. The numer of employees in each classification working in each department is given in the following employee-allocation matrix:

$$\text{Department} \begin{array}{c} A \\ B \\ C \\ D \end{array} \begin{bmatrix} \overset{\text{Unskilled}}{8} & \overset{\text{Classification}}{\underset{\text{Skilled}}{5}} & \overset{\text{Supervisor}}{1} \\ 20 & 10 & 3 \\ 15 & 9 & 2 \\ 5 & 10 & 1 \end{bmatrix}$$

(a) Give the matrix that when the employee-allocation matrix is multiplied by it, will yield a matrix containing the total number of employees in each classification.

(b) Determine the matrix that represents the total number of empoyees in each classification.

(c) Give the matrix that, when multiplied by the employee-allocation matrix, will yield a matrix containing the total number of employees in each department.

(d) Determine the matrix that represents the total number of employees in each department.

(e) Give the matrix that, when multiplied by the employee-allocation matrix, will yield the total weekly salaries paid by each department if unskilled employees are paid $100/week, skilled employees are paid $152/week, and supervisors are paid $250/week.

(f) Determine the matrix that represents the total weekly salaries paid by each department.

8. To make a unit of product X requires 2 hr in the molding department, 1 hr in the polishing department, and $\frac{1}{2}$ hr in the finishing department. To make a unit of product Y requires 1 hr in the molding department, 2 hr in the polishing department, and $\frac{3}{4}$ hr in the finishing department. Product Z requires 3 hr in the molding department, 2 hr in the polishing department, and 1 hr in the finishing department to make a unit. Yesterday the molding department was used for 44 hr, the polishing department was used for 42 hr, and the finishing department was used for 18 hr.

(a) Determine the system of linear equations that can be solved to give the number of units of each product manufactured yesterday.

(b) Solve the system of linear equations from part (a) by simultaneously finding the inverse of the coefficient matrix and the solution values of the system.

LINEAR INEQUALITIES AND LINEAR PROGRAMMING

5.1 INTRODUCTION

In Chap. 4, we considered situations in which constraints could be regarded as equations. The constraints of many practical problems must be expressed in inequalities rather than equalities. For example, if a businessman has available a maximum of $10,000, he cannot invest more than that amount; similarly, a person cannot work more than 24 hr in a day. Many other examples leading to inequalities are quite common in everyday situations.

In order to consider systems of linear inequalities, it is necessary that we consider first some of the basic properties of simple inequalities. One of the most fundamental properties of the real number system is that of *order*. Algebraically, real numbers are ordered such that two numbers are either equal or one is greater than the other. Symbolically we use the symbols $<$ and $>$, which mean *less than* and *greater than*, respectively. Thus $3<5$ is the statement that 3 is less than 5, and $6>2$ is the statement that 6 is greater than 2; the statement $a<b$ is the statement that a is less than b. It may be helpful to the reader to note that the point of each inequality sign is directed at the smaller number.

The ordering of the real numbers can be demonstrated quite easily by Fig. 5.1, which is comparable to the x axis presented in Chap. 1. With reference to the numbers on the axis in Fig. 5.1, the number a is less than the number b if it is to the left of b; if a is to the right of b, it is greater than b.

FIGURE 5.1

Thus we see that there are only three possibilities for two numbers a and b;

that is,
$$a < b$$
$$a = b$$
or
$$a > b$$

Quite often, we wish to express the relationship of two numbers by using the first two or last two possibilities. This is particularly true when we wish to consider a negative statement. For example, to say that a is *not greater than b* is the same as saying that a is *less than b* or a is *equal* to b, or simply, that a is *less than or equal to b*. We represent this ordering of less than or equal to symbolically by combining the $<$ and $=$ symbols to obtain the $\leq$ symbol; thus $a \leq b$ is read "a is less than or equal to b." Similarly, we use the symbol $\geq$ to mean "greater than or equal to."

Rules for Inequalities

Before we can successfully understand the use of inequalities, we need to understand the basic algebraic rules that govern them. Many of the rules for equalities are also valid for inequalities. For example, when $a = b$ and $b = c$, this implies that $a = c$ for any real numbers a, b, and c. The same ordering relationship is valid for inequalities. We shall use the combined symbols $\leq$ and $\geq$; however, the rules are equally valid for the strict inequalities $<$ and $>$.

Rule 5.1 If $a \leq b$ and $b \leq c$, then $a \leq c$.

Thus if Chrysler manufactures no more cars than Ford, and Ford manufactures no more cars than General Motors, then Chrysler manufactures no more cars than General Motors.

When we add equals to equals, we do not disturb the equality relationship. Similarly, when we add equals to unequals, we do not disturb the inequality relationship.

Rule 5.2 If $a \leq b$, then $a + c \leq b + c$.

Thus if the vice-president of a company makes no more salary than the president and they receive equal raises, the vice-president still receives no more salary than the president. Similarly, if the vice-president's increase in salary is *less than* the president's increase, the vice-president receives less salary than the president.

Rule 5.3 If $a \leq b$ and $c < d$, then $a + c < b + d$.

Sec. 5.1 / INTRODUCTION

Rule 5.2 is also valid for subtraction because c could be a negative number and the addition would be equivalent to subtracting a positive number. However, Rule 5.3 is *not* valid for subtraction because c and d are not equal. For example, for $2 \leq 3$ and $1 < 10$, it is obvious that $2 - 1 = 1$ is *not* equal to or less than $3 - 10 = -7$. Although c and d could be negative and it would appear to be equivalent to subtraction, Rule 5.3 is valid only if the two inequalities are valid and addition is performed.

Rules 5.2 and 5.3 are rules for addition. We have two comparable rules for multiplication of inequalities, but when dealing with multiplication of inequalities, we must be very careful. The following rule is the multiplication counterpart of Rule 5.2.

Rule 5.4 *If $a \leq b$, then $ac \leq bc$ when c is positive and $ac \geq bc$ when c is negative.*

Rule 5.4 says that the direction, or sense, of the inequality is the same if c is positive and is reversed if c is negative. For example, for $a = 3$, $b = 7$, and $c = 5$, we have $3 < 7$ and $3 \cdot 5 < 7 \cdot 5$, or $15 < 35$. If $c = -5$, we have $3 \cdot (-5) > 7 \cdot (-5)$, or $-15 > -35$. Thus we must be very careful when multiplying both sides of an inequality by a constant. The direction of the resulting inequality is dependent on the relationship between a and b and the sign of c.

The following rule is the multiplication counterpart of Rule 5.3.

Rule 5.5 *If $a \leq b$ and $c \leq d$, where a, b, c, and d are all positive, then $ac \leq bd$.*

For example, if we have three Volkswagens and four Fords and a Volkswagen costs less than a Ford, the total cost for the three Volkswagens is less than the total cost of the four Fords. However, we cannot tell whether four Volkswagens cost less than three Fords because this depends on the costs of the cars. Hence we cannot make a general statement that is always true unless the inequalities are pointing the same direction and each number is positive.

For the inequality $x \leq 2$, the solution set for x is the set of all values less than or equal to 2. This set is illustrated graphically by the shaded region in Fig. 5.2.

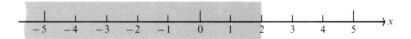

FIGURE 5.2

For the inequality containing two variables $x + y \leq 10$, the solution set is the set of all points (x,y) that are on or below the line $x+y=10$, as illustrated by the shaded region in Fig. 5.3. An easy way to determine on which side of the line the solution set lies is to select any arbitrary point to one side of the line and substitute it into the inequality. If the inequality is satisfied, the solution set is on the same side of the line as the arbitrary point. If it is not satisfied, then the solution set is on the opposite side of the line. Check the solution set for Fig. 5.3 with the point $(0,0)$.

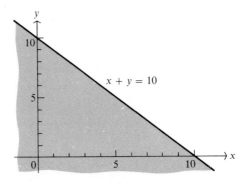

FIGURE 5.3

EXERCISES

1. Determine whether the following statements are true or false:

   (a) $6 > 4$
   (b) $5 < 3$
   (c) $-4 < -3$
   (d) $-8 > 2$
   (e) $4 \geqslant 0$
   (f) $-2 \leqslant -2$
   (g) $6 \leqslant -7$
   (h) $8 \geqslant 6$

2. Perform each of the following operations:

   (a) Add 6 to both sides of $5x - 6 < 14$.
   (b) Subtract 7 from both sides of $10x + 7 \geqslant 37$.
   (c) Multiply both sides of $2x \leqslant 10$ by $\frac{1}{2}$.
   (d) Multiply both sides of $\frac{1}{5}x \geqslant 10$ by 5.
   (e) Multiply both sides of $-4x \leqslant 32$ by $-\frac{1}{4}$.
   (f) Divide both sides of $6x \geqslant 36$ by 6.

3. Solve the following inequalities for x:

   (a) $3x + 5 \leqslant 20$
   (b) $2x - 3 \geqslant 9$
   (c) $6 - 4x < -10$
   (d) $2 + \frac{1}{3}x \geqslant 6$
   (e) $\frac{2}{3}x - 4 < 8$
   (f) $2x + 3 \leqslant 4x - 7$

4. A manufacturing firm can rent a warehouse that contains 5,000 sq ft of floor space. The product that they manufacture requires 4 sq ft of floor space per unit for

storage. Determine the inequality that must be satisfied by the number of units x that can be stored on the warehouse floor. Determine the greatest number of units that can be stored on the floor of the warehouse.

5. A grocer can purchase canned soup from a wholesaler for a discount if he buys at least 120 cans. However, the cans are boxed in cartons of 24 cans each. Determine the inequality that must be satisfied by the number of cartons x the grocer must purchase. Determine the least number of cartons the grocer can purchase and still receive the discount.

6. Draw graphs of the regions that represent the following inequalities:

(a) $y + 2x \leqslant 6$
(b) $3y + x < 9$
(c) $2y + 3x \geqslant 10$
(d) $y - 3x > -6$

7. A salesman makes $3 for every unit sold plus $370/month. Determine the minimum number of units per month he can sell in order to make at least $1,000/month.

8. A telephone installation person can install $2\frac{1}{2}$ phones per hour. Determine the maximum number of phones that a person can install in an 8-hr working day.

9. An office wants to recarpet the floor for a maximum cost of $15,000. Installation will cost $2,500 for the 1,200 sq ft office. Determine the maximum amount that can be paid per square foot of carpeting.

10. A cookie manufacturer puts 28 cookies in a box. For every 5,000 cookies made, an average of 16 are broken before getting to the packaging department. Determine the maximum number of boxes of cookies that can be packaged for every 5,000 cookies that are made.

5.2 SYSTEMS OF LINEAR INEQUALITIES

In business there are many situations where constraints are expressed by inequalities and where several inequalities are required to describe the constraints adequately. For example, a company may have a maximum amount of money available to invest in an existing business and may also require the business to be generating some minimum net profit per year. As a more specific example, a distributor of a certain product may have 650 units available for shipment, and, consequently, he cannot ship more than 650 or less than 0 units. Thus the two linear inequalities for the constraints, or the number of units x, that he can ship are

$$x \leqslant 650$$

and

$$x \geqslant 0$$

A solution to this system of inequalities is any value of x satisfying $0 \leqslant x \leqslant 650$, which is shown graphically by the shaded region in Fig. 5.4. Thus the *solution set* for the above inequalities is the set of x values satisfying $0 \leqslant x \leqslant 650$.

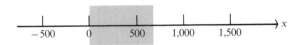

FIGURE 5.4

For a system of two linear inequalities in two variables such as

$$2x + 3y \leq 10$$
$$x + 2y \leq 6$$

we can plot graphs for the case where the left side of each equation is equal to its respective right side and then shade the areas that satisfy *both* inequalities; the shaded area in Fig. 5.5 represents the solution set for this system of inequalities. If we add the requirement that the variables cannot be negative, we have the following system of four inequalties in two variables:

$$2x + 3y \leq 10$$
$$x + 2y \leq 6$$
$$x \geq 0$$
$$y \geq 0 \qquad (5.1)$$

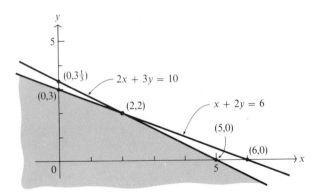

FIGURE 5.5

The solution set for this system of inequalities is exhibited in Fig. 5.6. The points on the boundaries of the solution set that represent the intersection point of any two straight lines forming the boundaries are called *extreme points*; thus the extreme points in Fig. 5.6 are $(0,0)$, $(0,3)$, $(2,2)$, and $(5,0)$.

Example 5.1 To make a widget requires 2 hr in the molding department and 3 hr in the polishing department. To make a gadget requires 5 hr in the molding department and 4 hr in the polishing department. The molding

Sec. 5.2 / SYSTEMS OF LINEAR INEQUALITIES

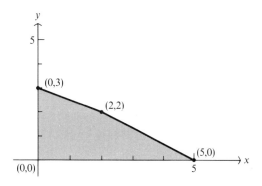

FIGURE 5.6

department has 10 hr available, and the polishing department has 12 hr available.

(a) Determine the system of linear inequalities that represent the solution set for the number of widgets and gadgets that can be made.
(b) Draw a graph of the solution set.
(c) Determine the extreme points of the solution set.

Solution (a)

$$2x + 5y \leqslant 10$$
$$3x + 4y \leqslant 12$$
$$x \geqslant 0$$
$$y \geqslant 0$$

where $x =$ number of widgets and $y =$ number of gadgets.

(b) The graph of the solution set is given in the shaded area of Fig. 5.7.
(c) The extreme points of the solution set are obtained by determining the points of intersection between appropriate pairs of boundary

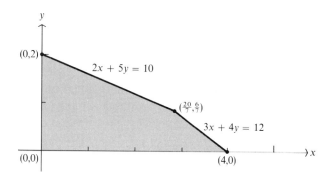

FIGURE 5.7

lines. Thus, in Fig. 5.7, we determine the point of intersection for the two lines for the equations

$$2x + 5y = 10$$

and

$$x = 0$$

which is the point $(0,2)$. Similarly, the lines for the two equations

$$2x + 5y = 10$$

and

$$3x + 4y = 12$$

have a point of intersection at $(\frac{20}{7}, \frac{6}{7})$. Also, the lines for the equations

$$3x + 4y = 12$$

and

$$y = 0$$

have a point of intersection at $(4,0)$. Finally the lines for the equations

$$x = 0$$

and

$$y = 0$$

intersect at $(0,0)$. The extreme points are shown in Fig. 5.7 and are $(0,2)$, $(\frac{20}{7}, \frac{6}{7})$, $(4,0)$, and $(0,0)$.

If the system of linear inequalities involves three variables, it is possible to graph the solution set; however, graphing is generally very difficult for most three-variable problems. For systems of linear inequalities involving more than three variables, graphing is out of the question and other techniques of solution must be utilized. Nevertheless, for systems containing three or more variables, it is still possible to convert the inequalities containing more than one variable to equations and to determine the extreme points to the solution set.

EXERCISES

1. For each of the following systems of inequalities, draw a graph of the solution set, and determine the extreme points of the solution set.

(a) $x + y \leq 7$
    $2x + y \leq 10$
(b) $x + 2y \leq 4$
    $x - y \leq 0$
(c) $x + 2y \leq 4$
    $2x - y \leq 6$
    $y \geq 0$
(d) $x + y \geq 4$
    $x - y \leq 0$
    $y \leq 4$

2. A wholesaler has 1,000 sq ft of storage space available in his warehouse and $2,000 available to purchase units of types A and B. A box of type A units costs $70 and requires 25 sq ft of storage space, and a box of type B units costs $50 and requires 30 sq ft of storage space. Let x and y represent the number of boxes of types A and B, respectively. Determine the system of linear inequalities that constrain the purchasing and storage ability of the wholesaler. Draw a graph of the solution set, and determine the extreme points.

3. The Specialty Manufacturing Company produces two models of barbecue grills: Super Deluxe and Deluxe. Each of the models is processed through three assembly departments. In department 1, which has 50 minutes (min) per day available, the Super Deluxe model requires 2 min of processing time per unit and the Deluxe model requires 5 min per unit; in department 2, which has 260 min per day available, the Super Deluxe model requires 20 min of processing time per unit and the Deluxe model also requires 20 min per unit; in department 3, which has 100 min of processing time per day available, the Super Deluxe model requires 10 min of processing time per unit and the Deluxe model requires 4 min per unit. Determine the system of linear inequalities that constrain the amount of processing time available from the departments. Draw a graph of the solution set, and determine the extreme points of the solution set.

4. The Big Spud Company processes potatoes into packages of french fries, hash browns, and flakes (for mashed potatoes). The company purchases potatoes from three sources. Due to the difference in size, length, and quality, the potatoes from the three sources differ in yield for the three products processed by Big Spud. It has been established that potatoes from source 1 yield 20 percent french fries, 20 percent hash browns, and 40 percent flakes; potatoes from source 2 yield 30 percent french fries, 10 percent hash browns, and 30 percent flakes; and potatoes from source 3 yield 20 percent french fries, 30 percent hash browns, and 30 percent flakes. The marketing department of Big Spud has determined that they can sell no more than the following amounts of each product: 1,800 lb of french fries, 1,600 lb of hash browns, and 3,000 lb of flakes. Let x, y, and z represent the amount of potatoes purchased from sources 1, 2, and 3, respectively. Determine the system of linear inequalities that constrain the amounts of the products processed by Big Spud.

5. A newspaper charges $3 for a block of black and white advertisement and $5 for a block of color advertisement. A small businessman wants to spend a maximum of $98 per week for newspaper advertising. For each block of color advertising, he wants at least three blocks of black and white. Determine the system of linear inequalities that constrain the amount of each type of advertising block to use. Draw a graph of the solution set, and determine the extreme points of the solution set.

6. A shirt manufacturer sells its long-sleeve shirts for $3.95 each and its short-sleeve shirts for $3.25 each. The long-sleeve shirts require $3\frac{1}{2}$ yards (yd) of fabric, and the short-sleeve shirts require 3 yd each. The manufacturer currently has 100 yd of fabric and wants revenue of at least $100 from the shirts made out of this fabric. Determine the system of inequalities that constrain the number of shirts that can be made and sold. Draw a graph of the solution set, and determine the extreme points of the solution set.

7. A fast-food restaurant serves regular hamburgers and double-meat hamburgers. The double burgers have 2 patties on them, and the regular burgers have 1 patty. The same buns are used for both types. Determine the system of inequalities that constrain the number of each type of hamburger that can be made if the restaurant has 50 hamburger patties and 40 buns. Draw a graph of the solution set, and determine the extreme points of the solution set.

5.3 LINEAR PROGRAMMING: GRAPHIC SOLUTION

In many business situations managers are confronted with the problem of establishing a plan or schedule for the allocation of available resources in order to optimize some objective value. The objective may be to maximize profit, maximize income, minimize cost, minimize loss, or any comparable goal. In real-world situations it is generally necessary that the manager who desires to optimize an objective value must operate within certain limitations, or constraints. Thus the objective value will be a function of the same variables that are restricted by resource limitations. This function is called the *objective function*, and the restrictions placed on the variables are called *constraints*. The variables are usually termed *decision variables* because the values of the variables constitute the decision, or schedule, for the allocation of resources. If the constraints and the objective function are *linear*, the decision problem is a problem in *linear programming*.

The systems of linear inequalities discussed in Sec. 5.2 constitute linear constraints. For example, suppose the ABC Manufacturing Company wishes to determine the maximum profit that can be obtained subject to certain constraints caused by manufacturing and resource limitations. ABC makes two types of toys: airplanes and trucks. Each type is made of aluminum, and the company has available only 250 ounces (oz) of aluminum for the next day. To manufacture one unit an airplane requires 10 oz of aluminum and a truck requires 25 oz of aluminum. An airplane and a truck each require 20 min of assembly time, and the assembly department has only 260 min of available time for the next day. Furthermore, an airplane requires 10 min to paint and a truck requires only 4 min to paint. The painting department has only 100 min of available time for the next day. ABC has determined that their profit is $1 per airplane and $2 per truck.

Stating the above problem in equation form, ABC wishes to maximize the profit function

$$x_0 = x + 2y \quad (5.2)$$

subject to the constraints

$$\begin{array}{rl} \text{Aluminum} & 10x + 25y \leqslant 250 \\ \text{Assembly} & 20x + 20y \leqslant 260 \\ \text{Painting} & 10x + 4y \leqslant 100 \\ x & \geqslant 0 \\ y & \geqslant 0 \end{array} \quad (5.3)$$

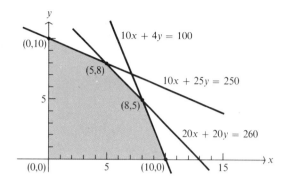

FIGURE 5.8

where x and y represent the number of airplanes and trucks, respectively, to be manufactured. The first three constraints given by the system of linear inequalities in (5.3) are for aluminum, assembly time, and painting time, respectively. The last two inequalities simply indicate that ABC cannot produce a negative number of airplanes and trucks. The solution set for the system in (5.3) is illustrated graphically by the shaded region in Fig. 5.8; any point in the solution set is a *feasible solution* of (5.3).

In order to maximize the objective function, we must determine the point (x,y) that yields the maximum value for x_0, the value of the objective function. By treating x_0 as a constant and plotting the straight line for increasing values of x_0 in Fig. 5.9, we obtain parallel lines that pass through the solution set until we choose a value of x_0 that is too large. In order to maximize the objective function value, we must select a value of x_0 that is large enough to have at least one point from the solution set common to the straight line representing the objective function such that the value of the objective function cannot be increased and still have a point common to both the objective-function line and the solution set. As can be seen in Fig. 5.9, this

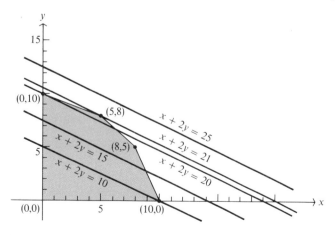

FIGURE 5.9

occurs for the objective function (5.2) when $x_0 = 21$, and the extreme point (5,8) is the only point in the solution set that is also on the line representing the objective function. Thus the solution to the linear programming problem given in (5.2) and (5.3) has been determined graphically. Hence the decision variables of the ABC Company indicate that they should manufacture five airplanes and eight trucks in order to produce a maximum profit of $21 and still satisfy each of the constraints given in (5.3).

If the goal is to *minimize* the objective function subject to the constraints, the value of x_0 is *decreased* until a point that is common to the objective-function line and the solution set is obtained and for which any further decrease in x_0 would render no points common to the two. Thus a point that satisfies the constraints and minimizes the objective function will be a point that is on the edge of the solution set and is nearest the point (or points) where the objective function loses "contact" with the solution set as the value of x_0 is *decreased*.

It can be proved (though the proof isn't given here) that if the solution set contains a point that yields the optimum (i.e., maximum or minimum) value of the objective function, there is at least one *extreme point* in the solution set that optimizes the objective-function value. This means that we need only consider extreme points to determine the solution if one exists. When solving linear programming problems graphically, it can be determined whether or not a solution exists by inspection of the solution set and the objective function. For example, there may be no points that satisfy the linear constraints; hence the solution set is empty. Or the solution set may not be bounded in the direction in which the value of the objective function is being changed; in this case, there is no *finite* optimum.

The graphic method of solving linear programming problems is quite straightforward and convenient for problems involving only two variables. However, the graphic method becomes very tedious for problems involving three variables and virtually impossible for problems involving more than three variables. Fortunately there are methods of solving linear programming problems that do not involve graphing the constraints, and we shall consider such methods later.

Example 5.2 The Chocolate Goodies Candy Company manufactures two kinds of chocolate candy: Mr. Goodnugget and Miss Snickles. A box of Mr. Goodnuggets sells for a profit of 40¢ per box, and a box of Miss Snickles sells for a profit of 50¢ per box. There are three main operations involved in processing the candy: blending, cooking, and packaging. The following table records the average time in minutes required by each box of candy for each of the three processing operations:

	Time, min		
	Blending	*Cooking*	*Packaging*
Mr. Goodnugget	2	5	3
Miss Snickles	4	4	1

Sec. 5.3 / LINEAR PROGRAMMING: GRAPHIC SOLUTION

During each production run, the blending equipment is available for at most 24 machine hours, the cooking equipment for a maximum of 30 machine hours, and the packaging equipment for at most 15 machine hours. The equipment can be used in making either type of candy at any time it is available. Determine how many boxes of each type of candy the Chocolate Goodies Candy Company should make in order to maximize their profit. Also determine the maximum profit.

Solution The objective function that must be maximized is

$$x_0 = 40x + 50y \tag{5.4}$$

where x is the number of boxes of Mr. Goodnuggets and y is the number of boxes of Miss Snickles. The variables x and y are subject to the following constraints expressed in minutes:

$$\begin{aligned} \text{Blending} \quad & 2x + 4y \leq 1{,}440 \\ \text{Cooking} \quad & 5x + 4y \leq 1{,}800 \\ \text{Packaging} \quad & 3x + y \leq 900 \end{aligned} \tag{5.5}$$

We must also have $x \geq 0$ and $y \geq 0$ because we cannot make a negative number of boxes of candy. Graphing the solution set and the objective function for different values of x_0 in Fig. 5.10, we obtain the point $(120, 300)$ of the solution set by plotting the line for $x_0 = 40x + 50y$ for different values of x_0 until the line has only one point in common with the solution set. The value of x_0 for which this occurs is 19,800. Thus the point $(120, 300)$ when substituted into $x_0 = 40x + 50y$ yields a maximum profit of 19,800¢ or \$198. Thus the Chocolate Goodies Candy Company should make 120 boxes of Mr. Goodnuggets and 300 boxes of Miss Snickles for a maximum profit of \$198.

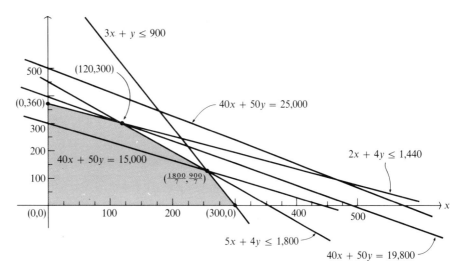

FIGURE 5.10

EXERCISES

1. The Ajax Metal Company manufactures paperclips and staples. The profit per case of paperclips is $3, and the profit per case of staples is $2. Both products require services from the shaping department and the packaging department, which for the next day have available only 30 and 20 min, respectively. A case of paperclips requires 3 min to shape and 1 min to package, and a case of staples requires 4 min to shape and 3 min to package. Determine the objective function for maximizing profit and the associated constraints. Graphically determine the optimum mixture of products and the associated maximum profit.

2. Ann's Hometown Bakery specializes in breads and rolls and also bakes two types of cakes: chocolate and white. A chocolate cake yields a profit of $3 per cake, and a white cake yields a profit of $4 per cake. A chocolate cake requires 15 oz of sugar, and a white cake requires 18 oz of sugar; the bakery has only 90 oz of sugar available today for use in cakes. There is a total of 15 min of mixer time available for the day; each chocolate cake requires 3 min of mixing, and each white cake requires 1 min of mixing. It requires 20 min to decorate a chocolate cake and 5 min to decorate a white cake; the bakery employs a decorator who is available at most for 130 min of work each day and who will have no other task for today. Determine the objective function for maximizing profit and the associated constraints. Graphically determine the optimum mixture of cakes and the associated maximum profit.

3. The Breezy Bull Ranch fattens beef cattle to sell to a large meat-packing plant. The manager of Breezy Bull Ranch believes that each steer should receive a minimum of 240 oz of nutritional ingredient A and a minimum of 150 oz of nutritional ingredient B each week. There are two grains available that contain both types of nutritional ingredients: One bushel of grain 1 contains 12 oz of ingredient A and 25 oz of ingredient B and costs $1.40 per bushel; one bushel of grain 2 contains 20 oz of ingredient A and 5 oz of ingredient B and costs $1.20 per bushel. Determine the objective function for minimizing cost and the associated constraints on nutritional ingredients. Graphically determine the optimum mixture of grains and the associated minimum cost.

4. A confectioner manufactures two kinds of candy bars: Sandies (packed with energy for growing kiddies) and Dandies (the "no-cal" nugget for weight watchers without willpower). Sandies sell at a profit of 40¢ per box, while Dandies bring in a profit of 50¢ per box. The candy is processed in three main operations: blending, cooking, and packaging. The following table records the average time (in minutes) required by each box of candy for each of the three processing operations:

	Time, min		
	Blending	Cooking	Packaging
Sandies	1	5	3
Dandies	2	4	1

During each production run, the blending equipment is available for a maximum of 12 machine hours, the cooking equipment for at most 30 machine hours, and the packaging equipment for no more than 15 machine hours. This machine time can be allocated to making either type of candy at all times that it is available. The confectioner wishes to know how many boxes of each kind he should make in order to maximize profit. Let x represent the number of boxes of Sandies and y represent the number of boxes of Dandies to be made. Specify the objective function for maximizing profit and the associated resource constraints. Graphically determine the optimal value of the objective function and the optimal number of boxes of each kind.

5. A candy bar manufacturer makes a candy bar that takes $\frac{1}{3}$ oz of chocolate, $\frac{1}{3}$ oz of caramel, and $\frac{1}{3}$ oz of peanuts and sells for 20¢ a bar. Another candy bar requires $\frac{2}{3}$ oz of chocolate and $\frac{1}{2}$ oz of caramel and sells for 15¢ a bar. For the next production run, 20 oz of chocolate, 18 oz of caramel, and 4 oz of peanuts are available. Specify the objective function for maximizing sales dollars and the associated resource constraints. Graphically determine the optimal value of the objective function and the optimal number of bars of each kind to make.

6. A drapery manufacturer makes drapes for homes and business. The drapes made for homes sell at a profit of $25 a set and require 160 yd of fabric and 10 hr to make. The drapes made for business sell at a profit of $40 a set and require 200 yd of fabric and 9 hr to make. The manufacturer has 8,000 yd of fabric and 450 hr available. Specify the objective function for maximizing profit and the associated resource constraints. Graphically determine the optimal value of the objective function and the optimal number of drapes of each kind to make.

5.4 INSPECTION OF EXTREME POINTS

In Sec. 5.3, it was noted that it has been proved that if the solution set contains a point that yields the optimum value of the objective function, there is *at least one extreme point* in the solution set that optimizes the objective function value. This limits considerably the number of points from the solution set that we need to consider. As a matter of fact, it reduces the number of candidate points from an infinite number (i.e., all points in the solution set) to a finite number. This property means that the maximum or minimum value of an objective function occurs at an extreme point. However, it does *not* necessarily mean that there is only one extreme point that yields the optimum value.

The above property clearly suggests a method for determining the optimum value of an objective function. (A more systematic method of inspecting extreme points is in the next chapter.) The method involves the following steps:

1. Determine the coordinates of the intersection points by treating the constraining inequalities as equations and solving for the points of intersection.
2. Determine whether each intersection point found in step 1 satisfies each of the constraints, and, for those that do, determine the value x_0 of the objective function for that point. Disregard those points that are infeasible (i.e., those that do not satisfy *all* constraints).
3. Select the optimum value of x_0.

In order to perform step 1 for an $m \times 2$ system of constraints, it is necessary that we determine solutions for 2×2 systems of equations. Since there are m equations in two variables, it is necessary that each equation be solved individually with each remaining equation. Using the constraints and objective function of Example 5.2, the number of boxes x and y of candy made by the Chocolate Goodies Candy Company must maximize $x_0 = 40x + 50y$ and satisfy the following constraints:

$$2x + 4y \leq 1{,}440$$
$$5x + 4y \leq 1{,}800$$
$$3x + y \leq 900$$
$$x \geq 0$$
$$y \geq 0$$

Replacing the inequalities with equalities, solving all possible pairs of equations for their intersection point, and eliminating the infeasible points, we obtain the following extreme points and their corresponding objective function values:

Extreme Point	$x_0 = 40x + 50y$
$(0, 0)$	0
$(0, 360)$	$18{,}000$
$(120, 300)$	$19{,}800 =$ maximum
$\left(\frac{1{,}800}{7}, \frac{900}{7}\right)$	$16{,}714$
$(300, 0)$	$12{,}000$

Thus the maximum profit is 19,800 (that is, $198), which is obtained by making 120 boxes of Mr. Goodnuggets and 300 boxes of Miss Snickles. Actually it was necessary that we examine 10 intersection points in order to determine the 5 extreme points.

In general, this method of examining the extreme points requires that all points of intersection for the constraining equations must be determined. The maximum possible number of points that must be determined for an $m \times n$ system can be established by the following formula:

$$_mC_n = \binom{m}{n} = \frac{m!}{n!(m-n)!} \qquad m \geq n \qquad (5.6)$$

where both $_mC_n$ and $\binom{m}{n}$ represent the number of points that must be considered and $m!$ is read m factorial and is defined by

$$m! = m(m-1)(m-2)\ldots(3)\cdot(2)\cdot(1)$$

When $m = 0$, $0!$ is defined to be equal to 1. For example,

$$5! = 5 \cdot 4 \cdot 3 \cdot 2 \cdot 1 = 120$$

and

$$8! = 8 \cdot 7 \cdot 6 \cdot 5 \cdot 4 \cdot 3 \cdot 2 \cdot 1 = 40{,}320$$

Thus, for the above example with $m = 5$ and $n = 2$, we determine

$$_5C_2 = \frac{5!}{2!(5-2)!} = \frac{5 \cdot 4 \cdot 3 \cdot 2 \cdot 1}{2 \cdot 1 \cdot 3 \cdot 2 \cdot 1} = 10$$

which means that initially we must consider 10 intersection points. For the example, we determine only 5 extreme points; however, in some situations, the number of extreme points may equal the number of intersection points.

The method of inspecting extreme points is applicable for any number of variables or constraints. When the system involves more than three variables, a major problem arises because of the awkwardness of solving an $n \times n$ system of equations for values of n greater than 3. It is difficult enough to solve one 4×4 system, not to mention solving several such systems. If $m = 5$, and $n = 4$, substitution into formula (5.6) yields

$$_5C_4 = \frac{5!}{4!1!} = 5$$

Thus we would need to determine five intersection points: one each for five different 4×4 systems.

Generally the systems of equations are somewhat simplified because of the requirement that each variable be nonnegative. Let us consider an example that contains five constraints and three variables.

Example 5.3 Use the method of inspecting extreme points to solve the linear programming problem for which we must minimize

$$x_0 = -2x_1 + 5x_2 - x_3 \tag{5.7}$$

subject to the constraints

$$
\begin{align}
\text{(a)} \quad & 4x_1 - 2x_3 \geq 2 \\
\text{(b)} \quad & 2x_1 + x_2 + 4x_3 \geq 12 \\
\text{(c)} \quad & x_1 \geq 0 \\
\text{(d)} \quad & x_2 \geq 0 \\
\text{(e)} \quad & x_3 \geq 0
\end{align}
\tag{5.8}
$$

Solution From formula (5.6), we determine for $m = 5$ and $n = 3$ that $_5C_3 = 5!/3!2! = 10$; thus we must determine 10 intersection points in order to find all extreme points. We shall change each constraint in (5.8) to an equation and begin with the first three equations. Solving

$$
\begin{align}
4x_1 - 2x_3 &= 2 \\
2x_1 + x_2 + 4x_3 &= 12 \\
x_1 &= 0
\end{align}
$$

Table 5.1

Equations from (5.8) Included in 3×3 System	Intersection Point	Feasibility Check	$x_0 = -2x_1 + 5x_2 - x_3$
(a), (b), (c)	$(0, 16, -1)$	No	
(a), (b), (d)	$(\frac{8}{5}, 0, \frac{11}{5})$	Yes	$-\frac{27}{5}$
(a), (b), (e)	$(\frac{1}{2}, 11, 0)$	Yes	54
(a), (c), (d)	$(0, 0, -1)$	No	
(a), (c), (e)	No solution	No	
(a), (d), (e)	$(\frac{1}{2}, 0, 0)$	No	
(b), (c), (d)	$(0, 0, 3)$	No	
(b), (c), (e)	$(0, 12, 0)$	No	
(b), (d), (e)	$(6, 0, 0)$	Yes	-12
(c), (d), (e)	$(0, 0, 0)$	No	

we obtain the intersection point $(0, 16, -1)$. Since the third coordinate (that is, x_3) is negative, this intersection point obviously does not satisfy the fifth constraint of (5.8); hence it is *not* an extreme point.

Next, we solve the 3×3 system composed of the first, second, and fourth equations of (5.8); thus solving

$$4x_1 \quad\quad\quad -2x_3 = 2$$
$$2x_1 + x_2 + 4x_3 = 12$$
$$x_2 \quad\quad = 0$$

we obtain the intersection point $(\frac{8}{5}, 0, \frac{11}{5})$. Checking this point for each constraint in (5.8), we determine that it is an extreme point and is a candidate for minimizing x_0. The results of each of the 3×3 systems and the check for feasibility are illustrated in Table 5.1. From Table 5.1, the minimum value of x_0 is -12 and the associated extreme point is $(6, 0, 0)$. Thus the solution to the linear programming problem that minimizes $x_0 = -2x_1 + 5x_2 - x_3$ is $x_1 = 6$, $x_2 = 0$, and $x_3 = 0$.

The method of inspecting extreme points has the advantage of being applicable to linear programming problems that contain any number of variables, whereas the graphic method is practically limited to problems that contain at most three variables. However, the method of inspecting extreme points does have the disadvantage of being rather cumbersome for large systems of inequalities. A method that is not as cumbersome is presented in Chap. 6.

EXERCISES

Use of the method of inspecting extreme points to determine the solution of each of the following exercises:

1. Sec. 5.3, Exercise 1

2. Sec. 5.3, Exercise 2

3. Sec. 5.3, Exercise 3

4. Sec. 5.3, Exercise 4

5. Sec. 5.3, Exercise 5

6. Sec. 5.3, Exercise 6

7. Maximize $x_0 = -x_1 + 5x_2 + 5x_3$ subject to the constraints

$$5x_1 - 3x_2 + 4x_3 \leq 10$$
$$3x_1 + 6x_2 - 2x_3 \leq 7$$
$$x_j \geq 0 \quad j = 1, 2, 3$$

8. Minimize $x_0 = 5x_1 - x_2 + 6x_3$ subject to the constraints

$$-2x_2 + x_3 \geq 2$$
$$2x_1 \quad\quad + 8x_3 \geq 31$$
$$x_j \geq 0 \quad j = 1, 2, 3$$

9. A toy manufacturer makes two types of toy trucks. The first type sells for a profit of $5 each and requires 2 hr in the molding department and 30 min in the painting department. The second type sells for a profit of $3 each and requires 1 hr in each department. Each day, the molding department has 40 hr available for making trucks, and the painting department has 25 hr available. Use the method of inspecting extreme points to determine the number of each type of truck to be made per day.

10. A painter can make a profit of $100 on an oil painting and $60 on an ink painting. It requires 50 hr to make an oil painting and 30 hr to make an ink painting. She paints approximately 600 hr per month and has outstanding orders for 10 ink paintings. The maximum number of oil paintings sold in a month is 5. Use the method of inspecting extreme points to determine the number of each type of painting to be done each month.

IMPORTANT TERMS AND CONCEPTS

Constraints
Decision variables
Extreme point
Feasible solution
Greater than

Greater than or equal to
Less than
Less than or equal to
Linear programming
Objective function

REVIEW PROBLEMS

1. A company sells its product for $2.50 per unit and wants next month's sales to be at least $8,500. Determine the inequality that must be satisfied by the number of units X the company must sell. Determine the least number of units that must be sold to meet next month's sales target.

2. Product X requires 3 oz of plastic, and product Y requires 5 oz of plastic. For next week's production, 525 oz of plastic are available. Determine the inequality that must be satisfied by the number of units of products X and Y manufactured. Draw a graph of the region that represents the number of units of products X and Y that can be manufactured.

3. A cookie manufacturer can make a case of peanut butter cookies with 1 lb of sugar and 2 lb of flour, a case of sugar cookies with 2 lb of sugar and 2 lb of flour, and a case of gingersnaps with 2 lb of sugar and 3 lb of flour. The manufacturer has 1,000 lb of sugar and 1,500 lb of flour available. Let X, Y and Z represent the number of cases produced of peanut butter cookies, sugar cookies, and gingersnaps, respectively. Determine the system of linear inequalities that constrain the amounts of the products produced.

4. The T Company makes three sizes, small, medium, and large, of t-shirts to sell to college bookstores. A small t-shirt requires $1\frac{1}{2}$ yd of fabric, 8 min in the sewing department, and 12 min in the lettering department. A medium t-shirt requires 2 yd of fabric, 8 min in the sewing department, and 15 min in the lettering department. A large t-shirt requires $2\frac{1}{2}$ yd of fabric, 10 min in the sewing department, and 15 min in the lettering department. Each day T Company has 1,500 yd of fabric, 6,000 min in the sewing department, and 11,250 min in the lettering department available for making college t-shirts. Let X, Y, and Z represent the number of small, medium, and large t-shirts, respectively. Determine the system of linear inequalities that constrain the number of t-shirts that can be made each day.

5. The math department has decided that a classroom should have at least 8 sq ft per desk and their rooms are 400 sq ft. Determine the inequality that must be satisfied to meet this requirement. Determine the maximum number of desks that should be in each classroom.

6. One unit of product A requires 4 oz of iron and 45 min in the molding department to manufacture. One unit of product B requires 6 oz of iron and 30 min in the molding department to manufacture. Each day there are 960 oz of iron and 9,450 min in the molding department available for the production of products A and B. Determine the system of linear inequalities that constrain the number of units of products A and B produced daily. Draw a graph of the solution set, and determine the extreme points of the solution set.

7. The Bright Star Company manufactures 75-watt and 100-watt light bulbs. A case of the 75-watt light bulbs yields a profit of \$12. The 100-watt light bulbs yield a profit of \$15 per case. In the production department, the 75-watt light bulbs require 2 hr per case, and the 100-watt light bulbs require 4 hr per case. Both types require 1 hr per case for packing. Each day there are 5,000 hr available in the production department and 2,000 hr in the packing department. Determine the objective function for maximizing profit and the associated constraints. Graphically determine the optimum mixture of light bulbs and the associated maximum profit.

8. A belt manufacturer produces leather and plastic belts. A leather belt requires 10 min in the cutting department, 5 min in the sewing department, and 2 min in the packing department. A plastic belt requires 6 min in the cutting department, 4 min in the sewing department, and 3 min in the packing department. The time available each day is 7,200 min in the cutting department, 5,760 min in the sewing department, and

2,880 min in the packing department. A leather belt yields a profit of $7, and a plastic belt yields a profit of $5. Determine the objective function for maximizing profit and the associated constraints. Graphically determine the optimum mixture of belts and the associated maximum profit.

9. Use the method of inspecting extreme points to determine the solution to Problem 7.

10. Use the method of inspecting extreme points to determine the solution to Problem 8.

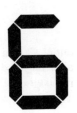

LINEAR PROGRAMMING: SOLUTIONS AND APPLICATIONS

6.1 THE SIMPLEX ALGORITHM

In Chap. 5, we considered the graphical solution to linear programming problems involving two variables. For those problems involving three or more variables, we considered the method of extreme point inspection, which was somewhat involved and awkward. In this chapter, we combine the gaussian elimination method from Chap. 4 with the extreme point inspection method from Chap. 5 and present a systematic procedure for examining extreme points. This procedure is known as the *simplex algorithm* and generally reduces greatly the number of extreme points we must consider. The simplex algorithm appears to be more difficult than the other methods previously considered. The primary reason for this is that it involves the use of the gaussian elimination method. Nevertheless, if you have a reasonable understanding of this method, you should have little difficulty with the material in this chapter.

Converting Inequalities to Equalities

In many instances, it is advantageous to convert an inequality into an equation. This is sometimes an advantage in that it permits us to convert a system of inequalities into a system of equations for which we can determine the solution in the manner presented in Chap. 4. For example, the inequality

$$2x + 3y \leq 10$$

can be converted to an equation by simply adding a new variable to the left term. Representing the new variable by z, we obtain

$$2x + 3y + z = 10$$

However, we note that z must be nonnegative because the original inequality would not be satisfied if it were otherwise. A nonnegative variable that is added to an inequality to convert it to an equation is called a *slack variable* because it takes up the existing slack due to the inequality relationship.

For an inequality that is "greater than or equal to," we need to subtract a nonnegative variable. Thus for

$$2x + 3y \geq 10$$

we subtract the nonnegative variable w in order to obtain the equation

$$2x + 3y - w = 10$$

A nonnegative variable that is subtracted from an inequality in order to convert it to an equation is called a *surplus variable* because it absorbs the surplus on the variable side of the inequality.

The introduction of slack or surplus variables does not completely eliminate inequalities from the system because each such variable is nonnegative and consequently is greater than or equal to zero. The advantage of introducing the slack and surplus variables lies in the conversion of each inequality involving more than one variable into an equation. After each inequality of this type is converted to an equation, the system can be treated like a system of linear equations, with the exception of the inequalities involving only one variable. These inequalities are handled separately from the system of equations; however, they are part of the constraints on the variables and are treated as such in determining any solution to the system of inequalities. For example, using this procedure to convert the system of inequalities given by

$$\begin{aligned} 3x + y &\leq 4 \\ 5x - 3y &\geq 7 \\ x &\geq 0 \\ y &\geq 0 \end{aligned}$$

we obtain

$$\begin{aligned} 3x + y + z &= 4 \\ 5x - 3y - w &= 7 \\ x &\geq 0 \\ y &\geq 0 \\ z &\geq 0 \\ w &\geq 0 \end{aligned}$$

where z and w are the slack and surplus variables, respectively. However, the first two inequalities have been converted to equations that do not have a unique solution but rather a multitude of solutions because they represent two equations in four unknowns.

Suppose we are confronted with determining the solution to the linear programming problem for which we must maximize

$$x_0 = 7x_1 + 3x_2 + 2x_3 + 9x_4 \qquad (6.1)$$

subject to

$$4x_1 + 2x_2 + x_3 + 8x_4 \leq 24$$
$$x_1 + x_2 + x_3 + x_4 \leq 9$$
$$x_j \geq 0 \quad j = 1, 2, 3, 4 \qquad (6.2)$$

In order to express this linear programming problem in a form with which we are familiar, we shall express Eq. (6.1) such that all variables are on the left side of the equation, and we shall introduce the slack variables x_5 and x_6 to express the two inequalities of (6.2) as equations. Performing these manipulations and combining the results, we have the following system of equations:

$$x_0 - 7x_1 - 3x_2 - 2x_3 - 9x_4 \qquad\qquad = 0$$
$$4x_1 + 2x_2 + x_3 + 8x_4 + x_5 \qquad = 24$$
$$x_1 + x_2 + x_3 + x_4 \qquad + x_6 = 9 \qquad (6.3)$$

In addition to the equations in the above system, we also must have $x_j \geq 0$ for $j = 1, 2, \ldots, 6$.

An Initial Basic Feasible Solution

In attempting to solve the system of equations given in (6.3), we note that the system involves three equations in seven variables. Thus there is an infinite number of solutions to the system because we can solve the system for the value of three variables each expressed in terms of the remaining four variables. We can solve the system immediately for a solution containing x_0, x_5, and x_6 in terms of x_1, x_2, x_3, and x_4 by solving the first, second, and third equations, respectively, for the respective variable. Thus, for given values of x_1, x_2, x_3, and x_4, we would obtain specific values for the three variables x_0, x_5, and x_6. One such solution that can be readily seen in (6.3) without solving for the three variables x_0, x_5, and x_6 in terms of the others is where the remaining four variables are all equal to zero; for this solution, we have $x_0 = 0$, $x_5 = 24$, and $x_6 = 9$, in addition to $x_1 = x_2 = x_3 = x_4 = 0$. We call this an *initial basic solution*, and x_0, x_5, and x_6 are known as *basic variables*. The remaining variables, for which the values were set equal to zero, are *nonbasic variables*.

In general, a *basic feasible solution* to a system of m equations in n variables is one that satisfies every equation in the system (i.e., it is feasible) and for which $n - m$ variables are set equal to zero and the remaining system of m equations in m variables is solved for values of the m variables. Furthermore, each of the variables must be nonnegative. Thus the basic variables for the above solution to (6.3) are x_0, x_5, and x_6, and they constitute the *basis* (i.e., the set of $m=3$ variables that were *not set* equal to zero); the nonbasic variables are x_1, x_2, x_3, and x_4.

Upon examination of (6.1) and (6.3), it is apparent that we must keep in mind that we are to maximize the objective-function value x_0. Therefore, it must always be present in the basis because we would never set it equal to zero even though it may assume such a value. Moreover, in attempting to maximize x_0, we need to examine different sets of basic variables, i.e., different *bases*. Returning to our example, if we examine the first equation of (6.3)

$$x_0 - 7x_1 - 3x_2 - 2x_3 - 9x_4 = 0$$

we note that an increase above zero in any of the four nonbasic variables x_1, x_2, x_3, and x_4 will cause x_0 to increase also in order to maintain the equality. For example, if x_1 is permitted to increase by one unit and the other nonbasic variables remain equal to zero, the objective-function row becomes

$$x_0 - 7(1) - 3(0) - 2(0) - 9(0) = 0$$

which yields $x_0 - 7 = 0$, or $x_0 = 7$. This indicates that an increase in x_1 will increase the value of x_0; hence x_0 would be nearer its maximum. From this, we note that we must introduce at least one nonbasic variable into the basis as long as there exists a negative coefficient in the row that yields the value of the objective function. (Unless we interchange rows in our system, this will always be the first row.) Since our basis can contain only three variables (that is, m variables), we must eliminate one of the variables that is currently in the basis. The variable that should be eliminated is the one that reaches zero first as the new basic variable increases.

From the previous discussion, the interpretation of the coefficients in the objective-function row is that each coefficient represents the increase (for negative coefficients) or decrease (for positive coefficients) in x_0 for a one-unit increase in the associated nonbasic variable. Our goal in solving a maximization problem should be to manipulate the objective-function row until the coefficients of all nonbasic variables are positive. For a minimization problem, we desire to obtain negative coefficients for all nonbasic variables.

The method that we shall use to achieve the optimum values for the objective function and the decision variables is known as the *simplex method*. This method examines in a systematic way the extreme points of the solution set until it determines an optimum value for the objective function. For the present, let us assume that a feasible solution exists and the optimal value of the objective function is finite. (These assumptions can be removed later.) The general steps involved in solving a linear programming problem for which the system of equations comparable to (6.3) contains m equations are given below.

Step 1. Select a set of m variables that yields an initial basic feasible solution. Eliminate the selected basic variables from the objective-function row except for the objective-function variable x_0.

Step 2. Check the objective-function row to determine whether there is a nonbasic variable in the current feasible solution that, if made positive, would improve the value of the objective function. If such a value exists, go to step 3; otherwise, stop.

Step 3. Determine how large the nonbasic variable found in step 2 can become until one of the m basic variables in the basic feasible solution becomes zero. Eliminate from the basis the basic variable that first becomes zero, and let the new basis contain the nonbasic variable found in step 2.

Step 4. Solve for these m variables, and set the remaining variables equal to zero in the next trial solution. Return to step 2.

Entering a Nonbasic Variable

In step 2 there may be more than one nonbasic variable that, if made positive, would improve the value of the objective function. If such is the case, we need a rule for deciding which nonbasic variable should be entered into the next trial basis. The rule will change slightly depending on whether we are maximizing or minimizing the objective function. First we shall introduce the rule for maximization and apply it to the problem in (6.3); later we shall consider the minimization problem.

Simplex Criterion I (Maximization) If there are nonbasic variables with negative coefficients in the objective-function row, when the objective function is expressed only in terms of nonbasic variables and x_0, select the nonbasic variable with the most negative coefficient, i.e., the variable with the best per-unit potential gain. If the coefficients of all nonbasic variables are positive or zero, an optimal solution has been obtained.

Applying the above criterion to system (6.3), we see that the nonbasic variable x_4 is the one that has the best per-unit potential gain for increasing the objective-function value; hence we should introduce x_4 into the new basis. However, we must determine which variable in the current basis should be eliminated. Since x_4 has the best per-unit potential gain, it is desirable to introduce as many units of x_4 as possible.

Removing a Basic Variable

The only limitation on the amount that x_4 can be increased is the requirement that no x_j value can be negative. Consequently, we must decrease the value of each current basic variable for which x_4 has a positive coefficient in the corresponding row. Thus we are assuming that the right-side values are also positive because the coefficients of basic variables are all equal to unity and

Sec. 6.1 / THE SIMPLEX ALGORITHM

negative values are not permitted for the variables. Hence, for each one-unit increase in x_4 in system (6.3), we decrease x_5 by eight units and decrease x_6 by one unit, so that rows 2 and 3, respectively, will be satisfied.

In order to determine the maximum amount that x_4 can assume and still have x_5 and x_6 nonnegative, we examine rows 2 and 3 of (6.3). In row 2

$$4x_1 + 2x_2 + x_3 + 8x_4 + x_5 = 24$$

we are introducing x_4 at a positive value while $x_1 = x_2 = x_3 = 0$ and we wish to force x_5 to 0. This row becomes

$$8x_4 + x_5 = 24$$

and when we set $x_5 = 0$ and solve for x_4, we obtain $x_4 = \frac{24}{8} = 3$. Thus, when x_4 has increased to a value of 3, the value of x_5 has decreased to 0. Similarly, for the third row of (6.3)

$$x_1 + x_2 + x_3 + x_4 + x_6 = 9$$

we have $x_1 = x_2 = x_3 = 0$ and wish to introduce x_4 at a positive level until x_6 becomes 0. This row becomes

$$x_4 + x_6 = 9$$

and when we set $x_6 = 0$ and solve for x_4, we have $x_4 = 9$. From these rows, we have determined the maximum value that x_4 can assume for *each* row. Hence, from row 2, x_4 can be no larger than 3 and x_5 can still be nonnegative. Similarly, from row 3, x_4 can be not larger than 9 and x_6 still remain nonnegative. Therefore, in order to have x_4 as large as possible and still keep all variables nonnegative, we cannot let x_4 be larger than the smallest of the maximum values for each row. Thus we let x_4 increase from 0 to 3, which causes x_5 to decrease from 24 to 0. This means that we eliminate x_5 from the current basis and introduce x_4 into the basis to obtain a new current basis. This completes step 3, and it remains to perform step 4 and solve for the values of x_0 and x_6 for the situation where $x_1 = x_2 = x_3 = x_5 = 0$ and $x_4 = 3$. The computations associated with step 3 are summarized in the following rule, and the calculations are given in Table 6.1

Simplex Criterion II (a) For all rows except the objective-function row, take the ratios of the current right-side constants to the coefficients of the entering variable x_j (ignore ratios with zero or negative denominators). (b) Select the minimum ratio, which will be the value of x_j in the new trial solution. The minimum ratio occurs for a variable x_k in the current solution. Eliminate x_k from the basis and set it equal to zero in the new trial solution.

Table 6.1 Computations of Criterion II for Entering x_4 into New Basis

Basic Variables	Current Solution	Coefficients of x_4	Ratios	Min	Next Solution
x_0	0	-9	Not Included		
x_5	24	8	3	3	$x_4 = 3, x_5 = 0$
x_6	9	1	9		

The Simplex Algorithm

Having completed step 3, where we determined that x_4 should enter the basis and x_5 should become a nonbasic variable, we need to solve the system for the basic variables. In order to do this, we shall use the gaussian elimination method of Sec. 4.2 and also retain the nonbasic variables on the left side of the equation so that we can readily inspect the objective-function row, as required by step 2. We shall repeat the four steps until we are able to stop with an optimal solution. The four steps and the two criteria constitute what is known as the *simplex algorithm* for solving linear programming problems of maximizing the objective function.

To further clarify the simplex algorithm for maximization problems, let us apply the algorithm to the linear programming problem given in (6.1) and (6.2) to obtain an optimal solution. We express the problem as the system of equations given in (6.3); that is,

$$\begin{aligned} x_0 - 7x_1 - 3x_2 - 2x_3 - 9x_4 &= 0 \\ 4x_1 + 2x_2 + x_3 + 8x_4 + x_5 &= 24 \\ x_1 + x_2 + x_3 + x_4 + x_6 &= 9 \end{aligned} \quad (6.4)$$

Performing step 1, we select x_0, x_5, and x_6 as the initial basic feasible solution. The objective-function row contains no basic variables except x_0; hence no elimination is required. Performing step 2, we find that any of the four nonbasic variables, if made positive, could improve the value of the objective function. Applying criterion **I** (maximization), we select the variable with the most negative coefficient, which is x_4 with a coefficient of -9. For step 3, we apply criterion **II**a and determine that the ratios for the second and third rows, respectively, of (6.4) are $\frac{24}{8}$ and $\frac{9}{1}$. Applying criterion **II**b, the minimum ratio is 3, which will be the value of x_4 in the new trial solution, and it occurs for the variable x_5 in the current solution. Thus we shall use the gaussian elimination method to eliminate x_5 from the basis and introduce x_4 into the new trial solution.

Since we have determined that x_4 should replace x_5 in our new trial solution, we need to apply the gaussian elimination method to change system (6.4) to an equivalent system in which the column of coefficients for x_4 is changed to appear as the column of coefficients of x_5 in (6.4). To do this, we divide the second row of (6.4) by the coefficient of x_4 (that is, by 8) to obtain

Sec. 6.1 / THE SIMPLEX ALGORITHM

the second row of (6.5). Next we multiply this row by 9 and add the result to the first row of (6.4) to obtain the first row of (6.5). Finally we subtract the second row of (6.5) from the third row of (6.4) to obtain the third row of (6.5):

$$\begin{aligned} x_0 - \tfrac{5}{2}x_1 - \tfrac{3}{4}x_2 - \tfrac{7}{8}x_3 + \tfrac{9}{8}x_5 &= 27 \\ \tfrac{1}{2}x_1 + \tfrac{1}{4}x_2 + \tfrac{1}{8}x_3 + x_4 + \tfrac{1}{8}x_5 &= 3 \\ \tfrac{1}{2}x_1 + \tfrac{3}{4}x_2 + \tfrac{7}{8}x_3 - \tfrac{1}{8}x_5 + x_6 &= 6 \end{aligned} \quad (6.5)$$

Setting the nonbasic variables x_1, x_2, x_3, and x_5 equal to zero, we have for the values of the basic variables $x_0 = 27$, $x_4 = 3$, and $x_6 = 6$. Thus we see that the value of the objective function x_0 has increased from 0 for our initial basic feasible solution to 27 for our current trial solution.

Returning to step 2, we see that if x_1, x_2, or x_3 were made positive, the objective-function value would improve. Applying criterion **I** (maximization), we choose to introduce x_1 into the basis, since $-\tfrac{5}{2}$ is the most negative coefficient. Performing step 3 and applying criterion **II**a, we determine the ratios of the right-side constants of (6.5) to the coefficients of x_1 for the second and third rows, obtaining 6 and 12, respectively. Applying criterion **II**b, we select the minimum ratio 6 and introduce x_1 into the new trial solution in place of x_4.

Going to step 4 and applying the gaussian elimination method, we shall solve for x_0, x_1, and x_6. This means that we must change the column of coefficients of x_1 in the new system to be the same as the column of coefficients of x_4 in (6.5). We do this first by dividing the second row of (6.5) by the coefficient of x_1 (that is, $\tfrac{1}{2}$) to obtain the second row of (6.6). Next we multiply the second row of (6.6) by $\tfrac{5}{2}$ and add the result to the first equation of (6.5) to obtain the first row of (6.6). Finally we multiply the second row of (6.6) by $\tfrac{1}{2}$ and subtract the result from the third equation of (6.5) to obtain the third equation of (6.6):

$$\begin{aligned} x_0 + \tfrac{1}{2}x_2 - \tfrac{1}{4}x_3 + 5x_4 + \tfrac{7}{4}x_5 &= 42 \\ x_1 + \tfrac{1}{2}x_2 + \tfrac{1}{4}x_3 + 2x_4 + \tfrac{1}{4}x_5 &= 6 \\ \tfrac{1}{2}x_2 + \tfrac{3}{4}x_3 - x_4 - \tfrac{1}{4}x_5 + x_6 &= 3 \end{aligned} \quad (6.6)$$

From (6.6), we obtain the values of the variables for the current trial solution; that is, $x_0 = 42$, $x_1 = 6$, $x_6 = 3$, and $x_2 = x_3 = x_4 = x_5 = 0$. Returning to step 2, we see that if x_3 were made positive, the objective-function value would improve. Applying criterion **I** (maximization), we choose to introduce x_3 into the basis because it has the only negative coefficient in the objective-function row. Performing step 3 and applying criterion **II**a, we determine the ratios of the right-side constants to the coefficients of x_3 for the second and third rows, obtaining 24 and 4, respectively. Applying criterion **II**b, we select the minimum ratio 4 and introduce x_3 into the new trial solution in place of x_6.

Applying the gaussian elimination method to (6.6) to introduce x_3 into and eliminate x_6 from the basis, we obtain the following system:

$$\begin{aligned} x_0 \phantom{{}+x_1} + \tfrac{2}{3}x_2 \phantom{{}+x_3} + \tfrac{14}{3}x_4 + \tfrac{5}{3}x_5 + \tfrac{1}{3}x_6 &= 43 \\ x_1 + \tfrac{1}{3}x_2 \phantom{{}+x_3} + \tfrac{7}{3}x_4 + \tfrac{1}{3}x_5 - \tfrac{1}{3}x_6 &= 5 \\ \tfrac{2}{3}x_2 + x_3 - \tfrac{4}{3}x_4 - \tfrac{1}{3}x_5 + \tfrac{4}{3}x_6 &= 4 \end{aligned} \qquad (6.7)$$

From (6.7), we have for the basic variables $x_0 = 43$, $x_1 = 5$, and $x_3 = 4$ and for the nonbasic variables $x_2 = x_4 = x_5 = x_6 = 0$. Returning to step 2, we see that all coefficients of nonbasic variables in the objective-function row are positive, and consequently there are no nonbasic variables that, if made positive, could improve the objective-function value. Thus we stop, and our current trial solution is the optimal solution for the linear programming problem given in (6.1) and (6.2).

We can check our answer at least partially by substituting the variable values into (6.3) and verifying that all equations are satisfied. If all equations of (6.3) are satisfied this verifies that the final solution is indeed a feasible solution; however, it does not verify that it is the solution that optimizes the objective function. If the solution does not satisfy the equations of (6.3), we have verified that the alleged solution is *not* a feasible solution and that we have made an error in our computations. The solution obtained by the simplex method will definitely be an optimal solution *provided* we have made *no errors* in our computations. Fortunately there is a way for checking to be sure that the final solution does in fact satisfy the equations of (6.3) *and* optimizes the objective function. (This checking method will be explained later.)

Tabular Representation Once the four steps and the associated two criteria are understood, the systems in (6.4) to (6.7) can be presented in a tabular, or matrix, form where we deal only with the coefficients of the variables and the right-side constants. Such a tabular representation is given in Table 6.2, where the coefficients in tableaux **I** to **IV** are those for systems (6.4) to (6.7), respectively. In the objective-function row of tableaux **I** to **III**, the most negative coefficient is indicated by an asterisk. Similarly, the coefficient that is circled in the column marked by an asterisk corresponds to the row in which the minimum ratio of the right-side constant to the coefficient of the new basic variable appears. The circled coefficient also is the one that is to be converted to unity in the gaussian elimination procedure. The steps for proceeding from tableaux **I** to **IV** in Table 6.2 are identical to those for proceeding from system (6.4) to (6.7), respectively; the only difference is in the way the systems are represented. In Table 6.2 the variables are listed at the top of their respective columns, whereas in systems (6.4) to (6.7) the variables are placed by their

Sec. 6.1 / THE SIMPLEX ALGORITHM

respective coefficients in each row. The tabular form of presentation is easier to write and more compact. However, when dealing with the tabular form, we should always keep in mind that each row represents an equation. We shall find this to be of considerable benefit.

Table 6.2

Tableau	Current Basis	x_0	x_1	x_2	x_3	x_4	x_5	x_6	Right-side Constants	Ratio
I	x_0	1	-7	-3	-2	-9^*	0	0	0	
	x_5	0	4	2	1	⑧	1	0	24	$\frac{24}{8}=3=\text{minimum}$
	x_6	0	1	1	1	1	0	1	9	$\frac{9}{1}=9$
II	x_0	1	$-\frac{5}{2}^*$	$-\frac{3}{4}$	$-\frac{7}{8}$	0	$\frac{9}{8}$	0	27	
	x_4	0	$(\frac{1}{2})$	$\frac{1}{4}$	$\frac{1}{8}$	1	$\frac{1}{8}$	0	3	$\frac{3}{1/2}=6=\text{minimum}$
	x_6	0	$\frac{1}{2}$	$\frac{3}{4}$	$\frac{7}{8}$	0	$-\frac{1}{8}$	1	6	$\frac{6}{1/2}=12$
III	x_0	1	0	$\frac{1}{2}$	$-\frac{1}{4}^*$	5	$\frac{7}{4}$	0	42	
	x_1	0	1	$\frac{1}{2}$	$\frac{1}{4}$	2	$\frac{1}{4}$	0	6	$\frac{6}{1/4}=24$
	x_6	0	0	$\frac{1}{2}$	$(\frac{3}{4})$	-1	$-\frac{1}{4}$	1	3	$\frac{3}{3/4}=4=\text{minimum}$
IV	x_0	1	0	$\frac{2}{3}$	0	$\frac{14}{3}$	$\frac{5}{3}$	$\frac{1}{3}$	43	
	x_1	0	1	$\frac{1}{3}$	0	$\frac{7}{3}$	$\frac{1}{3}$	$-\frac{1}{3}$	5	
	x_3	0	0	$\frac{2}{3}$	1	$-\frac{4}{3}$	$-\frac{1}{3}$	$\frac{4}{3}$	4	

Example 6.1 Use the tabular representation to determine the optimal solution of the following linear programming problem:

Maximize
$$x_0 = -x_1 + 7x_2 - 5x_3 + 14x_4$$

subject to
$$3x_1 + 4x_2 + 5x_3 + 6x_4 \leq 24$$
$$-x_1 + x_2 - 2x_3 + 2x_4 \leq 4$$
$$x_j \geq 0 \quad j = 1, 2, 3, 4$$

Solution First we employ slack variables to represent the problem as three equations:

$$\begin{aligned} x_0 + x_1 - 7x_2 + 5x_3 - 14x_4 &= 0 \\ 3x_1 + 4x_2 + 5x_3 + 6x_4 + x_5 &= 24 \\ -x_1 + x_2 - 2x_3 + 2x_4 + x_6 &= 4 \end{aligned} \quad (6.8)$$

Representing this system in tabular form, we obtain the following tableau:

Tableau	Current Basis	x_0	x_1	x_2	x_3	x_4	x_5	x_6	Right-side Constants	Ratio
I	x_0	1	1	-7	5	-14*	0	0	0	
	x_5	0	3	4	5	6	1	0	24	$\frac{24}{6}=4$
	x_6	0	-1	1	-2	②	0	1	4	$\frac{4}{2}=2=$ minimum

Since -14 is the most negative coefficient in the objective-function row, we shall introduce x_4 into the basis. We shall remove x_6 from the current basis because the minimum ratio occurs for the equation in which the basic variable x_6 appears. Thus we use the gaussian elimination method to change the column for x_4 to be like the column for x_6 in tableau I. Hence we obtain tableau II:

Tableau	Current Basis	x_0	x_1	x_2	x_3	x_4	x_5	x_6	Right-side Constants	Ratio
II	x_0	1	-6	0	-9*	0	0	7	28	
	x_5	0	6	1	⑪	0	1	-3	12	$\frac{12}{11}=$ minimum
	x_4	0	$-\frac{1}{2}$	$\frac{1}{2}$	-1	1	0	$\frac{1}{2}$	2	

In tableau II, we see that x_3 should be introduced into the basis in place of x_5. Thus we divide the second row by 11 and proceed according to the gaussian elimination method to convert the x_3 column to be like the x_5 column in tableau II. The results are given in tableau III:

Tableau	Current Basis	x_0	x_1	x_2	x_3	x_4	x_5	x_6	Right-side Constants	Ratio
III	x_0	1	$-\frac{12}{11}$*	$\frac{9}{11}$	0	0	$\frac{9}{11}$	$\frac{50}{11}$	$\frac{416}{11}$	
	x_3	0	$\left(\frac{6}{11}\right)$	$\frac{1}{11}$	1	0	$\frac{1}{11}$	$-\frac{3}{11}$	$\frac{12}{11}$	$\dfrac{12/11}{6/11}=2=$ minimum
	x_4	0	$\frac{1}{22}$	$\frac{13}{22}$	0	1	$\frac{1}{11}$	$\frac{5}{22}$	$\frac{34}{11}$	$\dfrac{34/11}{1/22}=68$

In tableau III, we determine that x_1 should be introduced into the basis in place of x_3. Again we apply the gaussian elimination method to change the x_1 column to be like the x_3 column in tableau III. The results are presented in tableau IV:

Tableau	Current Basis	x_0	x_1	x_2	x_3	x_4	x_5	x_6	Right-side Constants	Ratio
IV	x_0	1	0	1	2	0	1	4	40	
	x_1	0	1	$\frac{1}{6}$	$\frac{11}{6}$	0	$\frac{1}{6}$	$-\frac{1}{2}$	2	
	x_4	0	0	$\frac{7}{12}$	$-\frac{1}{12}$	1	$\frac{1}{12}$	$\frac{1}{4}$	3	

Sec. 6.1 / THE SIMPLEX ALGORITHM

From tableau **IV**, all coefficients in the objective-function row are nonnegative, which means that the current solution is the optimal solution to the linear programming problem in (6.8). Thus the optimal solution that maximizes the objective function is $x_1=2$, $x_2=0$, $x_3=0$, $x_4=3$, $x_5=0$, and $x_6=0$, which yields an objective-function value of $x_0=40$. These values can be verified by substitution into (6.8) to establish that they satisfy all constraints.

Minimization Problems

We have considered only maximization problems for the simplex method. Suppose the problem in Example 6.1 were to *minimize* the objective function subject to the same constraints. The objective-function row of (6.8) would be unchanged and would be given by

$$x_0 + x_1 - 7x_2 + 5x_3 - 14x_4 = 0$$

However, we must keep in mind that we wish to *minimize* x_0. Thus we wish to determine the nonbasic variables in the objective-function row that, if made positive, would cause x_0 to decrease and more closely approach a minimum value. If we were to increase the values of those nonbasic variables that have *positive* coefficients in the objective-function row, the value of x_0 would necessarily decrease in order to maintain equality. For example, if x_3 were increased from 0 to 1 and all other nonbasic variables remained 0, the objective-function value would become $x_0 + 5 = 0$, or $x_0 = -5$; hence x_0 has decreased by five units. However, if all coefficients of the nonbasic variables in the objective-function row were negative, x_0 could not decrease by letting a nonbasic variable be positive; actually it would increase. From this reasoning, we have the rule for *minimizing* the objective function in a linear programming problem.

Simplex Criterion I (Minimization) If there are nonbasic variables with positive coefficients in the objective-function row, when the objective function is expressed only in terms of nonbasic variables and x_0, select the nonbasic variable with the most positive coefficient, i.e., the variable with the best per-unit potential gain. If the coefficients of all nonbasic variables are negative or zero, an optimal solution has been obtained.

The remaining steps and criterion **II** for solving maximization problems are unchanged for minimization problems. The only difference between maximization and minimization problems is that in maximization problems we strive to force all coefficients in the objective-function row to be either zero or positive while in minimization problems we try to force all such coefficients to be either zero or negative. Table 6.3 presents the tableaux for minimizing the objective function of Example 6.1 subject to the original constraints. From tableau **II** in Table 6.3, we note that all the coefficients of the nonbasic variables in the objective-function row are nonpositive; thus the solution that minimizes the objective function is given by $x_1 = x_2 = x_4 = x_5 = 0$, $x_3 = \frac{24}{5}$, and $x_6 = \frac{68}{5}$, and the associated value of the objective function is $x_0 = -24$.

Table 6.3

Current Tableau	Basis	x_0	x_1	x_2	x_3	x_4	x_5	x_6	Right-side Constants	Ratio
I	x_0	1	1	-7	5*	-14	0	0	0	
	x_5	0	3	4	⑤	6	1	0	24	$\frac{24}{5}$
	x_6	0	-1	1	-2	2	0	1	4	
II	x_0	1	-2	-11	0	-20	-1	0	-24	
	x_3	0	$\frac{3}{5}$	$\frac{4}{5}$	1	$\frac{6}{5}$	$\frac{1}{5}$	0	$\frac{24}{5}$	
	x_6	0	$\frac{1}{5}$	$\frac{13}{5}$	0	$\frac{22}{5}$	$\frac{2}{5}$	1	$\frac{68}{5}$	

Unbounded and Infeasible Solutions

In applying the simplex algorithm, we assumed that a feasible solution existed and that the optimal value of the objective function was finite. (Later we shall see how to determine whether a feasible solution exists.) In order to determine whether the optimal value of the objective function is finite, we proceed according to the steps and criteria given previously. However, if in some tableau we apply criterion **II** and find that for the entering variable there is no positive coefficient in any row, then there exists an *unbounded* optimal solution. This follows because the entering variable can be made arbitrarily large and the value of x_0 increases without limits while the current basic variables remain nonnegative. Thus the simplex algorithm gives an indication that the optimal value of the objective function is not bounded.

In all problems considered so far, the constraints were expressed as "less than or equal to" constraints, with the right-side constant nonnegative, and the initial basic feasible solution has been selected to contain the slack variables. Suppose we wish to solve the following linear programming problem by the simplex method:

Maximize

$$x_0 = 3x_1 - 2x_2 + x_3$$

subject to

$$x_1 + 6x_2 + 3x_3 \geqslant 6$$
$$x_1 + 2x_2 + 4x_3 \leqslant 4$$
$$x_j \geqslant 0 \quad j = 1, 2, 3$$

Sec. 6.1 / THE SIMPLEX ALGORITHM

Using slack and surplus variables and expressing the above problem in equation form, we obtain

$$
\begin{aligned}
x_0 - 3x_1 + 2x_2 - x_3 &= 0 \\
x_1 + 6x_2 + 3x_3 - x_4 &= 6 \\
x_1 + 2x_2 + 4x_3 + x_5 &= 4
\end{aligned}
\qquad (6.9)
$$

where $x_j \geq 0$ for $j = 1, 2, 3, 4, 5$ and the goal is to maximize x_0.

In selecting the initial basic feasible solution required by step 1, generally we would choose x_0, x_4, and x_5 as the basic variables; however, x_4 has a negative coefficient. This can be made positive by multiplying both sides of the second equation of (6.9) by -1, but this yields $x_4 = -6$ if we let x_1, x_2, and x_3 be the nonbasic variables with values of zero. Clearly this violates the constraint that each variable, except x_0, must be nonnegative, and hence x_4 cannot be included in the basis with x_0 and x_5 if we require the basis to constitute a feasible solution to (6.9). In order to obtain a feasible basis containing x_0 and x_5, we must let x_4 be a nonbasic variable with a value of zero. This leaves us a choice between x_1, x_2, and x_3 as long as we do not force x_5 to be negative when eliminating the chosen variable from the last equation of (6.9). For example, if we were to choose x_1 as the variable to be introduced into the basis, we would apply the gaussian elimination method to (6.9) in order to eliminate x_1 from the first and third equations, which would result in (6.9) becoming

$$
\begin{aligned}
x_0 + 20x_2 + 8x_3 - 3x_4 &= 18 \\
x_1 + 6x_2 + 3x_3 - x_4 &= 6 \\
-4x_2 + x_3 + x_4 + x_5 &= -2
\end{aligned}
$$

The introduction of x_1 into the basis forces x_5 to become negative and hence infeasible.

If we choose to introduce x_2 into the basis with x_0 and x_5, we apply the gaussian elimination method to (6.9) to have x_2 appear in the second equation only. Solving for x_2 with x_1, x_3, and x_4 equal to zero, we obtain the following system:

$$
\begin{aligned}
x_0 - \tfrac{10}{3} x_1 - 2x_3 + \tfrac{1}{3} x_4 &= -2 \\
\tfrac{1}{6} x_1 + x_2 + \tfrac{1}{2} x_3 - \tfrac{1}{6} x_4 &= 1 \\
\tfrac{2}{3} x_1 + 3x_3 + \tfrac{1}{3} x_4 + x_5 &= 2
\end{aligned}
\qquad (6.10)
$$

The basis obtained in (6.10) is $x_0 = -2$, $x_2 = 1$, and $x_5 = 2$, which constitutes a basic feasible solution because x_0 is not restricted in sign like the other variables. Once an initial basic feasible solution is obtained, we have completed step 1 of the simplex method, and we can then proceed in the normal manner for the simplex method by applying the simplex algorithm to (6.10) for maximizing the objective-function value.

The solution to the linear programming problem in (6.9) is given in Table 6.4, where the first tableau is obtained from (6.10). We note that x_1 is the first variable to be introduced into the basis. We have already determined that x_5 and x_1 cannot be in the basis at the same time and so it should come as no surprise that x_5 is removed from the basis when x_1 is introduced into the basis. From the second tableau of Table 6.4, we obtain the maximum value of the objective function $x_0 = 8$, and the corresponding values of the variables are $x_1 = 3$, $x_2 = \frac{1}{2}$, and $x_3 = x_4 = x_5 = 0$.

Table 6.4

Tableau	Current Basis	x_0	x_1	x_2	x_3	x_4	x_5	Right-side Constants	Ratio
I	x_0	1	$-\frac{10}{3}*$	0	-2	$\frac{1}{3}$	0	-2	
	x_2	0	$\frac{1}{6}$	1	$\frac{1}{2}$	$-\frac{1}{6}$	0	1	6
	x_5	0	$\left(\frac{2}{3}\right)$	0	3	$\frac{1}{3}$	1	2	3 = minimum
II	x_0	1	0	0	13	2	5	8	
	x_2	0	0	1	$-\frac{1}{4}$	$-\frac{1}{4}$	$-\frac{1}{4}$	$\frac{1}{2}$	
	x_1	0	1	0	$\frac{9}{2}$	$\frac{1}{2}$	$\frac{3}{2}$	3	

If there were more than one "greater than or equal to" constraint in the previous problem, we would simply continue the process outlined until we obtained an initial basic feasible solution or until we exhausted all possible combinations of variables in the basis with x_0. If we exhaust all possible combinations of variables with x_0 in the basis, there does not exist a feasible solution; when this is the situation, the constraints are too strict and do not permit any simultaneous solution.

EXERCISES

1. For each of the following exercises, state the problem in equation form and solve by the simplex method. Compare your answers with those obtained by the graphical method.

    (a) Sec. 5.3, Exercise 1
    (c) Sec. 5.3, Exercise 3
    (e) Sec. 5.3, Exercise 5
    (b) Sec. 5.3, Exercise 2
    (d) Sec. 5.3, Exercise 4
    (f) Sec. 5.3, Exercise 6

2. For each of the following exercises, state the problem in equation form and solve by the simplex method. Compare your answers with those obtained by the method of inspecting extreme points.

(a) Sec. 5.4, Exercise 7   (b) Sec. 5.4, Exercise 8
(c) Sec. 5.4, Exercise 9   (d) Sec. 5.4, Exercise 10

3. Use the simplex method to solve the following linear programming problem:

Maximize
$$x_0 = x_1 - 3x_2 + 2x_3$$

subject to
$$x_1 + 6x_2 + 3x_3 \leq 6$$
$$2x_1 + 4x_2 + 8x_3 \leq 8$$
$$3x_1 - 3x_2 + 3x_3 \leq 9$$
$$x_j \geq 0 \quad j = 1, 2, 3$$

4. Use the simplex method to solve the following linear programming problem:

Maximize
$$x_0 = 5x_1 + 8x_2$$

subject to
$$x_1 + x_2 \leq 2$$
$$x_1 - 2x_2 \leq 0$$
$$-x_1 + 4x_2 \leq 1$$
$$x_j \geq 0 \quad j = 1, 2$$

5. Solve Exercise 4 with the objective function changed to $x_0 = 3x_1 + 4x_2$.

6. Solve Exercise 5 with the additional constraint $2x_1 + x_2 \leq 7$ added to the original constraints.

7. Solve Exercise 5 with the right-side constants changed to 3, 1, and 2 instead of 2, 0, and 1.

8. Solve Exercise 5 with the right-side constants changed to 7, 1, and 2 instead of 2, 0, and 1.

9. The Unique Manufacturing Company assembles three products. Each of the products is processed through two departments such that products A, B, and C require 3, 6, and 12 hr, respectively, from department X and 5, 15, and 30 hr, respectively, from department Y. The company makes a profit of $10, $12, and $4, respectively, for the three products A, B, and C. The current labor force of the company is such that there are 240 hr per week available in department X and 1,800 hr per week available in department Y. The company wishes to determine which product mix will maximize weekly profit.

(a) Determine the objective function and the constraints.
(b) Solve the linear programming problem by the simplex method.
(c) Interpret the meaning of the values of the decision variables and the slack variables for the optimal solution.

10. The U-Kamp Manufacturing Company makes two types of camping trailers: a standard model (the Kamper) and a luxury model (the Vacationer). The campers are sold to independent dealers at a profit of $200 per Kamper and $300 per Vacationer. A Kamper requires 30 man-hours for assembly, 20 man-hours for painting and finishing, and 10 man-hours for inspecting. A Vacationer requires 75 man-hours for assembly, 25 man-hours for painting and finishing, and 5 man-hours for inspecting. A production run generally has 15,000 man-hours available for assembly, 6,500 man-hours available for painting and finishing, and 2,500 for inspecting.

(a) Determine the objective function for maximizing profit and the associated constraints.
(b) Use the simplex method to solve for the maximum profit and optimal values of the decision and slack variables.
(c) Interpret the meaning of the values of the decision and slack variables for the optimal solution.

11. A furniture manufacturer makes three models of chairs. Models A, B, and C sell for a profit of $50, $80, and $100, respectively. The amount of vinyl required in each is 10 yd and 15 yd for models A and B, respectively. The amount of padding required in each is 5 lb, 2 lb, and 7 lb for models A, B, and C, respectively. The time required to make each model is 4 hr, 6 hr, and 10 hr for A, B, and C, respectively. The resources available are 250 yd of vinyl, 150 lb of padding, and 200 hr.

(a) Determine the objective function for maximizing profit and the associated constraints.
(b) Solve the linear programming problem by the simplex method.
(c) Interpret the meaning of the values of the decision and slack variables for the optimal solution.

12. A toy manufacturer makes three sizes of toy cars. The smallest sells for a profit of 50¢, the middle size sells for a profit of 90¢, and the largest sells for a profit of $1.40. The smallest requires 1 sheet of plastic and 2 oz of rubber. The middle size requires 2 sheets of plastic and $3\frac{1}{2}$ oz of rubber. The largest requires $4\frac{1}{2}$ sheets of plastic and $3\frac{1}{2}$ oz of rubber. For tomorrow's production, the manufacturer has 250 sheets of plastic and 300 oz of rubber.

(a) Determine the objective function for maximizing profit and the associated constraints.
(b) Solve the linear programming problem by the simplex method.
(c) Interpret the meaning of the values of the decision and slack variables for the optimal solution.

13. A manufacturer is required to decrease its pollution emission and is currently developing a plan to submit to the government for approval. Three pollution control devices can be used with their current machinery. For each hour used, device 1 decreases pollution 8 units, device 2 decreases pollution 10 units, and device 3 decreases pollution 6 units. On the machines to produce product A, the hourly cost in hundreds of dollars is $5 for device 1, $2 for device 2, and $4 for device 3. On product B machines, the hourly cost in hundreds of dollars is $2 for device 1, $6 for device 2, and $3 for device 3. The hourly cost in hundreds of dollars on product C machines is $3 for device 1, $5 for device 2, and $7 for device 3. If the government will approve the plan, the manufacturer will limit the weekly costs, in hundreds of dollars, to $750 for each product.

(a) Determine the objective function for minimizing pollution emission and the associated constraints.
(b) Solve the linear programming problem by the simplex method.
(c) Interpret the meaning of the values of the decision and slack variables for the optimal solution.

14. A carpet manufacturer makes three qualities of shag carpets that sell at a profit of $1, $2.50, and $3 per 6 sq yd for the low, medium, and high quality, respectively. Six sq yd require 15, 30, and 50 bolts of yarn for the low, medium, and high quality, respectively. The low quality carpet requires 1 hr in department A and $\frac{1}{2}$ hr in department B to make 6 sq yd. To make 6 sq yd of the medium quality requires 1 hr in each department. The high quality carpet requires $\frac{1}{2}$ hr in department A and $1\frac{1}{2}$ hr in department B. Tomorrow the manufacturer will have 1,000 bolts of yarn, 25 hr in department A, and 30 hr in department B.

(a) Determine the objective function for maximizing profit and the associated constraints.
(b) Solve the linear programming problem by the simplex method.
(c) Interpret the meaning of the values of the decision and slack variables for the optimal solution.

6.2 THE DUAL PROBLEM

In Sec. 6.1 we used the simplex algorithm to solve specific linear programming problems. It was noted that we could verify the feasibility of a solution, but we could not at that point verify the optimality of the solution. We now consider the problem of verifying the optimality of a solution. In order to do this, we make use of the property that for every linear programming problem there is a corresponding linear programming problem that is called its *dual*. The original linear programming problem is called the *primal* problem, and it is from this problem that we can construct the dual problem.

Let us consider the following pair of linear programming problems:

Maximize

$$x_0 = \sum_{j=1}^{n} c_j x_j \qquad (6.11)$$

subject to

Primal $\qquad \sum_{j=1}^{n} a_{ij} x_j \leq b_i \qquad i = 1, 2, \ldots, m \qquad (6.12)$

$\qquad x_j \geq 0 \qquad j = 1, 2, \ldots, n \qquad (6.13)$

and minimize

$$y_0 = \sum_{i=1}^{m} b_i y_i \qquad (6.14)$$

subject to

Dual
$$\sum_{i=1}^{m} a_{ij} y_i \geq c_j \qquad j=1,2,\ldots,n \qquad (6.15)$$

$$y_i \geq 0 \qquad i=1,2,\ldots,m \qquad (6.16)$$

We shall arbitrarily refer to (6.11) through (6.13) as the *primal problem* and to (6.14) through (6.16) as the *dual problem*.

For example, consider the following pair of linear programming problems:

Maximize

$$x_0 = 7x_1 + 3x_2 + 2x_3 + 9x_4 \qquad (6.17)$$

subject to

Primal
$$\begin{aligned} 4x_1 + 2x_2 + x_3 + 8x_4 &\leq 24 \\ x_1 + x_2 + x_3 + x_4 &\leq 9 \end{aligned} \qquad (6.18)$$

$$x_1 \geq 0 \quad x_2 \geq 0 \quad x_3 \geq 0 \quad x_4 \geq 0 \qquad (6.19)$$

and minimize

$$y_0 = 24 y_1 + 9 y_2 \qquad (6.20)$$

subject to

Dual
$$\begin{aligned} 4y_1 + y_2 &\geq 7 \\ 2y_1 + y_2 &\geq 3 \\ y_1 + y_2 &\geq 2 \\ 8y_1 + y_2 &\geq 9 \end{aligned} \qquad (6.21)$$

$$y_1 \geq 0 \quad y_2 \geq 0 \qquad (6.22)$$

We can view the dual problem in (6.20) to (6.22) as the primal problem flipped on its side. We note the following relationships between primal and dual problems:

1. The sense of optimization is reversed for the two problems.

Sec. 6.2 / THE DUAL PROBLEM

2. The coefficients of the primal objective function are the right-side constants for the constraints in the dual problem.
3. The direction of the inequalities in the constraints for the primal problem is reversed in the constraints for the dual problem.
4. The jth column of coefficients in the primal problem is the same as the jth row of coefficients in the dual problem.
5. The column of constants on the right side of the primal problem is the same as the row of coefficients of the objective function of the dual problem.

The real significance of the concept of duality is given in the following theorem:

Dual Theorem (a) If both the primal and dual problems possess feasible solutions, then the primal problem has an optimal solution x_j^* for $j = 1, 2, \ldots, n$, the dual problem has an optimal solution y_i^* for $i = 1, 2, \ldots, m$, and

$$x_0^* = \sum_{j=1}^{n} c_j x_j^* = \sum_{i=1}^{m} b_i y_i^* = y_0^* \qquad (6.23)$$

(b) If either the primal or dual problem possesses a feasible solution with a finite optimal objective-function value, then the other problem possesses a feasible solution with the same optimal objective-function value.

Optimal Dual Values By now you may be wondering how to obtain the solution to the dual problem. We could obtain it by applying the simplex algorithm. However, this would require the introduction of a surplus variable into each constraint of (6.15), that, in turn, would create some difficulty in obtaining an initial basic feasible solution. However, an initial feasible solution can be obtained as long as a feasible solution exists, but a considerable amount of effort would be required. Fortunately we do not need to solve the two problems independently in order to obtain their solutions. When the simplex method is used to obtain the optimal solution to the primal problem, the solution to the dual problem is given in the final tableau by simply selecting the values of the variables from their appropriate positions. The optimal values of the dual variables are determined from the *final* tableau of the primal problem in the following locations:

1. The minimum value of the objective function given in (6.14) is the right-side constant in the objective-function row of the primal problem.

2. The coefficients of the *slack* variables in the objective-function row are the optimal values of the dual variables.
3. The coefficient of the primal variable x_j in the objective-function row is the optimal value of the surplus variable introduced into the jth constraint of (6.15) for $j = 1, 2, \ldots, m$.

To illustrate the determination of optimal values of the dual problem, let us consider the primal and dual problems given in (6.17) to (6.22). The primal problem is expressed in equation form through introduction of the slack variables x_5 and x_6. (This problem was introduced at the beginning of this chapter, and its solution by the simplex method is represented in tabular form in Table 6.2.) First, we express the dual problem in equation form by introducing the surplus variables y_3, y_4, y_5, and y_6 into (6.20) to (6.22). Thus we obtain

$$
\begin{aligned}
y_0 \quad -24y_1 - 9y_2 &= 0 \\
4y_1 + y_2 - y_3 &= 7 \\
2y_1 + y_2 \quad - y_4 &= 3 \\
y_1 + y_2 \quad - y_5 &= 2 \\
8y_1 + y_2 \quad - y_6 &= 9
\end{aligned}
\qquad (6.24)
$$

$$y_1 \geq 0 \quad y_2 \geq 0 \quad y_3 \geq 0 \quad y_4 \geq 0 \quad y_5 \geq 0 \quad y_6 \geq 0 \qquad (6.25)$$

where we are to minimize the objective-function value y_0.

Using the final tableau of the primal problem (given in Table 6.5) and the directions given previously for determining the optimal values for the dual problem, from step 1 we determine that $y_0 = 43$; from step 2 we determine that $y_1 = \frac{5}{3}$ and $y_2 = \frac{1}{3}$; and from step 3 we determine the values of the surplus variables to be $y_3 = 0$, $y_4 = \frac{2}{3}$, $y_5 = 0$, and $y_6 = \frac{14}{3}$. It is apparent that (6.25) is satisfied; however, in order to verify that these values do, in fact, satisfy the equations in (6.24), you may wish to substitute them into the appropriate equations and see for yourself.

Table 6.5 Final Tableau from Table 6.2

Basis	x_0	x_1	x_2	x_3	x_4	x_5	x_6	Right-side Constants
x_0	1	0	$\frac{2}{3}$	0	$\frac{14}{3}$	$\frac{5}{3}$	$\frac{1}{3}$	43
x_1	0	1	$\frac{1}{3}$	0	$\frac{7}{3}$	$\frac{1}{3}$	$-\frac{1}{3}$	5
x_2	0	0	$\frac{2}{3}$	1	$-\frac{4}{3}$	$-\frac{1}{3}$	$\frac{4}{3}$	4

Verifying Optimal Values   One of the obvious advantages of the duality concept is that it permits us to verify without doubt that a primal solution is an optimal feasible solution. Recall that without the dual problem we can verify only that the constraints

Sec. 6.2 / THE DUAL PROBLEM

are satisfied, not whether the feasible solution is optimal. Now, with the dual solution, we can completely verify that we have an optimal feasible solution provided the primal variables (including the slack variables) satisfy the primal constraints, the dual variables (including the surplus variables) satisfy the dual constraints, and the value of the objective function for the primal problem is equal to the value of the objective function for the dual problem. This verification is ensured as a result of the dual theorem. Thus, if a solution obtained by the simplex method cannot be verified, we have an error in our computations that must be corrected before we can determine and verify an optimal feasible solution.

Example 6.2 For the following linear programming problem, determine its dual problem and the optimal values of the primal and dual variables:

Maximize

$$x_0 = 2x_1 + 4x_2 + x_3$$

subject to

$$x_1 + x_2 - 2x_3 \leq 15$$
$$3x_1 + x_2 + x_3 \leq 21$$
$$x_1 \geq 0 \quad x_2 \geq 0 \quad x_3 \geq 0$$

Solution The dual problem is minimize

$$y_0 = 15y_1 + 21y_2$$

subject to

$$y_1 + 3y_2 \geq 2$$
$$y_1 + y_2 \geq 4$$
$$-2y_1 + y_2 \geq 1$$
$$y_1 \geq 0 \quad y_2 \geq 0$$

Using slack variables to express the primal problem in equation form, we obtain

$$\begin{aligned} x_0 - 2x_1 - 4x_2 - x_3 &= 0 \\ x_1 + x_2 - 2x_3 + x_4 &= 15 \\ 3x_1 + x_2 + x_3 + x_5 &= 21 \end{aligned}$$

where we wish to maximize x_0 and all variables must be nonnegative. Using surplus variables to express the dual problem in equation form, we obtain

$$\begin{aligned} y_0 - 15y_1 - 21y_2 &= 0 \\ y_1 + 3y_2 - y_3 &= 2 \\ y_1 + y_2 - y_4 &= 4 \\ -2y_1 + y_2 - y_5 &= 1 \end{aligned}$$

where we wish to minimize y_0 and all variables except y_0 must be nonnegative.

Table 6.6

Current Tableau	Basis	x_0	x_1	x_2	x_3	x_4	x_5	Right-side Constants	Ratio
I	x_0	1	−2	−4*	−1	0	0	0	
	x_4	0	1	①	−2	1	0	15	$\frac{15}{1}=15=$ minimum
	x_5	0	3	1	1	0	1	21	$\frac{21}{1}=21$
II	x_0	1	2	0	−9*	4	0	60	
	x_2	0	1	1	−2	1	0	15	
	x_5	0	2	0	③	−1	1	6	$\frac{6}{3}=2=$ minimum
III	x_0	1	8	0	0	1	3	78	
	x_2	0	$\frac{7}{3}$	1	0	$\frac{1}{3}$	$\frac{2}{3}$	19	
	x_3	0	$\frac{2}{3}$	0	1	$-\frac{1}{3}$	$\frac{1}{3}$	2	

Applying the simplex algorithm to the primal problem, we obtain Table 6.6, and from this table we obtain the optimal values of the primal and dual variables:

Primal Values	Dual Values
$x_0 = 78$	$y_0 = 78$
$x_1 = 0$	$y_1 = 1$
$x_2 = 19$	$y_2 = 3$
$x_3 = 2$	$y_3 = 8$
$x_4 = 0$	$y_4 = 0$
$x_5 = 0$	$y_5 = 0$

These values can be verified by substitution into the equation form of the primal and dual problems, respectively.

Applying the Simplex Algorithm to a Dual Problem When we encounter a problem for which the constraints are like those given in (6.15), generally it is easier to solve its dual problem by the simplex method and determine the values of the primal variables from the final tableau of the dual problem. This can be done because of the reflective property that exists between the primal and dual problems; that is, the dual of a dual problem is simply the primal problem. Hence we may solve the dual problem by the simplex method and determine the values of its dual variables (i.e., the original primal variables) from the final tableau. For clarification, let us consider an example.

Sec. 6.2 / THE DUAL PROBLEM

Example 6.3 Solve the following linear programming problem by determining its dual problem by the simplex method and obtaining the primal variables from the final tableau of the dual problem.

Minimize

$$x_0 = 2x_1 + 3x_2$$

subject to

$$2x_1 + x_2 \geqslant 5$$
$$x_1 + 3x_2 \geqslant 10$$
$$2x_1 + 2x_2 \geqslant 12$$
$$x_1 \geqslant 0 \quad x_2 \geqslant 0$$

Solution The dual problem is maximize

$$y_0 = 5y_1 + 10y_2 + 12y_3$$

subject to

$$2y_1 + y_2 + 2y_3 \leqslant 2$$
$$y_1 + 3y_2 + 2y_3 \leqslant 3$$
$$y_1 \geqslant 0 \quad y_2 \geqslant 0 \quad y_3 \geqslant 0$$

Using the slack variables to express the dual problem in equation form, we obtain

$$\begin{aligned} y_0 - 5y_1 - 10y_2 - 12y_3 &= 0 \\ 2y_1 + y_2 + 2y_3 + y_4 &= 2 \\ y_1 + 3y_2 + 2y_3 + y_5 &= 3 \end{aligned}$$

where y_0 is to be maximized and all variables except y_0 must be nonnegative. Using slack variables to express the primal problem in equation form, we obtain

$$\begin{aligned} x_0 - 2x_1 - 3x_2 &= 0 \\ 2x_1 + x_2 - x_3 &= 5 \\ x_1 + 3x_2 - x_4 &= 10 \\ 2x_1 + 2x_2 - x_5 &= 12 \end{aligned}$$

where x_0 is to be minimized and all variables except x_0 must be nonnegative.

Table 6.7 presents the results of applying the simplex algorithm to the dual problem. From the final tableau of Table 6.7, we determine the dual variables to be $y_0 = 14$, $y_1 = 0$, $y_2 = \frac{1}{2}$, $y_3 = \frac{3}{4}$, $y_4 = 0$, and $y_5 = 0$; these values can be verified in the equations of the dual problem. The primal variables are determined from the objective-function row of the final tableau in Table 6.7 in the usual manner for determining dual variables; thus we obtain the values $x_0 = 14$, $x_1 = 4$, $x_2 = 2$, $x_3 = 5$, $x_4 = 0$, and $x_5 = 0$, which can be verified in the equations for the original problem. Hence we have obtained the optimal solution to the primal problem by solving the dual problem by the simplex method.

Table 6.7

Tableau	Current Basis	y_0	y_1	y_2	y_3	y_4	y_5	Right-side Constants	Ratio
I	y_0	1	-5	-10	-12^*	0	0	0	
	y_4	0	2	1	②	1	0	2	$\frac{2}{3} =$ minimum
	y_5	0	1	3	2	0	1	3	$\frac{3}{2}$
II	y_0	1	7	-4^*	0	6	0	12	
	y_3	0	1	$\frac{1}{2}$	1	$\frac{1}{2}$	0	1	$\frac{1}{1/2} = 2$
	y_5	0	-1	②	0	-1	1	1	$\frac{1}{2} =$ minimum
III	y_0	1	5	0	0	4	2	14	
	y_3	0	$\frac{5}{4}$	0	1	$\frac{3}{4}$	$-\frac{1}{4}$	$\frac{3}{4}$	
	y_2	0	$-\frac{1}{2}$	1	0	$-\frac{1}{2}$	$\frac{1}{2}$	$\frac{1}{2}$	

EXERCISES

1. For each of the following problems, determine the dual problem:

   (a) Maximize
   $$x_0 = 3x_1 + 2x_2$$
   subject to
   $$x_1 + 3x_2 \leq 4$$
   $$3x_1 + 2x_2 \leq 15$$
   $$x_1 \geq 0 \qquad x_2 \geq 0$$

   (b) Maximize
   $$x_0 = x_1 + 5x_2 - 4x_3$$
   subject to
   $$3x_1 + x_2 - 4x_3 \leq 12$$
   $$2x_1 + 2x_2 + x_3 \leq 25$$
   $$2x_1 - 2x_2 + 4x_3 \leq 10$$
   $$x_1 \geq 0 \qquad x_2 \geq 0 \qquad x_3 \geq 0$$

(c) Minimize

$$x_0 = 4x_1 + 5x_2$$

subject to

$$3x_1 + x_2 \geqslant 10$$
$$3x_1 - 4x_2 \geqslant 6$$
$$x_1 + 4x_2 \geqslant 25$$
$$x_1 \geqslant 0 \quad x_2 \geqslant 0$$

(d) Minimize

$$x_0 = 3x_1 + x_2 - 8x_3$$

subject to

$$4x_1 + x_2 + 3x_3 \geqslant 60$$
$$x_1 - 2x_2 + 8x_3 \geqslant 43$$
$$x_1 \geqslant 0 \quad x_2 \geqslant 0 \quad x_3 \geqslant 0$$

2. Determine the dual problem of the dual problem determined in

  (a) Exercise 1(a) (b) Exercise 1(c)

3. Determine the dual problem for Exercise 1, Sec. 5.3, and determine the optimal values of the dual variables from the final tableau obtained in solving Exercise 1(a), Sec. 6.1.

4. Determine the dual problem for Exercise 5, Sec. 5.4, and determine the optimal values of the dual variables from the final tableau obtained in solving Exercise 2(a), Sec. 6.1.

5. Determine the dual problem for Exercise 3, Sec. 6.1, and determine the optimal values of the dual variables from the final tableau obtained in solving it.

6. Determine the dual problem for Exercise 4, Sec. 6.1, and determine the optimal values of the dual variables from the final tableau obtained in solving it.

7. Determine the dual problem for Exercise 9, Sec. 6.1, and determine the optimal values of the dual variables from the final tableau obtained in solving it.

8. Determine the dual problem for Exercise 10, Sec. 6.1, and determine the optimal values of the dual variables from the final tableau obtained in solving it.

9. Solve Exercise 3, Sec. 5.3, by determining the dual problem, solving the dual problem by the simplex method, and determining the optimal values of the primal and dual variables from the final tableau obtained in solving the dual problem.

10. Solve Exercise 8, Sec. 5.4, by determining the dual problem, solving the dual problem by the simplex method, and determining the optimal values of the primal and dual variables from the final tableau obtained in solving the dual problem.

11. Determine the dual problem for Exercise 11, Sec. 6.1, and determine the optimal values of the dual variables from the final tableau obtained in solving it.

12. A veterinarian must provide 12 grams (g) of nutrient A, 16 g of nutrient B, and 20 g of nutrient C for each dog per day. One half lb of dog food 1 contains 4, 4, and 4 g of nutrients A, B, and C, respectively, and costs 70¢. A half lb of dog food 2 contains 3, 6, and 2 g of nutrients A, B, and C, respectively, and costs 60¢. A half lb of dog food 3 contains 4, 6, and 8 g of nutrients A, B, and C, respectively, and costs $1.20.

   (a) Determine the objective function for minimizing costs and the associated constraints.
   (b) Determine and solve the dual problem by the simplex method.
   (c) Determine the optimal values of the primal and dual variables from the final tableau obtained in solving the dual problem.

13. A television network sells 1-minute advertisements for $250 in prime time and $200 at other times. A company has estimated that 1 min of prime time will cause 80 people to purchase the product, and 1 min at other times will cause 65 people to purchase the product. The company wants a minimum of 500 purchases per week to be activated by these advertisements. Also, they have decided that they want at least 20 min of advertising per week with at least 5 min in prime time.

   (a) Determine the objective function for minimizing costs and the associated constraints.
   (b) Determine and solve the dual problem by the simplex method.
   (c) Determine the optimal values of the primal and dual variables from the final tableau obtained in solving the dual problem.

14. Determine the dual problem for Exercise 14, Sec. 6.1, and determine the optimal values of the dual variables from the final tableau obtained in solving it.

IMPORTANT TERMS AND CONCEPTS

Basic feasible solution
Basic solution
Basic variable
Basis
Dual problem
Dual theorem
Dual variables
Finite optimal solution
Initial basic feasible solution
Nonbasic variable
Primal problem

Primal variable
Simplex algorithm
Simplex criterion I (maximization)
Simplex criterion I (minimization)
Simplex criterion II
Simplex method
Slack variable
Surplus variable
Tableau
Unbounded optimal solution

REVIEW PROBLEMS

1. A manufacturer produces a portable black and white television, a portable color television, and a cabinet color television. A black and white television yields a profit

of $75 and requires 3 hr in department A, 2 hr in department B, and 1 hr in department C. A portable color television yields a profit of $120 and requires 2 hr in department A, 4 hr in department B, and 1 hr in department C. A cabinet color television yields a profit of $200 and requires 4 hr in department A, 6 hr in department B, and 2 hr in department C. The time available per week is 2,400 hr in department A, 3,000 hr in department B, and 900 hr in department C.

(a) Determine the objective function and the constraints.
(b) Solve the linear programming problem by the simplex method.
(c) Interpret the meaning of the values of the decision variables and the slack variables for the optimal solution.

2. A chemical plant manufactures three solutions. Solution X requires 2 units of chemical A, 1 unit of chemical B, 3 units of chemical C, and 2 units of chemical D. Solution Y requires 1 unit of chemical A, 4 units of chemical C, and 1 unit of chemical D. Solution Z requires 3 units of chemical A, 2 units of chemical B, 1 unit of chemical C, and 2 units of chemical D. There are 1,200 units of chemical A, 500 units of chemical B, 1,500 units of chemical C, and 800 units of chemical D available for producing solutions X, Y, and Z. Solution X yields a profit of $10 per unit, solution Y yields a profit of $8 per unit, and solution Z yields a profit of $15 per unit.

(a) Determine the objective function and the constraints.
(b) Solve the linear programming problem by the simplex method.
(c) Interpret the meaning of the values of the decision variables and the slack variables for the optimal solution.

3. A clock manufacturer produces alarm clocks, radio clocks, and kitchen clocks. An alarm clock yields a profit of $1, a radio clock yields a profit of $8, and a kitchen clock yields a profit of $5. An alarm clock requires 6 min in department A, 4 min in department B, and 2 min in department C. A radio clock requires 24 min in department A, 20 min in department B, and 6 min in department C. A kitchen clock requires 16 min in department A, 14 min in department B, and 5 min in department C. The time available is 4,500 min in department A, 4,000 min in department B, and 1,500 min in department C.

(a) Determine the objective function and the constraints.
(b) Solve the linear programming problem by the simplex method.
(c) Interpret the meaning of the values of the decision variables and the slack variables for the optimal solution.

4. A glass manufacturer makes 4 oz, 10 oz, and 12 oz glasses. A case of the 4 oz glasses requires 4 lb of glass, 1 hr in department A, and 30 min in department B. A case of 10 oz glasses requires 5 lb of glass, 2 hr in department A, and 45 min in department B. A case of 12 oz glasses requires $5\frac{1}{2}$ lb of glass, $2\frac{1}{2}$ hr in department A, and 1 hr in department B. Each week there are 6,000 lb of glass, 2,500 hr in department A, and 1,000 hr in department B available for production. The profit yield per case is $3 for 4 oz glasses, $4 for 10 oz glasses, and $4.50 for 12 oz glasses.

(a) Determine the objective function and the constraints.
(b) Solve the linear programming problem by the simplex method.
(c) Interpret the meaning of the values of the decision variables and the slack variables for the optimal solution.

5. Determine the dual problem for Problem 1, and determine the optimal values of the dual variables from the final tableau obtained in solving it.

6. Determine the dual problem for Problem 2, and determine the optimal values of the dual variables from the final tableau obtained in solving it.

7. Determine the dual problem for Problem 3, and determine the optimal values of the dual variables from the final tableau obtained in solving it.

8. Determine the dual problem for Problem 4, and determine the optimal values of the dual variables from the final tableau obtained in solving it.

7 PROBABILITY

7.1 INTRODUCTION

Probability theory originated in the seventeenth century. It was developed for use in predicting the long-term outcomes of games of chance, e.g., tossing coins, rolling dice, spinning roulette wheels, drawing cards. In each such game the outcome for a specific trial is uncertain; however, each possible outcome will appear with a certain *long-term regularity*. It is the *predictability* of this regularity that is the concern of probability theory.

In many areas, e.g., business, economics, engineering, biology, psychology, physics, chemistry, and meteorology (to name only a few), we encounter a similar type of uncertainty and long-term regularity. For example, a production manager for an electronics firm does not know *which* transistors from a production line will be defective, but in the long run he *does* know *what percent* will be defective; an economist cannot predict exactly *which* persons will spend less than $500 for clothing during a given year, but he *can* predict with some degree of certainty *how many* people will spend less than that amount; finally a geneticist cannot be certain that a specific offspring will have blue eyes, but she can predict *what percent* of offspring will have this characteristic. Such predictions are based on *probability*.

Statements such as "the probability of thundershowers is 40 percent," "the probability of team A winning the World Series is 0.7," and "the probability of obtaining three heads on three tosses of a coin is $\frac{1}{8}$" are commonplace in everyday life. The average person has an intuitive idea (regardless of its validity) of what we mean by each of the previous statements; however, such statements present us with at least two major problems: The first involves the correct interpretation of the statement, while the second concerns the calculation of the numerical value of probability. We consider the first problem briefly from the viewpoints of various definitions of probability; the remainder of the chapter is concerned primarily with the problem of calculating probabilities and related values. However, the correct interpretation of probability statements will be enhanced as we gain an understanding of the operating rules of probability.

The definition of probability is a very difficult problem, and thus several different definitions have evolved. We shall begin our treatment of probability by considering several approaches that lead to somewhat different interpretations of the meaning of probability. These include the *classical*, or *a priori*, theory approach, the *frequency*, or *a posteriori*, theory approach, and the *subjective* approach.

Classical, or a Priori, Probability

As mentioned previously, the theory of probability originated as a means of predicting long-term outcomes of games of chance. Such a beginning prompted the classical definition. For example, suppose we wish to find the probability that one toss of a fair coin will result in a head. To find this probability we can argue in the following manner: Since the coin is fair and there are only two ways the coin can fall, we would expect it to fall heads and tails with approximately equal frequency; that is, about one-half of the time it will fall heads, and one-half of the time it will fall tails. Hence the probability of the event of a head will be given the value $\frac{1}{2}$. This type of reasoning leads to the *classical* definition of probability.

Definition 7.1 If an event can occur in n mutually exclusive and equally likely ways and if m of these outcomes have an attribute A, then the probability of A is the fraction m/n.

The use of the definition is easy enough in simple cases, but it is not always obvious. Particular attention must be given to the terms *mutually exclusive* (to be defined and discussed later) and *equally likely*. For example, suppose we wish to compute the probability of getting two heads on two tosses of a fair coin. We might argue that there are three possible outcomes for the two tosses: two heads, two tails, or one head and one tail. Since there are three outcomes and only one has the desired attribute, an assumption that the outcomes are equally likely would yield a probability of $\frac{1}{3}$. However, this reasoning is in error because the three outcomes are *not* equally likely. The outcome of one head and one tail can occur in two ways; that is, the tail may appear on the first toss and the head on the second, or the head may appear on the first toss and the tail on the second. Therefore, there are four equally likely outcomes: *HH, HT, TH, TT*. The first of these has the desired attribute while the others do not. Thus the correct probability is $\frac{1}{4}$. We would have the same result if two fair coins were tossed simultaneously.

There are some rather serious drawbacks associated with the classical definition of probability: First the definition must be modified somewhat when the number of possible outcomes is infinite. Second an even more serious defect becomes apparent in the situation when we have an event that can occur in n ways but the ways are *not* equally likely. Finally the classical approach provides no means of answering such questions as "What is the probability that a rocket that will be launched tomorrow will reach the moon?" or "What is the probability of the Dallas Cowboys winning a specific ball game?" or "What is the probability that a specific automobile made by Ford Motor Company will get at least 17 mi/gal of gasoline?" All of these questions are of great interest; however, in order to be able to answer such questions, it is necessary that we modify our definition of probability so that such problems can be included within the framework of the theory.

Frequency, or a Posteriori, Probability

A coin that seems to be well balanced and symmetrical has been tossed 200 times, and the outcomes are listed in Table 7.1. We note that the relative frequencies (i.e., observed proportions) are quite close to long-run expected relative frequencies for a balanced coin, that is, $\frac{1}{2}$. This is not surprising if we consider the coin to be approximately "fair." In fact, we might be willing to use the relative frequencies listed in Table 7.1 as approximations for the probabilities of heads and tails on some future toss of that particular coin.

Table 7.1 Results of Tossing a Coin 200 Times

Outcome	Frequency	Observed Relative Frequency	Long-run Expected Relative Frequency
H	95	.475	.500
T	105	.525	.500
Total	200	1.000	1.000

For this experiment it seems reasonable to assume that there exists a number, say p, which *is* the probability of a head. If the coin appears symmetrical and well balanced, we might use Definition 7.1 to determine that p is approximately equal to $\frac{1}{2}$. This would only be an approximation, however, because we do not know for certain that the two cases, heads and tails, are exactly equally likely. As an alternative, we could toss the coin a *large* number of times, recording the results as in Table 7.1, and use the relative frequency of heads as an approximation of p. In such a case, we would have applied what is often called the *frequency-theory approach based on experience*.

Whether we use *relative frequency* or Definition 7.1 for the probability p seems unimportant in the previous case. On the other hand, suppose our coin is unbalanced so that we feel sure that the two cases, heads and tails, are not equally likely. For this situation we *can* postulate a value for p. However, the classical definition does not provide any way to *find* the value for p; hence we must use the frequency approach.

In many business situations, observations are taken that have an element of uncertainty or unpredictability associated with them. In such situations, it is often reasonable to postulate that the probability of a specific event is equal to a number p and to take the relative frequency of the event (obtained from a large number of trials) as the approximation of p. Although we term this the frequency-theory approach based on experience, the method is occasionally referred to as a *statistical probability*.

Let us assume that a series of observations (or experiments) can be made under quite uniform conditions; that is, a series of events can be repeated (the outcome being noted each time) under identical conditions, but there is an uncontrollable variation that is "random." Thus the observations (or outcomes) are *individually* unpredictable. In many cases, such observations fall into certain classes that have quite stable relative frequencies, and the relative frequency of a particular outcome is then used to approximate p. A question that arises is "How many times must the experiment be repeated to ensure a stable relative frequency?" The answer depends upon the stability of the entire series of experiments, and thus it is difficult to give a rigorous answer. The best we can do is to say that the experiment should be repeated until the relative frequencies that we desire to use as probabilities appear to be changing very little as the number of experiments increases. Because of this point, it is necessary to include a slight extension of the frequency-theory approach to probability; that is, the approach based on experience will not suffice entirely. Thus we include the concept of *the limit of the relative frequency of occurrence of an event as the number of experiments increases*.

In the *frequency-theory approach based on limits*, we are still concerned with the relative frequency m/n of an event. However, under this approach, we define and interpret the probability $P(A)$ of an event A as that number that the relative frequency of an event A approaches as the number of trials increases without bound. Thus we say that $P(A)$ is the limit of the relative frequency of the event A as n tends to infinity.

Subjective Probability

The two approaches to probability just discussed are not satisfactory for all situations. For example, they will not provide answers for such questions as "What is the probability of an investment returning a net profit of at least $10,000 during the first year?" They are of no use to us here because this type of question involves an event that cannot be repeated under similar conditions with any reasonable certainty. Thus for this type of situation we employ a different approach involving *subjective probability*. The subjective approach to probability is relatively new, but in recent years a deluge of attention has been devoted to the subject in related literature. We shall make no attempt to give an extensive treatment of subjective probability here. Rather we shall merely introduce the concept so that the reader may see how the subjective approach, as well as the classical and frequency-theory approaches, can be linked to the axiomatic development of probability theory. In the subjective approach, we interpret probability as a measure of personal confidence in a particular statement; and since the concept is difficult to treat in a general fashion, we shall merely exemplify it.

L. J. Savage, one of the pioneers of subjective probability, appeals to intuition for the definition of probability. He suggests that we associate the definition of probability with "the price we would pay for a particular contingency." For example, suppose we write a contract that promises a person $100 if an investment returns $10,000 during the first year. What would the person be willing to pay for the contract before he has seen what happens in the first year? If he would be willing to pay $50, we take this to mean that the probability of the investment returning $10,000 is $\frac{50}{100}$ *for him*. However, someone else may be willing to pay $75 for the contract; hence, for this person, the probability of the event is $\frac{75}{100}$. Under this approach, different people may assign different probabilities to the occurrence of the same event. However, even though this may be the case, the subjective approach assumes that the persons involved will be *reasonable*, not *arbitrary*, in their probability assignments.

In general, when probabilities are set forth in a situation where no objective (in the sense of the classical approach) or empirical (in the sense of the frequency approach) basis exists from which such probabilities are developed, we say that the probabilities are subjective. In this sense, subjective probabilities are necessarily expressions of personal judgment. Thus, even though a subjective probability may be based on *experience* with events *similar* to the one in question, it is *plausible* that different individuals may assign different subjective probabilities to the same event. This disadvantage, however, is more than offset by the tremendous advantage that the subjective approach can be applied to events that cannot occur or to events that can occur only once, as well as to events similar to those to which the classical and frequency approaches are applicable. Even so, the reader should be aware that even among experts in probability and statistics a great deal of controversy exists regarding the subjective approach.

Now that we have established a common understanding with regard to the meaning of the general term *probability*, our primary aim is to pursue the axiomatic development of probability theory to the extent that we are able to *establish* a set of operating rules and *use* such rules in the logical solution of a multiplicity of probability problems. However, before we can accomplish this task, it is necessary that we consider useful counting techniques and other related elementary concepts of probability. We pursue these subjects in Secs. 7.2 and 7.3.

EXERCISES

1. Explain the term *long term regularity*.

2. What two conditions must be satisfied before the definition of *classical*, or *a priori*, probability can be applied?

3. Explain the term *statistical probability*.

4. Explain each of the *objective* approaches to probability.

5. Explain the *subjective* approach to probability.

6. Big John's Pizza Parlor delivers orders via automobile. Bill Sharp orders a pizza to be delivered to his home. Bill's home is on Preston Lane, which is a dead-end street. There are three streets, Elm, Pine, and Pecan, that funnel into the open end of Preston Lane. Each street is equally accessible to the delivery car and is about the same total distance from the pizza parlor. Bill and his friends are going to try to guess which street the car will take to enter Preston Lane.

(a) Give your subjective probabilities for each street.
(b) Determine the probabilities for each street using the classical approach.
(c) For the last 10 deliveries, the car came down Elm 3 times, Pine 5 times, and Pecan 2 times. Determine the probabilities for each street using the frequency approach.

7.2 THE PRINCIPLES OF COUNTING

In discussing the axiomatic development of probability theory, we shall find that it is impossible to completely divorce such theory from the definitions associated with the approaches to probability previously described. In particular, we shall emphasize important points repeatedly by utilizing such approaches to determine values for elementary probabilities. However, in both the classical and frequency approaches, we are required to count possible outcomes, and in many situations this is a very difficult task. Thus, before proceeding with the axiomatic development, we shall consider certain useful counting techniques, that is, *permutations* and *combinations*.

Basic Counting Principle

First let us consider counting techniques that enable us to enumerate the number of elements of a given set. The most fundamental principle upon which most counting techniques are based is known as the *basic counting principle*.

Basic Counting Principle If event A can occur in m ways and if after event A has occurred event B can occur in n ways, then events A and B can occur together in m times n ways.

Suppose event A is getting from Chicago to Cleveland via one of 6 routes and event B is getting from Cleveland to New York via one of 8 routes, then the number of different routes for getting from Chicago to New York by going through Cleveland is $(6)(8)=48$.

The basic counting principle can be generalized to include more than two events.

General Counting Principle If k events can occur such that event A_1 can occur in n_1 ways, event A_2 can occur in n_2 ways after event A_1 has occurred,..., and event A_k can occur in n_k ways after events A_1 through A_{k-1} have occurred, then the events A_1 through A_k can occur together in $n_1 \cdot n_2 \cdots n_k$ ways.

For example, if a researcher has 10 volunteer subjects and he wishes to have each of 4 subjects take different personality tests, how many different test assignments will occur? For the first test there are 10 subjects. After the first test has been assigned to a subject, there are 9 subjects available for the second test. After the first and second tests have been assigned, there are 8 subjects available for the third test. Finally after the third test has been assigned, there are 7 subjects remaining to which the researcher can assign the fourth test. Hence there are $10 \cdot 9 \cdot 8 \cdot 7 = 5,040$ ways that the tests can be assigned to the subjects.

Permutations and Combinations The general problem in which permutations and combinations become expedient counting techniques is the one in which we desire to determine the number of ways in which a set of n items can be collected into groups of size r. If groupings are differentiated by *both* the items included in them *and* the *order* in which the items appear, such groupings are called *permutations*. However, if groupings are differentiated *only* by the items included in them and *not* by the *order* in which the items appear, they are known as *combinations*. Thus it follows that one combination may be *permuted* to obtain many possible permutations, and the general problem can be easily solved by use of the following relationship. For example, we have n distinguishable items with which to fill r positions. Applying the general counting principle, the first position can be filled in n ways, the second position can be filled in $n-1$ ways after the first position is filled, the third position can be filled in $n-2$ ways,..., and the rth position can be filled in $n-(r-1)=n-r+1$ ways after the first $r-1$ positions have been filled. Thus the number of possible permutations is $n(n-1)(n-2)\cdots(n-r+1)$. This result is presented in a more compact notation in the following rule.

Rule 7.1 For the general case where we have n distinguishable items and wish to determine the number of permutations that can be formed by grouping r of the n items at a time, we have

$$_nP_r = \frac{n!}{(n-r)!} = n \cdot (n-1) \cdot (n-2) \ldots (n-r+1) \qquad (7.1)$$

where $_nP_r$ is the number of permutations of n things taken r at a time (it is understood that the items are distinguishable unless otherwise noted) and $n!$ is read as "n factorial" and is evaluated by taking the product of all positive integers that are less than or equal to n.

As an example of the meaning of the factorial notation

$$6! = 6 \cdot 5 \cdot 4 \cdot 3 \cdot 2 \cdot 1 = 720$$

By definition, $0! = 1$. Also, $1! = 1 \cdot 0! = 1$.

We note further that if we are interested in groupings that are differentiated only by the items included in them and not by the order in which the items appear, we are concerned with combinations. From a group of n distinguishable items, we can first select r of the items and represent the number of ways we can make such a selection by $_nC_r$, which is read "the number of combinations of n items taken r at a time." Next, for each combination of r items, we can distribute them into r positions. This can be done in $_rP_r = r!$ ways. Applying the general counting principle, we have $_nC_r \cdot r!$ different arrangements. However, this is the same as determining the number of permutations, where order is considered. Thus we have $_nP_r = {_nC_r} \cdot r!$. So, solving for $_nC_r$, we have

$$_nC_r = \frac{_nP_r}{r!} = \frac{n!}{r!(n-r)!} \tag{7.2}$$

This result is stated more compactly in the following rule.

Rule 7.2 For the general case where we have n distinguishable items and wish to determine the number of combinations that can be formed by grouping r of the n items at a time, we have

$$_nC_r = \binom{n}{r} = \frac{n!}{r!(n-r)!} \tag{7.3}$$

where both $_nC_r$ and $\binom{n}{r}$ represent the number of combinations of n things taken r at a time.

As a special case of Rule 7.1, the number of permutations of n things taken n at a time is

$$_nP_n = \frac{n!}{(n-n)!} = \frac{n!}{1} = n! \tag{7.4}$$

Consider the problem of determining the number of combinations and permutations of the four letters A, B, C, and D when such letters are grouped three at a time. One way of solving the problem is simply to list and enumerate all possibilities, which is done in the following table for the cases of both combinations and permutations:

Combinations	*ABC*		*ABD*		*ACD*		*BCD*	
Corresponding permutations	*ABC* *BAC* *CAB*	*ACB* *BCA* *CBA*	*ABD* *BAD* *DAB*	*ADB* *BDA* *DBA*	*ACD* *CAD* *DAC*	*ADC* *CDA* *DCA*	*BCD* *CBD* *DBC*	*BDC* *CDB* *DCB*

Sec. 7.2 / THE PRINCIPLES OF COUNTING

Merely by counting, we have 4 combinations and 24 permutations. However, we also can solve the problem by use of (7.3) and (7.1); that is, for combinations

$$_nC_r = {_4C_3} = \frac{4!}{3!1!} = \frac{4 \cdot 3!}{3! \cdot 1!} = 4$$

and for permutations

$$_nP_r = {_4P_3} = \frac{4!}{(4-3)!} = \frac{4!}{1!} = 4 \cdot 3 \cdot 2 = 24$$

We can see also that each combination can be *permuted* to obtain $r! = 3! = 6$ different permutations; for example, *ABC, ACB, BAC, BCA, CAB,* and *CBA* are all *different* permutations that represent the *same* combination.

Example 7.1 How many different sets of officers can be obtained from a set of 10 people to fill the offices of president, vice-president, secretary, and treasurer?

Solution In this problem it is obvious that order (so to speak) is important. Thus, since the 4 positions to be filled are different (distinguishable), we can apply (7.1) to determine in how many different ways the 4 positions can be filled from the group of 10 people. The number of different sets of officers is

$$_{10}P_4 = \frac{10!}{(10-4)!} = \frac{10!}{6!} = \frac{10 \cdot 9 \cdot 8 \cdot 7 \cdot 6!}{6!} = 10 \cdot 9 \cdot 8 \cdot 7 = 5,040$$

Example 7.2 How many different marketing research teams of 5 employees can be obtained from a group of 10 employees?

Solution This is a combination problem because any group of 5 employees will give only one combination regardless of how they are arranged. Thus the number of different teams is

$$_{10}C_5 = \binom{10}{5} = \frac{10!}{5!5!} = \frac{10 \cdot 9 \cdot 8 \cdot 7 \cdot 6 \cdot 5!}{5!5!} = \frac{10 \cdot 9 \cdot 8 \cdot 7 \cdot 6}{5 \cdot 4 \cdot 3 \cdot 2 \cdot 1} = 252$$

Example 7.3 A committee of 3 students and 2 teachers is being selected. In how many ways can this be done if there are 6 students and 5 teachers from whom to choose?

Solution This requires use of combinations in conjunction with the basic counting principle. We use combinations to determine how many different groups of 3 students we can select from the 6 available students and how many different groups of 2 teachers we can select from the 5 available teachers. Thus we have

$$_6C_3 = \binom{6}{3} = \frac{6!}{3!3!} = \frac{6 \cdot 5 \cdot 4}{3 \cdot 2 \cdot 1} = 20$$

different groups of 3 students and

$$_5C_2 = \binom{5}{2} = \frac{5!}{2!3!} = \frac{5 \cdot 4}{2 \cdot 1} = 10$$

different groups of 2 teachers. Next we employ the basic counting principle since, for each group of students, we can match each of the 10 different groups of teachers. Thus the number of ways we can select 3 students and 2 teachers is

$$_6C_3 \cdot {_5C_2} = 20(10) = 200$$

Example 7.4

(a) How many different displays of 9 appliances can be made if the manager has 20 appliances to choose from and positions are ignored?
(b) If positions are considered?

Solution

(a) Since positions are ignored, we are concerned with combinations. Thus the number of different displays is

$$_{20}C_9 = \binom{20}{9} = \frac{20!}{9!11!} = \frac{20 \cdot 19 \cdot 18 \cdot 17 \cdot 16 \cdot 15 \cdot 14 \cdot 13 \cdot 12}{9 \cdot 8 \cdot 7 \cdot 6 \cdot 5 \cdot 4 \cdot 3 \cdot 2 \cdot 1} = 167{,}960$$

(b) This is a permutation problem. If 2 appliances exchange places, we have a different display even though we have an identical combination of appliances; hence the number of different displays is

$$_{20}P_9 = \frac{20!}{(20-9)!} = \frac{20!}{11!} = 20 \cdot 19 \cdot 18 \cdot 17 \cdot 16 \cdot 15 \cdot 14 \cdot 13 \cdot 12$$
$$= 60{,}949{,}324{,}800$$

Permutations for Groups of Indistinguishable Items

In the previous examples *all* the items are individually distinguishable from each other. However, suppose we are interested in the number of different permutations that can be obtained from a group of letters where such is not the case, say A, A, B, and B. Not all four letters of this group are distinguishable from each other because we have two groups of two letters each. Thus the permutations become

$$AABB \quad ABAB \quad ABBA$$
$$BAAB \quad BABA \quad BBAA$$

which are all the visibly different permutations that can be obtained. However, it is obvious that there are less than $_4P_4 (=24)$ permutations. Why? The reason is that no distinguishable difference exists between the two A's and between the two B's. If a distinguishable difference did exist, we would have a problem similar to those discussed previously in Examples 7.1 to 7.4. For example, if we had A_1, A_2, B_1, and B_2 instead of A, A, B, and B, then we would have four distinguishable items for which we could use (7.1).

Fortunately there exists a relationship for determining the number of different permutations in the problem just discussed. This relationship provides us with a means of determining the number of different permutations obtainable from a set of n items consisting of k subsets of items, where the items within each subset are not distinguishable from each other but the items in one subset are distinguishable from those in another subset. To obtain the relationship, we again rely on the general counting principle. We have n positions to fill and k types of items with which to fill them, r_i items of the ith type ($i = 1, 2, \ldots, k$). First we choose r_1 positions from the n positions for the first type of item; this can be done in $_nC_{r_1}$ ways. Second we choose r_2 positions from the remaining $n - r_1$ positions for the second type of item; this can be done in $_{n-r_1}C_{r_2}$ ways. We proceed similarly for the remaining types of items until, finally, we put the last r_k items of the kth type into the remaining $n - r_1 - r_2 - \cdots - r_{k-1} = r_k$ positions. Applying the general counting principle, this can be done in

$$_nC_{r_1} \cdot {}_{n-r_1}C_{r_2} \cdot {}_{n-r_1-r_2}C_{r_3} \cdots {}_{n-r_1-r_2-\cdots-r_{k-1}}C_{r_k}$$

$$= \frac{n!}{r_1!(n-r_1)!} \cdot \frac{(n-r_1)!}{r_2!(n-r_1-r_2)!} \cdots \frac{(n-r_1-r_2-\cdots-r_{k-1})!}{r_k!0!}$$

$$= \frac{n!}{r_1!r_2!\cdots r_k!}$$

ways. Thus the relationship is

$$_nP_{r_1, r_2, \ldots, r_k} = \frac{n!}{r_1!r_2!\cdots r_k!} \tag{7.5}$$

where n is the total number of items under consideration, k is the number of distinguishable subgroups, $r_i (i = 1, 2, \ldots, k)$ is the number of nondistinguishable items in subgroup i, $r_1 + r_2 + \cdots + r_k = n$, and $_nP_{r_1, r_2, \ldots, r_k}$ is the number of permutations obtainable from n items consisting of the k distinguishable subgroups.

Example 7.5 How many different permutations can be obtained when arranging a group of nine numbers if the nine numbers are made up of three 4's, four 5's, and two 8's?

Solution Using (7.5), we have

$$_9P_{3,4,2} = \frac{9!}{3!4!2!} = 1{,}260$$

Thus 1,260 *different* nine-digit numbers can be constructed from the group of nine numbers.

Example 7.6 Five employees are being considered for promotion to 5 positions, i.e., 2 foremen, 2 supervisors, and 1 manager. How many sets of promotions are possible if the sets are considered as being different when employees exchange *types* of positions?

Solution We are concerned merely with the number of permutations obtainable from 5 employees when they are to be divided into three sets of sizes, 2, 2, and 1. The number of different permutations is

$$_5P_{2,2,1} = \frac{5!}{2!2!1!} = \frac{5 \cdot 4 \cdot 3 \cdot 2!}{2!2!1!} = 30$$

Example 7.7 How many different management teams can be selected from 8 employees under the rules given in Example 7.6?

Solution This problem is different from the one in Example 7.6 in that we have more potential employees than positions on the team. However, if we view the state of "not being selected on the team" as a fourth type of position, we shall simply divide the 8 players into 4 distinguishable groups of sizes, 2 (foremen), 2 (supervisors), 1 (manager), and 3 (nonpromoted). Thus the number of different management teams is

$$_8P_{2,2,1,3} = \frac{8!}{2!2!1!3!} = \frac{8 \cdot 7 \cdot 6 \cdot 5 \cdot 4 \cdot 3 \cdot 2 \cdot 1}{2 \cdot 1 \cdot 2 \cdot 1 \cdot 1 \cdot 3 \cdot 2 \cdot 1} = 1{,}680$$

It can be noted that (7.3) is a special case of (7.5); that is, since $r + (n - r) = n$, when we find the number of combinations of n things taken r at a time, *essentially* we are finding the number of permutations obtainable from n items when they are to be divided into *two* distinguishable sets of size r and $n - r$. Thus

$$_nP_{r,n-r} = \frac{n!}{r!(n-r)!} = {_nC_r} \tag{7.6}$$

We can use this same reasoning also to note that the number of combinations of n things taken r at a time is equal to the number of combinations of n things taken $(n - r)$ at a time; that is,

$$_nC_r = \frac{n!}{r!(n-r)!} = {_nC_{n-r}} \tag{7.7}$$

For example, the number of combinations of 5 items taken 2 at a time is 10; that is,

$$_5C_2 = \frac{5!}{2!(5-2)!} = \frac{5!}{2!3!} = \frac{5 \cdot 4 \cdot 3!}{2!3!} = 10$$

and the number of combinations of 5 items taken 3 at a time is also 10; that is,

$$_5C_3 = \frac{5!}{3!(5-3)!} = \frac{5!}{3!2!} = \frac{5 \cdot 4 \cdot 3!}{3!2!} = 10$$

Sec. 7.2 / THE PRINCIPLES OF COUNTING

EXERCISES

1. Explain the difference between *permutations* and *combinations*.

2. Calculate the following:

   (a) $_5P_3$ (b) $_5C_3$ (c) $_{10}P_6$
   (d) $_9C_5$ (e) $_{11}P_{1,8,2}$ (f) $_{12}P_{3,5,4}$

3. Give the number of distinct sets of 8 letters that can be selected from the set A, B, C, D, E, F, G, H, I, J, K, and L

   (a) When order is *not* considered
   (b) When order *is* considered

4. Calculate how many different three-digit numbers can be constructed when zero is always considered and any digit

   (a) May appear only once
   (b) May appear no more than twice
   (c) May appear up to three times

5. How many different management teams (1 president, 2 vice-presidents, 2 plant managers, 2 production managers, 1 purchasing manager, 1 sales manager, 2 planning managers) can be selected from 11 employees when the team is considered different if the employees exchange *types* of positions?

6. How many different management teams (2 plant managers, 2 production managers, 2 quality control managers, 3 sales managers, 2 planning managers) can be selected from 15 employees when the team is considered different if the employees exchange *types* of positions?

7. Calculate the numbers of different ways that a set of 20 students may be divided into

   (a) 4 groups of 5 students
   (b) 5 groups of 4 students
   (c) 2 groups of 10 students

8. A manager is considering 8 applicants for 2 job vacancies.

   (a) How many different pairs of employees can be hired?
   (b) If the 2 jobs are different, how many different pairs can be hired when position is considered?

9. A chair manufacturer has 7 different upholstery materials. If he is going to put a different material on each of 3 chairs, how many different sets of chairs can he make?

10. How many different groups of 4 employees can a manager send to a workshop if he has 6 employees to choose from?

11. A jewelry store is trying to make a display and has 8 rings from which to choose. How many different displays can the store make if they use 4 rings and their position in the display is considered?

12. A fast food restaurant has 9 employees. Each day they need 4 cooks and 2

people to serve customers. How many different groups of employees can work on one day if the job performed is considered?

13. The Elmore Park city council consists of 9 members. A delegation of 3 is to attend a court hearing in the state capitol. Of the 9 members there are 6 men and 3 women. Also, there are 3 Republicans, 4 Democrats, and 2 Independents.

(a) Determine how many different delegations are possible.
(b) Determine how many delegations containing 2 men are possible.
(c) Determine how many delegations containing 1 Republican and 1 Independent are possible.
(d) Given that the chairman of the council must be in the delegation, determine how many delegations are possible.

14. The Golden H Hamburger House sells hamburgers with mustard, catsup, pickles, onion, relish, tomato, or lettuce. How many different hamburgers can be made using

(a) Any 2 of the extras
(b) Any 3 of the extras
(c) Either mustard or catsup with any 2 of the remaining extras
(d) Either mustard or catsup and either pickles or relish with any 2 of the remaining extras

15. Crocker's Pancake House makes 10 different flavors of pancakes that are served with either butter or margarine and one of 5 flavors of syrup. How many different orders are there?

16. How many six-digit license plates can be made using exactly 4 of the 9 digits 1 through 9 and exactly 2 of the 26 letters in the English alphabet

(a) If repetition is permitted
(b) If repetition is not permitted

7.3 ELEMENTARY PROBABILITY CONCEPTS

In order to understand the operating rules of probability, we must consider first certain fundamental concepts and definitions.

Definition 7.2 A random experiment is a process, or operation, that results in one of a number of possible outcomes such that the outcome is determined only by chance and such that it is impossible to predict the exact outcome.

Example 7.8 Consider the process of tossing a fair die. This process is a random experiment because the possible outcomes of 1, 2, 3, 4, 5, and 6 are determined strictly by chance and it is impossible to predict exactly which of the six values will occur on any given toss.

Example 7.9 Consider the process of selecting a card from a well-shuffled deck of ordinary playing cards. Since the cards are well shuffled, the selection of any particular card is strictly by chance; furthermore, it is impossible to predict the exact card that will be obtained on any given selection. Hence this process is a random experiment.

Definition 7.3 A population (or sample space) of events, denoted S, is the set of all possible outcomes that can occur as the result of a given random experiment.

In Example 7.8 the sample space is given by $S = \{1, 2, 3, 4, 5, 6\}$.

Definition 7.4 A random event is an event, or occurrence, that is generated by a random experiment.

In Example 7.8 the occurrence of a 4 on the face of the die represents a random event if the die has been tossed without any restrictions. Likewise, in Example 7.9 the occurrence of the ace of spades represents a random event if both the selection is made from an ordinary (unmarked) deck of playing cards without the selector observing the faces of the cards and the cards have been thoroughly shuffled prior to the selection.

Definition 7.5 A random variable is a variable, the values of which are determined by the outcome of a random event and can be ordered according to numerical value.

Note that the second part of Definition 7.5 is somewhat restrictive because it represents the essence of the difference between the possible outcomes of a random experiment and the possible values of a random variable. For instance, in Example 7.8 the number appearing on the face of the die *is* the random variable of interest. The *random variable* receives its value as the result of a random event (our tossing the die and observing the number that is facing upward). Furthermore, the possible values $(1, 2, \ldots, 6)$ can be ordered according to numerical value, that is, $6 > 2, 3 > 1$, and so forth.

However, although the *suit and rank* of the card selected in Example 7.9 are the result of a random event, they cannot be construed strictly as a random variable in the sense of Definition 7.5 because (for example) it does not make sense numerically to speak of the 4 of clubs being larger or smaller than the 4 of hearts. Note that we can create a random variable by arbitrarily assigning numerical values to the various possible combinations of suit and rank.

Definition 7.6 A simple event is a subset of the sample space containing only one element of the sample space.

In Example 7.8 each of the six events {1}, {2}, {3}, {4}, {5}, and {6} is a simple event. In Example 7.9 a simple event is the event {ace of spades}.

Definition 7.7 A *compound event* is an event that can be represented as the union of two or more simple events.

In Example 7.8 a compound event is that of obtaining an even number on the face of the die as the result of only one toss. This particular event can be represented as the union of the three simple events {2}, {4}, or {6}; that is, if either of the simple events {2}, {4}, or {6} is obtained, the compound event {*even number*} will have occurred. In Example 7.9, a compound event is that of drawing a card from the heart suit. This event can be represented by the union of the 13 simple events {A}, {2}, {3}, {4}, {5}, {6}, {7}, {8}, {9}, {10}, {J}, {Q}, or {K} from the heart suit; for, if either of these simple events occurs, the compound event {*heart*} will necessarily have occurred.

Definition 7.8 A *sample* is a union of simple events (or observations) that is taken from a population of events.

Definition 7.9 A *simple random sample* is a sample selected by a random process such that any simple event remaining in the population at any given instant has the same chance of appearing in the sample as any other remaining simple event.

This definition implies that there are two conditions required for simple random sampling; that is, each simple event in the population has an equal chance of being selected and each possible sample of a specific size has the same chance of being selected as any other possible sample of the same size.

The population for Example 7.8 is the union of the six simple events {1}, {2}, {3}, {4}, {5}, and {6}; and a sample of size 2 from this population could be {1,3} while a sample of size 3 could be {1,4,6}. The population for Example 7.9 is the entire deck of 52 cards (that is, the 52 simple events); and a sample of size 3 from this population could be {ace of clubs, king of hearts, 7 of diamonds}.

In Example 7.9, the sample obtained by randomly selecting a card from the well-shuffled deck of 52 cards is a simple random sample because any given card will have the same opportunity of being selected as any other card. If the experiment of this example is such that more than one draw is required, we have the option of replacing the card in the deck and reshuffling before the next draw or keeping the selected card out of the deck. In the former case the procedure is called *sampling with replacement* while in the latter case it is called *sampling without replacement*. In either case, each of the *remaining* cards has the same chance of being selected on the next draw. Hence, in this example, a simple random sample is obtained whether we sample with replacement or without replacement.

In Example 7.8 the sample obtained by tossing the die (if it is a fair die) is a simple random sample because each number has the same chance of appearing in the sample. Furthermore, any sample of size greater than 1 will be obtained *essentially* by replacing the event in the original population owing to the nature of the experiment of tossing a fair die (i.e., each simple event has the same chance of occurring on any repeated toss). Hence, in this example, the sampling procedure is necessarily sampling with replacement.

Definition 7.10 A probability sample is a sample obtained by a random process wherein each event has a known (not necessarily equal) probability of occurring.

In Example 7.8 if the die were not a fair die (i.e., known but different probabilities for each possible value), the resulting sample would be a probability sample rather than a simple random sample. It should be noted that any simple random sample is also a probability sample because the probabilities are equal and thus *known*; however, because of the very nature of the definition, a probability sample is not necessarily a simple random sample. Thus it is evident that a simple random sample is a more restrictive classification than a probability sample.

Definition 7.11 Two events A and B are mutually exclusive if $A \cap B = \emptyset$.

The name *mutually exclusive* is attached to two such events because the definition implies that the occurrence of one event *excludes* the occurrence of the other. Thus, by definition, any two different simple events are mutually exclusive.

Definition 7.12 A set of events $A_1, A_2, \ldots, A_k$ is exhaustive for a population S if

$$A_1 \cup A_2 \cup \cdots \cup A_k = S$$

In Example 7.8 consider one toss of the die. Let $A = \{1,2\}$, $B = \{3,4\}$, $C = \{2,3\}$, and $D = \{4,5,6\}$. For example, event A will have occurred if either a 1 or 2 is obtained on a given toss; and because $A \cap B = \emptyset$ the events A and B are mutually exclusive. However, events A and C are not mutually exclusive because $A \cap C = \{2\}$.

The events A, B, and D are exhaustive events because $A \cup B \cup D = \{1,2,3,4,5,6\} = S$.

EXERCISES

1. Explain the relationships between a *random experiment*, a *random event*, and a *random variable*, and give an example of each.

2. Define the terms *simple event* and *compound event*, and give an example of each.

3. Define the terms *population* and *sample*, and give an example of each.

4. Define *simple random sampling*, and give an example.

5. What two conditions are required by the definition of *simple random sampling*?

6. Define *probability sampling*, and give an example that is not a *simple random sample*.

7. What two conditions are required by the definition of *probability sampling*?

8. (a) Is a *probability sample* always a *simple random sample*? Explain.
   (b) Is a *simple random sample* always a *probability sample*? Explain.

9. Define *mutually exclusive events*, and give an example.

10. Define a set of *exhaustive events*, and give an example.

11. Give an example of two events that are *mutually exclusive* but not *exhaustive*.

12. Give an example of two events that are *exhaustive* but not *mutually exclusive*.

13. Give an example of two events that are neither *mutually exclusive* nor *exhaustive*.

14. Give an example of three events that are not *mutually exclusive* but are *exhaustive*.

15. A card is to be drawn from a deck. Consider the following events:

    (a) Black card, heart
    (b) Heart, diamond, club
    (c) Face card, not a face card, club

Determine whether or not the events are *mutually exclusive* or *exhaustive*. Which of these events are *simple events* and which are *compound events*?

7.4 POSTULATES OF PROBABILITY

Now we begin the axiomatic development of probability theory. In mathematics any set of operating rules is based generally on an explicitly stated set of assumptions, and such is the case in probability. In the development of probability theory, these assumptions are known as the *postulates of probability*, and they are as follows.

Let S be a sample space (or population of events) and let A represent *any event* in S. Then, if we represent the probability of the event A by the function $P(A)$, P becomes a *probability function* having for its domain the set of subsets of the sample space S if the following three postulates are satisfied.

Sec. 7.4 / POSTULATES OF PROBABILITY

Postulate 7.1 The range of P is the set of real numbers such that

$$0 \leqslant P(A) \leqslant 1 \tag{7.8}$$

Postulate 7.2 If $A_i \cap A_j = \emptyset$, then

$$P(A_i \cup A_j) = P(A_i) + P(A_j) \tag{7.9}$$

Postulate 7.3

$$P(S) = 1 \tag{7.10}$$

Although these postulates are motivated by the classical and frequency approaches to probability, they provide a basis for the development of a much more powerful and generally applicable approach; that is, they can be used to develop an idealized model of a real-world situation. However, it is important to note that because the axiomatic approach does not always provide an expedient means for determining specific values for probabilities of mutually exclusive events, often we use one of the other approaches to probability to determine such values. Thus the axioms are applicable to all probabilities regardless of which approach is used to obtain them. Furthermore, even when the situation is such that use of the postulates will lead directly to specific probabilities for mutually exclusive events, often it is easier to use the basic counting principles associated with the other two approaches.

Postulate 7.2 can be generalized to apply to any set of *mutually exclusive* events (say, $A_1, A_2, \ldots, A_m$); that is,

$$P(A_1 \cup A_2 \cup \cdots \cup A_m) = P(A_1) + P(A_2) + \cdots + P(A_m) \tag{7.11}$$

Example 7.10 Suppose we have 10 different companies, of which 2 manufacture only automobiles, 3 manufacture only airplanes, and 5 manufacture only electronic components. From this set we wish to select randomly one company such that each company has an equal chance of being selected. Hence, using the classical definition of probability and letting auto, plane, or component represent, respectively, the set of companies that make the corresponding product, the probabilities of drawing a specific type of manufacturer are $P(\text{auto}) = .2$, $P(\text{plane}) = .3$, and $P(\text{component}) = .5$ [that is, there are 10 possible outcomes, 2 of which have the attribute auto so that $P(\text{auto}) = \frac{2}{10} = .2$].

For a better understanding of how these probabilities can be obtained by using the postulates, consider Table 7.2. Since there are 10 companies and we are sampling such that each company has an equal chance of being selected, Postulates 7.2 and 7.3 and Eq. (7.11) can be used to solve for the common probability (equal chance) of the occurrence of each company; thus each manufacturer has a probability of .1 of being selected. Furthermore, since the 10 *simple* events of selecting specific manufacturers are *necessarily* mutually exclusive (because they manufacture *only* one type of product), we may add the probabilities of individual events to obtain the probability of a compound event, e.g., the compound event of selecting an automobile manufacturer.

Table 7.2

Simple Event	Probability	Compound Event	Probability
$Auto_1$	.1	Automobile	
$Auto_2$	.1	manufacturer	.2
$Plane_1$	.1		
$Plane_2$	.1	Airplane	
$Plane_3$	.1	manufacturer	.3
$Component_1$	.1	Electronic-	
$Component_2$	.1	component	.5
$Component_3$	.1	manufacturer	
$Component_4$	.1		
$Component_5$	.1		
	1.0		1.0

Similarly, Postulate 7.2 and Eq. (7.11) can be used to determine the probabilities of two or more *properly defined* (i.e., mutually exclusive) compound events. For example, by Postulate 7.2, the probability of selecting a company that manufactures a transportation vehicle is

$$P(\text{transportation vehicle}) = P(\text{auto} \cup \text{plane}) = P(\text{auto}) + P(\text{plane})$$
$$= .2 + .3 = .5 \tag{7.12}$$

By the same postulate, the probability of selecting a company that manufactures a flight-related product is

$$P(\text{flight-related product}) = P(\text{plane} \cup \text{component})$$
$$= P(\text{plane}) + P(\text{component}) = .3 + .5 = .8$$

Note that in both of these cases we have been able to obtain the probability of one *compound* event by adding the probabilities of two other *compound* events. This is possible because (and only because) in each case the two compound events involved in the addition are *mutually exclusive*, which leads us to a very important point: Postulate 7.2 does *not* apply in the case of compound events that are *not* mutually exclusive. Suppose we wish to find $P(\text{transportation vehicle} \cup \text{flight-related product})$. Using Postulate 7.2, we would obtain

$$P(\text{transportation vehicle} \cup \text{flight-related product})$$
$$= P(\text{transportation vehicle}) + P(\text{flight-related product})$$
$$= .5 + .8 = 1.3$$

However, Postulate 7.1 tells us that this result is impossible because the probability of an event *can never be greater than* 1.0. Although we shall discuss the proper approach to this type of problem later, we mention it here only to emphasize the vital role that the term *mutually exclusive* plays with regard to the postulates.

Let us develop one other point in this example: Through use of Postulate 7.2, the probability of selecting a company that does not make automobiles can be obtained by the following calculation:

$$P(\text{not auto}) = P(\text{plane} \cup \text{component}) = P(\text{plane}) + P(\text{component})$$
$$= .3 + .5 = .8 \qquad (7.13)$$

Furthermore, by Postulates 7.2 and 7.3, the probability of selecting a company that manufactures either automobiles, airplanes, or components is

$$P(S) = P(\text{plane} \cup \text{auto} \cup \text{component})$$
$$= P(\text{plane}) + P(\text{auto}) + P(\text{component})$$
$$= .3 + .2 + .5 = 1$$

This is necessarily true because the events are both *exhaustive* and *mutually exclusive*. However, knowing that the three compound events are mutually exclusive and exhaustive, we can also find the probability of selecting a company that does not make automobiles by combining Postulates 7.2 and 7.3 and solving for the unknown sum; that is, since

$$P(\text{plane}) + P(\text{component}) + P(\text{auto}) = 1$$

we have

$$P(\text{plane}) + P(\text{component}) = 1 - P(\text{auto})$$

Substituting from (7.13),

$$P(\text{not auto}) = 1 - P(\text{auto}) = 1 - .2 = .8$$

The last result of Example 7.10 leads us to a rule that is quite useful in many probability problems. If we let the event A' represent the complement of event A, that is, the failure of event A to occur, then $S = A \cup A'$. However, $A \cap A' = \emptyset$ so we may apply Postulates 7.2 and 7.3 to obtain $P(S) = P(A \cup A') = P(A) + P(A') = 1$. Solving for $P(A)$, we obtain the following rule.

Rule 7.3

$$P(A) = 1 - P(A') \qquad (7.14)$$

For Example 7.10, we can calculate other probabilities through use of this rule; that is,

$$P(\text{auto} \cup \text{plane}) = 1 - P(\text{component}) = 1 - .5 = .5$$

and

$$P(\text{plane} \cup \text{component}) = 1 - P(\text{auto}) = 1 - .2 = .8$$

Example 7.11 A large producer of cosmetics is interested in conducting a market-research study concerning the effects of advertising in drugstores. For purposes of this study, a sample will be selected from the drugstores in a given geographical area, the distribution of which is given in Table 7.3.

Table 7.3

Population of City	No. of Drugstores	Probability
Under 5,000	424	.0424
5,000 to 15,000	843	.0843
15,000 to 30,000	1,427	.1427
30,000 to 50,000	2,012	.2012
50,000 to 100,000	2,384	.2384
100,000 and over	2,910	.2910
Total	10,000	1.0000

The probabilities given in Table 7.3 are those of choosing a drugstore in a city of the given population when randomly selecting a drugstore from the 10,000 drugstores in the geographical area. Let X be the random variable that has for its value the population of the city in which the selected drugstore is located. Then, by the classical approach,

$$P(X < 5,000) = .0424$$
$$P(5,000 \leq X < 15,000) = .0843$$
$$P(15,000 \leq X < 30,000) = .1427$$
$$P(30,000 \leq X < 50,000) = .2012$$
$$P(50,000 \leq X < 100,000) = .2384$$
$$P(X \geq 100,000) = .2910$$

and these events represent an arbitrarily defined set of mutually exclusive and exhaustive events. Thus Postulates 7.2 and 7.3 can be used to determine the probabilities of the occurrence of certain other compound events, as follows. The probability of randomly selecting a drugstore from a city with a population of 30,000 or larger is

$$P(X \geq 30,000) = P(30,000 \leq X < 50,000 \cup 50,000 \leq X < 100,000 \cup X \geq 100,000)$$
$$= P(30,000 \leq X < 50,000) + P(50,000 \leq X < 100,000)$$
$$+ P(X \geq 100,000)$$
$$= .2012 + .2384 + .2910 = .7306$$

Sec. 7.4 / POSTULATES OF PROBABILITY

This means that if we sample repeatedly with replacement from the population of drugstores, we expect an average of 7,306 drugstores per 10,000 to be located in a city with a population of at least 30,000.

The probability of randomly selecting a drugstore from a city with a population of between 15,000 and 100,000 is

$$P(15,000 \leq X < 100,000) = P(15,000 \leq X < 30,000 \cup 30,000 \leq X$$
$$< 50,000 \cup 50,000 \leq X < 100,000)$$
$$= P(15,000 \leq X < 30,000) + P(30,000 \leq X < 50,000)$$
$$+ P(50,000 \leq X < 100,000)$$
$$= .1427 + .2012 + .2384 = .5823$$

The reader should satisfy himself, however, that we cannot obtain $P(X \geq 30,000 \cup 15,000 \leq X < 100,000)$ simply by adding $P(X \geq 30,000)$ and $P(15,000 \leq X < 100,000)$ because the two events are *not mutually exclusive*.

EXERCISES

1. State the three postulates of probability. What conditions must hold for these postulates to be valid?

2. The results of a survey of 50 randomly selected retail outlets for a certain product classified 30 outlets as small, 15 as medium, and 5 as large in volume.

(a) Calculate the probabilities that a randomly selected retail outlet for this product will be small, medium, or large by applying the frequency-theory (a posteriori) approach based on experience.

(b) Use the results of part (a) to illustrate each of the three postulates of probability.

3. A test consisting of 6 complex problem-solving tasks was given to each of 100 employees with the following distribution of correct answers:

Correct Answers	f
1	20
2	16
3	17
4	14
5	18
6	15

(a) Use the frequency-theory (a posteriori) approach based on experience to calculate the probabilities associated with each event.

(b) Use the results of part (a) to illustrate each of the three postulates of probability.

4. A manufacturing company has the following number of employees in each department:

Department	No. of Employees
Production	63
Marketing	14
Finance	18
Purchasing	5

Calculate the probability that a randomly selected employee

    (a) Works in the production or marketing department
    (b) Does not work in the production department
    (c) Works in the finance or purchasing department
    (d) Does not work in the marketing or finance department
    (e) Does not work in the purchasing, finance, or production department

5. A retail store has categorized their accounts receivable according to how many days it has been since the date of purchase. The results are:

No. of Days Since Purchases	No. of Accounts
Under 20	1,825
20 to 35	1,180
35 to 60	815
60 to 90	645
90 to 110	385
Over 110	150

Calculate the probability that a randomly selected account is

    (a) From a purchase made 20 to 35 days ago   .236
    (b) From a purchase made less than 60 days ago .764
    (c) From a purchase made 35 to 90 days ago .292
    (d) Not from a purchase made over 90 days ago   .893
    (e) Not from a purchase made under 60 days ago .236

6. The accounting department also categorized their accounts receivable according to the balance due. The results are:

Balance due in Dollars	No. of Accounts
Under 25	370
25 to 50	1,600
50 to 100	1,250
100 to 200	785
200 to 500	695
500 to 1,000	225
Over 1,000	75

Sec. 7.5 / JOINT, MARGINAL, AND CONDITIONAL PROBABILITIES

Calculate the probability that a randomly selected account is

(a) Under $100
(b) For $50 to $200
(c) Not for $100 to $500
(d) Not $50 or over
(e) Not $500 or over

7. In an experiment on taste sensitivity, sugar solutions of 4 different concentrations are presented to a subject. What is the probability that he can, by chance alone, correctly arrange them in order from most concentrated to least?

8. The Federal Aviation Administration gives multiple choice questions on the written examination for a private pilot's license. A question is such that there are 5 answers of which 1 is correct and is worth five points. Another answer is not completely correct and is worth three points. The remaining answers are all incorrect and worth zero points. A person taking the exam randomly selects the answers.

(a) Calculate the probabilities associated with the possible scores on each question.
(b) Determine the probability of getting, by chance alone, five points. Three points. At least three points. At most three points. Zero points.

9. Park City has ten candidates for mayor: 5 Democrats, 3 Republicans, and 2 Independents. If the chances of any of the candidates being elected are equally likely, determine the probability that the new mayor is

(a) A Democrat or an Independent   (b) A Republican or a Democrat
(c) Not a Republican   (d) Not a Democrat
(e) A Democrat or not an Independent

7.5 JOINT, MARGINAL, AND CONDITIONAL PROBABILITIES

The theory and related applications of *joint*, *marginal*, and *conditional* probabilities play a very important role in probability theory. We shall begin our discussion using an example that is designed specifically to illustrate the meaning of these terms. Suppose that we are concerned with a population of 1,000 individuals, each of whom is classified according to both educational level achieved and annual salary (the results of such a classification are given in Table 7.4). Suppose also that we are to choose one individual from the population by simple random sampling. For example, as indicated in the table, the event A_1 will occur if we select an individual who has at most eight years of education regardless of salary earned. Similarly, B_3 will occur if we select an individual who earns more than $15,000 per year regardless of educational background.

Joint Probabilities

Now suppose we were to draw an individual who has at most eight years' education *and* who earns more than $15,000/year. In such a case, *both* events A_1 and B_3 will have occurred. Thus the event $A_1 \cap B_3$ has occurred. Any event that is the result of the intersection of two or more events is called a *joint event*.

Table 7.4

	Annual Salary in Dollars (X)			
Level of Education (Years, Y)	Event B_1 $0 < X \leq 5,000$	Event B_2 $5,000 < X \leq 15,000$	Event B_3 $X > 15,000$	Total
Event A_1, $0 < Y \leq 8$	210	80	10	300
Event A_2, $8 < Y \leq 12$	400	150	50	600
Event A_3, $Y > 12$	30	50	20	100
Total	640	280	80	1,000

It is clear that when there are n equally likely possible outcomes of a chance event and such outcomes may be classified according to two criteria (as in Table 7.4), the probability of the occurrence of a joint event may be obtained by the classical approach. For example, we obtain $P(A_1 \cap B_3)$ by taking the ratio of the number of individuals having both attributes of interest to the total number of individuals; that is,

$$P(A_1 \cap B_3) = \frac{10}{1,000}$$

Now, in cross-classifying the individuals according to education and annual salary, we have arbitrarily divided the 1,000 possible outcomes into 9 *mutually exclusive joint events*. This is obvious because one random selection will produce an individual having the attributes associated with *only* one cross classification. It follows, then, that if we compute and add the probabilities of *all* the joint events, we obtain 1; that is,

$$P(A_1 \cap B_1) + P(A_1 \cap B_2) + \cdots + P(A_3 \cap B_3) = \frac{210}{1,000} + \frac{80}{1,000} + \cdots + \frac{20}{1,000}$$

$$= \frac{1,000}{1,000} = 1 \qquad (7.15)$$

Marginal Probabilities The fact that the joint events are mutually exclusive leads to another concept: We may be interested in only one of the criteria of classification, say, education, while indifferent to the other, say, annual salary. In such a case, we may be interested, for example, in obtaining the probability that a randomly selected individual will have more than 12 years of education. We can obtain this probability immediately through the classical approach as

$$P(A_3) = \frac{100}{1,000}$$

Sec. 7.5 / JOINT, MARGINAL, AND CONDITIONAL PROBABILITIES

However, it is enlightening to apply Postulate 7.2. Since B_1, B_2, and B_3 are exhaustive, we have $S = B_1 \cup B_2 \cup B_3$ which yields $A_3 = A_3 \cap S = (A_3 \cap B_1) \cup (A_3 \cap B_2) \cup (A_3 \cap B_3)$. Thus, since the joint events are mutually exclusive, we have

$$P(A_3) = P[(A_3 \cap B_1) \cup (A_3 \cap B_2) \cup (A_3 \cap B_3)]$$
$$= P(A_3 \cap B_1) + P(A_3 \cap B_2) + P(A_3 \cap B_3)$$
$$= \frac{30}{1,000} + \frac{50}{1,000} + \frac{20}{1,000} = \frac{100}{1,000} \quad (7.16)$$

which is called a *marginal probability*. (Note that it was obtained by dividing the number in the proper *margin* by the total possible number of outcomes.) The term *marginal* is used whenever one or more criteria of classification are ignored. In general, *we can obtain the marginal probability of an event by summing the probabilities of all mutually exclusive joint events that satisfy the event in question.*

Rule 7.4 If the event A can occur jointly with all or any subset of k mutually exclusive and exhaustive events $B_1, B_2, \ldots, B_k$, then the marginal probability of the event A can be obtained by

$$P(A) = P(A \cap B_1) + P(A \cap B_2) + \cdots + P(A \cap B_k) \quad (7.17)$$

In subsequent sections we shall be concerned with other general rules that are useful in determining both marginal and joint probabilities. First, however, we shall treat the concept of *conditional probabilities*.

Conditional Probabilities

Often we are interested in knowing the probability of an event occurring when we possess or can assume information regarding the occurrence of another event. For example, with the information that a green ball has already been drawn on the only previous selection, we may wish to know the probability of selecting a green ball from an urn that originally contained six green balls and four blue balls. If we let the events of selecting colors green and blue be denoted by G and B, respectively, and the subscript of each correspond to the draw to which we have reference, the above *conditional probability* may be denoted by $P(G_2 | G_1)$. This probability statement is read as "the probability of selecting a green ball on the second draw given that a green ball was obtained on the first draw." In general, for any events A and B, we shall have $P(A | B)$, "the probability of event A occurring given that event B has occurred." This probability is actually the *probability of a conditional event* (that is, the event described by "the occurrence of event A given that event B has occurred"). However, the probability of a conditional event is generally termed a *conditional probability*.

To determine conditional probabilities, we *can* construct the remaining population (perhaps only a subpopulation) from which we shall be sampling under the given conditions, and then we may proceed by determining the probabilities in the usual manner of enumeration of possibilities. As a numerical example, suppose we have randomly selected an individual from the population given in Table 7.4 and have been told that he has no more than eight years of education (i.e., event A_1 has occurred). Under this *conditional* information, how do we assess the probability that the individual earns more than \$15,000/year? That is, what is $P(B_3|A_1)$? The fact that we desire information regarding the conditional event $(B_3|A_1)$ tells us that we are concerned only with the subpopulation of individuals having the attribute denoted by the event A_1, and thus we limit our consideration to these 300 individuals. Since we know that the randomly selected individual came from this subpopulation, we know that the total number of possible outcomes has been reduced to 300; and, of these 300, 10 have the attribute of interest, that is, B_3. Thus, simply by limiting consideration to the proper subpopulation and enumerating outcomes (the classical approach), we obtain

$$P(B_3|A_1) = \frac{10}{300} \tag{7.18}$$

There are many conditional probability situations in which the above type of reasoning will lead us conveniently to a correct answer. However, because there are many other situations in which it is very difficult, perhaps even impossible, to count cases (so to speak), we require a more general approach for obtaining conditional probabilities.

Rule 7.5 Let A and B be two events in a sample space S such that $P(B) > 0$. The conditional probability of event A given event B is

$$P(A|B) = \frac{P(A \cap B)}{P(B)} \tag{7.19}$$

It also follows that as long as $P(A) > 0$,

$$P(B|A) = \frac{P(A \cap B)}{P(A)} \tag{7.20}$$

For the case concerning the information in Table 7.4, we have

$$P(B_3 \cap A_1) = \frac{10}{1,000} \quad \text{and} \quad P(A_1) = \frac{300}{1,000} \tag{7.21}$$

Thus we can apply (7.20) to obtain

$$P(B_3|A_1) = \frac{P(B_3 \cap A_1)}{P(A_1)} = \frac{10/1,000}{300/1,000} = \frac{10}{300} \tag{7.22}$$

Sec. 7.5 / JOINT, MARGINAL, AND CONDITIONAL PROBABILITIES

The result is identical to that of (7.18), obtained by reducing consideration to the appropriate subpopulation and using the classical approach of counting cases.

In the previous example, the desired conditional probability is easy to obtain regardless of which method is used; however, such is not always the case. Some situations will arise for which the method involving direct enumeration in the appropriate subpopulation is much better suited, and some situations will occur for which the general formula (7.19) is most expedient. Unfortunately, which method to use for determining conditional probabilities is not always obvious; in fact, it is practically impossible to set forth general rules for choosing the proper method. Thus the reader should develop the habit of always considering the complications that may be involved in both methods before selecting one of them. The following examples are designed for two purposes: to enhance development of such a habit and to further clarify situations involving determination of conditional probabilities.

Example 7.12 Suppose we must conduct a survey among the employees of Essex Shoe Company. The 20 employees of the company can be categorized in the following table:

	M, Men	W, Women	Total
S, Salaried	4	3	7
H, Hourly	8	5	13
Total	12	8	20

The survey sample is to be chosen by simple random sampling without replacement. If we randomly select an individual and observe that we have selected a man, what is the probability that we have selected one who is salaried, that is, what is $P(S|M)$?

Solution Having the information of the table available, it makes very little difference which method we choose for calculating the conditional probability. $P(S|M)$ can be obtained by considering *only* the population of *men*. Thus the subpopulation of interest is the total of 12 men, 4 salaried and 8 hourly. Hence we obtain immediately $P(S|M) = \frac{4}{12}$. However, we can count cases to obtain the relevant joint and marginal probabilities to be

$$P(S \cap M) = \frac{4}{20} \quad \text{and} \quad P(M) = \frac{12}{20}$$

Thus Eq. (7.19) gives us

$$P(S|M) = \frac{P(S \cap M)}{P(M)} = \frac{4/20}{12/20} = \frac{4}{12}$$

The reader may wish to verify that either method will yield the following probabilities for other conditional events of Example 7.12:

$$P(H|M) = \tfrac{8}{12} \quad P(S|W) = \tfrac{3}{8} \quad P(M|S) = \tfrac{4}{7} \quad P(M|H) = \tfrac{8}{13}$$

$$P(H|W) = \tfrac{5}{8} \quad P(W|S) = \tfrac{3}{7} \quad P(W|H) = \tfrac{5}{13}$$

Example 7.13 Suppose a firm is going to randomly select 2 accountants to go to an accounting convention. The firm has 4 female accountants and 6 male accountants. Determine the probability of the second person chosen being a female given that a female was the first person chosen; that is find $P(F_2|F_1)$.

Solution In this type of problem, selection of the appropriate method is critical. Let us illustrate by again calculating the desired probability through use of both the methods of direct enumeration and formula (7.19). First note how easy it is to count cases (direct enumeration). To determine $P(F_2|F_1)$ by this method, we simply consider the subpopulation that remains after F_1 has occurred. Thus under the condition of having selected a female accountant on the first selection, the subpopulation from which we select the second person consists of the original 6 males and the 3 remaining female accountants. Since we are selecting a simple random sample, the probability of any specific accountant being obtained on the second selection is $\tfrac{1}{9}$. There are 3 female accountants that constitute mutually exclusive events for the second selection; thus we may count cases (or apply Postulate 7.2) to determine immediately

$$P(F_2|F_1) = \tfrac{3}{9} = \tfrac{1}{3} \qquad (7.23)$$

Now consider the excessive labor that is expended if we choose to use the method involving formula (7.19), which requires that we solve

$$P(F_2|F_1) = \frac{P(F_1 \cap F_2)}{P(F_1)}$$

Clearly, the marginal probability of event F_1 can be obtained easily (by counting cases) as $P(F_1) = \tfrac{4}{10}$. However, it is very difficult to obtain the joint probability $P(F_1 \cap F_2)$; for in order to count cases, we must use permutations. There are $_{10}P_2$ ways of selecting 2 females from the accountants. (Order is clearly important; so we must consider each accountant as a distinguishable person if we are to include all possible selections.) Thus our sample space consists of $_{10}P_2$ mutually exclusive events. Since we are employing simple random sampling, each of these events has equal probability of occurring, that is, $1/(_{10}P_2)$. Thus, to obtain $P(F_1 \cap F_2)$, we must determine how many of these $_{10}P_2$ events satisfy the event $F_1 \cap F_2$. Since there are 4 female accountants available, we obtain $_4P_2$ events that satisfy the event $F_1 \cap F_2$. Thus, by counting cases, we have

$$P(F_1 \cap F_2) = \frac{_4P_2}{_{10}P_2} = \frac{4!/2!}{10!/8!} = \frac{12}{90} = \frac{2}{15}$$

and finally

$$P(F_2|F_1) = \frac{P(F_1 \cap F_2)}{P(F_1)} = \frac{2/15}{4/10} = \frac{10}{30} = \frac{1}{3}$$

Although this answer is identical to (7.23), the solution is excessively laborious. However, the method involving formula (7.19) should not be dismissed completely because it represents a more generally applicable method than that of counting cases, particularly for certain practical problems.

Although it can hardly be called a rule, the calculations of Example 7.13 suggest that often it is expedient initially to try to count cases in problems similar to the one above; however, it is also necessary that extreme caution be exercised in order to avoid oversimplification. One of the reasons why we were able to count cases so easily in Example 7.13 is that we were considering the events in logical order; but what would we do if for some reason we were required to calculate $P(F_1|F_2)$? In this case, apparently it is impractical to attempt to solve the problem by reducing consideration to a subpopulation and counting cases. Furthermore, evaluation of the formula

$$P(F_1|F_2) = \frac{P(F_1 \cap F_2)}{P(F_2)}$$

is even more difficult than in Example 7.13 because the calculation of $P(F_2)$ is not completely straightforward. Thus, in this instance, neither method is expedient. In subsequent sections, however, we shall see that the problem can be solved quite easily through the use of another approach.

Example 7.14 The joint probabilities in the following table have been calculated from the information of the table presented in conjunction with Example 7.12. Using these probabilities and Rules 7.4 and 7.5, find the probabilities

(a) $P(M)$ (b) $P(W)$ (c) $P(S)$ (d) $P(H)$
(e) $P(S|M)$ (f) $P(H|M)$ (g) $P(S|W)$ (h) $P(H|W)$
(i) $P(M|S)$ (j) $P(W|S)$ (k) $P(M|H)$ (l) $P(W|H)$

	M	W
S	$\frac{4}{20}$	$\frac{3}{20}$
H	$\frac{8}{20}$	$\frac{5}{20}$

Solution

(a) $P(M) = P(M \cap S) + P(M \cap H) = \frac{4}{20} + \frac{8}{20} = \frac{12}{20}$

(b) $P(W) = P(W \cap S) + P(W \cap H) = \frac{3}{20} + \frac{5}{20} = \frac{8}{20}$

(c) $P(S) = P(M \cap S) + P(W \cap S) = \frac{4}{20} + \frac{3}{20} = \frac{7}{20}$

(d) $P(H) = P(M \cap H) + P(W \cap H) = \frac{8}{20} + \frac{5}{20} = \frac{13}{20}$

(e) $P(S|M) = \frac{P(M \cap S)}{P(M)} = \frac{4/20}{12/20} = \frac{4}{12}$

(f) $P(H|M) = \frac{P(M \cap H)}{P(M)} = \frac{8/20}{12/20} = \frac{8}{12}$

(g) $P(S|W) = \frac{P(W \cap S)}{P(W)} = \frac{3/20}{8/20} = \frac{3}{8}$

(h) $P(H|W) = \frac{P(W \cap H)}{P(W)} = \frac{5/20}{8/20} = \frac{5}{8}$

(i) $P(M|S) = \frac{P(M \cap S)}{P(S)} = \frac{4/20}{7/20} = \frac{4}{7}$

(j) $P(W|S) = \frac{P(W \cap S)}{P(S)} = \frac{3/20}{7/20} = \frac{3}{7}$

(k) $P(M|H) = \frac{P(M \cap H)}{P(H)} = \frac{8/20}{13/20} = \frac{8}{13}$

(l) $P(W|H) = \frac{P(W \cap H)}{P(H)} = \frac{5/20}{13/20} = \frac{5}{13}$

EXERCISES

In all exercises that involve sampling, assume that simple random sampling is used unless stated otherwise.

1. Define *conditional probability*, and give an example.

2. Define *joint probability*, and give an example.

3. The following is a wage classification of the employees of the Ramjet Company:

Wage	Division A	B	C
W1	100	900	1,000
W2	150	700	200
W3	250	400	200
W4	500	500	100

    (a) State what the event $(B \cup W3)$ represents.
    (b) Calculate $P(B \cup W3)$.
    (c) State what the event $(A \cap W2)$ represents.
    (d) Calculate $P(A \cap W2)$.
    (e) State what the event $(C|W1)$ represents.
    (f) Calculate $P(C|W1)$.

Sec. 7.5 / JOINT, MARGINAL, AND CONDITIONAL PROBABILITIES

(g) State what the probability $1 - P(B \cap W4)$ represents.
(h) Calculate $1 - P(B \cap W4)$.

4. The following is a classification of the positions held by the employees of a company:

Position	Female	Male
Executive	1	7
Staff	12	18
Manager	6	16
Hourly	151	189

(a) State what the event (Manager ∩ Female) represents.
(b) Calculate $P(\text{Manager} \cap \text{Female})$.
(c) State what the event (Executive ∪ Staff) represents.
(d) Calculate $P(\text{Executive} \cup \text{Staff})$.
(e) State what the event (Female | Manager) represents.
(f) Calculate $P(\text{Female} | \text{Manager})$.
(g) State what the event (Hourly) represents.
(h) Calculate $P(\text{Hourly})$.

5. The following is a classification of the number of parts produced at each plant in one day:

Plant	A	B	C	D
P1	100	80	90	150
P2	90	110	30	80
P3	60	70	90	140
P4	120	90	130	40

Calculate the probability that a randomly selected part

(a) Is produced by $P1$ or $P4$
(b) From $P3$ is part A
(c) Is from $P4$ given that it is part D
(d) Is from $P2$ and is part D
(e) Is part C
(f) Is from $P4$ given that it is part C
(g) Is either from $P3$ or is part B
(h) Is not both from $P1$ and part D

6. Using the data in Exercise 3, determine the marginal probabilities for each division and for each wage classification.

7. A group of students at Parkhurst Junior College consists of 200 boys and 300 girls. One-third of the girls and one-fourth of the boys are freshmen and the remainder are sophomores. If a student is chosen at random, determine the probability that

(a) The student is a freshman boy

(b) The student is either a freshman or a boy
(c) The student is either a sophomore or a girl
(d) The student is a sophomore if we know that he is a boy
(e) The student is a girl if we know she is a freshman

8. Two senatorial candidates are randomly selected from a group of candidates consisting of 5 Democrats and 3 Republicans.

   (a) Determine the probability that the second candidate is a Republican if the first one is a Democrat.
   (b) Determine the probability that the second candidate is a Republican.
   (c) Determine the probability that the first candidate is a Republican if the second one is a Democrat.

9. The Dixie Discount Supermarket is planning a sale of canned vegetables. The cans were damaged to the extent that all labels were removed. However, it is known that there are 200 cans of peas, 300 cans of green beans, 400 cans of sweet corn, and 100 cans of beets. You are the first shopper to select a can.

   (a) Determine the probability of selecting a can of corn.
   (b) Determine the probability of not getting a can of beets.
   (c) Are the two events "corn" and "green beans" mutually exclusive?
   (d) Determine the probability of getting a can of green beans on your second selection if you got beets on the first selection.

10. The group of patients at the Golden Acres Rest Home is $\frac{1}{2}$ male and $\frac{1}{2}$ female. Furthermore, $\frac{1}{10}$ of all male patients play checkers, and $\frac{1}{20}$ of the female patients play checkers. For a randomly selected patient, determine the probabilities for each of the following events:

    (a) The patient is a male checker player.
    (b) The patient is a female who does not play checkers.
    (c) The patient is a checker player.
    (d) The patient is not a checker player.
    (e) The patient is a male if not a checker player.

7.6 MULTIPLICATION RULE

In Example 7.13 we have demonstrated how difficult it is in some situations to compute joint probabilities through the classical approach. However, there exists a rule, known as the *multiplication rule for probabilities*, that provides another (and often more expedient) means of computing joint probabilities in such situations.

Rule 7.6 For any events A and B

$$P(A \cap B) = P(A)P(B|A)$$

and (7.24)

$$P(A \cap B) = P(B)P(A|B)$$

Sec. 7.6 / MULTIPLICATION RULE

The first equality in (7.24) is obtained simply by solving Eq. (7.20) for $P(A \cap B)$, while the second is obtained by solving (7.19) for $P(A \cap B)$. Thus (as the reader was warned previously) it is obvious that it is necessary to view (7.19), (7.20), and the two equalities of (7.24), *not as formulas*, but as *relations* involving joint, marginal and conditional probabilities. In a sense, the relationship is circular; that is, for any given problem we cannot use either of the relationships to solve for a desired probability until we have employed some other approach to arrive at values for other required probabilities.

Example 7.15 Suppose the firm in Example 7.13 is going to randomly select 2 accountants to go to an accounting convention. The firm has 4 female (F) accountants and 6 male (M) accountants. Use Rule 7.6 to find the following joint probabilities:

(a) $P(F_1 \cap F_2)$ (b) $P(F_1 \cap M_2)$ (c) $P(M_1 \cap F_2)$ (d) $P(M_1 \cap M_2)$

where the subscripts 1 and 2 represent the order of selection.

Solution We obtain the marginal probabilities $P(F_1)$ and $P(M_1)$ by considering the problem of selecting only one person from the original population; thus

$$P(F_1) = \frac{2}{5} \quad \text{and} \quad P(M_1) = \frac{3}{5}$$

We obtain relevant conditional probabilities by considering only the appropriate conditional populations and counting cases, as in Example 7.13. Hence

$$P(F_2 | F_1) = \frac{1}{3} \quad P(M_2 | F_1) = \frac{2}{3}$$

$$P(F_2 | M_1) = \frac{4}{9} \quad P(M_2 | M_1) = \frac{5}{9}$$

Using this information, we find the required joint probabilities:

(a) $P(F_1 \cap F_2) = P(F_1)P(F_2|F_1) = \frac{2}{5} \cdot \frac{1}{3} = \frac{2}{15}$

(b) $P(F_1 \cap M_2) = P(F_1)P(M_2|F_1) = \frac{2}{5} \cdot \frac{2}{3} = \frac{4}{15}$

(c) $P(M_1 \cap F_2) = P(M_1)P(F_2|M_1) = \frac{3}{5} \cdot \frac{4}{9} = \frac{4}{15}$

(d) $P(M_1 \cap M_2) = P(M_1)P(M_2|M_1) = \frac{3}{5} \cdot \frac{5}{9} = \frac{5}{15}$

Note that the four joint events are mutually exclusive and exhaustive; thus, as is to be expected, the sum of their probabilities is 1.

In determining probabilities of joint events for events occurring in a natural sequence, such as in Example 7.15, it is sometimes convenient to represent the probabilities in a *tree diagram*, as illustrated in Fig. 7.1, where each branch of the tree represents a possible outcome at that particular point and the numbers on each branch represent the probability of that particular event. The probability of being at the end of a particular branch is simply the product of the probabilities on the path that was traveled to get there. However, for more complicated events, tree diagrams become impractical.

Note also that we could have determined $P(F_1 \cap F_2)$, for example, by

$$P(F_1 \cap F_2) = P(F_2) P(F_1 | F_2)$$

However, with a bit of common sense and a little reasoning it is quite clear that this is *not* the relation to use in this particular instance because it requires that we obtain $P(F_2)$ and $P(F_1|F_2)$, and neither of these tasks is easy in Example 7.15. Let us pursue this area in a more general context, and, in so doing, we shall be able to summarize the operating rules that we have treated so far.

Let us consider *all* marginal, joint, and conditional probabilities that might be of interest in Example 7.15 and see if a logical approach can be developed for deciding which methods and/or relations should be used to solve for each probability. The probabilities are

Marginal: $P(F_1), P(M_1)$ (7.25)

Conditional: $P(F_2|F_1), P(F_2|M_1), P(M_2|F_1), P(M_2|M_1)$ (7.26)

Joint: $P(F_1 \cap F_2), P(F_1 \cap M_2), P(M_1 \cap F_2), P(M_1 \cap M_2)$ (7.27)

Marginal: $P(F_2), P(M_2)$ (7.28)

Conditional: $P(F_1|F_2), P(M_1|F_2), P(F_1|M_2), P(M_1|M_2)$ (7.29)

The events are grouped in this particular fashion for a specific reason: The nature of the problem is such that the grouping represents the logical order of solution. Why? Let us answer this by working backward through the grouping.

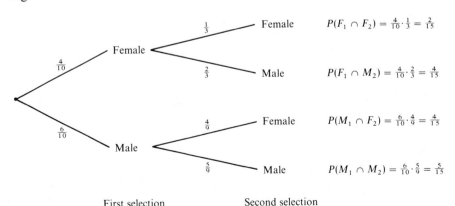

FIGURE 7.1

Sec. 7.6 / MULTIPLICATION RULE

1. It is risky to rely on the intuition necessary to obtain the conditional probabilities of (7.29) by counting cases. (The reader should be sure he/she agrees with and understands this statement.) Thus we must use formula (7.19) for conditional probabilities; however, this will require that we know the marginal *and* joint probabilities of (7.28) and (7.27), respectively.
2. It is not practical (again, intuition is risky) to count cases to obtain the marginal probabilities of (7.28). Hence we require the use of formula (7.17), which necessarily entails the joint probabilities of (7.27).
3. Although we *can* count cases (using permutations, as in Example 7.13) to obtain the joint probabilities of (7.27), it is much easier to employ the rule for multiplication of probabilities (7.24); however, this requires the conditional *and* marginal probabilities of (7.26) and (7.25), respectively.
4. We can determine the conditional probabilities of (7.26) simply by reducing consideration to the appropriate subpopulations and by counting cases.
5. It is very easy to obtain the marginal probabilities of (7.25) by counting cases in the original population.

As an illustration, consider the case where we desire $P(M_1|F_2)$. We cannot count cases, and so we use (7.19) to obtain

$$P(M_1|F_2) = \frac{P(M_1 \cap F_2)}{P(F_2)} \qquad (7.30)$$

But, by (7.17),

$$P(F_2) = P(F_1 \cap F_2) + P(M_1 \cap F_2) \qquad (7.31)$$

and, by (7.24),

$$P(F_1 \cap F_2) = P(F_1)P(F_2|F_1) \qquad (7.32)$$

and

$$P(M_1 \cap F_2) = P(M_1)P(F_2|M_1) \qquad (7.33)$$

Combining (7.30) to (7.33), we have

$$P(M_1|F_2) = \frac{P(M_1 \cap F_2)}{P(F_2)} = \frac{P(M_1 \cap F_2)}{P(F_1 \cap F_2) + P(M_1 \cap F_2)}$$

$$= \frac{P(M_1)P(F_2|M_1)}{P(F_1)P(F_2|F_1) + P(M_1)P(F_2|M_1)} \qquad (7.34)$$

Now each of the probabilities in the last term of (7.34) can be found by counting cases, and they are

$$P(M_1) = \frac{6}{10} \qquad P(F_2|M_1) = \frac{4}{9} \qquad P(F_1) = \frac{4}{10} \qquad P(F_2|F_1) = \frac{3}{9} \qquad (7.35)$$

Substituting the values of (7.35) into (7.34), we obtain finally

$$P(M_1|F_2) = \frac{(6/10)(4/9)}{(4/10)(3/9) + (6/10)(4/9)} = \frac{24/90}{36/90} = \frac{2}{3} \qquad (7.36)$$

It may be somewhat disconcerting to the reader to note that $P(M_1|F_2)$ turns out to be exactly what we would have obtained by reasoning intuitively in the following manner: We have selected one accountant without noting whether the person was male or female, and on the second selection we selected a female accountant. This means that even though one of the accountants cannot be selected, there remain 9 accountants (a subpopulation): 6 of these are male, and 3 are female. Thus, intuitively, the accountant we selected on the first selection has 6 equally likely chances out of 9 of being male; hence $P(M_1|F_2) = \frac{2}{3}$. However, even though the correct value for the probability of interest could have been obtained intuitively in this instance, the reasoning process described in the previous paragraph is valuable for at least two major reasons: First it will provide valid results in problems that are so complicated that the intuitive process breaks down completely; second, and perhaps even more important, it illustrates the logical approach to general probability problems involving formulas and rules that are circular.

Independence From the above illustration, it should be evident that the probability of the occurrence of a specific event will change as the conditional information regarding another event changes. For example,

$$P(F_2|F_1) = \frac{1}{3} \neq P(F_2|M_1) = \frac{4}{9}$$

In this case, the probabilities *depend* upon the given condition, or the events are *dependent*. When this is *not* the case, we have *independent events*, which are defined as follows:

Definition 7.13 Let A and B be two events in a sample space S. The events are independent if any of the following is satisfied:

$$P(A|B) = P(A) \qquad (7.37)$$
$$P(B|A) = P(B) \qquad (7.38)$$

or

$$P(A \cap B) = P(A)P(B) \qquad (7.39)$$

It can be shown that if any one of the relationships in Definition 7.13 is satisfied, both of the other two relationships will also be satisfied; thus, if we wish to check as to whether or not two events are independent, we need test only for one of the three conditions. Likewise, if any one of the three relationships is *not* satisfied, neither of the two remaining relationships will be satisfied. The proof of this is left as an exercise for the reader.

In addition to providing a means for *checking* the independence of two events, in the situation where we know that the events under consideration are independent, relation (7.39) allows a simpler solution for joint probabilities than the general multiplication rule (Rule 7.6). However, if there is any doubt regarding dependence, Rule 7.6 should always be employed because it will give the correct answer whether or not the events are independent.

Let us consider several examples that demonstrate the use of Definition 7.13.

Example 7.16 We have received an order containing 6 good parts (G) and 4 bad parts (B) from which we wish to select a simple random sample of size 2 *with replacement*. Determine whether or not two events B_1 and G_2 are independent.

Solution We need to test to see if $P(G_2) = P(G_2 | B_1)$. Knowing that we are sampling *with* replacement, we can determine the appropriate marginal and conditional probabilities by considering only the population from which we shall sample. Thus

$$P(G_2) = \frac{3}{5} \quad \text{and} \quad P(G_2 | B_1) = \frac{3}{5}$$

Hence we have

$$P(G_2 | B_1) = \frac{3}{5} = P(G_2)$$

and, by (7.38), the two events are independent (that is, if A and B represent the outcomes of the first and second draws, respectively).

The reader should apply the reasoning just described to be satisfied that, under the given sampling scheme, the two events are *independent* for each of the following: B_1 and B_2; G_1 and B_2; and G_1 and G_2.

Example 7.17 We have received an order containing 6 good parts (G) and 4 bad parts (B) from which we wish to select a simple random sample size 2 *without replacement*. Determine whether or not the two events B_1 and G_2 are independent.

Solution Since we are sampling *without* replacement, we can determine the appropriate marginal and conditional probabilities in the following manner. We reduce consideration to the appropriate subpopulation and count cases to obtain

$$P(G_2 | B_1) = \frac{2}{3}$$

Then we employ (7.17) and (7.24) to obtain

$$P(G_2) = P(G_1 \cap G_2) + P(B_1 \cap G_2) = P(G_1)P(G_2|G_1) + P(B_1)P(G_2|B_1)$$

$$= \left(\frac{6}{10}\right)\left(\frac{5}{9}\right) + \left(\frac{4}{10}\right)\left(\frac{6}{9}\right)$$

$$= \frac{30}{90} + \frac{24}{90} = \frac{54}{90} = \frac{3}{5}$$

Hence, by testing for (7.38), we have

$$P(G_2|B_1) = \frac{2}{3} \neq P(G_2) = \frac{3}{5}$$

and thus the two events are *not* independent.

Again the reader should be satisfied that, under the given sampling scheme, the two events are *dependent* for each of the following: B_1 and B_2; G_1 and B_2; and G_1 and G_2.

Example 7.18 Both Adams Construction Company and Biltrite, Inc., are bidding on the same contracts for each of two customers, customers 1 and 2. Assuming that (1) each company has an equal chance of winning each contract, (2) the two companies are the only bidders, and (3) the contract awards are independent events, determine (a) the probability that Adams Construction Company wins both contracts and (b) the probability that Biltrite, Inc., wins only one contract.

Solution Letting A and B represent the events of Adams Construction Company and Biltrite, Inc., respectively, receiving the contract and the subscript denote the customer number, we obtain

$$P(A_1) = \frac{1}{2} \qquad P(A_2) = \frac{1}{2} \qquad P(B_1) = \frac{1}{2} \qquad P(B_2) = \frac{1}{2}$$

Thus, using the independence property of the contract awards and (7.39), we find

(a) $P(A \text{ wins both contracts}) = P(A_1 \cap A_2) = P(A_1)P(A_2) = (\frac{1}{2})(\frac{1}{2}) = \frac{1}{4}$

(b) $P(B \text{ wins only one contract}) = P[(B_1 \cap A_2) \cup (A_1 \cap B_2)]$

$$= P(B_1 \cap A_2) + P(A_1 \cap B_2)$$
$$= P(B_1)P(A_2) + P(A_1)P(B_2)$$
$$= (\frac{1}{2})(\frac{1}{2}) + (\frac{1}{2})(\frac{1}{2})$$
$$= \frac{1}{4} + \frac{1}{4} = \frac{1}{2}$$

The tree diagram for all possible outcomes is given in Fig. 7.2.

Sec. 7.6 / MULTIPLICATION RULE

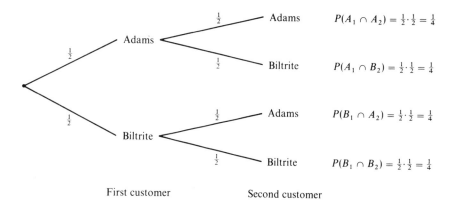

First customer   Second customer

FIGURE 7.2

EXERCISES

In all exercises that involve sampling, assume that simple random sampling is used unless stated otherwise.

1. Define *independent events*.

2. Events E_1 and E_2 are independent. Does this imply that
   (a) $P(E_2 | E_1) = 0$
   (b) $P(E_1 \cap E_2) = 0$
   (c) $P(E_1 | E_2) = P(E_1)$
   (d) $P(E_1 \cup E_2) > 0$
   (e) $P(E_1 \cap E_2) = P(E_1)P(E_2)$

3. When is $P(A | B)$ not equal to $P(A \cap B)/P(B)$?

4. If $P(A | B) = \frac{1}{2}$ and $P(B) = \frac{2}{3}$, what is $P(A \cap B)$?

5. If $P(A \cap B) = \frac{1}{6}$ and $P(A | B) = \frac{1}{5}$, what is $P(B)$?

6. If $P(A) = \frac{1}{4}$, $P(B) = \frac{1}{3}$, and $P(B | A) = \frac{2}{3}$, what is $P(A | B)$?

7. A shipment contains 7 red shirts and 8 white shirts. Use Rule 7.6 to determine the following joint probabilities when selecting a simple random sample of size 2 *without replacement* and draw a tree diagram:
   (a) $P(R_1 \cap R_2)$ (b) $P(R_1 \cap W_2)$
   (c) $P(W_1 \cap R_2)$ (d) $P(W_1 \cap W_2)$

8. For the shipment given in Exercise 7, use Rule 7.6 to determine the following joint probabilities when selecting a sample of size 3 *without replacement* and draw a tree diagram:
   (a) $P(R_1 \cap R_2 \cap R_3)$ (b) $P(R_1 \cap R_2 \cap W_3)$

(c) $P(R_1 \cap W_2 \cap R_3)$ (d) $P(R_1 \cap W_2 \cap W_3)$
(e) $P(W_1 \cap R_2 \cap R_3)$ (f) $P(W_1 \cap R_2 \cap W_3)$
(g) $P(W_1 \cap W_2 \cap R_3)$ (h) $P(W_1 \cap W_2 \cap W_3)$
(i) $P(R_1 \cap R_2)$ (j) $P(R_1 \cap W_2)$
(k) $P(W_1 \cap R_2)$ (l) $P(W_1 \cap W_2)$

9. Rework Exercise 7 when selecting a simple random sample of size 2 *with replacement*.

10. Rework Exercise 8 when selecting a simple random sample of size 3 *with replacement*.

11. The probabilities of a customer purchasing products A and B are $\frac{1}{3}$ and $\frac{1}{2}$, respectively. The probability of a customer purchasing B if he or she has purchased A is $\frac{1}{4}$.

(a) Determine whether or not the two events A and B are independent.
(b) Calculate the probability of purchasing A after B has been purchased.

12. The probabilities of machines X, Y, and Z working are $\frac{7}{8}$, $\frac{3}{4}$, and $\frac{9}{10}$, respectively. Assume that whether or not one machine works is independent of the others working.

(a) Calculate the probability that all three machines work.   .590625
(b) Calculate the probability of X and Y working and Z not working.   .065625
(c) Calculate the probability of Y working and X and Z not working.   .009375
(d) Calculate the probability that all three machines do not work.   .003125

13. A manager has been observing the absentee rate of two employees who work together. The probability of Betty missing work on any given day is .2, and the probability of Mary being absent is .15. However, if Betty is absent on one day, the probability of Mary being absent the next day is .25. Calculate the probability of Betty being absent today if Mary was not at work yesterday.

14. The Federal Aviation Administration has found that the probability of a person passing the examination for a private pilot's license on the first try is .70. The probability that an individual will pass on the second try is .75, and the probability of passing on the third try if failing on the previous two is .60. Determine the probability that a person

(a) Fails all three tests   .03
(b) Fails both of the first two tests   .075
(c) Requires no more than two tests   .475
(d) Requires at least two tests   .525

15. John Scott, an insurance salesman, figures that the probability that he can consumate a sale on the first contact with a client is .50. However, if the client fails to buy insurance on the first contact, the probability increases to .65 on the second contact.

(a) Determine the probability that the client will buy insurance from Mr. Scott.
(b) Determine the probability that he will not buy insurance.

16. The Meinhart Electronics Company is considering bidding on a government contract. Its subjective probability of winning the contract is $\frac{1}{2}$ if Miller Instruments,

Inc. does not submit a bid; however, the probability they will get the contract if Miller Instruments does submit a bid is $\frac{1}{3}$. The probability of Miller Instruments submitting a bid is $\frac{2}{5}$.

(a) Determine the probability of Miller Instruments submitting a bid and Meinhart Electronics winning the contract.
(b) Determine the probability of Miller Instruments not submitting a bid and Meinhart Electronics winning the contract.
(c) Determine the probability of Meinhart Electronics winning the bid.

17. Each of two groups of six employees in a machine shop contains exactly 2 college students. If 2 employees are selected from each group,

(a) Determine the probability that all 4 are college students
(b) Determine the probability that exactly 3 are college students
(c) Determine the probability that at most 2 are college students

7.7 ADDITION RULE

Postulate 7.2 is applicable for determining the probability of the event $A \cup B$ *only* if events A and B are *mutually exclusive*. If A and B are not mutually exclusive events, that is, $A \cap B \neq \emptyset$, then $P(A \cup B)$ can be determined by using the classical approach. That is, $P(A \cup B)$ is the ratio of the total number of outcomes that satisfy $A \cup B$ to the total number of outcomes in S. However, the number of outcomes in $A \cup B$ is equal to the number of outcomes in A plus the number of outcomes in B minus the number of outcomes common to both, i.e., the number in $A \cap B$. We must subtract the number of outcomes in $A \cap B$ because we have counted them twice: once in A and once in B. Thus, to determine $P(A \cup B)$ when events A and B are not mutually exclusive, we use the *addition rule for probabilities*:

Rule 7.7 For any two events A and B

$$P(A \cup B) = P(A) + P(B) - P(A \cap B) \qquad (7.40)$$

Postulate 7.2 [Eq. (7.9)] is a special case of Rule 7.7 because $P(A \cap B) = 0$ if A and B are mutually exclusive events. However, it should be noted that if we know that $P(A \cap B) = 0$, we *do not necessarily* know that A and B are mutually exclusive events. Such a situation will be illustrated in Sec. 7.10.

Since the events in probability are sets of outcomes, it is occasionally helpful to view the occurrence of events by using Venn diagrams. If we view the events A and B as sets in the Venn diagram of Fig. 7.3 and recall that the probability of an event is the ratio of the number of outcomes satisfying an event to the total number of possible outcomes, Rule 7.7 becomes clearer. That is, we see that the reason for subtracting $P(A \cap B)$ on the right side of (7.40) is that the number of outcomes satisfying $A \cap B$ have been included in both the outcomes in A and the outcomes in B; thus those outcomes have been counted twice. Furthermore, if the events A and B are mutually

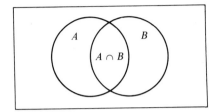

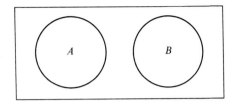

FIGURE 7.3   **FIGURE 7.4**

exclusive, the events may be represented by the Venn diagrams of Fig. 7.4, which illustrates that $P(A \cap B) = 0$.

Example 7.19   The 1,000 current loans of the Certainteed Loan Company are distributed according to the breakdown given below:

	P, Personal	B, Business
H, High risk	260	40
M, Medium risk	300	100
L, Low risk	140	160

Suppose an auditor is going to examine the loan company records to determine if the finance firm for which he works will supply the loan company with more money to lend to its clients. The auditor will base his decision on one loan record selected randomly from the files of the loan company. If the loan is either a business loan or a low-risk personal loan, the company will receive more money for lending purposes; otherwise, it will not receive more money. Find the probability that the loan company (a) *will* receive more money and (b) *will not* receive more money.

Solution   Since the auditor is randomly selecting a record from the files, each individual record has an equal chance of being selected. Thus, from the table that is given, we can compute the joint probabilities

$$P(H \cap P) = .260 \quad P(H \cap B) = .040$$
$$P(M \cap P) = .300 \quad P(M \cap B) = .100$$
$$P(L \cap P) = .140 \quad P(L \cap B) = .160$$

(a) To solve the first problem, we can use Rule 7.7; that is,

$$P(\text{money}) = P(L \cup B) = P(L) + P(B) - P(L \cap B) \tag{7.41}$$

However, we must determine $P(L)$ and $P(B)$, and so we use Rule 7.4 to obtain

$$P(L) = P(L \cap P) + P(L \cap B)$$
$$= .140 + .160 = .300 \tag{7.42}$$

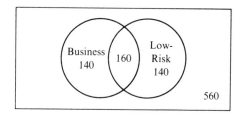

FIGURE 7.5

and

$$P(B) = P(H \cap B) + P(M \cap B) + P(L \cap B)$$
$$= .040 + .100 + .160 = .300 \tag{7.43}$$

Therefore, combining (7.41) to (7.43), we have

$$P(\text{money}) = .300 + .300 - .160 = .440$$

Viewing the problem through the use of Venn diagrams, we can represent the loans as business loans versus personal loans and as low-risk loans versus high- and medium-risk loans. Thus, in the Venn diagram of Fig. 7.5, we represent the number of loans in each category when classified in the above manner. Using the number in each category, we have

$$P(\text{money}) = P(L \cup B) = \frac{140 + 160 + 140}{1{,}000} = \frac{440}{1{,}000} = .440$$

(b) To solve the second problem we consider two alternatives: First we can apply Rule 7.3 to obtain

$$P(\text{no money}) = 1 - P(\text{money}) = 1 - .440 = .560$$

Second we can consider the event of "no money" as being satisfied if we select either the event $H \cap P$ or the event $M \cap P$. These two events are mutually exclusive, and hence

$$P(\text{no money}) = P[(H \cap P) \cup (M \cap P)]$$
$$= P(H \cap P) + P(M \cap P)$$
$$= .260 + .300 = .560$$

Either of these two alternatives can be used in conjunction with Fig. 7.5 to determine that there are 560 satisfactory outcomes out of a possible 1,000 outcomes that result in the company not receiving more money.

We have stated previously that generally there are several correct ways of solving a probability problem, which allows sufficient means for checking a solution. If we employ two approaches and do not arrive at the same solution, we must conclude that our "logic" (that is, common sense) is in error for at least one of the methods. Let us consider another example for which there are at least two alternative methods of solution.

Example 7.20 Suppose we have a shipment containing 6 good (G) transistors and 4 bad (B) transistors from which we wish to select a simple random sample of size 2 *with replacement*. Determine the probability of selecting a good transistor on the first draw or a bad transistor on the second draw, that is, $P(G_1 \cup B_2)$.

Solution In our first approach, we use Rules 7.6 and 7.7 to obtain

$$P(G_1 \cup B_2) = P(G_1) + P(B_2) - P(G_1 \cap B_2)$$
$$= P(G_1) + P(B_2) - P(G_1)P(B_2|G_1)$$
$$= P(G_1) + P(B_2) - P(G_1)P(B_2)$$
$$= \frac{3}{5} + \frac{2}{5} - \left(\frac{3}{5}\right)\left(\frac{2}{5}\right)$$
$$= \frac{19}{25}$$

(From Example 7.16, note that G_1 and B_2 are independent.)

In our second approach, let us consider the four mutually exclusive and exhaustive joint events $B_1 \cap B_2, B_1 \cap G_2, G_1 \cap B_2$, and $G_1 \cap G_2$. Since the event $G_1 \cup B_2$ is satisfied by the occurrence of any one of the events $B_1 \cap B_2, G_1 \cap B_2$, or $G_1 \cap G_2$, we may apply Postulate 7.2, obtaining

$$P(G_1 \cup B_2) = P[(B_1 \cap B_2) \cup (G_1 \cap B_2) \cup (G_1 \cap G_2)]$$
$$= P(B_1 \cap B_2) + P(G_1 \cap B_2) + P(G_1 \cap G_2) \quad (7.44)$$

However, applying (7.24) to each of the three terms of (7.44), we have

$$P(B_1 \cap B_2) + P(G_1 \cap B_2) + P(G_1 \cap G_2)$$
$$= P(B_1)P(B_2|B_1) + P(G_1)P(B_2|G_1) + P(G_1)P(G_2|G_1)$$
$$= \left(\frac{2}{5}\right)\left(\frac{2}{5}\right) + \left(\frac{3}{5}\right)\left(\frac{2}{5}\right) + \left(\frac{3}{5}\right)\left(\frac{3}{5}\right)$$
$$= \frac{4}{25} + \frac{6}{25} + \frac{9}{25}$$
$$= \frac{19}{25} \quad (7.45)$$

This second approach allows an even easier solution. Since the only one of the mutually exclusive and exhaustive joint events that does *not* satisfy $G_1 \cup B_2$ is $B_1 \cap G_2$, we *could* apply Rule 7.3 to obtain

$$P(G_1 \cup B_2) = 1 - P(B_1 \cap G_2)$$
$$= 1 - P(B_1)P(G_2|B_1)$$
$$= 1 - \left(\frac{2}{5}\right)\left(\frac{3}{5}\right)$$
$$= \frac{19}{25}$$

Sec. 7.7 / ADDITION RULE

EXERCISES

In all exercises that involve sampling, assume that simple random sampling is used unless stated otherwise.

1. Events A, B, and C are defined as follows:

$$A = \{1, 3, 5, 6\} \quad B = \{2, 4, 6\} \quad C = \{1, 3, 7\}$$

   Draw a Venn diagram representing the three sets.
   - (a) Find $P(A \cup B)$.
   - (b) Find $P(A \cap B)$.
   - (c) Find $P(A \cup C)$.
   - (d) Find $P(A \cap C)$.
   - (e) Find $P(B \cap C)$.
   - (f) Find $P[A \cup (B \cup C)]$.
   - (g) Find $P[A \cap (B \cup C)]$.
   - (h) Find $P[(A \cup C) \cap (A \cup B)]$.
   - (i) Are A and B mutually exclusive?
   - (j) Are B and C mutually exclusive?

2. Events A, B, and C are independent and are not mutually exclusive:

$$P(A) = \frac{1}{2} \quad P(B) = \frac{2}{3} \quad P(C) = \frac{1}{3}$$

   - (a) Find $P(A \cap B \cap C)$.
   - (b) Find $P(A \cup C)$.
   - (c) Find $P(A \cup B \cup C)$.
   - (d) Find $P(A \cap C')$.
   - (e) Find $P[(A \cap C) \cup B']$.
   - (f) Find the probability that at least one of the three events will occur.

3. Events E_1 and E_2 are mutually exclusive. Does this imply that
   - (a) $P(E_2 | E_1) = 0$
   - (b) $P(E_1 \cap E_2) = 0$
   - (c) $P(E_2 \cup E_1) = P(E_2) + P(E_1)$
   - (d) $P(E_1 \cup E_2) = 0$
   - (e) $P(E_2 \cap E_1) = P(E_2) P(E_1)$

4. If a fair die is rolled, what is the probability of the outcome being the following:
   - (a) 3 or more
   - (b) 5 or less
   - (c) Greater than or equal to 4 but less than or equal to 2
   - (d) Greater than 3 but less than 5

5. If a pair of fair dice is tossed, what is the probability that the total number of spots facing upward is as follows:
   - (a) 2
   - (b) 3
   - (c) 4
   - (d) 5
   - (e) 6
   - (f) 7
   - (g) 8
   - (h) 9

(i) 10   (j) 11   (k) 12   (l) {2, 3, 12}
(m) {7, 11}   (n) {4, 10}   (o) {5, 9}   (p) {6, 8}
(q) {2, 3, 7, 11, 12}   (r) {4, 5, 6, 8, 9, 10}

6. The orders of a particular company have been classified as follows:

	Region			
Amount	A	B	C	D
I	50	100	50	10
J	100	50	75	25
K	75	30	100	75
L	75	20	25	140

(a) State what $P(K|C)$ represents.
(b) Calculate $P(K|C)$ by counting cases.
(c) Compute $P(K|C)$ indirectly by using the appropriate probability rules. [Compare the results with your answer to part (b) and check to see that they are equal.]
(d) State what $P(J \cup D)$ represents.
(e) Find $P(J \cup D)$ by counting cases.
(f) Compute $P(J \cup D)$ indirectly by using the appropriate probability rules. [Compare the result with your answer to part (e) and check to see that they are equal.]
(g) State what $P(L \cap A)$ represents.
(h) Compute $P(L \cap A)$ by counting cases.
(i) Calculate $P(L \cap A)$ indirectly by using the appropriate probability rules. [Compare the result with your answer to part (h) and check to see that they are equal.]

7. A marketing research team took a survey of the number of customers in each age group. The results of one week are:

	Age Group				
Product	Under 12	12–21	22–35	35–60	over 60
A	230	170	200	30	20
B	50	80	100	120	110
C	70	110	220	150	90
D	10	60	120	180	130

(a) State what $P(B \cup \text{over } 60)$ represents.
(b) Calculate $P(B \cup \text{over } 60)$ by counting cases.
(c) Compute $P(B \cup \text{over } 60)$ indirectly by using the appropriate probability rules.
(d) State what $P(A|\text{under } 12)$ represents.
(e) Find $P(A|\text{under } 12)$ by counting cases.

Sec. 7.7 / ADDITION RULE

(f) Compute $P(A|\text{under 12})$ indirectly by using the appropriate probability rules.

(g) State what $P(B \cap 35\text{–}60)$ represents.

(h) Find $P(B \cap 35\text{–}60)$ by counting cases.

(i) Compute $P(B \cap 35\text{–}60)$ indirectly by using the appropriate probability rules.

8. A company is going to send 3 managers to a special training session. The 3 managers are to be randomly selected from 10 managers in department A, 6 in department B, and 12 in department C.

(a) Compute the probability that all 3 managers will be from department A.

(b) Compute the probability that no one will be selected from department B.

(c) Compute the probability that a manager will be selected from A, B, and C in that order.

9. An automobile company obtained the following results from a recent poll:

		Preferred Car Size		
		Small	Medium	Large
A	Single	20	5	2
B	Married (No children)	25	12	5
C	Married (Has children)	5	18	8

Compute the following probabilities by using the appropriate rules.

(a) $P(A \cap \text{Medium})$

(b) $P(\text{Large}|C)$

(c) $P(B|\text{Medium})$

(d) $P(A \cup \text{Small})$

(e) $P(C \cap \text{Small})$

(f) $P(\text{Small}|C)$

10. The Middlesex Community Hospital has two physicians on call for emergencies. Due to their professional commitments and other demands on their time, the probability that a physician will be available for an emergency is $\frac{9}{10}$. The availability of one physician is independent of the other. Determine the probabilities for the following events:

(a) Both physicians will be available.

(b) At least one physician will be available.

(c) Neither physician will be available.

11. D & D Enterprises employs two salesmen who are assigned to all prospective clients. Joe Flishman is assigned to $\frac{2}{3}$ of all prospective clients and signs a sales contract with 40 percent of all contacts. However, Harold Thomas signs 60 percent of all his contacts. For a given prospective client, determine the probability of the following events.

(a) Joe gets him to sign a contract.

(b) Harold gets him to sign a contract.

(c) He signs a contract.

(d) If he signs a contract, Joe is the one who signs him.
(e) If he signs a contract, Harold is the one who signs him.
(f) He does not sign a contract.
(g) Either Joe signs him or Harold does not sign him.

7.8 PROBABILITY DISTRIBUTIONS

Probability distributions consist of the possible values of a *random* variable and the corresponding *probability* for each value (or group of values). In each of the examples in this chapter we have actually been using a probability distribution (or at least a portion of one). In no case can we make exact probability statements about specific events without having the probability distribution of the population or at least the portion of the distribution related to the events concerning the probability statements. (Both the truth of this statement and the importance of probability distributions are more apparent in statistics.)

Example 7.21 Consider the experiment of tossing a fair die. Let X be the random variable that has for its possible values the number of spots showing on the top side of the die. The probability distribution for the random variable X is

x	$P(X=x)$
1	$\frac{1}{6}$
2	$\frac{1}{6}$
3	$\frac{1}{6}$
4	$\frac{1}{6}$
5	$\frac{1}{6}$
6	$\frac{1}{6}$
	1.00

Using this probability distribution and appropriate postulates, we can determine the probability of any event concerning the variable X. For example,

$$P(X \text{ is odd}) = P(X=1 \cup X=3 \cup X=5) = \frac{1}{6} + \frac{1}{6} + \frac{1}{6} = \frac{1}{2}$$

$$P(X \text{ is even}) = P(X=2 \cup X=4 \cup X=6) = \frac{1}{6} + \frac{1}{6} + \frac{1}{6} = \frac{1}{2}$$

$$P(X \geq 3) = P(X=3 \cup X=4 \cup X=5 \cup X=6) = \frac{1}{6} + \frac{1}{6} + \frac{1}{6} + \frac{1}{6} = \frac{2}{3}$$

$$P(2 \leq X \leq 4) = P(X=2 \cup X=3 \cup X=4) = \frac{1}{6} + \frac{1}{6} + \frac{1}{6} = \frac{1}{2}$$

$$P(X < 5) = P(X=1 \cup X=2 \cup X=3 \cup X=4) = \frac{1}{6} + \frac{1}{6} + \frac{1}{6} + \frac{1}{6} = \frac{2}{3}$$

Sec. 7.8 / PROBABILITY DISTRIBUTIONS

Example 7.22 Let X be the outcome of the arbitrary classification of a loan in the files of the loan company mentioned in Example 7.19. The probability distribution for X is

Random Event	x	$P(X = x)$
$H \cap P$	0	.26
$H \cap B$	1	.04
$M \cap P$	2	.30
$M \cap B$	3	.10
$L \cap P$	4	.14
$L \cap B$	5	.16
		1.00

Now we can use this probability distribution to make a probability statement concerning any random event from the population of events.

Example 7.23 Suppose we select samples from a group of teenage customers that contains 6 girls (G) and 4 boys (B). If we select a sample of size 1, we have the following probability distribution, where X represents the number of girls in the sample.

Random Event	x	$P(X = x)$
B	0	$\frac{4}{10}$
G	1	$\frac{6}{10}$
		1.00

However, if we select a sample of size 2 *without* replacement, we have the following probability distribution, where X represents the number of girls in the outcome of the two draws:

Random Event	x	$P(X = x)$
$B_1 \cap B_2$	0	$\frac{2}{15}$
$B_1 \cap G_2$ or $G_1 \cap B_2$	1	$\frac{8}{15}$
$G_1 \cap G_2$	2	$\frac{5}{15}$
		1.00

Finally, if we select a sample of size 2 *with* replacement, the probability distribution becomes

Random Event	x	$P(X=x)$
$B_1 \cap B_2$	0	$\frac{4}{25}$
$B_1 \cap G_2$ or $G_1 \cap B_2$	1	$\frac{12}{25}$
$G_1 \cap G_2$	2	$\frac{9}{25}$
		1.00

In the last two tables of Example 7.23 it should be noted that we obtain *different* probability distributions for what appear to be identical situations. However, the two cases differ in the techniques used for selecting the sample. Thus, as was cautioned at the beginning of the chapter, it is evident that the probabilities associated with various events necessarily depend on how the experiment is being conducted.

In Examples 7.21 to 7.23, we referred to the tabular association of the possible values of a random variable X with corresponding probabilities as a probability distribution. However, in many situations we can make such an association more conveniently by using a mathematical function. In such a case, we represent a probability distribution notationally by an entity termed a *probability function* (in the discrete case) or a *probability density function* (in the continuous case). Consider an illustration: In Example 7.21 we represented the probability distribution by a table; however, we can also represent such distribution by a function, namely

$$f(x) = \frac{1}{6} \qquad x = 1, 2, 3, 4, 5, 6 \tag{7.46}$$

where $f(x)$ is the probability function of the *discrete* random variable X and x represents any possible value which X may assume. Whenever we describe a probability function (or a probability density function) mathematically, we *always* include the possible values that X may assume, as in the right side of (7.46). In this case, the material ($x = 1, 2, 3, 4, 5, 6$) to the right of the description of the probability function tells us that we are considering a *discrete* random variable because it can assume only one of *six* possible values.

As another illustration, we can use a probability function to represent the answer to the question "What is the probability that x heads will appear if

Sec. 7.8 / PROBABILITY DISTRIBUTIONS

three coins are tossed?" Representing the answer in functional notation, we have

$$f(x) = \frac{_3C_x}{2^3} \quad x = 0, 1, 2, 3 \tag{7.47}$$

However, we obtain the same results when we represent the probability distribution in tabular form:

x	$P(X = x)$
0	$\frac{1}{8}$
1	$\frac{3}{8}$
2	$\frac{3}{8}$
3	$\frac{1}{8}$
	1.00

You are encouraged to use a tree diagram to verify the values in the above table.

In each of the above illustrations, we note that $f(x)$ *corresponds to* $P(X = x)$ *for each possible value* x; furthermore, we note that the events represented by the possible values of X are mutually exclusive and exhaustive. Hence the properties of a probability function for a discrete random variable X are as follows:

1. The function is nowhere negative; that is $f(x) \geq 0$ for any possible value of X.
2. The function $f(x)$ represents the probability that X assumes the possible value x.
3. $\sum f(x) = 1$, *where the summation is over all possible values of X.*

It should be noted that a function of a discrete random variable is a probability function *if and only if* the function satisfies all three of these requirements. [Observe that the variable X actually may describe the attribute of interest, as was the case for the die-tossing illustration of (7.46), or it may be simply a code, as in Example 7.22.]

We shall now consider another probability-associated concept. In many instances, it is necessary to evaluate probabilities such as $P(X < 8)$ and $P(3 < X \leq 7)$. In these cases (as well as in many other situations), it is convenient to use a new function, known as the *cumulative probability distribution function*, or, more simply, the *cumulative distribution*. For a discrete probability function $f(x_i)$, $i = 1, 2, \ldots$, the cumulative distribution $F(x)$ is defined by

$$F(x) = \sum f(x_i) \tag{7.48}$$

where the summation is over those values of i such that $x_i \leq x$.

We note that

$$F(x) = P(X \leq x)$$

where x is any possible value that X may assume;

$$P(a < X \leq b) = F(b) - F(a)$$

where a and b are any two possible values of X such that $a < b$; and

$$F(x) = 1$$

where the appropriate summation is over all possible values of X.

Example 7.24 Suppose 100 employees are equally distributed in critical categories 1 to 5. Using simple random sampling, find the following: (a) the probability distribution in both tabular and functional form; (b) the corresponding cumulative probability distribution; (c) $P(X \leq 4)$; and (d) $P(X > 3)$.

Solution

(a) The probability distribution in tabular form is

x	$P(X = x)$
1	.2
2	.2
3	.2
4	.2
5	.2
	1.00

The functional form is

$$f(x) = \frac{1}{5} \quad (x = 1, 2, 3, 4, 5)$$

(b) The corresponding cumulative probability distribution is

Sec. 7.8 / PROBABILITY DISTRIBUTIONS

x	$F(x)$
1	.2
2	.4
3	.6
4	.8
5	1.0

(c) $P(X \leq 4) = F(4) = .8$
(d) $P(X > 3) = P(3 < X \leq 5) = F(5) - F(3) = 1.0 - .6 = .4$

EXERCISES

In all exercises that involve sampling, assume that simple random sampling is used unless stated otherwise.

1. A pair of fair dice is to be rolled:
   (a) List each possible outcome and the probability of each outcome.
   (b) What is the probability of the outcome being less than 5?
   (c) What is the probability of the outcome being even?
   (d) What is the probability of the outcome being both even and less than 5?
   (e) What is the probability of the outcome being even or less than 5?

2. Four fair coins are to be tossed at one time:
   (a) List each possible outcome and the probability of each outcome.
   (b) What is the probability of the outcome being two or more heads?
   (c) What is the probability of the outcome being an even number of heads? (Zero is to be considered an even number.)
   (d) What is the probability of the outcome being both an even number of heads and two or more heads?
   (e) What is the probability of the outcome being an even number of heads or two or more heads?

3. Two boxes contain five items each, numbered 1 to 5. An item is to be drawn at random from each box and the sum of the numbers on the two items is the variable X:
   (a) List each of the possible values of X and the probability associated with each value of X.
   (b) What is $P(X \geq 6)$?
   (c) What is $P(X \text{ is odd})$?
   (d) What is $P(X \geq 6 \cap X \text{ is odd})$?
   (e) What is $P(X \geq 6 \cup X \text{ is odd})$?

4. If a box contains 7 good and 6 defective parts and a simple random sample of size 2 is to be drawn, construct the probability distribution for the variable X, where X represents

(a) The number of good parts obtained and sampling is with replacement
(b) The number of good parts obtained and sampling is without replacement
(c) The number of defective parts obtained and sampling is with replacement
(d) The number of defective parts obtained and sampling is without replacement

5. A group of students contains 6 business, 4 psychology, and 2 geography majors, and a simple random sample of size 3 is to be drawn. Construct a probability distribution for the variable X, where X represents

(a) The number of business majors obtained and sampling is with replacement
(b) The number of business majors obtained and sampling is without replacement
(c) The number of psychology majors obtained and sampling is with replacement
(d) The number of psychology majors obtained and sampling is without replacement
(e) The number of geography majors obtained and sampling is with replacement
(f) The number of geography majors obtained and sampling is without replacement
(g) The number of students obtained that are either psychology or geography majors and sampling is with replacement
(h) The number of students obtained that are either psychology or geography majors and sampling is without replacement

6. Construct the probability function of the variable X that represents the total number of spots showing on the top surfaces for the experiment of tossing three fair dice.

7. What are the three properties of the probability function of a discrete random variable? (Verify that the function obtained in Exercise 6 is, in fact, a probability function.)

8. Construct a cumulative distribution for parts (a) and (b) of Exercise 5.

9. Construct a cumulative distribution for Exercise 6 and find

(a) $P(X \geq 12)$ (b) $P(X \leq 8)$

10. A department store has 3 employees in the shoe department, 10 employees in ladies' clothing, 8 employees in men's clothing, and 4 employees in home furnishings. Three employees are to be randomly selected to learn a new sales technique. Construct a probability distribution for the variable X, where X represents

(a) The number of employees selected from the shoe department
(b) The number of employees selected from ladies' clothing
(c) The number of employees selected from men's clothing
(d) The number of employees selected from home furnishings

11. The OK Manufacturing Company has a record of producing 30 percent defects. From each production batch 5 items are inspected.

(a) Construct a probability distribution for the number of defects inspected from each batch.
(b) Construct a cumulative distribution for part (a).

(c) Find $P(X \geq 4)$.
(d) Find $P(X \leq 2)$.

12. Construct a cumulative distribution for parts (c) and (d) of Exercise 10 and find

    (a) The probability that at most 2 employees are selected from men's clothing
    (b) The probability that at least 2 employees are selected from men's clothing
    (c) The probability that at most 1 employee is selected from home furnishings
    (d) The probability that at least 1 employee is selected from home furnishings

13. A recent survey shows that among the families in Dallas, 32 percent have no children, 24 percent have 1 child, 18 percent have 2 children, 13 percent have 3 children, 9 percent have 4 children, and 4 percent have 5 or more children.

    (a) For a randomly selected family in Dallas, list the possible number of children they may have and the associated probability.
    (b) Determine the probability that they will have 3 or more children.
    (c) Given that they have at least 2 children, determine the probability that they will have at least 4 children.
    (d) Determine the probability that they will have at least 1 child but less than 4 children.

7.9 MATHEMATICAL EXPECTATION

Occasionally we are interested in knowing the average value for a random variable. For example, we may wish to know the average profit, cost, income, or payoff for an investment. If we know the probability distribution for the associated random variable, we can find the average of the random variable. This average is called the *mathematical expectation* of the random variable X, which is simply a special weighted average of the possible values of X. The proper weights are the probabilities corresponding to the possible values (or groups of values) of X.

Definition 7.14 If $x_1, x_2, \ldots, x_n$ are the n possible values of the discrete random variable X, the mathematical expectation of X (or the mean of the probability distribution of X) is

$$E(X) = x_1 P(X = x_1) + x_2 P(X = x_2) + \cdots + x_n P(X = x_n)$$
$$= \sum_{i=1}^{n} x_i P(X = x_i) \qquad (7.49)$$

where $P(X = x)$ is the probability that the random variable X assumes the value x. The mathematical expectation is often referred to as an average, or "expected," value. Note that since $f(x_i) = P(X = x_i)$, we can also write (7.49) as $E(X) = \sum_{i=1}^{n} x_i f(x_i)$.

Example 7.25 Suppose Ms. Jones owns a steakhouse that specializes in only three types of steaks, and, furthermore, the only other selection on the menu is hamburgers. The price (including beverage, dessert, and all trimmings) of each of the four items on the menu and the probability that a randomly selected customer will buy each are as follows:

Item	x(Price)	P(X = x)
Sirloin	$8.50	.30
T-bone	9.25	.25
Rib eye	6.95	.35
Hamburger	2.45	.10
		1.00

What is the expected value of customer expenditure?

Solution

$$E(X) = \$8.50(.30) + \$9.25(.25) + \$6.95(.35) + \$2.45(.10) = \$7.54$$

It should be noted that the expected value of $7.54 is not one of the possible values of x. Furthermore, it is not necessary that it be a possible value since it represents a weighted average.

Example 7.26 Ms. Jones (Example 7.25) wishes to know the average *profit* per customer. In order to find the expected profit, she has determined her profit on each of the four items, and the random variable now becomes profit per item rather than total price per item:

Item	x(Profit)	P(X = x)
Sirloin	$2.45	.30
T-bone	2.43	.25
Rib eye	2.47	.35
Hamburger	1.25	.10
		1.00

What is the expected profit per customer?

Solution

$$E(X) = \$2.45(.30) + \$2.43(.25) + \$2.47(.35) + \$1.25(.10) = \$2.3320$$

Thus, if Ms. Jones has 100 customers during an evening, she expects to make $100 \times \$2.3320 = \$233.20$ profit for that evening. Even with the same number of customers, her profit will vary from one evening to another owing to the different orders she will receive; that is, the expected profit is an *average*. Hence $233.20 is expected as the average profit for every 100 customers over a *long period of time*.

Sec. 7.9 / MATHEMATICAL EXPECTATION

Mathematical expectations, or expected values, are of considerable importance in decision making. As an illustration, in Example 7.26 expected values could be used in determining the effects of price increases, the total profit increases due to various price changes, and the effects on profits due to increases in costs (i.e., decreases in profits). The calculation of mathematical expectations is often simplified by use of the following rules, which are derived immediately from (7.49):

Rule 7.8 If c is a constant,

$$E(c) = c \tag{7.50}$$

Rule 7.9 If c is a constant and $g(X)$ is a function of X,

$$E[cg(X)] = cE[g(X)] \tag{7.51}$$

Rule 7.10 If $g_i(X)$ is a function of X for $i = 1, 2, \ldots, n$,

$$E\left[\sum_{i=1}^{n} g_i(X)\right] = \sum_{i=1}^{n} E[g_i(X)] \tag{7.52}$$

Using these three rules, we can obtain the following rule:

Rule 7.11 If a and b are constants,

$$E(aX + b) = aE(X) + b \tag{7.53}$$

Rule 7.11 shows what effect a linear transformation (i.e., a transformation of $mX + b$) on a random variable X has on the expected value (mean) of the variable.

Example 7.27 Suppose Brown Lumber Company sells four grades of $\frac{1}{4}$-in. den paneling. Furthermore, suppose the probability that a customer who is purchasing paneling will select a specific grade and the corresponding prices are as given in the table below:

Grade	Price per Panel (x)	$P(X = x)$
A	$5.00	.2
B	4.50	.4
C	4.00	.3
D	3.50	.1
		1.0

Now suppose the company from which Brown Lumber Company buys paneling plans to increase its price by 10 percent and the city decides to collect a tax of 25¢ for each panel sold. Brown plans to increase the price of paneling accordingly, that is, $1.1X + .25$.

(a) Find the expected (average) price per panel prior to the price increase.
(b) Use Rule 7.11 to find the expected (average) price per panel after the price increase.
(c) Use the individual new prices to find the expected (average) price per panel after the price increase.

(In each case assume that relevant probabilities will not be changed by the price increases.)

Solution

(a) We use (7.49) to find the average price per panel prior to the price increase:

$$E(X) = 5.00(.2) + 4.50(.4) + 4.00(.3) + 3.50(.1) = \$4.35$$

(b) Using the result of (a) and Rule 7.11, we have

$$E(1.1X + .25) = 1.1E(X) + .25 = 1.1(4.35) + .25 = \$5.035$$

(c) First we determine each of the individual prices after the price increase:

Grade	Price per Panel(x)	$P(X=x)$
A	$\$.25 + 1.1(5.00) = \$5.75$	.2
B	$.25 + 1.1(4.50) = 5.20$	.4
C	$.25 + 1.1(4.00) = 4.65$	.3
D	$.25 + 1.1(3.50) = 4.10$	.1
		1.0

Then

$$E(X) = 5.75(.2) + 5.20(.4) + 4.65(.3) + 4.10(.1) = \$5.035$$

[Note that the results of (b) and (c) are identical.]

EXERCISES

In all exercises that involve sampling, assume that simple random sampling is used unless stated otherwise.

1. Define the term *mathematical expectation*.

Sec. 7.9 / MATHEMATICAL EXPECTATION

2. The following is the probability distribution of the discrete variable Y:

y	$P(Y=y)$
1	$\frac{1}{20}$
2	$\frac{4}{20}$
3	$\frac{9}{20}$
4	$\frac{3}{20}$
5	$\frac{2}{20}$
6	$\frac{1}{20}$

What is $E(Y)$?

3. The Moore Manufacturing Company figures that the present values and probabilities of each of the possible outcomes of introducing a new product are as follows:

Outcome	Present Value	Probability
A	$105,000	.75
B	1,400,000	.10
C	−780,000	.15

What is the mathematical expectation of the present value of introducing the product?

4. Marilyn Gray wishes to invest in some business properties. She figures that the present values of monetary returns and the related probabilities for a new plant are as follows:

Outcome	Present Value	Probability
A	$850,000	.85
B	450,000	.10
C	−1,800,000	.05

The present values and probabilities of buying a chain of stores are as follows:

Outcome	Present Value	Probability
D	$750,000	.75
E	200,000	.25

If Marilyn must choose between these alternatives on the basis of expected monetary returns, should she buy the plant or the stores?

5. An urn contains 7 black balls, 4 red balls, 3 white balls, and 1 blue ball, and a player is to draw 1 ball. If it is black, he wins $1; if it is red, he wins $2; if it is white, he wins $3; and if it is blue, he pays $25. What is the mathematical expectation of this game [that is, if we let X represent "the amount won," what is $E(X)$]?

6. If $E(X) = \$45$, find

   (a) $E(1.5X + \$5)$ (b) $E(.5X + \$10)$

7. Given the following probability distribution

x	$P(X=x)$
10	.11
11	.00
12	.35
13	.44
14	.08
15	.02

   (a) Find $E(X)$.   (b) Find $E(.6X - 2)$.

8. A company will accept an investment opportunity if it has an expected monetary value greater than $10,000. An investment opportunity has the following present values and probabilities of each of the possible outcomes:

Outcome	Present Value	Probability
A	$500,000	.40
B	750,000	.10
C	−100,000	.30
D	−200,000	.20

   Should the investment opportunity be accepted?

9. The profits per week and probabilities of expanding plant capacity to make component parts are as follows:

Outcome	Present Value	Probability
A	$1,000	.3
B	500	.5
C	−500	.2

   The profits per week and probabilities of purchasing the component parts are as follows:

Sec. 7.10 / CONTINUOUS PROBABILITY DISTRIBUTIONS

Outcome	Present Value	Probability
A	$5,000	.02
B	750	.40
C	250	.30
D	−150	.15
E	−1,500	.13

If a company must choose between these alternatives on the basis of expected monetary returns, should it expand capacity or purchase the parts?

10. A book salesman makes a commission of $1.25 on each book sold. The following table gives the probabilities for the number of books sold to a customer:

No. of Books Sold	Probability
0	.38
1	.25
2	.15
3	.10
4	.06
5	.04
6	.02

(a) Compute the expected number of books to be sold to a customer.
(b) Compute the expected commission from calling on a customer.

11. A manager has kept a record of the number of employees absent each day for the past 10 years. The resulting probability is as follows:

No. of Absentees	Probability
5	.16
6	.26
7	.22
8	.19
9	.12
10	.05

Compute the expected number of absentees on any given day.

7.10 CONTINUOUS PROBABILITY DISTRIBUTIONS

In Sec. 7.8, we discussed probability distributions for discrete variables and listed the properties that a function must satisfy in order to be classed as a probability function. The case where the random variable of interest is continuous can be treated in a somewhat analogous manner; however, shortly we shall see that a very important distinction exists in associated interpretations.

In many applications of probability, the random variable of interest can take on any real value within an interval (or intervals). For example, suppose we are interested in the probability that a typical lightbulb made by Westinghouse will function properly for a period in excess of 2,000 hr of continuous use. Let the random variable X represent the length of useful life of a lightbulb made by Westinghouse; in such a case, we seek $P(X > 2{,}000)$. Theoretically, the variable X can take on any positive real value and is thus continuous.

Before we can compute probabilities for a continuous random variable, we must consider certain properties that its probability density function must satisfy. In general, we say that the function $f(x)$ is a probability density function having for its domain the set of all possible values that the continuous random variable X can assume if and only if the following conditions are satisfied:

1. The function is nowhere negative; that is, $f(x) \geq 0$ for all possible values of X.
2. The probability $P(a \leq X \leq b)$, where a and b are any two possible values of X such that $a \leq b$, is determined by the area under the curve $f(x)$ between the values a and b.
3. The total area under the curve $f(x)$ is equal to unity (1).

Example 7.28 Suppose we are firing a rifle at a 1-ft long piece of string that is stretched horizontally in a plane perpendicular to our line of fire. Furthermore, suppose the probability density function of the random variable X, which is the distance that our bullet hits from the left end of the string, is given by

$$f(x) = \begin{cases} 1 & 0 \leq x \leq 1 \\ 0 & \text{otherwise} \end{cases}$$

Find $P(\frac{1}{2} \leq X \leq \frac{2}{3})$.

Solution We are given that $f(x)$ is a probability density function; hence $P(\frac{1}{2} \leq X \leq \frac{2}{3})$ may be determined by finding the area between $f(x)$ and the x axis and bounded by $\frac{1}{2}$ on the left and $\frac{2}{3}$ on the right. The area of interest is displayed by the cross-hatched region in Fig. 7.6. Since the region of interest is a rectangle, we can determine the appropriate area by using plane geometry. Thus

$$P(\tfrac{1}{2} \leq X \leq \tfrac{2}{3}) = \text{base} \times \text{height} = (\tfrac{2}{3} - \tfrac{1}{2})(1) = \tfrac{1}{6}$$

The determination of $P(a \leq X \leq b)$ *can be* a very complicated process in the case of continuous random variables. Finding the numerical value for such probability requires the use of integral calculus, except for very simple functions as in Example 7.28. However, for several special probability density functions, the probabilities have been computed and recorded in tables so that we shall not be required to determine the area by use of integral calculus.

As mentioned previously, there is another property of continuous random

Sec. 7.10 / CONTINUOUS PROBABILITY DISTRIBUTIONS

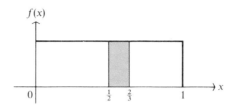

FIGURE 7.6

variables that demands consideration. In the discrete case we have seen that a probability function provides information regarding the probability that the random variable assumes a particular value, but in the continuous case this is not true. That is, in the continuous case, a probability density function, when evaluated at a given possible value of the *continuous* random variable X, *does not* provide the probability that the random variable assumes that particular value (as is the case for a discrete variable). The reasoning is as follows.

In considering the properties of a probability density function for a continuous variable, we note that the probability $P(a \leqslant X \leqslant b)$ is determined by the area between the function and the x axis and bounded by the values of a and b on the x axis, where $a \leqslant b$. What is $P(X=c)$, where c is any possible value of X? To determine $P(X=c)$, we must find the area between $f(x)$ and the x axis and bounded on the x axis by the values c and c; hence this yields only a line that *has no width* and therefore zero area. Thus, if X is a continuous random variable, we have $P(X=c)=0$, where c is any possible value of X. (Note that this is contrary to the results that were obtained for *discrete* random variables.) It *does not* mean that we *cannot* obtain a value of c for the variable, but, rather, that we have an extremely small chance of doing so. This does not appear unreasonable if we note that for a continuous variable we have an infinite (noncountable) number of possible values. *The above difference in computing probabilities makes it desirable to have separate names for the mathematical function that represents a discrete probability distribution (a "probability function") and the one that represents a continuous probability distribution (a "probability density function").*

Finally it should be noted that the *expected value* of a continuous distribution is defined in a manner which parallels that for the discrete case; however, since such definitions involve the use of integral calculus, we shall not include them here.

EXERCISES

In all exercises that involve sampling, assume that simple random sampling is used unless stated otherwise.

1. What three conditions must be satisfied for a function to be the probability density function of a continuous random variable?

2. Given that the probability density function for the random variable X is

$$f(x) = \begin{cases} \frac{1}{2} & 0 < x < 2 \\ 0 & \text{otherwise} \end{cases}$$

determine

(a) $P(X \leq 1)$
(b) $P(\frac{1}{2} < X < \frac{3}{4})$
(c) $P(\frac{1}{2} \leq X < 2)$
(d) $P(X \geq 1.5)$
(e) $P(X = 1)$
(f) $P(X \geq 2.4)$
(g) $P(X < 0)$
(h) Whether or not the three conditions of Exercise 1 are satisfied

3. Given that the probability density function for the random variable X is

$$f(x) = \begin{cases} 2x & 0 < x < 1 \\ 0 & \text{otherwise} \end{cases}$$

determine

(a) $P(X < \frac{1}{2})$
(b) $P(X \geq \frac{3}{4})$
(c) $P(\frac{1}{2} < X < \frac{3}{4})$
(d) $P(X > 2)$
(e) $P(X = \frac{1}{2})$
(f) Whether or not the three conditions of Exercise 1 are satisfied

4. A store makes deliveries within a 25-mi radius. The probability density function for the random variable X is

$$f(x) = \begin{cases} 1/25 & 0 < x < 25 \\ 0 & \text{otherwise} \end{cases}$$

where x is the number of miles from the store to the point of delivery.
Determine:

(a) $P(X \leq 10)$
(b) $P(X \geq 40)$
(c) $P(5 < X < 25)$
(d) $P(10 < X < 15)$
(e) $P(X = 30)$
(f) $P(X \geq 15)$

7:11 SUMMARY

The major objectives of this chapter will have been accomplished if the student possesses a good understanding of the following formulas and related concepts:

1. *Postulates of probability*: Let S be a sample space and let A represent *any event* in S; then, if we represent the probability of event A by the function $P(A)$, P is a *probability function* on the sample space S if the following three postulates are satisfied.

Postulate 7.1 The range of P is the set of real numbers such that

$$0 \leq P(A) \leq 1$$

Postulate 7.2 If $A_i \cap A_j = \emptyset$, then

$$P(A_i \cup A_j) = P(A_i) + P(A_j)$$

Postulate 7.3

$$P(S) = 1$$

2. For any set of mutually exclusive events $A_1, A_2, \ldots, A_k$,
$P(A_1 \cup A_2 \cup \cdots \cup A_k) = P(A_1) + P(A_2) + \cdots + P(A_k)$.
3. $P(A \cap B) = P(A)P(B|A) = P(B)P(A|B)$.
4. $P(A \cup B) = P(A) + P(B) - P(A \cap B)$.
5. $P(A|B) = \dfrac{P(A \cap B)}{P(B)}$, where $P(B) \neq 0$.
6. Events A and B are independent if and only if any one of the following conditions is satisfied:

$$P(A|B) = P(A)$$
$$P(B|A) = P(B)$$
$$P(A \cap B) = P(A)P(B)$$

7. If A and B are mutually exclusive events, $P(A \cap B) = 0$; however, $P(A \cap B) = 0$ *does not* imply that A and B are mutually exclusive.
8. $P(A) = 1 - P(A')$.
9. $E(X) = x_1 P(X = x_1) + x_2 P(X = x_2) + \cdots + x_n P(X = x_n)$.
10. If c is constant, $E(c) = c$.
11. If c is constant and $g(X)$ is a function of X,

$$E[cg(X)] = cE[g(X)]$$

12. If $g_i(X)$ is a function of X for each $i = 1, 2, \ldots, n$,

$$E\left[\sum_{i=1}^{n} g_i(X)\right] = \sum_{i=1}^{n} E[g_i(X)]$$

13. If a and b are constants,

$$E(aX + b) = aE(X) + b$$

14. The function $f(x)$ is a probability function for the *discrete* random variable X if the following three conditions are satisfied:
    (a) $f(x) \geq 0$ for any possible value x.
    (b) $f(x) = P(X = x)$.
    (c) $\sum f(x) = 1$ where the summation is over all possible values of X.
15. The function $f(x)$ is a probability density function for the *continuous*

random variable X if the following three conditions are satisfied:
(a) $f(x) \geq 0$ for every value of x.
(b) $P(a \leq X \leq b)$ is the area under the curve $f(x)$ between the constant values a and b, where a and b are any two values of X such that $a \leq b$.
(c) The total area under the curve $f(x)$ is equal to unity (1).

IMPORTANT TERMS AND CONCEPTS

Addition rule of probabilities
Basic counting principle
Classical, or a priori, probability
Combinations
Complement of an event
Compound event
Conditional event
Conditional probability
Continuous probability distribution
Cumulative probability distribution
Cumulative probability distribution function
Exhaustive
Expected value
Frequency, or a posteriori, probability
General counting principle
Independence
Joint event
Joint probability
Long-term regularity
Marginal probability
Mathematical expectation
Multiplication rule for probabilities

Mutually exclusive events
Permutations
Population
Postulates of probability
Probability density function
Probability distribution
Probability distribution function
Probability distribution mean
Probability function
Probability sample
Random event
Random experiment
Random variable
Sample
Sample space
Sampling with replacement
Sampling without replacement
Simple event
Simple random sample
Statistical probability
Subjective probability
Tree diagram
Venn diagram

REVIEW PROBLEMS

1. A company is considering 15 applications. Calculate the number of distinct groups of 3 people to be hired when

(a) The position is not considered
(b) The position is considered

2. A survey of 80 customers of a manufacturing firm classified 36 as wholesalers, 24 as retailers, and 20 as manufacturers.

(a) Calculate the probabilities that a randomly selected customer of this manufacturer will be a wholesaler, a retailer, or a manufacturer by applying the frequency-theory (a posteriori) approach based on experience.
(b) Use the results of part (a) to illustrate each of the three postulates of probability.

3. The following is a job classification of the employees of a company:

	Department			
Classification	A	B	C	D
Unskilled (U)	100	210	305	0
Skilled (S)	43	110	170	3
Manager (M)	6	17	22	1
Staff (T)	0	0	0	6
Executive (E)	1	2	2	2

(a) State what the event ($S \cap B$) represents.
(b) Calculate $P(S \cap B)$.
(c) State what the event ($T \cup B$) represents.
(d) Calculate $P(T \cup B)$.
(e) State what the event ($U|C$) represents.
(f) Calculate $P(U|C)$.
(g) State what the event ($M \cup C$) represents.
(h) Calculate $P(M \cup C)$.
(i) State what the event ($B|M$) represents.
(j) Calculate $P(B|M)$.
(k) State what the probability $1 - P(S \cup D)$ represents.
(l) Calculate $1 - P(S \cup D)$.

4. The following is a classification of accounts receivable for the stores of a chain:

	Store			
Amount of Account	A	B	C	D
Under $50 ($R1$)	320	380	260	310
$50 to $100 ($R2$)	390	380	280	320
$100 to $500 ($R3$)	210	230	390	340
$500 to $1000 ($R4$)	120	100	230	270
$1000 to $5000 ($R5$)	60	55	147	188
$5000 and over ($R6$)	10	5	3	2

(a) State what $P(R2 \cup B)$ represents.
(b) Calculate $P(R2 \cup B)$. $P(R2) = 1370 + P(B) = 1150 ; 1370 + 1150 - 380 = 2140$
(c) State what $P(C|R5)$ represents.
(d) Calculate $P(C|R5)$. $\dfrac{147}{450}$

(e) State what $P(R6|A)$ represents.
(f) Calculate $P(R6|A)$. $\frac{10}{110}$
(g) State what $P(R1 \cap B)$ represents.
(h) Calculate $P(R1 \cap B)$. $\frac{3}{80}$
(i) State what $P(R2 \cup R3 \cup R4)$ represents.
(j) Calculate $P(R2 \cup R3 \cup R4)$.
(k) State what the probability $1 - P(R5 \cup R6)$ represents.
(l) Calculate $1 - P(R5 \cup R6)$.

5. A manufacturing company packaged 8 boxes of green cups, 6 boxes of blue cups, and 4 boxes of yellow cups and forgot to label the boxes as to color. A simple random sample of size 3 is to be drawn without replacement. Construct a probability distribution for the variable X, where X represents

(a) The number of cases of green cups obtained
(b) The number of cases of blue cups obtained
(c) The number of cases of yellow cups obtained
(d) The number of cases obtained that are either green or blue cups
(e) The number of cases obtained that are either green or yellow cups
(f) The number of cases obtained that are either blue or yellow cups

6. Construct a cumulative distribution for each part of Exercise 5.

7. Use part (b) of Problem 6 and find (a) $P(X \leq 2)$; (b) $P(X > 0)$.

8. Use part (f) of Problem 6 and find (a) $P(1 \leq X \leq 3)$; (b) $P(0 \leq X \leq 1)$.

9. The present values and probabilities of each of the possible outcomes of an opening day special for a new store are as follows:

Outcome	Present Value	Probability
A	20,000	.15
B	12,000	.40
C	2,000	.30
D	−10,000	.15

What is the mathematical expectation of the present value of having this special on opening day?

10. The present values and probabilities of each of the possible outcomes of adopting a new marketing strategy are as follows:

Outcome	Present Value	Probability
A	$750,000	.25
B	150,000	.60
C	−800,000	.15

The present values and probabilities of a second possible marketing strategy are as follows:

Outcome	Present Value	Probability
D	$1,500,000	.10
E	200,000	.78
F	−1,000,000	.12

If the company must choose between these alternatives on the basis of expected monetary returns, which marketing strategy should they use?

11. Given that the probability density function for the random variable X is

$$f(x) = \begin{cases} \dfrac{x}{8} & 0 < x < 4 \\ 0 & \text{otherwise} \end{cases}$$

determine:

(a) $P(X \leq 2)$
(b) $P(X > 3)$
(c) $P(X \geq 4)$
(d) $P(1 \leq X \leq 3)$
(e) $P(X = 1)$
(f) $P(2 \leq X \leq \frac{7}{2})$
(g) $P(X < 1)$
(h) $P(X > \frac{5}{2})$

SPECIAL PROBABILITY DISTRIBUTIONS AND APPLICATIONS

8.1 INTRODUCTION

Nearly every aspect of a business is affected by uncertainty. Consequently, businesspeople will find the concepts and applications of probability extremely important in assisting them in their decision making. For example, they may wish to know the chances of a new product achieving a certain level of sales for its first year on the market. Or they may need to know the probability of a production process resulting in faulty items. Or they may need to know the chances of getting a new contract for $800,000 for which they must spend $125,000 in bidding expenses. Or they may need to know the odds of snow during the coming winter in order to plan for business investments in the skiing industry. More specifically, they may need to determine probabilities for many similar repetitive situations such as periodically checking on the performance process to see if the process is in control. Or they may wish to make the decision of accepting or rejecting a shipment of units based on the results of a sample.

In order to make decisions for similar or repetitive situations, it is generally quite helpful to have available a function or equation that will yield the probability for specific values. Some special probability distributions that arise quite frequently will be considered in this chapter. For most of these special distributions, we first develop the probability function for a specific example and then derive the probability function for the general problem. For the general solution, we consider characteristics of the distribution such as the mean and cumulative probability function. Moreover, we investigate the proper use of appropriate tables, thus making the task of finding probabilities for specific problems a relatively simple one.

In our investigations we consider a probability distribution as a mathematical model of a real-world situation. For each such model, we have a set of assumptions that characterize a large class of experiments. Although the choice of a particular probability function as a model for a specific situation is somewhat arbitrary, it should not be completely so. Before a specific model is used to describe a particular situation, we should attempt to justify the use of the model and to ensure that our assumptions are realistic for the problem under consideration.

8.2 BINOMIAL DISTRIBUTION

The *binomial distribution* is probably the most frequently used discrete distribution in probability. It is concerned with repeated trials of a given event when the outcomes of the trials are independent. Furthermore, for each trial, the event has *only two mutually exclusive* outcomes such as heads or tails, good or bad, male or female, defective or nondefective, dead or alive, satisfied or dissatisfied, operative or inoperative, and success or failure. Such a trial is known as a *Bernoulli trial*, or, more specifically, a trial generated by a *Bernoulli process*.

Suppose we toss a coin three times and wish to determine the probability function of X, where X is the number of heads to be obtained on the three tosses. Let us suppose, *strictly for illustrative purposes*, that the coin is *unfair*, so that, for example, the probability of obtaining a head on a given toss is equal to .4 and hence the probability of obtaining a tail is .6. We shall begin the solution of this problem by considering the probability distribution associated with one toss of our unfair coin; then we shall determine the probability distribution associated with three tosses of the coin.

The probability distribution for the experiment of one toss of our unfair coin is given in functional form in Eq. (8.1), where x represents the number of heads on a given toss:

$$f(x) = (.4)^x (.6)^{1-x} \qquad x = 0, 1 \tag{8.1}$$

We note that when we substitute the appropriate value of x into $f(x)$, we obtain the probability distribution in tabular form:

x	$P(X = x)$
0	.6
1	.4
	1.00

The probability function given by (8.1) can be generalized for the case where p is the probability that $X = 0$, and $(1 - p)$ is the probability that $X = 1$. The more general probability function is

$$f(x) = p^x (1-p)^{1-x} \qquad x = 0, 1 \tag{8.2}$$

which is known as a *Bernoulli probability function*.

Let us examine certain properties of the events that occur when we toss our coin three times: First we note that there are *only two* possible mutually exclusive outcomes *per toss* (i.e., heads or tails). Second, since we are essentially sampling *with* replacement, the outcome of a given toss will not be influenced by any previous outcome and thus the outcome of a given trial is independent of the outcomes of any previous trials; i.e., the probabilities for a given toss are not changed by the presence or absence of knowledge concerning outcomes of previous trials. Third the probability of each possible outcome remains unchanged from trial to trial. Although these three properties *appear* to be reasonable for our coin-tossing problem, we cannot prove or guarantee that they are correct; therefore, we shall *assume* such properties.

To determine the probability function for three tosses of a coin, we must find the probability for each possible value of a random variable X, where X now represents the number of heads obtained on three tosses of the coin; thus, in this particular example, the possible values of X are 0, 1, 2, 3. Let H and T represent the outcomes of head and tail, respectively, for a given toss; and let the position in the arrangement represent the number of the toss (that is, HTH represents the event of obtaining heads on the first and third tosses and a tail on the second toss). Utilizing the above described assumptions, we determine the required probabilities:

$$P(X=0) = P(TTT) = P(T)P(T)P(T)$$
$$= (.6)(.6)(.6)$$
$$= (.6)^3 = .216$$

$$P(X=1) = P(HTT \cup THT \cup TTH) = P(HTT) + P(THT) + P(TTH)$$
$$= (.4)(.6)(.6) + (.6)(.4)(.6) + (.6)(.6)(.4)$$
$$= 3(.4)(.6)^2 = .432 \tag{8.3}$$

$$P(X=2) = P(HHT \cup HTH \cup THH) = P(HHT) + P(HTH) + P(THH)$$
$$= (.4)(.4)(.6) + (.4)(.6)(.4) + (.6)(.4)(.4)$$
$$= 3(.4)^2(.6) = .288$$

$$P(X=3) = P(HHH)$$
$$= (.4)(.4)(.4)$$
$$= (.4)^3 = .064$$

Hence the probability distribution in tabular form is

x	$P(X=x)$
0	.216
1	.432
2	.288
3	.064
	1.000

Sec. 8.2 / BINOMIAL DISTRIBUTION

Several points are so vital to the development of the functional form of probability distributions that they should be spelled out in detail: *First* the four compound events

$$X=0 \quad X=1 \quad X=2 \quad X=3 \tag{8.4}$$

are *mutually exclusive* [i.e., for example, it is impossible on one repeat of the experiment of tossing three coins to obtain both one head ($X=1$) and two heads ($X=2$) simultaneously] and *exhaustive*; thus their probabilities will sum to *unity*. *Second* the joint events

$$HHH \quad HHT \quad HTH \quad THH \quad HTT \quad THT \quad TTH \quad TTT \tag{8.5}$$

each of which satisfies one (and only one) of the events of (8.4), are obviously mutually exclusive. This being true, we can obtain the probability of any of the four compound events of (8.4) simply by adding the probabilities of *all* joint events of (8.5) that satisfy the compound event of interest; for example, $P(X=2) = P(HHT \cup HTH \cup THH) = P(HHT) + P(HTH) + P(THH)$. *Third* the individual tosses of the coin are independent, and thus we can compute the probability for each joint event of (8.5) simply by taking the product of the probabilities of the marginal events included in such a joint event. However, in so doing, we see that if, when order is ignored, the overall makeup of the two joint events is the same (for example, *HHT,* and *HTH*), their probabilities of occurrence will be equal.

Having made these observations, we see that we can determine the probability of a given *compound event* by finding the probability for *any* specific joint event that satisfies the compound event of interest and multiplying the probability of that joint event by the number of joint events that satisfy the compound event of interest. (That is, when we add the probabilities of the mutually exclusive joint events, actually we are adding *equal* numbers and hence we may multiply.) For example, to determine *P(X = 1),* we can choose one of the joint events, say *HTT,* and compute $P(HTT) = P(H)P(T)P(T) = (.4)(.6)(.6) = (.4)(.6)^2$; next we can determine how many such joint events satisfy the event $X=1$. Since order is important, this may be done either by enumeration or by use of permutations, but the use of permutations is much easier for complex problems. For our example, we note that we have two groups of outcomes with two in one group (tails) and one in the other group (heads). Thus, using Eq. (7.5), we have

$$_3P_{2,1} = \frac{3!}{2!1!} = {_3C_1}$$

Having thus obtained the number of joint events and the probability of each joint event that satisfies the compound event $X=1$, we combine this information to obtain

$$P(X=1) = {_3C_1}(.4)(.6)^2 \tag{8.6}$$

We *could* repeat the above process for each of the possible values of X. However, we prefer to have a function of x such that we can determine $P(X=x)$ (that is, the probability that the random variable X assumes the value of x) by substituting x (that is, the appropriate possible value of X) into the function. To determine the form of such a function, let us consider for our example the more general case of $P(X=x)$, without x being specified (except that it must be one of the possible values of X). We note that in order for the event $X=x$ to be satisfied, we must obtain x heads and $3-x$ tails; hence we need consider only one specific joint event that yields exactly x heads. For simplicity, let us consider that event for which the first x outcomes are heads and the last $3-x$ outcomes are tails; the probability for this joint event is $(.4)^x(.6)^{3-x}$. Now to find the number of different orders for the outcomes that have exactly x heads and $3-x$ tails, we determine.

$$_3P_{x,3-x} = \frac{3!}{x!(3-x)!} = {_3C_x}$$

Finally we obtain the desired probability function by multiplying the above probability by the number of satisfying events. (Again note that we are actually adding the *equal* probabilities of *mutually exclusive events*.) The desired probability function is

$$P(X=x) = f(x) = {_3C_x}(.4)^x(.6)^{3-x} \qquad x=0,1,2,3 \qquad (8.7)$$

Note that by substituting the possible values of X into $f(x)$, we obtain the same probabilities as in (8.3):

$$\begin{aligned} P(X=0) &= f(0) = {_3C_0}(.4)^0(.6)^{3-0} = (1)(1)(.216) = .216 \\ P(X=1) &= f(1) = {_3C_1}(.4)^1(.6)^{3-1} = (3)(.4)(.36) = .432 \\ P(X=2) &= f(2) = {_3C_2}(.4)^2(.6)^{3-2} = (3)(.16)(.6) = .288 \\ P(X=3) &= f(3) = {_3C_3}(.4)^3(.6)^{3-3} = (1)(.064)(1) = .064 \end{aligned} \qquad (8.8)$$

Thus we have generalized the solution to the initial problem of finding the probability function of X, where X is the number of heads that appear on three tosses of an unfair coin; the function is given in (8.7).

In the general problem of the nature just described, we usually speak of the two mutually exclusive events of interest as being either *successes* or *failures*: A *success* denotes the *occurrence* of the outcome of interest, while a *failure* denotes the nonoccurrence of the outcome of interest; thus, in the previous problem, the event of obtaining a head is a success while that of obtaining a tail is a failure. It should be clear that the use of the terms success and failure has nothing to do with personal feelings; for example, if we were interested in finding the probability of rolling three losing numbers in six tosses of a pair of dice, we would refer to the event of obtaining a losing number as a success, which, obviously, would be contrary to personal desire.

Sec. 8.2 / BINOMIAL DISTRIBUTION

We now consider the more general problem of determinimg the probability function of X, where X is the number of *successes* in n trials of an experiment for which there exist only two mutually exclusive outcomes, success and failure, with the probabilities of each outcome being p and $1-p$, respectively. Furthermore, we assume that the outcomes of the trials are independent events and the probabilities of success and failure remain unchanged from trial to trial.

To determine the probability function of X, first we find the probability of a specific joint event that gives x successes and $n-x$ failures, and then we find the number of such events and multiply the probability by the number of events. This product will give us a general function for $P(X=x)$. Let us begin by considering that specific joint event for which the first x trials result in successes (S) and the last $n-x$ trials result in failures (F). We have

$$P\bigg(\underbrace{SS\ldots S}_{\substack{x \\ \text{Successes}}} \underbrace{FF\ldots F}_{\substack{n-x \\ \text{Failures}}}\bigg)$$

$$= \underbrace{P(S)P(S)\ldots P(S)}_{x \text{ times}} \underbrace{P(F)P(F),\ldots,P(F)}_{n-x \text{ times}}$$

$$= \underbrace{pp\ldots p}_{x \text{ values}} \underbrace{(1-p)(1-p)\ldots(1-p)}_{n-x \text{ values}}$$

$$= p^x(1-p)^{n-x} \tag{8.9}$$

Moreover, we have a group of n items (outcomes), of which there are x in one group (successes) and $n-x$ in the other group (failures), and we wish to determine the number of different orders in which these outcomes may occur. Hence, again using (7.5), we have

$$_nP_{x,n-x} = \frac{n!}{x!(n-x)!} = {_nC_x} \tag{8.10}$$

Multiplying the results of (8.9) and (8.10), we have

$$P(X=x) = f(x) = {_nC_x} p^x(1-p)^{n-x} \qquad x=0,1,2,\ldots,n \tag{8.11}$$

We must check that the function in (8.11) is actually a valid probability function. [The requirements for $f(x)$ to be a probability function were given in Sec. 7.8. They are: $f(x) \geq 0$ for all x; $f(x) = P(X=x)$; and $\sum f(x) = 1$, where the summation is over all possible values of X.] Since each term in the product is nonnegative, we can see immediately that the first requirement is satisfied. Also, $f(x) = P(X=x)$ because we have derived the function for exactly that purpose. Finally we need to determine only whether or not $\sum f(x) = 1$. Thus

$$\sum_{x=0}^{n} f(x) = \sum_{x=0}^{n} {_nC_x} p^x(1-p)^{n-x} \tag{8.12}$$

must be equal to unity in order for $f(x)$ to qualify as a valid probability function. Recall from the binomial expansion in algebra that

$$(a+b)^n = \sum_{x=0}^{n} {}_nC_x a^x b^{n-x}$$

We note that the right side of Eq. (8.12) is the binomial expansion of $[p+(1-p)]^n = 1^n = 1$. Thus $f(x)$ in (8.11) is, in fact, a valid probability function.

Since the terms in such functions are those of the *binomial expansion* of $[p+(1-p)]^n$, the probability function in (8.11) is called the *binomial probability function* while the corresponding probability distribution is called simply the *binomial distribution*. We shall follow the policy of denoting the function by

$$b(x;n,p)$$

where x is any possible value that "number of successes" (X) can assume, n is the number of trials, and p is the probability of a success on any specific trial. Functionally we denote the binomial probability function as

$$b(x;n,p) = {}_nC_x p^x (1-p)^{n-x} \qquad x = 0, 1, \ldots, n \qquad (8.13)$$

In $b(x;n,p)$, x is a variable while n and p are parameters (i.e., they are fixed constants for a given problem). Whenever we need to refer to the binomial distribution, we shall always use the above notation, including the term ${}_nC_x$. Since often it is difficult to see the reason for such a term intuitively, the reader may wish to refer often to the development associated with (8.11), in which we have taken advantage of the fact that ${}_nP_{x,n-x} = {}_nC_x$; we use this last term because it has come to be known as the *binomial coefficient*.

In analyzing $b(x;n,p)$, we see that the distribution is symmetrical when $p = .5$. For a better understanding of the graphic shape of the binomial distribution for fixed n and different values of p, refer to Fig. 8.1, where the probabilities are represented by the height of the lines above the respective x values. This representation reflects the fact that the variable of interest is a *discrete* random variable.

Example 8.1 Suppose we have a group of employees containing 3 foremen and 7 workers from which we wish to select *with replacement* a simple random sample of size $n = 3$. Find the probability of selecting (a) no workers, (b) 1 worker, (c) 2 workers, and (d) 3 workers.

Solution The problem is one to which (8.11) is applicable. Let X represent the number of workers (i.e., successes) in the sample. We have that $p = \frac{7}{10}$ because this is the probability of success for one trial; furthermore, since we are sampling with replacement, the probabilities remain unchanged from trial to trial; finally we have that $n = 3$. Thus we obtain answers to (a) to (d)

Sec. 8.2 / BINOMIAL DISTRIBUTION

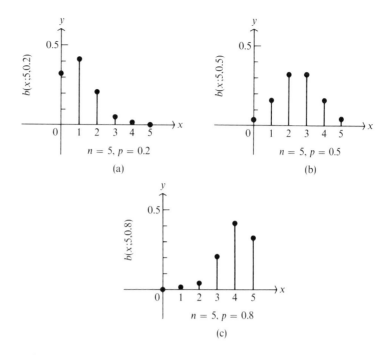

FIGURE 8.1

simply by substituting appropriate values into (8.11):

(a) $P(X=0) = b(0;3,.7) = {}_3C_0(.7)^0(.3)^{3-0} = .027$
(b) $P(X=1) = b(1;3,.7) = {}_3C_1(.7)^1(.3)^{3-1} = .189$
(c) $P(X=2) = b(2;3,.7) = {}_3C_2(.7)^2(.3)^{3-2} = .441$
(d) $P(X=3) = b(3;3,.7) = {}_3C_3(.7)^3(.3)^{3-3} = .343$

Note that the four events are mutually exclusive and exhaustive; hence, as is to be expected, their probabilities sum to unity (1).

Example 8.2 Suppose that from past history a shoe manufacturer knows that 20 percent of his orders will be for more than 100 pairs of shoes. What is the probability that the next six orders he receives will contain three orders of more than 100 pairs of shoes if, in fact, each order can be considered as the outcome of a Bernoulli process?

Solution Since we are interested in orders of more than 100 pairs of shoes, we shall call such an order a *success*; conversely, we shall call any order less than or equal to 100 a *failure*. We know that 20 percent of the orders are successes; hence $p = .2$. Furthermore, $n = 6$ and $x = 3$; thus

$$P(X=3) = b(3;6,.2) = {}_6C_3(.2)^3(.8)^{6-3} = .08192$$

Example 8.3 Suppose a bakery has 5 ovens available for baking and the minimum number of ovens required to meet the demands of its customers is

4. If the probability that an oven is inoperative on a given day is .1 and the failure of ovens can be considered as a Bernoulli process, determine the probability that the bakery will be able to meet the demands of its customers on a given day.

Solution To fulfill the demands of its customers on a given day, the bakery must have either 4 or 5 ovens in operation. Thus let X represent the number of successes, i.e., operative ovens on a given day; then $p=.9$ $n=5$, and

$$\begin{aligned} P(\text{meet demands}) &= P(X=4 \cup X=5) \\ &= P(X=4) + P(X=5) \\ &= {}_5C_4(.9)^4(.1)^{5-4} + {}_5C_5(.9)^5(.1)^{5-5} \\ &= .32805 + .59049 \\ &= .91854 \end{aligned}$$

Cumulative Binomial Probabilities

In many practical situations we are required to determine $P(X \leqslant x)$, where X is distributed according to the binomial distribution. From Example 8.3, we see that computing very many of the values of $b(x;n,p)$ could become quite laborious, particularly for large values of n. For example, suppose we want to know the probability of obtaining 40 or fewer successes out of 90 trials of a Bernoulli process with $p=.3$. In order to obtain such a probability by straightforward application of the binomial probability function, we would have to evaluate and add 41 probability terms of the nature of (8.13); that is, recalling that the terms following the vertical line in a probability statement are conditions imposed on events,

$$\begin{aligned} P_b(x \leqslant 40 \mid n=90, p=.3) \\ = P_b(X=0 \mid n=90, p=.3) + P_b(X=1 \mid n=90, p=.3) + \cdots \\ + P_b(X=40 \mid n=90, p=.3) \\ = {}_{90}C_0(.3)^0(.7)^{90} + {}_{90}C_1(.3)^1(.7)^{89} + \cdots + {}_{90}C_{40}(.3)^{40}(.7)^{50} \end{aligned}$$

(Hereafter the subscript b will be used to denote probabilities for the binomial distribution, which will provide a means of distinguishing binomial probabilities from other probabilities that have similar forms.) Obviously, if it is possible to do so, we would like to avoid performing this very difficult task, and so in order to circumvent such a problem, we utilize the *cumulative binomial distribution*, denoted by $B(k;n,p)$. The relation between the cumulative binomial distribution and the binomial distribution is naturally

$$B(k;n,p) = P_b(X \leqslant k \mid n,p) = \sum_{x=0}^{k} b(x;n,p) \qquad (8.14)$$

Information regarding the cumulative binomial distribution is given in Table A.2 in the Appendix for various possible values of x, n, and p;

however, before employing *any* table of probabilities in order to simplify solutions of problems, we must understand the proper use of that specific table. In the case of Table A.2, we must make certain that we understand precisely what probabilities are given. For example, we *could* have a table that presents any one of the following types of probabilities: $P(X < x), P(X \leq x)$, $P(X > x)$, or $P(X \geq x)$. All such information is interesting; hence different tables may be constructed for different purposes. However, from a table giving one of the previous types of probabilities, we can determine any of the other probabilities listed, as well as many other required probabilities. Recall that for any variable X, $P(a < X \leq b) = F(b) - F(a)$, where (as given in Chap. 7) $F(x)$ is the cumulative probability distribution of X. To be sure, we shall have great need for evaluating this kind of probability for the binomial distribution, and to do so, we use Table A.2.

For the binomial distribution $b(x;n,p)$, Table A.2 yields the probability $P_b(X \geq x \mid n,p)$ for various possible combinations of x, n, and p. To find $P_b(X \geq x \mid n,p)$ in Table A.2, in the body of the table we must locate the probability that corresponds to appropriate values of x, n, and p. Then we answer any probability question regarding the binomial distribution in *terms* of the probabilities given in Table A.2; for example, to find cumulative binomial probabilities $B(k;n,p)$, we simply find

$$B(k;n,p) = P_b(X \leq k \mid n,p) = 1 - P_b(X \geq k+1 \mid n,p) \qquad (8.15)$$

Table A.2 is what is generally called a *greater than or equal to* probability table, and it may be used to make any desired probability statement concerning events that follow a binomial distribution. Let us consider exemplary probabilities that can be obtained from Table A.2, limiting our consideration to the binomial distribution with $n = 12$ and $p = .42$. First we locate the page that gives information for $n = 12$ and $p = .42$; then we note that regardless of the probability question we wish to answer, we must phrase such a question in terms of one or more probabilities of the form $P(X \geq x)$ because this is the type of information available in Table A.2.

1. Find $P_b(X \geq 5 \mid n = 12, p = .42)$. To determine this information, we read the probability in Table A.2 that corresponds to $n = 12$, $p = .42$, and $x = 5$; thus, $P_b(X \geq 5 \mid n = 12, p = .42) = .6175$.
2. Find $P_b(X > 5 \mid n = 12, p = .42)$. We note that $P_b(X > 5 \mid n = 12, p = .42) = P_b(X \geq 6 \mid n = 12, p = .42) = .3889$.
3. Find $P_b(X < 9 \mid n = 12, p = .42)$. We note that $P_b(X < 9 \mid n = 12, p = .42) = 1 - P_b(X \geq 9 \mid n = 12, p = .42) = 1 - .0218 = .9782$.
4. Find $P_b(X \leq 4 \mid n = 12, p = .42)$. We note that $P_b(X \leq 4 \mid n = 12\ p = .42) = 1 - P_b(X \geq 5 \mid n = 12, p = .42) = 1 - .6175 = .3825$.
5. Find $P_b(X = 8 \mid n = 12, p = .42)$. We note that $P_b(X = 8 \mid n = 12, p = .42) = P_b(X \geq 8 \mid n = 12, p = .42) - P_b(X \geq 9 \mid n = 12, p = .42) = .0760 - .0218 = .0542$.

Table A.2 does not provide probabilities for values of p larger than .5; however, Table A.2 may be used to determine probabilities for binomial distributions where $p > .5$. In order to clarify the reasoning behind the proper procedure for using the table, let us consider the following example. Suppose we flip an unfair coin five times, and suppose, furthermore, that we choose to call the occurrence of a head a success, where the probability of a head on one toss is $p = .7$. Then, naturally, the occurrence of a tail becomes a failure, and the probability of a failure on one toss is $1 - p = 1 - .7 = .3$. Let X represent "number of successes" and Y represent "number of failures"; then the relevant probabilities for number of successes (heads) and number of failures (tails) are as given in Table 8.1, where we note that each possible value of number of heads corresponds to one of the possible values of number of tails. For example, every time we obtain three ($x = 3$) heads out of five tosses, we necessarily obtain two ($n - x = 2$) tails out of five tosses; thus, as indicated in Table 8.1, $P(3$ heads out of 5 trials $|P(H) = .7) = P(2$ tails out of 5 trials $|P(T) = .3)$.

Table 8.1

(1) No. Successes Heads, x	(2) $P(X = x)$	(3) No. Failures Tails, y	(4) $P(Y = y)$
0	$_5C_0(.7)^0(.3)^5 = .0024$	5	$_5C_5(.3)^5(.7)^0 = .0024$
1	$_5C_1(.7)^1(.3)^4 = .0284$	4	$_5C_4(.3)^4(.7)^1 = .0284$
2	$_5C_2(.7)^2(.3)^3 = .1323$	3	$_5C_3(.3)^3(.7)^2 = .1323$
3	$_5C_3(.7)^3(.3)^2 = .3087$	2	$_5C_2(.3)^2(.7)^3 = .3087$
4	$_5C_4(.7)^4(.3)^1 = .3601$	1	$_5C_1(.3)^1(.7)^4 = .3601$
5	$_5C_5(.7)^5(.3)^0 = .1681$	0	$_5C_0(.3)^0(.7)^5 = .1681$

The equivalence also holds true for probabilities of other corresponding events. The general result is as follows: We have arbitrarily designated *heads* as a success; thus, for example, we may seek $P_b(X = 3 | n = 5, p = .6)$, where X is the number of successes. However, we can just as easily designate *tails* as a success rather than a failure; then [since three heads with $P(H) = .6$ is the same event as two tails with $P(T) = .4$] we need find only $P_b(X = 2 | n = 5, p = .4)$.

In general, when we wish to find $P_b(X = x | n,p)$ when $p > .5$, we simply restate the problem in terms of $n - x$ successes with a probability of success of $1 - p$ on a given trial; that is,

$$P_b(X = x | n,p) = P_b(X = n - x | n, 1 - p) \qquad (8.16)$$

More generally, for cumulative probabilities we have

$$P_b(X \geq x | n,p) = P_b(X \leq n-x | n, 1-p) \tag{8.17}$$

Thus, to solve any binomial probability problem involving a value of $p > .5$, we merely interchange the roles of (1) success with failure, (2) p with $1-p$, and (3) x with $n-x$, utilizing the concept of (8.16). For example, suppose we wish to find

$$P_b(X \geq 5 | n=12, p=.58)$$

We see [from (8.17)] that

$$P_b(X \geq 5 | n=12, p=.58) = P_b(X \leq 12-5 | n=12, p=1-.58)$$
$$= P_b(X \leq 7 | n=12, p=.42)$$

and then

$$P_b(X \leq 7 | n=12, p=.42) = 1 - P_b(X \geq 8 | n=12, p=.42)$$
$$= 1 - .0760 = .9240$$

The general formula of (8.17) is representative of only one of many types of problems that need to be solved for the case of $p > .5$. In fact, there are so many such problems that it is fruitless to attempt to set down a general formula for each case; therefore the wise reader will exert every effort to *learn the line of reasoning* leading to the statement of (8.17) rather than attempting to *memorize* the relation per se.

In order to determine exact probabilities for the binomial distribution, we must use Table A.2 or a comparable table. However, in many practical situations we may have to deal with values of n that are larger than those given in Table A.2 or with values of p that are not available in Table A.2; in these situations, generally we are willing to settle for an approximation of such probabilities. However, in the situation where it is *mandatory* that we have extreme accuracy, we may wish to consult certain publications that include very extensive tables for the binomial distribution, such as those published by the Harvard University Computation Laboratory [1],* U.S. Army Ordnance Corps [2], National Bureau of Standards [3], Owen [4], and Romig [5]†.

Recalling the definition of the mean of a probability distribution of a

---

*See the references at the end of this chapter.

†The tables given in the Harvard University publication and the U.S. Army Ordnance Corps pamphlet are the most extensive.

discrete random variable X (Definition 7.14), we may determine the *mean of the binomial distribution* as

$$E(X) = \sum_{x=0}^{n} xf(x) = \sum_{x=0}^{n} x\,_nC_x\, p^x(1-p)^{n-x}$$

For clarity, let us begin with the special case $n=3$. We have

$$E(X) = \sum_{x=0}^{3} x\,_3C_x\, p^x(1-p)^{3-x}$$

$$= 0\,_3C_0\, p^0(1-p)^3 + 1\,_3C_1\, p^1(1-p)^2 + 2\,_3C_2\, p^2(1-p)^1$$
$$+ 3\,_3C_3\, p^3(1-p)^0$$
$$= 0 + 3p(1-p)^2 + 6p^2(1-p) + 3p^3$$
$$= 3p\left[(1-p)^2 + 2p(1-p) + p^2\right]$$
$$= 3p\left[(1-p) + p\right]^2 = 3p(1)^2 = 3p$$

For the general case, we obtain

$$E(X) = \sum_{x=0}^{n} x\,_nC_x p^x(1-p)^{n-x}$$

$$= \sum_{x=0}^{n} x\frac{n!}{x!(n-x)!} p^x(1-p)^{n-x} \qquad (8.18)$$

However, since the first term yields 0 for $x=0$, we may begin our summation with $x=1$ and write (8.18) as

$$E(x) = \sum_{x=1}^{n} x\frac{n!}{x!(n-x)!} p^x(1-p)^{n-x}$$

$$= \sum_{x=1}^{n} \frac{n!}{(x-1)!(n-x)!} p^x(1-p)^{n-x}$$

Factoring out np from each term, we have

$$E(X) = np \sum_{x=1}^{n} \frac{(n-1)!}{(x-1)!(n-x)!} p^{x-1}(1-p)^{n-x}$$

If we let $y = x - 1$, the previous equation becomes

$$E(X) = np \sum_{y=0}^{n-1} \frac{(n-1)!}{y!(n-1-y)!} p^y(1-p)^{n-1-y}$$

$$= np \sum_{y=0}^{n-1} {}_{n-1}C_y\, p^y(1-p)^{n-1-y}$$

Sec. 8.2 / BINOMIAL DISTRIBUTION

We recognize the summation of the last equation to be that for a binomial distribution with $n-1$ trials; thus the sum is necessarily 1, and we have for the mean for the binomial distribution

$$E(X) = np \qquad (8.19)$$

EXERCISES

1. Define the term *Bernoulli process* and give an example.

2. Define the terms *success* and *failure* and give an example.

3. What is the *binomial coefficient*?

4. Suppose a car rental agency has available 6 station wagons, 9 sedans, and 5 compact cars from which we wish to select with replacement a simple random sample of size $n=5$. Determine the probability of selecting

   (a) 3 sedans   (b) 5 sedans   (c) 2 sedans
   (d) 1 sedan   (e) 0 sedans

5. (a) Using only the results of Exercise 4, determine the probability of selecting 4 sedans.
   (b) Use the binomial distribution to find the probability of selecting 4 sedans in Exercise 4 and check to see that it equals the probability determined in part (a) of this exercise.
   (c) Construct the probability distribution function of the experiment defined in Exercise 4 in tabular form using the answers to that exercise and part (b) of this one.

6. Suppose we have a group of employees containing 13 males and 12 females from which we wish to select with replacement a simple random sample of size $n=15$. Determine the probability of selecting

   (a) 8 males   (b) 5 females
   (c) 6 males   (d) 12 females

7. The guidance system of a Hawkeye missile contains 5 redundant subsystems, 3 of which must fail before the guidance system fails. If the probability of any one of the subsystems failing on any given flight is .01, what is the probability of the guidance system failing on a particular flight?

8. An assembly line of Ford Motor Company has 8 painting booths, at least 6 of which must be in operation in order to meet the production schedule on any given day. The probability that a mechanical failure will make a booth inoperative on a given day is .02, and the probability that the absence of an operator will make a booth inoperative on a given day is .01. If the failure of booths can be considered as a Bernoulli process and the causes of failure are independent, what is the probability that a failure of painting booths will not prevent the assembly line from achieving scheduled production on a given day?

9. Use Table A.2 (in the Appendix) to find

(a) $P_b(X \geq 5 | n=15, p=.06)$ (b) $P_b(X \leq 16 | n=19, p=.73)$
(c) $P_b(X = 8 | n=12, p=.95)$ (d) $P_b(X > 7 | n=14, p=.59)$
(e) $P_b(8 \leq X \leq 13 | n=13, p=.46)$ (f) $P_b(3 < X < 8 | n=10, p=.88)$
(g) $P_b(9 \leq X \leq 11 | n=18, p=.69)$ (h) $P_b(12 \leq X \leq 15 | n=50, p=.72)$

10. Use Table A.2 to find
    (a) $P_b(X \geq 7 | n=11, p=.28)$ (b) $P_b(X \geq 4 | n=16, p=.37)$
    (c) $P_b(X < 9 | n=13, p=.08)$ (d) $P_b(X \leq 6 | n=9, p=.48)$
    (e) $P_b(X > 12 | n=18, p=.25)$ (f) $P_b(X = 10 | n=50, p=.13)$
    (g) $P_b(X > 3 | n=8, p=.81)$ (h) $P_b(X < 14 | n=17, p=.64)$

11. Fifteen percent of all items produced in department A are defective. If a batch is rejected when a sample of 10 items has 4 or more defects in it, determine the probability that a batch will be rejected.

12. The Choco Candy Company makes 15 percent of its candy bars too small. For a sample of size 20, use Table A.2 to find

    (a) The probability that less than 4 candy bars are too small
    (b) The probability that at least 15 candy bars are too small
    (c) The probability that between 4 and 8 candy bars are too small
    (d) The probability that at least 7 candy bars are too small
    (e) The probability that 1 or 2 candy bars are too small
    (f) The probability that at most 1 candy bar is too small
    (g) The probability that more than 8 candy bars are too small

13. Forty percent of all customers who enter Nieman-Marcus do not purchase anything. If 100 customers enter the store in an afternoon, use Table A.2 to find

    (a) The probability that 65 will make a purchase
    (b) The probability that at most 50 will make a purchase
    (c) The probability that at least 71 will make a purchase
    (d) The probability that 25 will not make a purchase
    (e) The probability that less than 47 will not make a purchase
    (f) The probability that more than 53 will not make a purchase

14. Kresge & Co. has determined that 8 percent of all purchases are returned. If a sample of 13 is taken, use Table A.2 to find

    (a) The probability that 5 will be returned
    (b) The probability that none will be returned
    (c) The probability that at least 4 will be returned
    (d) The probability that between 2 and 6 will be returned
    (e) The probability that no more than 2 will be returned
    (f) The probability that more than 4 will be returned
    (g) The probability that less than 2 will be returned

8.3 POISSON DISTRIBUTION

In Sec. 8.2 we were concerned with Bernoulli processes, i.e., processes consisting of distinct trials in which the probability of success is constant from trial to trial. In this section we shall study another situation in which the probability of success is *constant*; however, we cannot view the process as consisting of a series of distinct trials. The probability function for such a problem is named the *Poisson distribution*, after the famous French mathematician Siméon Poisson. This distribution is generally used to compute probabilities for random variables that represent a number of objects (or events) distributed over a *fixed* time or space.

The Poisson distribution has been shown to be a very good model to use for determining probabilities associated with, for example, the following random variables:

1. The number of calls coming into a telephone switchboard during a *fixed* interval of time
2. The number of defects in a manufactured part
3. The number of small organisms contained in a given volume of fluid
4. The number of automobile deaths in a city during a *fixed* interval of time
5. The number of red cells in a blood specimen of a *fixed* volume
6. The number of customers arriving at a service facility in a *fixed* interval of time
7. The number of sales of a particular product during a given period
8. The number of typing errors on a page.

Of course, the Poisson distribution is a satisfactory model in many more areas.

When is the Poisson distribution a feasible tool? The process under consideration must satisfy certain assumptions before we are justified in even considering the Poisson distribution for use as a model in a realistic situation. These assumptions are:

1. The number of occurrences of the phenomenon under consideration (hereafter termed a *success*) is independent from unit (of time or space) to unit.
2. The probability of a single occurrence in a very small unit is proportional to the size of the unit.
3. The probability of two or more occurrences in a very small unit is so small that it can be ignored. (8.20)

Through use of the assumptions of (8.20), it is possible to derive the Poisson probability function for x, where x is the number of successes in a given space or time interval. The Poisson probability function is

$$p(x;\lambda) = \frac{e^{-\lambda}\lambda^x}{x!} \qquad x = 0, 1, 2, \ldots \qquad (8.21)$$

where λ is the *mean of the distribution* (i.e., the average number of successes in a given space or time interval) and e is the *constant* 2.7182818. As can be seen from (8.21), the Poisson distribution for the number of occurrences x depends on (or is completely characterized by) the *average number of occurrences per stated unit of space or time*.

To better visualize the Poisson distribution, consider Bill Cartwright, a vacuum cleaner salesman. If Mr. Cartwright averages one sale per day (i.e., $\lambda = 1$) and the number of sales is distributed according to the Poisson distribution, the probabilities of selling 0, 1, 2, or 3 vacuum cleaners on a given day are as follows:

$$p(0;1) = \frac{e^{-1}1^0}{0!} = .368$$

$$p(1;1) = \frac{e^{-1}1^1}{1!} = .368$$

$$p(2;1) = \frac{e^{-1}1^2}{2!} = .184$$

$$p(3;1) = \frac{e^{-1}1^3}{3!} = .061$$

Similarly, if Mr. Cartwright averages 1.5 or 2 sales per day, the probabilities are as follows:

$$p(0;1.5) = \frac{e^{-1.5}(1.5)^0}{0!} = .223 \qquad p(0;2) = \frac{e^{-2}(2)^0}{0!} = .135$$

$$p(1;1.5) = \frac{e^{-1.5}(1.5)^1}{1!} = .335 \qquad p(1;2) = \frac{e^{-2}(2)^1}{1!} = .270$$

$$p(2;1.5) = \frac{e^{-1.5}(1.5)^2}{2!} = .251 \qquad p(2;2) = \frac{e^{-2}(2)^2}{2!} = .270$$

$$p(3;1.5) = \frac{e^{-1.5}(1.5)^3}{3!} = .125 \qquad p(3;2) = \frac{e^{-2}(2)^3}{3!} = .180$$

For a better understanding of the graphic shape of the Poisson distribution for different values of λ, see Fig. 8.2, where the probabilities are represented by the height of the lines above the respective x values. Thus we see that the variable of interest for the Poisson distribution is also a discrete random variable.

In many problems it is no easy task to satisfy ourselves that the assumptions in (8.20) hold true for a given situation; in fact, often this requires either experience in the problem area or experimental evidence. (For example, it is sometimes obvious that the independence requirement is not satisfied.) If we are in doubt as to whether or not the Poisson distribution is appropriate for a specific situation, we may gather data from the process in question and compare the observed data with computed Poisson probabilities. From such comparison, we can determine if the Poisson model provides a satisfactory

Sec. 8.3 / POISSON DISTRIBUTION

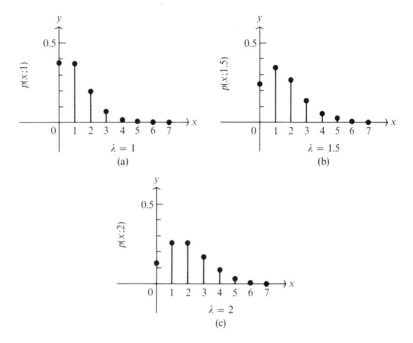

FIGURE 8.2

model for the pattern of occurrences. However, such comparisons require that we are able to determine Poisson probabilities; hence let us consider the method for determining Poisson probabilities from Table A.3 in the Appendix.

The cumulative probability distribution for the Poisson distribution is

$$P(k;\lambda) = \sum_{x=0}^{k} p(x;\lambda) = P_p(X \leq k \mid \lambda) \tag{8.22}$$

(Note that we use the subscript p to denote a *Poisson* probability.) When the values are multiplied by 1,000, we have precisely the values given in Table A.3. That is, Table A.3 presents cumulative sums for $P_p(X \leq k \mid \lambda)$, where X is the random variable representing the number of occurrences (successes) of the phenomenon of interest (in a given unit of time or space); k is any possible value of X; and λ is the average number of occurrences (successes) of the phenomenon of interest in the given unit of time or space. In order to determine $P_p(X \leq k \mid \lambda)$ in Table A.3, from the body of the table we read the probability that corresponds to k and λ.

For a better understanding of both the kinds of realistic probability problems in which the Poisson distribution is helpful and the use of Table A.3 for determining Poisson probabilities, let us consider several examples.

Example 8.4 The Atlantic Wire Company has an automatic machine that produces *many* spools of wire per hour; the machine produces an average of

3 defective spools per hour. Assuming a Poisson model, find the probability of the machine producing the following number of defective spools of wire in 1 hr:

(a) Exactly 3 (b) 3 or less (c) Less than 3
(d) 3 or more (e) More than 3

Solution We are given that $\lambda = 3$ defects per hour; therefore, we determine the appropriate probabilities from Table A.3, which are

(a) $P_p(X=3 \mid \lambda=3) = P_p(X \leq 3 \mid \lambda=3) - P_p(X \leq 2 \mid \lambda=3) = .647 - .423 = .224$
(b) $P_p(X \leq 3 \mid \lambda=3) = .647$
(c) $P_p(X < 3 \mid \lambda=3) = P_p(X \leq 2 \mid \lambda=3) = .423$
(d) $P_p(X \geq 3 \mid \lambda=3) = 1 - P_p(X < 3 \mid \lambda=3) = 1 - P_p(X \leq 2 \mid \lambda=3) = 1 - .423 = .577$
(e) $P_p(X > 3 \mid \lambda=3) = 1 - P_p(X \leq 3 \mid \lambda=3) = 1 - .647 = .353$

Example 8.5 For the machine in Example 8.4 what is the probability that the machine will produce seven defective spools in a period of 2 hr?

Solution We recall that the Poisson distribution for the number of occurrences x depends on the average number of occurrences per *stated* unit of space or time; thus, $\lambda = 6$ defects per 2 hr, and the required probability is

$$P_p(X=7 \mid \lambda=6) = P_p(X \leq 7 \mid \lambda=6) - P_p(X \leq 6 \mid \lambda=6)$$
$$= .744 - .606 = .138$$

Example 8.6 Acme Electronics Company produces a special-purpose transistor that will have an operating life of 2,000 hr or more (as claimed by company officials). From past history it is known that about 4 out of every 100 transistors will fail before 2,000 hr. Assuming the Poisson model, what is the probability that a *large* randomly selected group of Acme transistors will have zero failures in the first 2,000 hr of operation? One or more failures?

Solution Since the average number of failures is 4 out of every 100 transistors, we have $\lambda = .04$ failures per 2,000 hr; hence the probability of zero failures is

$$P_p(X=0 \mid \lambda=.04) = .961$$

Furthermore, the probability of 1 or more failures is

$$P_p(X \geq 1 \mid \lambda=.04) = 1 - P_p(X=0 \mid \lambda=.04) = 1 - .961 = .039$$

Poisson Approximation of Binomial Probabilities

From these simple examples it should be obvious that the Poisson distribution is a probability model that is directly applicable in many different types of problems. However, in addition to this direct utility, one of the primary uses of the Poisson distribution is in approximating probabilities for the *binomial distribution when n is very large and p is very small*. (Generally it is considered feasible to use such an approximation when $p \leqslant .1$ and $np \leqslant 5$.) Such use is made possible by the fact that the Poisson distribution closely approximates the binomial distribution for large n and small p. This fact provides a much needed tool because generally tables for determining binomial probabilities do not include probabilities for large values of n.

In order to approximate a binomial probability, we use the mean of the binomial distribution np as the *mean of the Poisson distribution* λ and determine the appropriate probability from the table of Poisson probabilities.

Example 8.7 A company buys parts from a certain vendor knowing that the vendor's manufacturing process produces 0.005 percent defective parts. What is the probability that the company will receive more than 3 defective parts from a lot of 20,000 parts?

Solution A binomial model appears to fit this problem reasonably well; thus

$$P_b(X > 3 \mid 20{,}000, .00005) = 1 - P_b(X \leqslant 3 \mid n = 20{,}000, .00005)$$

$$= 1 - \sum_{x=1}^{3} b(x; 20{,}000, .00005)$$

This cannot be located in any available table, and so we shall use the Poisson approximation with $\lambda = np = 20{,}000(.00005) = 1$; thus

$$\sum_{x=0}^{3} b(x; 20{,}000, .00005) \approx P_p(X \leqslant 3 \mid \lambda = 1) = .981$$

where (*throughout the text*) $\approx$ should be read "approximately equal to." Finally

$$P_b(X > 3 \mid 20{,}000, .00005) \approx 1 - .981 = .019$$

Recall from (8.21) that the Poisson distribution depends on only one parameter λ, which we have called the *mean of the distribution*. Our reason for so labeling the parameter is that application of Definition 7.14 for the expected value of a distribution to the Poisson distribution yields

$$E(X) = \sum_{x=0}^{\infty} x \frac{e^{-\lambda} \lambda^x}{x!} = \lambda \qquad (8.23)$$

EXERCISES

1. What three assumptions must be satisfied before we can even consider applying the Poisson distribution to a given process?

2. What characteristic completely characterizes the Poisson distribution?

3. An average of 5 customers per hour enter Zorba's Men's Store. Assuming a Poisson model, determine the probability that

   (a) 6 or less customers will enter the store during a 1-hr period.
   (b) Exactly 11 customers will enter the store in a given 1-hr period.
   (c) At least 20 customers will enter the store in a given 2-hr period.
   (d) At least 15 but not more than 20 customers will enter the store in a given 2-hr period.

4. The random-access data file of an IBM 360 computer produces an average of 1 error in 100 hr of operation. Assuming the Poisson model, what is the probability that the file will produce

   (a) 1 or more errors in a job that will require it to be in operation for 50 hr
   (b) Less than 10 errors during 1,000 hr of operation
   (c) More than 7 but less than 13 errors during 500 hr of operation

5. A high-speed power loom of the Birmingham Textile Mills produces an average of 5 flaws in each 100 yd of fabric woven. Assuming a Poisson model, what is the probability that the loom will produce

   (a) More than 10 flaws in a production run of 100 yd of fabric
   (b) Less than 4 flaws in 50 yd of fabric
   (c) At least 7 but not more than 12 flaws in 200 yd of fabric

6. Under what specific conditions is it generally considered feasible to use the Poisson distribution to approximate probabilities for the binomial distribution?

7. Determine the following binomial probabilities:

   (a) $P_b(X \leq 4 \mid n = 5,000;\ p = .0004)$ (b) $P_b(X > 1 \mid n = 100,000;\ p = .00003)$
   (c) $P_b(X < 3 \mid n = 50;\ p = .05)$ (d) $P_b(2 \leq X \leq 5 \mid n = 500;\ p = .008)$
   (e) $P_b(X > 2 \mid n = 1,000;\ p = .003)$

8. The A-1 Gear Company has a machine that produces 0.5 percent defective nylon gears. What is the probability that a lot of 600 of these gears will contain 2 or less defective gears?

9. The Perry fire department receives an average of 8 alarms on one day of the weekend. Assuming a Poisson model, what is the probability that the fire department will receive

   (a) 8 alarms on Saturday
   (b) At least 15 alarms on a weekend
   (c) Less than 3 alarms on Sunday
   (d) Less than 6 alarms on a weekend
   (e) More than 2 alarms but less than 9 alarms on Saturday
   (f) 5 alarms during a 12-hr period on a weekend

10. The Citizens' National Bank has an average of 24 customers between 12:00 and 1:00 each day. Assuming the Poisson model, determine the probability that

    (a) Exactly 17 customers will arrive between 12 and 1
    (b) Less than 3 customers will arrive between 12:00 and 12:15
    (c) More than 10 customers will arrive between 12:30 and 1:00

(d) 6 customers will arrive between 12:10 and 12:30
(e) 2 or more customers will arrive in a 5-min interval between 12 and 1

11. The D&D store sells an average of 6 dresses in a 2-hr period. Assuming a Poisson model, determine the probability that

(a) Exactly 1 dress is sold in a 30-min period
(b) More than 4 dresses are sold within a 2-hr period
(c) At least 18 dresses are sold in an 8-hr work day
(d) Less than 6 dresses are sold in a 4-hr period
(e) More than 1 but less than 5 dresses are sold in 1 hr
(f) At the most 25 dresses are sold in an 8-hr work day

12. A car salesman sells an average of 18 cars during a 6-day week. Assuming a Poisson model, determine the probability that

(a) He sells exactly 15 cars one week
(b) He sells more than 3 cars one day
(c) He does not sell any cars one day
(d) He sells between 15 and 20 cars one week
(e) He sells less than 12 cars one week
(f) He sells more than 20 cars one week

8.4 HYPERGEOMETRIC DISTRIBUTION

Recall that the development of the binomial distribution required that we consider a random process where (1) each trial results in one of only two mutually exclusive outcomes (termed success and failure), (2) the probability of success is constant from trial to trial, (3) each trial is independent of all other trials, and (4) the number of trials is fixed. The binomial distribution was derived through consideration of repeated tosses of an unfair coin; in such a case, essentially we were sampling *with replacement*.

Let us consider the problem of determining the probability distribution for a similar situation, i.e., the situation of sampling *without replacing* from a finite population where each observation (trial) will result in one of only two mutually exclusive outcomes. As an example of such a situation, consider a set that contains 10 batteries: 6 good and 4 bad. Suppose we wish to find the probability of obtaining 4 good and 2 bad batteries when we select a simple random sample of size 6 *without* replacement. We shall determine such probability by using the classical approach and the properties of simple random sampling.

From simple random sampling, we know that when order is disregarded, there are $_{10}C_6$ possible samples of size 6 that can be selected from a population of size 10; thus the total number of possible outcomes for the experiment of interest is $_{10}C_6$. The next step is to determine how many of these outcomes satisfy the event of 4 good and 2 bad batteries. Since 4 good batteries can be selected from 6 in $_6C_4$ ways and 2 bad batteries can be selected from 4 in $_4C_2$ ways, the total number of ways to obtain 4 good and 2 bad batteries is $_6C_4 \cdot _4C_2$. Thus, for the case of sampling *without replacement*, we have from the classical definition of probability that the probability of selecting 4 good and 2 bad batteries from a population of 10 batteries (6 good and 4 bad) is

$$\frac{_6C_4 \cdot _4C_2}{_{10}C_6} = \frac{(6!/4!2!)(4!/2!2!)}{10!/6!4!} = \frac{(6 \cdot 5 \cdot 4!/4!2!)(4 \cdot 3 \cdot 2!/2!2!)}{10 \cdot 9 \cdot 8 \cdot 7 \cdot 6!/6!4!} = \frac{3}{7}$$

From this example, we see the following properties: (1) we can classify the outcomes for each trial into one of only two mutually exclusive events (good and bad); (2) *the probability of success (or failure) changes on each trial*; (3) the trials *are dependent* upon outcomes of previous trials; and (4) the number of trials is fixed. Properties 2 and 3 become obvious if we use the conditional population approach to evaluate relevant conditional probabilities. For example, $P(G_2|B_1) = \frac{6}{9} \neq P(G_2|G_1) = \frac{5}{9}$, which means that the probability for the second trial is dependent upon the outcome of the first trial. Furthermore, the fact that the probability of selecting a good battery changes from trial to trial, for example, is evident from the following:

$$P(G_2|B_1) = \tfrac{6}{9} \neq \tfrac{6}{10} = P(G_1)$$

The binomial distribution cannot be used as a model in this situation because of properties 2 and 3. Since such properties are a direct result of the fact that we are sampling *without replacement*, we note that it would be appropriate to use the binomial distribution as a model if we were sampling *with replacement*.

The conditions of the previous example may be generalized as follows:

1. The population from which we are sampling contains a items having a characteristic termed *success* and b items having a characteristic termed *failure*; thus there are a total of $a+b$ items in the population.
2. We are to select *without* replacement a simple random sample of size n.

We wish to determine the probability of obtaining x successes in a sample of size n. Since we are not concerned with order, we can determine the total number of possible different samples of size n to be $_{a+b}C_n$. These samples are mutually exclusive, and, owing to the simple-random-sampling technique, each sample is equally likely to occur. Using the classical definition of probability, now we need to find the number of samples that satisfy the event of x successes and $n-x$ failures. Again, since we are not concerned with order, we have $_aC_x$ different ways of obtaining x successes and $_bC_{n-x}$ different ways of obtaining $n-x$ failures. Moreover, for each of the different ways of obtaining x successes, we have $_bC_{n-x}$ ways of obtaining $n-x$ failures. Hence we have a total of $_aC_x \cdot _bC_{n-x}$ different ways of obtaining x successes and $n-x$ failures. Therefore, we have

$$P_h(X=x|n,a,b) = \frac{_aC_x \cdot _bC_{n-x}}{_{a+b}C_n} \qquad (8.24)$$

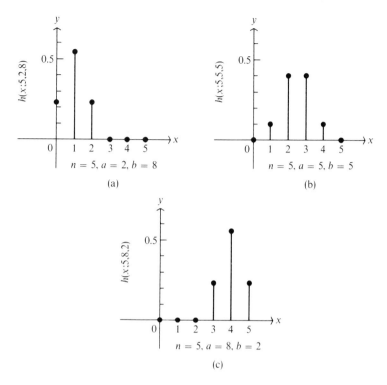

FIGURE 8.3

where $P_h(X = x \mid n,a,b)$ is the probability of obtaining x successes out of n draws from a finite population containing two mutually exclusive groups of items, a of which are termed successes and b of which are termed failures. Relation (8.24) obviously holds true *only for the case of simple random sampling without replacement.*

A random variable X having the probability function given by (8.24) is called a *hypergeometric random variable,* and the probability function is termed the *hypergeometric probability function.* This being the case, we employ the subscript h to denote a *hypergeometric* probability. For a more specific functional notation, we write (8.24) as

$$h(x; n, a, b) = \frac{{}_aC_x \cdot {}_bC_{n-x}}{{}_{a+b}C_n} \qquad x = 0, 1, 2, \ldots, a \qquad (8.25)$$

The graphic shape of the hypergeometric distribution is represented in Fig. 8.3 for fixed n and several values of a and b. Again the probabilities are represented by the height of the line, which reflects that the variable of interest is a discrete random variable.

Lieberman and Owen [6] have created an extensive table that provides both cumulative and individual probabilities for the hypergeometric distribution; however, such a table is too voluminous to be included. Nevertheless, individual values of $h(x;n,a,b)$ can be determined by substituting appropriate values for the binomial coefficient (i.e., combinations) into (8.24). Table A.4 in the Appendix gives the binomial coefficients for values of n to 20. In order to determine $_nC_r$, we simply locate the value in Table A.4 that corresponds to n and r; for example, to evaluate $_{11}C_5$, we read the value in Table A.4 that corresponds to $n=11$ and $r=5$, thus obtaining $_{11}C_5=462$.

Example 8.8 The Johnson Manufacturing Company has received a shipment of 20 parts that were ordered from a manufacturer and the order specifies 8 parts of type A and 12 parts of type B. Assuming that the correct order was received, what is the probability of selecting without replacement a simple random sample of size 3 containing 2 parts of type A and 1 part of type B?

Solution This problem satisfies the requirements for the *hypergeometric* model; namely, the population can be classified into mutually exclusive groups, and the sampling technique is that of simple random sampling without replacement. Furthermore, we have $a=8$, $b=12$, $x=2$, and $n=3$; therefore, using (8.25), we have

$$P_h(X=2 \mid n=3, a=8, b=12) = h(2;3,8,12) = \frac{_8C_2 \cdot _{12}C_1}{_{20}C_3}$$

From Table A.4, $_8C_2=28$, $_{12}C_1=12$, and $_{20}C_3=1{,}140$. Thus

$$P_h(X=2 \mid n=3, a=8, b=12) = \frac{(28)(12)}{1{,}140} = \frac{28}{95}$$

Example 8.9 We wish to conduct a market-research study involving 27 electronic-component manufacturers, of which 12 are classified as "large" and 15 are classified as "small." If we select a simple random sample of size 4 without replacement from the group of 27 manufacturers, what is the probability of selecting 2 large and 2 small manufacturers?

Solution Again this problem satisfies the requirements for the hypergeometric model. Therefore, $x=2$, $a=12$, $b=15$, and $n=4$ gives

$$P_h(X=2 \mid n=4, a=12, b=15) = h(2;4,12,15) = \frac{_{12}C_2 \cdot _{15}C_2}{_{27}C_4}$$

Using Table A.4, we obtain $_{12}C_2=66$ and $_{15}C_2=105$. We see that $_{27}C_4$ is not included in Table A.4. Thus we compute manually $_{27}C_4=17{,}550$, and hence we have

$$P_h(X=2 \mid n=4, a=12, b=15) = \frac{(66)(105)}{17{,}550} = \frac{77}{195}$$

A very useful feature of the hypergeometric distribution is its recursion formula, which allows us [once we have computed $h(x;n,a,b)$ for a specific

Sec. 8.4 / HYPERGEOMETRIC DISTRIBUTION

value x] to find probabilities for a value of $x+1$ without employing (8.25). The recursion formula is

$$h(x+1;n,a,b) = \frac{(n-x)(a-x)}{(x+1)(b-n+x-1)} h(x;n,a,b) \qquad (8.26)$$

Example 8.10 Acme Electronics Company has an inventory of 18 boxes (each containing 100 units) of transistors such that 7 boxes are of type A and 11 boxes are of type B. Consider the orders for the next 5 boxes of transistors to be a simple random sample of size 5 without replacement. Determine the probability that the orders contain either 2 or 3 boxes of type A. (Use the recursion formula where applicable.)

Solution We have $n=5$, $a=7$ and $b=11$. Thus

$$P_h(X=2 \cup X=3 \mid n=5, a=7, b=11)$$
$$= P_h(X=2 \mid n=5, a=7, b=11) + P_h(X=3 \mid n=5, a=7, b=11)$$
$$= h(2;5,7,11) + h(3;5,7,11) \qquad (8.27)$$

However, using Table A.4, we have

$$h(2;5,7,11) = \frac{{}_7C_2 \cdot {}_{11}C_3}{{}_{18}C_5} = \frac{(21)(165)}{8,568} = .4044 \qquad (8.28)$$

Also, by (8.26), we have

$$h(3;5,7,11) = \frac{(5-2)(7-2)}{(3)(11-5+3)} \cdot h(2;5,7,11) = \frac{15}{27}(.4044) = .2247 \qquad (8.29)$$

Finally, substituting the results of (8.28) and (8.29) into (8.27), we have

$$P_h(X=2 \cup X=3 \mid n=5, a=7, b=11) = .4044 + .2247 = .6291$$

Binomial Approximations for Hypergeometric Probabilities

We have noted that there are two reasons why the binomial model is not applicable in situations for which the hypergeometric probability function has been developed: First the probability of success is *not* constant from trial to trial; and second, the outcomes of various trials are *not* independent. However, if the population size $a+b$ is large and $n/(a+b)$ is small, the probability of success changes very little from trial to trial and the outcomes are practically independent. Thus, if these conditions are satisfied, we would expect the binomial distribution to be a fairly good approximation of the hypergeometric distribution, and it turns out that such is the case. For example, suppose a dealer has 5,000 items in inventory such that 2,000 are of type A and 3,000 are of type B. If we randomly select 5 items from the inventory without replacement, the probability that we have 3 of type A in our sample can be approximated by use of the binomial distribution with

$$p = \frac{2{,}000}{5{,}000} = \frac{2}{5} = .4 \quad \text{and} \quad n = 5$$

Thus

$$P_h(X = 3 \mid n = 5, a = 2{,}000, b = 3{,}000) = h(3; 5, 2{,}000, 3{,}000)$$
$$\approx P_b(X = 3 \mid n = 5, p = .4)$$
$$= P_b(X \geq 3 \mid n = 5, p = .4) - P_b(X \geq 4 \mid n = 5, p = .4)$$
$$= .3174 - .0870 = .2304$$

You should verify that computing this probability using the hypergeometric function yields a value of 0.23044.

The mean for the hypergeometric distribution can be determined in the same manner as for the binomial distribution. We have

$$E(X) = \frac{na}{a+b} \tag{8.30}$$

where n is the sample size, a is the number of successes in the population, and b is the number of failures in the population.

EXERCISES

1. What is the essential difference between situations in which the hypergeometric distribution can be used as a model and situations in which the binomial distribution can be used as a model?

2. Suppose a bin of parts contains 11 usable and 4 nonusable parts. Determine the probability of selecting without replacement a simple random sample of size 4 containing

(a) Exactly 3 usable parts
(b) At least 3 usable parts
(c) At most 3 usable parts
(d) At least 1 but not more than 3 usable parts
(e) Exactly 4 usable parts

3. Determine the following probabilities:

(a) $P_h(X = 3 \mid n = 5, a = 6, b = 9)$ (b) $P_h(X = 0 \mid n = 7, a = 14, b = 3)$
(c) $P_h(X = 9 \mid n = 10, a = 9, b = 10)$ (d) $P_h(X = 4 \mid n = 7, a = 11, b = 14)$
(e) $P_h(X = 2 \mid n = 6, a = 21, b = 4)$

4. Independent Research Associates wishes to conduct an employee-attitude survey in a small firm with 17 employees, 9 of whom are college graduates. If a simple

Sec. 8.4 / HYPERGEOMETRIC DISTRIBUTION

random sample of size 5 is selected without replacement, determine the probabilities that the sample contains

(a) 2 college graduates
(b) 3 college graduates
(c) 2 or more college graduates
(d) Less than 3 college graduates

5. The marketing department of D&D Enterprises wishes to conduct a survey of the size of the orders placed by 19 customers, 9 of whom are classified as "wholesale" and 10 of whom are classified as "retail." If a simple random sample of size 10 is selected without replacement, determine the probabilities that

(a) Exactly 3 retail customers are selected
(b) More than 3 retail customers are selected
(c) At most 3 retail customers are selected
(d) At least 4 but not more than 6 retail customers are selected

6. Under what circumstances may the binomial distribution be used as an approximation of the hypergeometric distribution? Why?

7. Repeat the probabilities requested in Exercise 4 with a large firm of 10,000 employees, 2,000 of whom are college graduates.

8. Repeat the probabilities requested in Exercise 5 with a firm that has 4,000 customers, 1,200 of whom are classified as wholesale and 2,800 of whom are classified as retail.

9. Twenty-five customers have placed their purchases on layaway. For these, 10 payments are overdue. Determine the probability of selecting without replacement a simple random sample of size 5 containing

(a) Exactly 1 payment overdue
(b) At least 4 payments overdue
(c) Less than 4 payments overdue
(d) No overdue payments

10. The Fix-It Repair Shop has 5 toasters and 9 irons to repair. If the repairman randomly selects 3 appliances to repair, determine the probability that he selects

(a) Exactly 3 toasters
(b) Exactly 3 irons
(c) At most 1 toaster
(d) At least 1 toaster
(e) Exactly 1 iron

11. Repeat the probabilities requested in Exercise 10 with 6,000 toasters and 9,000 irons.

12. An apartment building of 20 apartments has 12 single tenants and 8 married couples. If a survey of 6 apartments is conducted, determine the probability of selecting

(a) Exactly 4 apartments with single tenants
(b) At most 2 apartments with married couples

(c) At least 5 apartments with married couples
(d) No apartments with married couples
(e) At least 1 apartment with a single tenant and at most 4 apartments with single tenants

13. Repeat the probabilities requested in Exercise 12 for the entire apartment complex of 2,000 apartments, of which 800 are occupied by married couples.

8.5 MARKOV CHAINS

In Sec. 8.2 we considered the binomial probability distribution for which the individual trials were Bernoulli trials and the outcome of each trial was independent of the outcome of any preceding trials. Furthermore, the probability of success was constant from one trial to the next. In this section, we consider a slight generalization of Bernoulli trials. That is, we permit more than two possible outcomes per trial and we permit the trials to have a certain amount of dependence. This change allows us to consider a large number of random variables in business because many variables are observed at regular time intervals over a given period of time. Examples of such situations include consumers switching from one brand to another as the result of advertising, special promotion, price, dissatisfaction, etc.; voters switching allegiance from one political party to another as a result of platforms, candidates, issues, etc.; and managers making different decisions due to changing conditions.

A *Markov chain* is a sequence of experiments in which the outcomes of each trial depend at most on the outcomes of the preceding trial; the outcomes are called *states*. The process of a Markov chain is described by a *probability matrix*. In Chap. 4 we studied matrices, so the manipulation of a matrix will not be new; however, the elements in the probability matrix are of a special character. The *transition matrix* of a Markov chain is an $n \times n$ matrix **P** whose elements p_{ij}, called transition probabilities, are such that

(1) $\quad 0 \leq p_{ij} \leq 1$

(2) $\quad p_{i1} + p_{i2} + \cdots + p_{in} = 1 \quad$ for each $i = 1, 2, \ldots, n$

Furthermore, the element p_{ij} in the ith row and jth column of **P** represents the probability of state (or outcome) j occurring on the next trial given that we are currently in state i.

As an example of a Markov chain, let us consider a problem related to politics. Suppose that of all voters who voted Democratic (D) in the last election, 60 percent will vote Democratic in the next election, 30 percent will vote Republican, and 10 percent will vote Independent. Of those who voted Republican (R) in the last election, 20 percent will vote Democratic in the next election, 60 percent will vote Republican, and 20 percent will vote Independent. Of those who voted Independent (I), 30 percent will vote

Democratic in the next election, 20 percent will vote Republican, and 50 percent will vote Independent. The transition matrix for this situation is

$$\begin{array}{c} \text{Last election} \end{array} \begin{array}{c} \text{States} \\ D \\ R \\ I \end{array} \overbrace{\begin{bmatrix} D & R & I \\ .6 & .3 & .1 \\ .2 & .6 & .2 \\ .3 & .2 & .5 \end{bmatrix}}^{\text{Next election}} = \mathbf{P}$$

Thus the probability in the first row and first column $p_{11} = .6$ represents the conditional probability that a voter will vote Democratic in the next election given that he voted Democratic in the last election. In other words, the probability is conditional and is dependent upon what the voter did in the last election. If we wish to determine the probability that a voter who voted Democratic in the last election will vote Democratic in the second election from now, we simply make use of our rules for conditional probabilities, in particular Rule 7.4. Letting the subscript of the state represent the number of the future election with which we are concerned, we may apply Rule 7.4 to determine the probability of a Democratic voter in the last election voting Democratic in two elections hence. We see that the voter could vote for any of the three parties in the first election hence. Thus we wish to determine

$$P(D_2) = P(D_1 \cap D_2) + P(R_1 \cap D_2) + P(I_1 \cap D_2)$$

Applying the multiplication rule and using the conditional probabilities from the transition matrix, we have

$$P(D_2) = P(D_2|D_1)P(D_1) + P(D_2|R_1)P(R_1) + P(D_2|I_1)P(I_1)$$
$$= (.6)(.6) + (.2)(.3) + (.3)(.1) = .45$$

Hence the probability that a Democratic voter in the last election will vote Democratic in two elections hence is .45.

We note that the above probability can be obtained simply by matrix multiplication. That is, the first row of the matrix $\mathbf{P}$ times its first column yields the probability. In other words, the element in the first row and first column of $\mathbf{PP}$ (or $\mathbf{P}^2$) is the desired probability. If we square the matrix $\mathbf{P}$, we have

$$\mathbf{P}^2 = \mathbf{PP} = \begin{bmatrix} .6 & .3 & .1 \\ .2 & .6 & .2 \\ .3 & .2 & .5 \end{bmatrix} \begin{bmatrix} .6 & .3 & .1 \\ .2 & .6 & .2 \\ .3 & .2 & .5 \end{bmatrix} = \begin{bmatrix} .45 & .38 & .17 \\ .30 & .46 & .24 \\ .37 & .31 & .32 \end{bmatrix}$$

The matrix obtained by squaring the transition matrix is called the *two-step* transition matrix; it yields the probabilities of the voter's actions for two elections hence. For example, the probability .46 in the second row and second column is the probability that one who voted Republican in the last election will vote Republican in two elections hence. Similarly, the other probabilities may also be obtained.

In general, the element a_{ij} of the matrix obtained by taking the nth power of **P** represents the probability that an experiment starting in the ith state will be in the jth state after n trails. Furthermore, $\mathbf{P}^n$ is known as the *n-step transition matrix*.

Example 8.11 The behavior patterns of an assembly worker indicate that if he picks up a part with his left hand, the probability he will pick up the next part with his left hand is $\frac{4}{5}$. If he picks up a part with his right hand, the probability he will pick up the next part with his left hand is $\frac{2}{5}$.

Solution

(a) The transition matrix is

$$\mathbf{P} = \text{This part} \begin{Bmatrix} \text{Left hand} \\ \text{Right hand} \end{Bmatrix} \begin{matrix} \overbrace{\text{Left hand} \quad \text{Right hand}}^{\text{Next part}} \\ \begin{bmatrix} \frac{4}{5} & \frac{1}{5} \\ \frac{2}{5} & \frac{3}{5} \end{bmatrix} \end{matrix}$$

(b) To determine the four-step transition matrix, we shall determine the two-step transition and then square it.

$$\mathbf{P}^2 = \begin{bmatrix} \frac{4}{5} & \frac{1}{5} \\ \frac{2}{5} & \frac{3}{5} \end{bmatrix} \begin{bmatrix} \frac{4}{5} & \frac{1}{5} \\ \frac{2}{5} & \frac{3}{5} \end{bmatrix} = \begin{bmatrix} \frac{18}{25} & \frac{7}{25} \\ \frac{14}{25} & \frac{11}{25} \end{bmatrix}$$

$$\mathbf{P}^4 = \begin{bmatrix} \frac{18}{25} & \frac{7}{25} \\ \frac{14}{25} & \frac{11}{25} \end{bmatrix} \begin{bmatrix} \frac{18}{25} & \frac{7}{25} \\ \frac{14}{25} & \frac{11}{25} \end{bmatrix} = \begin{bmatrix} \frac{422}{625} & \frac{203}{625} \\ \frac{406}{625} & \frac{219}{625} \end{bmatrix}$$

EXERCISES

Which of the matrices in Exercises 1 to 8 satisfy the definition for transition matrices?

1. $\begin{bmatrix} 1 & 0 \\ 0 & 1 \end{bmatrix}$

2. $\begin{bmatrix} \frac{1}{2} & \frac{1}{2} \\ \frac{1}{2} & 1 \end{bmatrix}$

3. $\begin{bmatrix} \frac{3}{2} & -\frac{1}{2} \\ \frac{1}{4} & \frac{3}{4} \end{bmatrix}$

4. $\begin{bmatrix} \frac{1}{2} & \frac{1}{2} & 0 \\ 0 & 0 & 1 \\ \frac{1}{2} & 0 & \frac{1}{2} \end{bmatrix}$

5. $\begin{bmatrix} \frac{1}{4} & \frac{1}{4} & \frac{1}{4} \\ 1 & 0 & 1 \\ \frac{2}{3} & \frac{1}{3} & 0 \end{bmatrix}$

6. $\begin{bmatrix} 1 & 0 & 0 \\ 0 & 1 & 0 \\ 0 & 0 & 1 \end{bmatrix}$

7. $\begin{bmatrix} \frac{1}{3} & \frac{1}{3} & \frac{1}{3} \\ \frac{1}{4} & -\frac{1}{2} & \frac{1}{4} \\ \frac{1}{2} & 0 & \frac{1}{2} \end{bmatrix}$

8. $\begin{bmatrix} \frac{1}{2} & \frac{1}{4} & 0 & \frac{1}{4} \\ 0 & \frac{1}{3} & \frac{1}{3} & \frac{1}{3} \\ \frac{1}{4} & \frac{1}{4} & \frac{1}{4} & \frac{1}{4} \\ \frac{1}{5} & \frac{1}{5} & \frac{1}{10} & \frac{1}{2} \end{bmatrix}$

Sec. 8.5 / MARKOV CHAINS

For each of the transition matrices in Exercises 9 to 11, determine the two-step and three-step transition matrices.

9. $\begin{bmatrix} \frac{1}{2} & \frac{1}{2} \\ \frac{1}{3} & \frac{2}{3} \end{bmatrix}$ 10. $\begin{bmatrix} .6 & .4 \\ .3 & .7 \end{bmatrix}$ 11. $\begin{bmatrix} \frac{1}{2} & 0 & \frac{1}{2} \\ \frac{1}{3} & \frac{1}{3} & \frac{1}{3} \\ 1 & 0 & 0 \end{bmatrix}$

12. The children at Kollin's Kiddie Kollege are given milk for lunch. If a child had chocolate milk for lunch the previous day, the probability that he will have chocolate milk for lunch today is .5. However, if he had plain milk the day before, the probability that he will have chocolate milk today is .7.

    (a) Determine the transition matrix for the children's selection
    (b) Determine the proportion of the time children will choose chocolate milk.
        [Hint: Determine P^n for large values of n such that it is essentially unchanged when multiplied by P.]

13. A study was made for the city of Chicago. It was determined that the breakdown of the educational level of parents and their children is as follows:

		Highest educational level of children		
		College	High school	Elementary school
Highest educational level of parents	College	.80	.15	.05
	High school	.50	.35	.15
	Elementary school	.20	.60	.20

    (a) Determine the transition matrix.
    (b) Determine the probability that the grandchild of a college graduate is a college graduate.
    (c) Determine the probability that the grandchild of a high school graduate is a college graduate.
    (d) Determine the probability that the grandchild of an elementary school graduate is a college graduate.
    (e) Determine the probability that the great-grandchild of a high school graduate is a college graduate.

14. Weather Bureau records for Texas show that the probability is .8 that a sunny day will be followed by a sunny day, .1 that a sunny day will be followed by a cloudy day, and .1 that a sunny day will be followed by a rainy day. Likewise, the probabilities that a cloudy day will be followed by a sunny, cloudy, or rainy day are .6, .2, and .2, respectively. The probabilities that a rainy day will be followed by a sunny, cloudy, or rainy day are .6, .3, and .1, respectively.

    (a) Determine the transition matrix for the Texas weather
    (b) Determine the probability that if it is sunny today it will be rainy two days from now.
    (c) In order to have a successful picnic and outing at the lake, we must have

either sunny or cloudy weather. If it is raining today, determine the probability of having a successful picnic two days from now.

15. A market analysis of three brands of toothpaste gives the following transition matrix of probabilities of customers shifting each year from one brand to another. Assume that all customers stay with one of the three brands.

$$\mathbf{P} = \begin{matrix} & \begin{matrix} A & B & C \end{matrix} \\ \begin{matrix} A \\ B \\ C \end{matrix} & \begin{bmatrix} .7 & .1 & .2 \\ .2 & .5 & .3 \\ .2 & .1 & .7 \end{bmatrix} \end{matrix}$$

(a) Determine the two-step transition matrix
(b) Determine the probability that those using brand A today will be using brands B or C 2 years hence.
(c) Determine the probability that those using brand C today will still be using brand C 2 years hence.

8.6 GAME THEORY

In numerous situations, the mathematical expectation of each of the outcomes associated with several events may be used as a basis for deciding among alternatives. For example, one may attempt to maximize expected sales, gains, or profits or minimize expected costs or losses. Occasionally it is possible to view a business, economic, political, or sociological situation as a game among two or more parties. *Game theory* is the mathematical analysis of events in business, warfare, or social situations where there is conflict of interest. Examples of such situations include collective bargaining, competition for market share when there is a small number of competitors, politicians competing to get elected, countries at war against each other, and investment alternatives. The problem is much simpler when there are only two parties involved; thus we shall restrict our consideration to that type of problem.

The types of competitive activities described above are of a very serious nature. However, since they are competitive, we label them *games*. Furthermore, when we permit only two parties or persons or players to be involved, we have what we call a *two-person game*. The games that we shall consider are those in which the gains equal the losses, that is, when one player wins an amount a the other player loses an equal amount. Such games are called *zero-sum games*. The objective of game theory is to determine the *optimal* strategies for each player; that is, strategies that are most profitable to each player. The payoff associated with these optimal strategies is the *value* of the game.

A two-person game between players A and B is defined by a *payoff matrix*. Each player will have a number of strategies available to him. The payoff is dependent upon the strategies of all players. Thus it is apparent why such games are occasionally referred to as games of strategy and matrix

Sec. 8.6 / GAME THEORY

games. An example of a payoff matrix where players A and B each have available two strategies is given below

$$\text{Player } A \begin{array}{c} \\ 1 \\ 2 \end{array} \overset{\begin{array}{cc} \text{Player } B \\ \text{I} \quad\quad \text{II} \end{array}}{\begin{bmatrix} 4 & -1 \\ -3 & 5 \end{bmatrix}}$$

The elements of the payoff matrix are the payoffs and they are expressed as payments to player A. For example, the element 4 in the first row and first column represents the amount player A will receive (and consequently the amount player B will lose) if he chooses strategy 1 and player B chooses strategy **I**. The element -3 in the second row and first column represents the amount player A will win (i.e., he loses 3) if he chooses strategy 2 and player B chooses strategy **I**.

A game for which player A has m strategies and player B has n strategies is called an $m \times n$ game, since the payoff matrix is of order $m \times n$. Thus a 3×3 game may be described by the following payoff matrix:

$$\text{Player } A \begin{array}{c} \\ 1 \\ 2 \\ 3 \end{array} \overset{\begin{array}{ccc} \text{Player } B \\ \text{I} \quad\quad \text{II} \quad\quad \text{III} \end{array}}{\begin{bmatrix} -5 & -7 & 5 \\ 6 & 0 & 3 \\ 4 & -3 & -4 \end{bmatrix}}$$

From the viewpoint of player B, in the above payoff matrix, strategy **II** is always better than strategy **I** because each payoff is more favorable for him regardless of the strategy chosen by player A. Thus player B can eliminate strategy **I** from consideration and consider only the reduced payoff matrix.

$$\text{Player } A \begin{array}{c} \\ 1 \\ 2 \\ 3 \end{array} \overset{\begin{array}{cc} \text{Player } B \\ \text{II} \quad\quad \text{III} \end{array}}{\begin{bmatrix} -7 & 5 \\ 0 & 3 \\ -3 & -4 \end{bmatrix}}$$

When a player has a strategy for which the payoff is as good as or better than a second strategy for each possible strategy of his opponent, that strategy is said to *dominate* the second strategy. Hence, anytime a strategy is dominated by another strategy, we may eliminate the one that is dominated.

Further considering the reduced payoff matrix, we see that strategy 2 for player A dominates strategy 3; thus we may eliminate strategy 3 and obtain the following 2×2 payoff matrix:

$$\begin{array}{c} \\ \text{Player } A \end{array} \begin{array}{c} \text{Player } B \\ \begin{array}{cc} \text{II} & \text{III} \end{array} \\ \begin{array}{c} 1 \\ 2 \end{array} \begin{bmatrix} -7 & 5 \\ 0 & 3 \end{bmatrix} \end{array}$$

Now comparing strategy **III** versus strategy **II** for player *B*, we see that **II** dominates **III**; hence we may eliminate **III** and obtain the following 2×1 payoff matrix:

$$\begin{array}{c} \\ \text{Player } A \end{array} \begin{array}{c} \text{Player } B \\ \text{II} \\ \begin{array}{c} 1 \\ 2 \end{array} \begin{bmatrix} -7 \\ 0 \end{bmatrix} \end{array}$$

Furthermore, examining the 2×1 payoff matrix, we see that strategy 2 for player *A* dominates strategy 1. Hence we eliminate strategy 1 and obtain the 1×1 payoff matrix that leaves each player with only one strategy and a payoff of zero. Thus the value of the game is zero. Furthermore, whenever the value of the game is zero we say that the game is *fair game*.

The optimum pair of strategies (2,**II**) for the above game is called a *saddle point* because the associated value in the payoff matrix is the minimum in its row *and* maximum in its column. Of course, the goal in game theory is to obtain the optimal strategies for each player. Thus the saddle point simultaneously represents the maximum of the minimum gains for player *A* in each row and the minimum of the maximum losses for player *B* in each column. When a game has a saddle point, the optimal strategies for both players are *pure strategies*; that is, each player will always use the same strategy every time the game is played regardless of what his opponent does. When a player uses two or more different strategies he is using a *mixed strategy*.

For the 3×3 game considered above, we can determine the saddle point more easily by simultaneously determining the maximum of the minimum payoffs for the rows and minimum of the maximum payoffs for the columns, as illustrated below.

$$\begin{array}{cc} & \text{Player } B \\ & \begin{array}{ccc} \text{I} & \text{II} & \text{III} \end{array} \quad \text{Row minimum} \\ \text{Player } A \begin{array}{c} 1 \\ 2 \\ 3 \end{array} & \begin{bmatrix} -5 & -7 & 5 \\ 6 & 0 & 3 \\ 4 & -3 & -4 \end{bmatrix} \quad \begin{array}{l} -7 \\ 0 \text{ (maximum)} \\ -4 \end{array} \end{array}$$

$$\text{Column maximum} \quad \begin{array}{ccc} 6 & 0 & 5 \end{array}$$
$$\text{(minimum)}$$

The maximum of the row minimums is equal to the minimum of the column maximums, which reflects that the game has a saddle point (2,**II**) and the value of the game is zero. Games for which saddle points exist are called *strictly determined* games. However, not all games are strictly determined, since not all games have saddle points.

For those games that are not strictly determined, we must use another means of obtaining the optimal strategies. In strictly determined games, the optimal strategy for each player is a pure strategy; that is, one that is always used regardless of the opponent's actions. In games that are not strictly determined, we view the game as one that will be played repeatedly and the players are permitted to change their strategies from game to game. Thus each player wishes to prevent the development of a systematic pattern of choosing his strategies for fear that his opponent will discover his pattern. Consequently, for an $m \times n$ game, this can be accomplished by having player A select a set of probabilities $p_1, p_2, \ldots, p_m$ that will determine the frequencies with which he plays his m strategies and, likewise, having player B select a set of probabilities $q_1, q_2, \ldots, q_n$ that will determine the frequency with which he plays his n strategies. Then each player can select his strategies at random according to his set of probabilities. Strategies generated in this manner are called *randomized strategies* and include the one-play, pure strategies.

The associated payoff for the randomized strategies will vary from game to game due to the chance of selection of the strategies. However, to determine the optimal strategy for a long sequence of games, we shall consider the average payoff (i.e., the expected payoff) per game. To assist us in this consideration, we must rely on a theorem that we shall state and use; however, we shall not attempt to prove it. This theorem is known as the *Basic Theorem of Game Theory* and is given below

---

Theorem 8.1 For every matrix game, there exist optimal strategies for players A and B and a number v, called the value of the game, such that if A uses his optimal strategy his expected payoff from B will be greater than or equal to v for every strategy of B, and such that if B uses his optimal strategy the expected payoff to A will be less than or equal to v for every strategy of A.

This theorem assures us that every game has a solution when randomized strategies are used. Furthermore, it assures that if player A chooses to employ the ith pure strategy (i.e., the ith strategy of the m available), then there is an optimal randomized strategy $q_1, q_2, \ldots, q_n$ that player B can use such that the expected payoff is less than or equal to v. In other words, for the payoff matrix $A = [a_{ij}]$, the expected payoff is such that

$$a_{i1}q_1 + a_{i2}q_2 + \cdots + a_{in}q_n \leq v$$

Thus, since player A may choose any pure strategy for a given game, the following system of inequalities must be satisfied:

$$a_{11}q_1 + a_{12}q_2 + \cdots + a_{1n}q_n \leqslant v$$
$$a_{21}q_1 + a_{22}q_2 + \cdots + a_{2n}q_n \leqslant v$$
$$\vdots \qquad \vdots \qquad \vdots \qquad \vdots$$
$$a_{m1}q_1 + a_{m2}q_2 + \cdots + a_{mn}q_n \leqslant v \tag{8.31}$$

We assume that $v > 0$; however, if it is not positive then we can add a sufficiently large constant to each element of A such that the newly obtained payoff matrix will have only positive payoff values. Thus the value of the game will be positive. The only effect that adding a constant to each payoff element will have on the value of the game is that it will increase it by the amount added without changing the optimal strategies. It is left as an exercise for the reader to show that, when the elements of A are all positive, the value of the game is positive.

Since $v > 0$, we may divide both sides of each inequality in the system (8.31) by v. Performing this division for each inequality and substituting $x_i = q_i/v$ in order to simplify the notation, we have the following system of inequalities:

$$a_{11}x_1 + a_{12}x_2 + \cdots + a_{1n}x_n \leqslant 1$$
$$a_{21}x_1 + a_{22}x_2 + \cdots + a_{2n}x_n \leqslant 1$$
$$\vdots \qquad \vdots \qquad \vdots \qquad \vdots$$
$$a_{m1}x_1 + a_{m2}x_2 + \cdots + a_{mn}x_n \leqslant 1 \tag{8.32}$$

Because the q_i values constitute a set of probabilities, it follows that

$$x_1 + x_2 + \cdots + x_n = \frac{q_1 + q_2 + \cdots + q_n}{v} = \frac{1}{v} \tag{8.33}$$

If the probabilities chosen by player B are optimal, they must minimize v. Consequently, to determine the optimal probabilities, we may simply determine the values of $x_1, x_2, \ldots, x_n$ that minimize v subject to the constraints given by the system of inequalities (8.32). However, minimizing v is equivalent to maximizing $x_1 + x_2 + \cdots + x_n$ in Eq. (8.33). Letting $x_0 = x_1 + x_2 + \cdots + x_n$, we may express our problem as the following linear programming problem:

Maximize

$$x_0 = x_1 + x_2 + \cdots + x_n$$

Sec. 8.6 / GAME THEORY

subject to

$$a_{11}x_1 + a_{12}x_2 + \cdots + a_{1n}x_n \leq 1$$
$$a_{21}x_1 + a_{22}x_2 + \cdots + a_{2n}x_n \leq 1$$
$$\vdots \qquad \vdots \qquad \vdots$$
$$a_{m1}x_1 + a_{m2}x_2 + \cdots + a_{mn}x_n \leq 1$$
$$x_j \geq 0 \qquad j = 1, 2, \ldots, n \quad (8.34)$$

Similar reasoning for player A and his strategies leads to the following linear programming problem, where $y_i = p_i/v$:

Minimize

$$y_0 = y_1 + y_2 + \cdots + y_m$$

subject to

$$a_{11}y_1 + a_{21}y_2 + \cdots + a_{m1}y_m \geq 1$$
$$a_{12}y_1 + a_{22}y_2 + \cdots + a_{m2}y_m \geq 1$$
$$\vdots \qquad \vdots \qquad \vdots \qquad \vdots$$
$$a_{1n}y_1 + a_{2n}y_2 + \cdots + a_{mn}y_m \geq 1 \quad (8.35)$$
$$y_i \geq 0 \qquad i = 1, 2, \ldots, m$$

The linear programming problem expressed in (8.35) is the dual of the problem expressed in (8.34). Hence, if we employ the simplex algorithm of Chap. 6 to solve (8.34), we will also have determined the solutions to the dual problem (8.35).

Example 8.12 Determine the value of the game and the optimal strategies for players A and B for the payoff matrix

$$\begin{bmatrix} 1 & 6 & 3 \\ 2 & 4 & 8 \\ 3 & -3 & 3 \end{bmatrix}$$

Solution We first determine that the payoff matrix does not contain a saddle point, because if it did the optimal strategy for each player would be the pure strategy corresponding to the saddle point. Setting the problem up as a linear programming problem similar to (8.34), we obtain

Maximize

$$x_0 = x_1 + x_2 + x_3$$

subject to

$$x_1 + 6x_2 + 3x_3 \leq 1$$
$$2x_1 + 4x_2 + 8x_3 \leq 1$$
$$3x_1 - 3x_2 + 3x_3 \leq 1$$
$$x_1 \geq 0 \quad x_2 \geq 0 \quad x_3 \geq 0$$

Applying the simplex algorithm given in Sec. 6.1, we obtain Table 8.2.

Table 8.2

Tableau	Current Basis	x_0	x_1	x_2	x_3	x_4	x_5	x_6	Right-Side Constant	Ratio
I	x_0	1	-1^*	-1	-1	0	0	0	0	
	x_4	0	1	6	3	1	0	0	1	1
	x_5	0	2	4	8	0	1	0	1	$\frac{1}{2}$
	x_6	0	③	-3	3	0	0	1	1	$\frac{1}{3}$ = minimum
II	x_0	1	0	-2^*	0	0	0	$\frac{1}{3}$	$\frac{1}{3}$	
	x_4	0	0	7	2	1	0	$-\frac{1}{3}$	$\frac{2}{3}$	$\frac{2}{21}$
	x_5	0	0	⑥	6	0	1	$-\frac{2}{3}$	$\frac{1}{3}$	$\frac{1}{18}$ = minimum
	x_1	0	1	-1	1	0	0	$\frac{1}{3}$	$\frac{1}{3}$	
III	x_0	1	0	0	2	0	$\frac{1}{3}$	$\frac{1}{9}$	$\frac{4}{9}$	
	x_4	0	0	0	-5	1	$-\frac{7}{6}$	$\frac{4}{9}$	$\frac{5}{18}$	
	x_2	0	0	1	1	0	$\frac{1}{6}$	$-\frac{1}{9}$	$\frac{1}{18}$	
	x_1	0	1	0	2	0	$\frac{1}{6}$	$\frac{2}{9}$	$\frac{7}{8}$	

From tableau **III** of Table 8.2, we have the optimal values of $x_0 = \frac{4}{9}$, $x_1 = \frac{7}{18}$, $x_2 = \frac{1}{18}$, and $x_3 = 0$. Thus, from (8.33), we have $x_0 = \frac{4}{9} = 1/v$, or the value of the game is $v = \frac{9}{4}$. Substituting for x_1, x_2, and x_3, we get

$$\frac{q_1}{v} = \frac{7}{18} \qquad \frac{q_2}{v} = \frac{1}{18} \qquad \frac{q_3}{v} = 0$$

Substituting $v = \frac{9}{4}$, we obtain

$$q_1 = \frac{7}{8} \qquad q_2 = \frac{1}{8} \qquad q_3 = 0$$

which means that player B would use his first strategy in $\frac{7}{8}$ of all games he plays, his second strategy in $\frac{1}{8}$ of all games, and his third strategy would never be used. The lack of use of the third strategy is apparent since the first column dominates the third column from player B's viewpoint. Furthermore, the use of the first two strategies would be *randomly* generated in a 7:1 ratio in favor of the first strategy in order to avoid the development of a systematic rule by player B.

Sec. 8.6 / GAME THEORY

Player A's strategies are represented in the linear programming problem similar to (8.35). For the payoff matrix of this example, this problem is given by

Minimize
$$y_0 = y_1 + y_2 + y_3$$

subject to
$$y_1 + 2y_2 + 3y_3 \geqslant 1$$
$$6y_1 + 4y_2 - 3y_3 \geqslant 1$$
$$3y_1 + 8y_2 + 3y_3 \geqslant 1$$
$$y_1 \geqslant 0 \qquad y_2 \geqslant 0 \qquad y_3 \geqslant 0$$

However, since this is the dual of the problem that we have already solved, we can determine the optimal values for y_1, y_2, and y_3 from tableau **III** of Table 8.2. Thus we have $y_1 = 0$, $y_2 = \frac{1}{3}$, and $y_3 = \frac{1}{9}$. Substituting for y_1, y_2, and y_3, we have

$$\frac{p_1}{v} = 0 \qquad \frac{p_2}{v} = \frac{1}{3} \qquad \frac{p_3}{v} = \frac{1}{9}$$

Substituting $v = \frac{9}{4}$ and solving for the p_i values, we have

$$p_1 = 0 \qquad p_2 = \frac{3}{4} \qquad p_3 = \frac{1}{4}$$

Thus player A will not employ his first strategy, but will use his second strategy $\frac{3}{4}$ of the time and his third strategy $\frac{1}{4}$ of the time. Of course, the use of the second and third strategies will be randomly generated in a 3:1 ratio in order to prevent player A from establishing a systematic pattern for his strategies.

EXERCISES

In Exercises 1 to 8 determine whether the game has a saddle point. For each game with a saddle point, determine the optimal strategies and the value of the game

1. $\begin{bmatrix} 3 & 1 \\ -1 & 7 \end{bmatrix}$
2. $\begin{bmatrix} -2 & 0 \\ -3 & 2 \end{bmatrix}$
3. $\begin{bmatrix} 5 & 4 \\ -5 & 1 \end{bmatrix}$

4. $\begin{bmatrix} 2 & 1 \\ 0 & 1 \end{bmatrix}$
5. $\begin{bmatrix} 4 & -6 & -3 \\ 3 & 1 & 2 \end{bmatrix}$
6. $\begin{bmatrix} 9 & 2 & 0 \\ 5 & 4 & 7 \end{bmatrix}$

7. $\begin{bmatrix} 3 & -1 \\ 5 & 2 \\ -2 & 5 \end{bmatrix}$
8. $\begin{bmatrix} -2 & 0 & -4 & -1 \\ 2 & 3 & -3 & 1 \\ 0 & 2 & -1 & 1 \end{bmatrix}$

In Exercises 9 to 12, determine the value of the game and the optimal randomized strategies for both players

9. $\begin{bmatrix} 5 & -1 \\ 3 & 4 \end{bmatrix}$ 10. $\begin{bmatrix} 5 & 0 \\ 4 & 3 \end{bmatrix}$ 11. $\begin{bmatrix} 3 & 0 \\ 2 & 1 \\ 1 & 3 \end{bmatrix}$

12. $\begin{bmatrix} -1 & 2 & 1 \\ 3 & -1 & 1 \end{bmatrix}$

13. Gene Bates, a young businessman, wishes to invest some of his money. He has researched the alternatives among stocks, bonds, and banks for the different economic conditions. The payoff matrix is given below for the percentage expected return on his investment from the time period he wished to commit his funds. Determine the percentage of his money that he should put in each type of investment regardless of the economic situation.

	Recession	Normal	Inflation
Stocks	−70	30	70
Bonds	30	20	40
Banks	10	20	30

14. The Miller Electronics Company plans to introduce a new electronic calculator. Their competitor, Fairfield Corporation, is also planning to introduce a similar calculator. The price of the new calculator is going to be either $8 or $10, depending on how they are sold. It is believed that Fairfield Corporation is planning to use the same prices. The payoff matrix given below represents the amount (in thousands of dollars) of profit for Miller. Determine the optimal strategy for each firm and the value of the game

		Fairfield	
		$8	$10
Miller	$8	−80	40
	$10	40	0

15. Bob and Tom each simultaneously show a nickel, dime, or quarter. If the total amount shown is even, Bob wins Tom's coin. Otherwise, Tom wins Bob's coin.

   (a) Construct the payoff matrix.
   (b) Determine optimal strategies for Bob and Tom.
   (c) Determine the value of the game.

16. Sue and Bill play a game in which each simultaneously shows one or two fingers. If there is a match, Sue wins $1 per finger. If there is no match, Bill wins $1 per finger.

   (a) Determine the payoff matrix.
   (b) Determine the optimal strategies for each player.
   (c) Determine the value of the game.

IMPORTANT TERMS AND CONCEPTS	Basic theorem of game theory
	Bernoulli
	    probability function
	    process
	    trial
	Binomial probability function
	Cumulative
	    binomial distribution
	    hypergeometric distribution
	    Poisson distribution
	Dominance
	Fair game
	Game theory
	Games
	    fair
	    strictly determined
	    two-person
		value
		zero-sum
		Hypergeometric probability function
		Markov chains
		Matrix
		    n-step transition
		    payoff
		    probability
		    transition
		Poisson probability function
		Saddle point
		States
		Strategy
		    mixed
		    pure
		    randomized

REFERENCES

1. Harvard University Computation Laboratory: *Tables of the Cumulative Binomial Probability Distribution*, Harvard University Press, Cambridge, Mass., 1955.
2. U.S. Army Ordnance Corps: "Tables of Cumulative Binomial Probabilities," *Ordinance Corps Pamphlet* ORDP 20-1 (September 1952).
3. National Bureau of Standards: "Tables of the Binomial Probability Distribution," in *Applied Mathematics*, ser. 6, U.S. Government Printing Office, Washington, D.C., 1950.
4. Owen, D. B.: *Handbook of Statistical Tables*, Addison-Wesley Publishing Company, Inc., Reading, Mass., 1962.
5. Romig, H. G.: 50–100 *Binomial Tables*, John Wiley & Sons, Inc., New York, 1953.
6. Lieberman, G. J., and D. B. Owen: *Tables of the Hypergeometric Probability Distribution*, Stanford University Press, Stanford, Calif., 1961.

REVIEW PROBLEMS

1. Records show that 8 percent of the parts produced are defective. Each day 100 parts are sampled. Use Table A.2 to find the probability

(a) That exactly 5 defects are found
(b) That more than 97 of the parts sampled are good
(c) That at least 2 defects are found
(d) That between 94 and 97, inclusive, good parts are found
(e) That no defects are found

2. An office has 10 females and 5 males. If 3 are randomly selected to serve on a committee, determine the probability that

(a) All 3 are females
(b) There is exactly 1 male on the committee
(c) There is at least 1 female on the committee
(d) There is no more than 1 male on the committee
(e) All 3 are males

3. A manufacturer produced 6 widgets and 14 gadgets from which we wish to select with replacement a simple random sample of size $n=4$. Determine the probability of selecting

(a) 2 widgets
(b) 3 widgets
(c) 4 gadgets
(d) Less than 2 gadgets
(e) At least 3 widgets

4. The production department has 7 cutting machines, of which at least 4 must be operating to meet the production schedule on any given day. The probability that a machine will fail is .3, and the failure of one machine is independent of all other machines failing. If the failure of cutting machines can be considered as a Bernoulli process, determine the probability that the cutting machines will meet the scheduled production on a given day.

5. Twenty-five percent of the purchases from a company are for more than $1,000. A simple random sample of size $n=20$ is to be taken from their sales slips. If the amount of a sale greater than $1,000 can be considered a Bernoulli process, determine the probability of selecting

(a) 3 receipts for an amount greater than $1,000
(b) At least 18 receipts for $1,000 or less
(c) Less than 5 receipts for more than $1,000
(d) Between 15 and 18, inclusive, receipts for $1,000 or less
(e) 2 or 3 receipts for more than $1,000

6. A store has 24 employees of which 4 are managers. If 5 employees are to be randomly selected to attend a special training program, what is the probability that

(a) All 5 will not be managers
(b) 2 will be managers
(c) Less than 2 will not be managers
(d) At least 4 will not be managers
(e) Less than 3 will be managers

7. A secretary makes an average of 2 mistakes for every 5 pages typed. Assuming a Poisson model, determine the probability that

(a) Exactly 1 mistake will be made on a page

(b) Less than 4 mistakes will be made on a 10 page manuscript
(c) More than 3 mistakes will be made on a 5 page manuscript
(d) At least 4 mistakes will be made on a 15 page manuscript
(e) More than 25 mistakes will be made on a 50 page manuscript

8. A telephone operator receives an average of 3 calls per minute. Assuming a Poisson model, determine the probability that

(a) No phone calls are received in 1 min
(b) Exactly 9 calls are received in a 5-min period
(c) Less than 3 calls are received in a 3-min period
(d) More than 25 calls are received in a 7-min period
(e) Between 5 and 10 calls are received in a 2-min period

9 The catalog department of a retail store receives an average of 10 orders in a 4-hr period. Assuming a Poisson model, determine the probability that

(a) Exactly 8 orders are received in a 4-hr period
(b) Exactly 5 orders are received in a 2-hr period
(c) More than 30 orders are received in an 8-hr day
(d) Between 15 and 21 orders are received in an 8-hr day
(e) Between 2 and 6 orders are received in a 2-hr period

10. A bakery delivered 4 banana cream pies and 5 chocolate cream pies. If a simple random sample of size 3 is to be selected for serving first, determine the probability that

(a) All 3 are banana cream
(b) 2 are banana cream
(c) Less than 2 are banana cream
(d) At least 1 is chocolate cream

11. Repeat the probabilities requested in Problem 6 with a firm that has 8,000 employees, of whom 1,200 are managers.

9 MATHEMATICS OF FINANCE

9.1 INTRODUCTION

The concepts of *mathematics of finance* are applicable to most business situations and to most of the transactions performed by an average person. In a society where a great deal of the purchasing is done on a credit basis, the concepts of *simple interest, compound interest, interest rate, maturity value, annuity,* and *amortization* are particularly applicable; the last four concepts are basically variations of the concepts of simple and compound interest.

In order to study such concepts, we consider first the concept of a *sequence*. The reader may already be familiar with sequences from the study of algebra; if so, the considerations contained in Sec. 9.2 can either serve as a review or be omitted.

9.2 SEQUENCES

A *sequence* is a list of elements arranged in a specific order according to some definite pattern or law of formation. This law determines which element is first, which is second, etc. For instance, the first 10 positive integers 1, 2, 3,..., 10, arranged in increasing order, constitute a sequence. Other examples of sequences are the following:

1. The positive odd integers from 1 to 10, arranged in increasing order: 1, 3, 5, 7, 9
2. The reciprocals of the positive even integers from 1 to 10, arranged in decreasing order: $\frac{1}{2}, \frac{1}{4}, \frac{1}{6}, \frac{1}{8}, \frac{1}{10}$
3. The integer-valued positive multiples of 10 from 1 to 50, arranged in increasing order: 10, 20, 30, 40, 50
4. The first 10 positive integers that are divisible only by themselves and 1: 1, 2, 3, 5, 7, 11, 13, 17, 19, 23

The previous examples of sequences are such that the elements are

Sec. 9.2 / SEQUENCES

numbers; however, the elements may be virtually anything; for example, the first letters of the names of the seven weekdays arranged in their natural order constitute a sequence: S, M, T, W, T, F, S. Although such arrangements do constitute sequences, we shall be concerned here only with sequences of numbers.

In each of the previous examples of sequences we were given the rule that established the pattern, or law of formation. In many situations we are given the first few terms of a sequence and asked to determine the next term or the next few terms. Usually, the goal is to determine the pattern, or law of formation, on the basis of only a few terms. For example, we may be asked to determine the next two numbers of the sequence for the following successions of numbers:

$$1, 2, 3, 4, \ldots \tag{9.1}$$

$$2, 4, 6, 8, \ldots \tag{9.2}$$

$$1, \tfrac{1}{2}, \tfrac{1}{3}, \tfrac{1}{4}, \ldots \tag{9.3}$$

The sequence in (9.1) *appears* to be the set of positive integers in increasing order for which the next two terms of the sequence would be 5 and 6. Similarly, the sequence in (9.2) *appears* to be the even positive integers in increasing order for which the next two terms of the sequence would be 10 and 12. The sequence in (9.3) *appears* to be the reciprocals of the positive integers in decreasing order for which the next two terms would be $\tfrac{1}{5}$ and $\tfrac{1}{6}$.

The alert reader will note that we have stated that each sequence *appears* to follow a specific pattern, and, actually, this is about all we can say about sequences. We would all agree that the terms in each of the previous sequences certainly satisfy the law of formation as stated for each sequence; however, the simple law that we have selected may not be the *proper one*. For instance, the first eight terms of the sequence in (9.1) would be

$$1, 2, 3, 4, 5, 6, 7, 8, \ldots$$

if the law of formation is the set of positive integers in increasing order. However, another sequence for which the first four terms are the same as the sequence in (9.1) could be

$$1, 2, 3, 4, 1, 2, 3, 4, \ldots$$

This sequence can be generated by taking consecutive groups of the first four positive integers in increasing order; thus, with a little effort, one can easily construct sequences that become

$$1, 2, 3, 4, 2, 4, 6, 8, \ldots$$
$$1, 2, 3, 4, 0, 1, 2, 3, \ldots$$

or many others.

From this discussion, it is apparent that it is not always possible to determine the proper law of formation for a sequence from only the first few terms unless they constitute *all* of the terms of the sequence. However, we did see that in many instances we can construct *a simple rule* for forming a sequence. In most practical situations, the rule is stated in a form such that any term in the sequence can be determined from some initial value (or values) and the order of appearance of the term in the sequence. We denote the nth term in the sequence by the term a_n.

It is most desirable to express a_n as a function of the initial values and the position n in which it appears in the sequence. For example, if the correct rule for the sequence in (9.1) is the set of positive integers in increasing order, we can represent the nth term by the function $a_n = n$, which yields for the first term $a_1 = 1$, for the second term $a_2 = 2$, for the third term $a_3 = 3$, and so on. Similarly, if the correct rule for the sequence in (9.2) is the set of even positive integers in increasing order, we can represent the nth term by the function $a_n = 2n$, which yields for the first term $a_1 = 2$, for the second term $a_2 = 4$, for the third term $a_3 = 6$, and so on. If the proper rule for the sequence in (9.3) is the set of reciprocals of the positive integers arranged in decreasing order, we can represent the nth term by the function $a_n = 1/n$, which yields for the first term $a_1 = 1$, for the second term $a_2 = \frac{1}{2}$, for the third term $a_3 = \frac{1}{3}$, and so on.

There are two specific types of sequences that are of primary concern in the study of the mathematics of finance: One is known as an *arithmetic progression*, and the other is known as a *geometric progression*. We shall first study these two sequences and then apply the results to some particular problem areas.

Arithmetic Progressions

First let us consider the type of sequence for which the first term is a given value a_1 and for which each term thereafter is determined by adding the same fixed value d to the preceding term. The term a_1 is appropriately called the *first term*, and the value d is called the *common difference*; this type of sequence is called an *arithmetic progression*. For example, the sequence 1, 3, 5, 7, 9 is an arithmetic progression in which each term after the first is obtained by adding 2 to the preceding term; thus the first term is $a_1 = 1$, and the common difference is $d = 2$. Similarly, the sequence 0.8, 0.7, 0.6, 0.5, 0.4 is an arithmetic progression in which the first term is $a_1 = 0.8$ and the common difference is $d = -0.1$.

The rule for an arithmetic progression states that we begin with a first term a_1 and add the common difference d to this term to obtain the second term; to the second term, we add d to obtain the third term; and we continue this process until we obtain the terms in the sequence. In order to express the nth term a_n as a function of an initial value or values and the order that it appears in the sequence, let us consider several consecutive terms of the

Sec. 9.2 / SEQUENCES

sequence. Starting with the first term a_1 and performing the above process, we obtain

$$a_1 = a_1$$
$$a_2 = a_1 + d$$
$$a_3 = a_2 + d = (a_1 + d) + d = a_1 + 2d$$
$$a_4 = a_3 + d = (a_1 + 2d) + d = a_1 + 3d$$
$$\vdots \quad \vdots \quad \vdots \quad \vdots$$

To obtain each term, we simply add d to the preceding term. Thus the coefficient of d is one less than the number (or order) of the term in the sequence. Hence we see that for any positive integer n,

$$a_n = a_1 + (n-1)d \tag{9.4}$$

Usually we are concerned only with a finite number of terms; furthermore, generally we wish to know the sum of the terms in the sequence. We shall denote the sum of the first n terms in the sequence by S_n; this sum of n terms, each expressed in terms of a_1, n, and d will be

$$S_n = a_1 + (a_1 + d) + (a_1 + 2d) + \cdots + [a_1 + (n-1)d]$$

or, expressed in terms of a_n and with the right side of the above in reverse order,

$$S_n = a_n + (a_n - d) + (a_n - 2d) + \cdots + (a_1 + d) + a_1$$

Adding corresponding sides of the two equations, we obtain

$$2S_n = (a_1 + a_n) + (a_1 + a_n) + (a_1 + a_n) + \ldots + (a_1 + a_n)$$

Solving for S_n, we have

$$S_n = \frac{n}{2}(a_1 + a_n) \tag{9.5}$$

or, substituting the right side of (9.4) for a_n,

$$S_n = \frac{n}{2}[2a_1 + (n-1)d] \tag{9.6}$$

The choice of whether to use (9.5) or (9.6) to obtain the sum of the first n terms of an arithmetic progression depends upon whether a_n or d, respectively, is given; if both are given, it does not matter which equation is used. It is convenient to note that Eq. (9.5) is simply n times the average of the first and last terms of the arithmetic progression.

Example 9.1 Determine the first six terms and their sum for an arithmetic progression with $a_1 = 5$ and $d = 2$.

Solution Using Eq. (9.4) and the values for a_1 and d, we determine $a_n = 5 + (n-1)2$; thus, for the six terms, we obtain

$$5, 7, 9, 11, 13, 15$$

To determine the sum of the six terms, we can use either (9.5) or (9.6) because we were given d and also have computed a_6. Using Eq. (9.5), we obtain

$$S_6 = \tfrac{6}{2}(5 + 15) = 60$$

Using Eq. (9.6), we obtain

$$S_6 = \tfrac{6}{2}[2(5) + 5(2)] = 60$$

Both of the sums in Example 9.1 can be verified by summing the six individual terms in the sequence. For this kind of problem, such a check is feasible; however, if there are very many terms in the sequence, it is much simpler to substitute the appropriate values into either (9.5) or (9.6).

Example 9.2 Determine the first 12 terms and their sum for an arithmetic progression with $a_1 = 10$ and $d = 0.05$.

Solution Using Eq. (9.4) and the values for a_1 and d, we determine $a_n = 10 + (n-1)0.05$. Substituting the different values for n, we obtain

$$10.00, 10.05, 10.10, 10.15, 10.20, 10.25, 10.30, 10.35, 10.40, 10.45, 10.50, 10.55$$

Using the values for a_1, d, and n, we substitute into Eq. (9.6) and obtain

$$S_{12} = \tfrac{12}{2}[2(10) + (12-1)0.05]$$
$$= 123.30$$

Example 9.3 The profits, in thousands of dollars, of the Carbo Chemical plant have been 3, 7, 11, 15, 19, 23, 27, 31 for the past 8 years. Since the profits over the past 8 years have followed an arithmetic progression, determine the first term, the common difference, and the sum of the profits.

Solution By inspection, we determine the first term to be $a_1 = 3$. The common difference can be obtained simply by taking the difference between any two consecutive terms; thus $d = 7 - 3 = 4$. Since there are eight terms in the sequence and we know a_1 and d, we can use Eq. (9.6) to determine S_8; thus

$$S_8 = \tfrac{8}{2}[2(3) + 7(4)] = 136$$

We could have determined the sum using the values for the first and last terms and the number of terms in the progression; that is, using Eq. (9.5),

$$S_8 = \tfrac{8}{2}(3+31) = 136$$

Geometric Progressions Now let us consider the second type of sequence that is of considerable importance: *geometric progression*. A geometric progression is formed by starting with a first term a_1 and obtaining each term thereafter by multiplying the preceding term by a fixed number r, called the *common ratio*. For example, the sequence 1, 2, 4, 8, 16, 32 is a geometric progression in which each term is obtained by multiplying the preceding term by 2; thus the first term is $a_1 = 1$, and the common ratio is $r = 2$. Likewise, the sequence 2.0, 0.2, 0.02, 0.002, 0.0002 is a geometric progression in which the first term is $a_1 = 2$ and the common ratio is $r = 0.1$. Another sequence that is a geometric progression is 1, -3, 9, -27, 81, where the first term is $a_1 = 1$ and the common ratio is $r = -3$.

Beginning with a first term a_1 and obtaining each term by multiplying the preceding term by the common ratio r, we obtain the following general terms in the sequence:

$$a_1 = a_1$$
$$a_2 = a_1 r$$
$$a_3 = a_2 r = (a_1 r)r = a_1 r^2$$
$$a_4 = a_3 r = (a_1 r^2)r = a_1 r^3$$
$$a_5 = a_4 r = (a_1 r^3)r = a_1 r^4$$
$$\vdots \quad \vdots \quad \vdots \quad \vdots$$

Letting a_n represent the nth term, we note that the exponent of r is one less than n for each term; thus the nth term in a geometric progression is given by

$$a_n = a_1 r^{n-1} \tag{9.7}$$

In order to determine the sum of the first n terms of a geometric progression, we denote this sum by S_n and obtain

$$S_n = a_1 + a_1 r + a_1 r^2 + \cdots + a_1 r^{n-1}$$

Multiplying both sides of this equation by the common ratio r, we obtain

$$rS_n = a_1 r + a_1 r^2 + a_1 r^3 + \cdots + a_1 r^n$$

We note that every term on the right side of the second equation (except the last term) has a corresponding identical term in the first equation; the same is true for the first equation when compared with the second, except that the first term does not have an identical counterpart. Thus, subtracting the second equation from the first, we obtain

$$S_n - rS_n = a_1 - a_1 r^n$$

since all terms cancel except the first term of the first equation and the last term of the second equation. Solving the resulting equation for S_n, we obtain

$$S_n = \frac{a_1(1-r^n)}{1-r} \qquad (9.8)$$

provided that $r \neq 1$. Equation (9.8) can be written in another form if we expand the numerator and substitute r times both sides of Eq. (9.7). That is,

$$S_n = \frac{a_1 - a_1 r^n}{1-r} = \frac{a_1 - a_n r}{1-r}$$

since $a_n r = a_1 r^n$.

Example 9.4 Determine the first nine terms and their sum for a geometric progression with $a_1 = 5$ and $r = 2$.

Solution Using Eq. (9.7), we can determine the nth term in the progression by the equation

$$a_n = 5 \cdot 2^{n-1}$$

Substituting the different values for n, we obtain the terms

$$5 \cdot 2^0, 5 \cdot 2^1, 5 \cdot 2^2, 5 \cdot 2^3, 5 \cdot 2^4, 5 \cdot 2^5, 5 \cdot 2^6, 5 \cdot 2^7, 5 \cdot 2^8$$

Expanding these terms, we obtain the numerical values

$$5, 10, 20, 40, 80, 160, 320, 640, 1{,}280$$

Substituting the values of a_1, n, and r into Eq. (9.8), we obtain

$$S_9 = \frac{5(1-2^9)}{1-2} = 2{,}555$$

(The reader may wish to verify this sum by adding the nine terms in the sequence.)

Example 9.5 The number of sales Mr. Jones has made over the past 4 months can be represented by a geometric progression with $a_1 = 100$ and $r = 1.05$. Determine the number of sales for each month and their sum.

Sec. 9.2 / SEQUENCES

Solution Using Eq. (9.7), we determine the nth term of the progression to be

$$a_n = 100 \cdot (1.05)^{n-1}$$

Substituting the four integer values for n, we obtain the number of sales each month

$$100, 105, 110.25, 115.7625$$

The total number of sales during the 4 months can be obtained by use of Eq. (9.8). Substituting the appropriate values, we obtain

$$S_4 = 100 \cdot \frac{1-(1.05)^4}{1-1.05}$$
$$= 100 \cdot \frac{-0.21550625}{-.05}$$
$$= 431.0125$$

Example 9.6 Determine the first term, the common ratio, and the sum of the following geometric progression:

$$4, -12, 36, -108, 324, -972$$

Solution By inspection, the first term is determined to be $a_1 = 4$. The common ratio can be obtained by taking the ratio of any term to its immediately preceding term. Thus we determine

$$r = \frac{36}{-12} = -3$$

Substituting these values into Eq. (9.8), we determine the sum of the six terms to be

$$S_6 = 4 \cdot \frac{1-(-3)^6}{1-(-3)} = 4 \cdot \frac{1-729}{4} = -728$$

EXERCISES

1. For each of the following sequences, determine a simple rule or formula that characterizes the sequence and give the next three terms:

(a) 0, 2, 4, 6, 8,...
(b) 3, 5, 7, 9, 11,...
(c) 3, 7, 11, 15, 19,...
(d) 1, $\frac{1}{4}$, $\frac{1}{7}$, $\frac{1}{10}$, $\frac{1}{13}$,...
(e) $\frac{1}{16}$, $\frac{1}{8}$, $\frac{1}{4}$, $\frac{1}{2}$, 1,...
(f) 3, $\frac{9}{2}$, 6, $\frac{15}{2}$, 9,...
(g) 3, 6, 12, 24, 48,...
(h) 2, 6, 18, 54, 162,...

2. For each of the following functions, determine the first five terms of the sequence

(a) $a_n = 2n + 4$
(b) $a_n = 2n^2 + 1$
(c) $a_n = 1/n^2$
(d) $a_n = n^2 + 3n + 1$
(e) $a_n = (5n + 2)/(n + 1)$
(f) $a_n = [n(n + 1)]/2$

3. Determine which of the following sequences are arithmetic progressions and find the common difference for each such progression:

(a) 57, 44, 31, 28
(b) 9.8, 8.4, 7.2, 6.0
(c) 9, 13, 17, 21
(d) $11a + 12b$, $6a + 5b$, $a - 2b$
(e) $5b$, $10b$, $15b$, $20b$
(f) 2, 4, 8, 16

4. Given $a_1 = 2$ and $d = 5$, determine a_8 and S_8.

5. Given $a_1 = 7$ and $d = -2$, determine a_{10} and S_{10}.

6. Given $a_1 = 10$ and $d = 10$, determine a_{15} and S_{15}.

7. Determine which of the following sequences are geometric progressions and find the common ratio for each such progression:

(a) 6, 3, 1, $\frac{1}{2}$
(b) 100, 50, 25, 5
(c) 5, 25, 125, 625
(d) c, cx, cx^2, cx^3
(e) $(1.03)^2$, $(1.03)^5$, $(1.03)^8$
(f) $(1.05)^{-4}$, $(1.05)^{-1}$, $(1.05)^2$

8. Given $a_1 = 2$ and $r = 5$, determine a_5 and S_5.

9. Given $a_1 = 300$ and $r = 3$, determine a_6 and S_6.

10. Given $a_1 = 100$ and $r = 1.1$, determine a_4 and S_4.

11. If the profit of Cox Electronics increases at the rate of 5 percent/year and the present profit is $300,000, what will the profit be 6 years from now? What will be the total profit earned during the 6-year period?

12. Jean Allen is offered a position at a salary that starts at $10,000/year and increases yearly by $500. If she continues to work under this salary schedule, what will her salary be at the end of 7 years? What is the total salary she will receive during the 7-year period?

13. Travelers Inn is presently valued at $80,000 and will depreciate as follows: $2,900 the first year, $2,800 the second year, $2,700 the third year, and so on. Based on these estimates, what will be the amount of depreciation for the fifteenth year and what will the property be worth after the fifteenth year?

14. Each month an employer puts $3 plus $2 times the number of months the employees have worked that calendar year into a fund to give them as a Christmas bonus.

(a) Determine the amount that will be placed into the fund at the end of October for an employee who has worked since March 1.
(b) Determine the amount to be given to the employee in part (a) at Christmas.
(c) Determine the amount that will be placed in the fund at the end of October for an employee who has worked since Jan. 1.

Sec. 9.3 / SIMPLE AND COMPOUND INTEREST

(d) Determine the amount to be given to the employee in part (c) at Christmas.

15. The cost of maintaining a machine is expected to be $15 the first year and increases by $10 each year. What is the expected maintenance cost during the fifth year? If the machine has a life of 20 years, what is the expected total maintenance cost for the machine?

16. Sales have been increasing by 100 units times the number of years from 1965 when sales were 150,000 units. What are the expected sales for 1978? What are the total expected sales for 1978 through 1983?

9.3 SIMPLE AND COMPOUND INTEREST

In many business situations funds are borrowed or loaned for various lengths of time. Generally, a *short-term* loan is considered to be for less than 1 year, and a *long-term* loan is considered to be for 1 year or longer. In any event, money has a *time value* because a dollar today is worth more than a dollar a year from now. This time value is given in terms of interest charges or receipts.

Simple Interest

The amount paid for the use of a quantity of money over a period of time is called *interest*, symbolically denoted by I. If interest is paid only for the use of the original quantity and not on any accrued interest, it is termed *simple interest*. Interest is usually stated as a percentage per period of the original quantity, with the period usually being 1 year; this percentage is called the *rate of interest*, symbolically denoted by r, which is expressed in decimal form. For example, $100 borrowed at an 8 percent/year rate of interest will cost $8 interest. Thus, denoting the quantity borrowed by P, called the *principal*, we have $P=\$100$, $r=.08$, and $I=\$8$ for a period of 1 year.

If the time period for the previous example is other than 1 year, the interest would be multiplied by the number of years or parts of a year. Denoting by n the number of interest periods, the interest can be determined by

$$I = Prn \tag{9.9}$$

Example 9.7 A salesman borrows $4,000 at a 7 percent/year simple interest rate for 3 years. Determine the amount of interest he must pay.

Solution We are given $P=\$4,000$, $r=.07$, and $n=3$. Substituting into (9.9), we determine the interest to be

$$I = (4,000)(.07)3 = \$840$$

The total sum or *maturity value*, denoted by S, to be repaid at the end of the time period is simply the principal plus the interest, that is,

$$S = P + I = P + Prn$$

or
$$S = P(1 + rn) \qquad (9.10)$$

Example 9.8 Determine the total sum to be repaid by the salesman in Example 9.7.

Solution Using Eq. (9.10), we obtain

$$S = (4,000)[1 + (.07)3]$$
$$= (4,000)(1.21)$$
$$= \$4,840$$

Simple interest is most commonly used for short-term loans, and since short-term loans are for a duration of less than 1 year, the term of the loan is expressed as a fractional part of a year. Furthermore, for computational convenience, each month is viewed as consisting of 30 days, and consequently, 1 year is viewed as consisting of 360 days.

Example 9.9 A college professor borrows $2,000 from her bank at 8 percent/year simple interest on a 6-month note. Determine the interest and the maturity value of the note.

Solution Using Eq. (9.9) with $P = \$2,000$, $r = .08$, and $n = \frac{180}{360}$, we obtain

$$I = 2,000(.08)\left(\frac{180}{360}\right) = \$80$$

The maturity value is determined simply by adding the interest to the principal to obtain

$$S = 2,000 + 80 = \$2,080$$

The maturity value can also be determined by use of Eq. (9.10); however, the previous simple addition is somewhat easier because the interest is already computed.

Compound Interest Simple interest is the interest paid only for the use of the principal. If the interest is payable at the end of each time period in which the interest rate is stated and it is left to accrue interest for the remaining length of the loan, this interest is called *compound interest*. An example of such interest is the typical savings account for which the interest is payable at the end of each period, called the *conversion period*, and for which the interest is added to the principal to accrue more interest. The total amount payable at the end of the term is the principal plus the interest. However, the interest is compound interest; thus the total amount payable is called the *compound amount*, or *compound maturity value*.

Sec. 9.3 / SIMPLE AND COMPOUND INTEREST

The compound maturity value at the end of the first period is

$$S_1 = P + Pr = P(1+r)$$

At the end of the second period, the compound maturity value is

$$S_2 = S_1 + S_1 r = S_1(1+r)$$
$$= P(1+r)(1+r) = P(1+r)^2$$

and at the end of the third period, it is

$$S_3 = S_2 + S_2 r = S_2(1+r)$$
$$= P(1+r)^2(1+r) = P(1+r)^3$$

Noting that

$$S_1 = P(1+r) \qquad S_2 = P(1+r)^2 \quad \text{and} \quad S_3 = P(1+r)^3$$

we see that the values constitute the terms of a geometric progression with a common ratio of $(1+r)$. Thus the nth term of the progression is

$$S_n = P(1+r)^n \qquad (9.11)$$

which represents the compound maturity value of a principal P that is borrowed or invested at an interest rate of r per period for n periods and that is compounded at the end of each period. The value S_n is commonly called the *future value* of the amount P that is being compounded at a rate r for n periods.

Example 9.10 Mary Porter, a career woman, borrows $3,000 from a bank at a 7 percent/year interest rate compounded annually for 3 years. Assuming that no payments are made, determine the compound maturity value.

Solution Substituting $P = \$3,000$, $r = .07$, and $n = 3$ into Eq. (9.11), we have

$$S_3 = 3,000(1 + .07)^3$$
$$= 3,000(1.225)$$
$$= \$3,675$$

The term $(1+r)^n$ in (9.11) becomes awkward and laborious to compute for large values of n. Since the term is frequently used, it has been computed for a wide range of values for r and n. Table A.5 in the Appendix presents the values of this term for values of n from 1 to 50 for each of the values of r from 0.01 to 0.08. The use of Table A.5 is not absolutely necessary, since the use of modern electronic calculators and computers will generally obviate the need for the tables used in this chapter and will give greater flexibility in the rates and periods considered. However, the table is included for those who do not have easy access to calculators and for comparison purposes.

The rate r and the number of periods in the term $(1+r)^n$ are generally stated in percent per year and years, respectively. However, in order for (9.11) to be valid: it is necessary only that the time period for which the interest rate is expressed and for which the number of periods is given be the same period of time. If this is the case, Eq. (9.11) can be employed without change for the compound maturity value.

Example 9.11 Mr. Joe Friendly, a local insurance salesman, deposited $6,000 with a savings-and-loan firm. The money is to accrue interest at the rate of 6 percent/year compounded semiannually. Determine the compound maturity value at the end of $2\frac{1}{2}$ years if no further deposits or withdrawals are made.

Solution Since the interest rate and the number of periods must be expressed in semiannual terms, we obtain $r=.06/2=.03$ and $n=(2.5)(2)=5$. Substituting these semiannual values into (9.11) and locating the appropriate value from Table A.5, we obtain the compound maturity value

$$S_5 = 6,000(1+.03)^5$$
$$= 6,000(1.03)^5$$
$$= 6,000(1.159)$$
$$= \$6,954$$

Example 9.12 The Charge-Card Company extends to customers who use credit cards issued by the company the privilege of carrying a charge account with the company. This privilege costs the customer the rate of 1.5 percent/month on the unpaid balance. The interest charges are added to the unpaid balance at the end of each month unless the account is paid in full. Mary Smith used her charge card to purchase a $1,000 diamond ring. Suppose Mary makes no payments on her account. Determine the amounts that she would owe at the end of 6 months and at the end of 1 year.

Solution Since the interest accrues interest, the problem involves *compound interest*. The monthly interest rate is 1.5 percent, and the interest is compounded monthly. Thus, for 6 months, Mary would owe

$$S_6 = 1,000(1+.015)^6$$

Since an interest rate of 1.5 percent is not available in Table A.5, we use our calculator or logarithms to determine $(1.015)^6 = 1.093$. Thus, at the end of 6 months, Mary would owe

$$S_6 = 1,000(1.093) = \$1,093$$

For 1 year, or 12 months, Mary would owe

$$S_{12} = 1,000(1+.015)^{12}$$

Sec. 9.3 / SIMPLE AND COMPOUND INTEREST

Again using our calculator or logarithms, we obtain $(1.015)^{12} = 1.196$. Thus, for 1 year, Mary would owe

$$S_{12} = 1,000(1.196) = \$1,196$$

From Example 9.12, we see that an interest rate of 18 $[=(12)(1.5)]$ percent/year compounded monthly has cost \$196 for the use of \$1,000 for 1 year, which is equivalent to a rate of 19.6 percent/year simple interest.

Equation (9.11) has been used exclusively to determine S_n for given values of P, r, and n. The equation can be treated simply as an equation in four unknowns and can thus be used to solve for any one unknown when the values of the remaining three unknowns are given. For example, if we wish to determine the principal that must be invested to yield a compound maturity value of S_n at the end of n periods for an interest rate of r per period, we simply solve (9.11) for P to obtain

$$P = S_n(1+r)^{-n} \tag{9.12}$$

The value P, when viewed in this context, is termed the *present value* of the compound maturity value S_n. Table A.6 in the Appendix presents values of $(1+r)^{-n}$ for various values of r and n.

Similarly, if we wish to determine the number of periods that a principal P must be invested at a compound interest rate of r per period to yield a compound maturity value of S_n, we solve Eq. (9.11) for n; thus we have

$$\frac{S_n}{P} = (1+r)^n$$

Taking the common logarithm of both sides, we obtain

$$\log\left(\frac{S_n}{P}\right) = n\log(1+r)$$

Solving for n and using the rule for the logarithm of a ratio, we obtain

$$n = \frac{\log(S_n/P)}{\log(1+r)}$$

or

$$n = \frac{\log S_n - \log P}{\log(1+r)} \tag{9.13}$$

In order to solve for r, we begin with (9.11) to obtain

$$\frac{S_n}{P} = (1+r)^n$$

Raising both sides of the equation to the $1/n$ power, we obtain

$$\left(\frac{S_n}{P}\right)^{1/n} = (1+r)$$

Solving for r, we have

$$r = \left(\frac{S_n}{P}\right)^{1/n} - 1 \qquad (9.14)$$

[The use of Eq. (9.14) to solve for r generally will require the use of a calculator or logarithms to determine the value of $(S_n/P)^{1/n}$. The reader is not encouraged to memorize Eqs. (9.12) to (9.14), but rather to be able to solve (9.11) for each individual term.]

Example 9.13 Mr. Cash is confronted with a situation that requires him to have available a total of $6,000 in the not-too-distant future. He is faced with three alternatives:

1. He can buy a 2-year note at an interest rate of 6 percent/year that is compounded annually.
2. He can invest $5,000 with an organization at 8 percent/year compounded annually.
3. He can invest $5,000 with a private individual for a fixed period of 2 years.

Assuming that no additional money is withdrawn or deposited, determine

(a) What principal Mr. Cash must invest for alternative 1 to yield the required $6,000 compound maturity value
(b) How long Mr. Cash's $5,000 would need to remain invested to yield the required $6,000 for alternative 2
(c) What rate of interest Mr. Cash should charge to yield the required $6,000 if the interest is compounded annually for alternative 3

Solution

(a) Using (9.12), we have

$$P = 6{,}000(1 + .06)^{-2}$$
$$= 6{,}000(1.06)^{-2}$$

From Table A.6 we obtain the value for $(1.06)^{-2} = 0.890$; thus, in order for Mr. Cash to receive $6,000, he must invest

$$P = 6{,}000(0.890) = \$5{,}340$$

(b) Using (9.13), we have

$$n = \frac{\log(6{,}000) - \log(5{,}000)}{\log(1 + .08)}$$
$$= \frac{3.7782 - 3.6990}{.0334}$$
$$= \frac{.0792}{.0334}$$
$$= 2.37$$

Sec. 9.3 / SIMPLE AND COMPOUND INTEREST

Thus Mr. Cash must leave his $5,000 invested for 2.37 years in order to receive a total of $6,000 maturity value.

(c) Using (9.14), we have

$$r = \left(\frac{6,000}{5,000}\right)^{1/2} - 1$$
$$= (1.2)^{1/2} - 1$$
$$= 1.0954 - 1$$
$$= 0.0954$$

Hence Mr. Cash must charge 9.54 percent interest/year compounded annually in order to have a $6,000 maturity value.

EXERCISES

1. Determine the simple interest for

   (a) $P = \$1{,}500$, $r = .07$, and $n = 90$ days
   (b) $P = 6{,}000$, $r = .06$, and $n = 6$ months
   (c) $P = \$375$, $r = .08$, and $n = 9$ months
   (d) $P = \$1{,}800$, $r = .075$, and $n = 200$ days

2. Determine the principal for the simple interest loan with

   (a) $I = \$100$, $r = .06$, and $n = 90$ days
   (b) $I = \$75$, $r = .05$, and $n = 4$ months
   (c) $I = \$250$, $r = .08$, and $n = 180$ days
   (d) $I = \$280$, $r = .07$, and $n = 1$ year

3. Determine the simple interest rate per year for

   (a) $I = \$360$, $P = \$4{,}000$, and $n = 6$ months
   (b) $I = \$240$, $P = \$2{,}000$, and $n = 9$ months
   (c) $I = \$300$, $P = \$3{,}800$, and $n = 240$ days
   (d) $I = \$120$, $P = \$1{,}000$, and $n = 1$ year

4. Determine the term of the simple interest loan for

   (a) $I = \$80$, $P = \$1{,}000$, and $r = .08$
   (b) $I = \$120$, $P = \$2{,}000$, and $r = .08$
   (c) $I = \$50$, $P = \$1{,}500$, and $r = .05$
   (d) $I = \$210$, $P = \$4{,}000$, and $r = .06$

5. Determine the maturity value for each of the loans in

   (a) Exercise 1   (b) Exercise 2
   (c) Exercise 3   (d) Exercise 4

6. Using logarithms where necessary, determine the compound maturity value for the following, with an annual conversion period:

   (a) $P = \$408$, $r = .05$, and $n = 3$
   (b) $P = \$5{,}216$, $r = .07$, and $n = 4$

(c) $P = \$1,850$, $r = .055$, and $n = 5$
(d) $P = \$3,100$, $r = .075$, and $n = 2$

7. Determine the following values for the given information where the interest is compounded annually:

   (a) Find P when $S_4 = \$3,000$ and $r = .06$.
   (b) Find r when $S_5 = \$701$ and $P = \$500$.
   (c) Find n when $S_n = \$1,800$, $P = \$1,200$, and $r = .07$.

8. Determine the compound maturity value of $1,500 invested for 19 years at a rate of 8 percent/year compounded semiannually. What is the amount of interest due?

9. The Airtight Bank pays its depositors an annual interest rate of 5.5 percent compounded quarterly. The Goldbrick Bank pays its depositors an annual rate of 5.2 percent compounded monthly. For a $1,000 investment for 1 year, determine the compound maturity value from each bank and determine which bank offers the most favorable option.

10. Betty Cash borrowed $1,500 from the Best National Bank on a 90-day note at 7.5 percent interest. Determine the maturity value of the note if the interest is simple interest. Determine the compound maturity value of the note if the interest is compounded monthly.

11. The sales of the Electric Power Company are increasing at the rate of 8 percent/year. How many years will it take for the sales to double? Triple?

12. Mr. Goodrisk borrowed $2,000 from Friendly State Bank today and will receive an additional $3,000 in 2 years. He will repay $1,000 at the end of 3 years, another $2,000 at the end of 4 years, and a final payment at the end of 6 years. If Friendly State Bank charges 7 percent/year compounded annually, determine the amount of the final payment.

13. The Bluestreak Motor Company has been steadily increasing its business over the past 10 years. If 10 years ago the sales were $300,000 and today's sales are $600,000, what is the annual growth rate compounded annually?

14. The Slick Oil Company has produced 2.36 million barrels of crude oil this year, and they have been increasing production each year by 6 percent over the previous year. How many million barrels of crude oil did Slick produce 10 years ago?

15. The Fishy Pet Shop has 12 fish they use for breeding fish to sell. If their reproduction rate is 9 fish each per month, how many fish are available for sale during a 3-month period? A 6-month period? A year?

16. A farmer needs to borrow $8,000 to plant his crop. Three banks have given the following proposals:

    (a) Bank A will loan him the $8,000 to be repaid in 8 months at an annual interest rate of 15 percent.
    (b) Bank B will loan him the $8,000 to be repaid in 8 months at an annual interest rate of 12 percent compounded monthly.
    (c) Bank C will loan him the $8,000 with $4,000 plus all accrued interest at the end of 4 months and the remaining $4,000 and its accrued interest at the end of 8 months. The annual interest rate is 18 percent and is compounded monthly.

Which bank should the farmer borrow the $8,000 from?

17. Mr. Jones withdrew $2,500 from his savings account. How much did he have to put in his savings account 4 years ago in order to have this amount at an interest rate of 6.4 percent compounded quarterly?

9.4 ANNUITIES

In the preceding section we considered the maturity value of a single initial amount, called the *principal*, which was invested at a periodic interest rate r for a length of n periods. Let us now consider the situation where equal periodic payments are made and each payment is left to accrue interest until the final payment is made; such a sequence of payments is called an *annuity*. If we are interested in determining the amount that will be available at the time of the last payment, we are seeking the *future value* of the annuity. However, if we are interested in determining how large a sum we must deposit now at a compound interest rate in order to receive a fixed amount each period, we are interested in the *present value* of the annuity.

Ordinary Annuities

First let us consider the problem of determining the future value of an annuity. More specifically, we shall consider only *ordinary annuities*, i.e., annuities for which the payment is made (i.e., put in) at the end of the period. Also, we shall consider only those situations for which the time of payment is the same as the time when interest is computed. Thus, for ordinary annuities, the payment will be made and the interest will be computed at the end of the period.

Suppose we make n equal periodic payments of p dollars that are to earn compound interest at a rate of r per period. If we begin making payments today and continue making payments for n periods, the compound amount of each payment is as shown in Table 9.1. The compound amount of payment for each period represents the future value of that payment for n periods from now, and it is computed by applying Eq. (9.11) to each respective payment. If we wish to know the total future value of the annuity, we simply add the future values for the individual payments. Letting A_n represent the compound maturity value or future value of the annuity, we obtain

$$A_n = p + p(1+r)^1 + \cdots + p(1+r)^{n-3} + p(1+r)^{n-2} + p(1+r)^{n-1}$$

In order to obtain a more compact form of the sum, we note that the compound maturity value of the payments constitutes a geometric progression for which the first term is p and the common ratio is $(1+r)$. Thus A_n is simply the sum of n terms of a geometric progression, the equation for which is given by (9.8). Thus, substituting the previous values for the first term and the common ratio, we obtain the equation for determining the future value of an annuity, that is,

$$A_n = p \frac{1-(1+r)^n}{1-(1+r)}$$

Table 9.1

Payment No.	Amount of Payment	No. of Periods at Interest Rate r	Compound Amount of Payment
1	p	n−1	$p(1+r)^{n-1}$
2	p	n−2	$p(1+r)^{n-2}$
3	p	n−3	$p(1+r)^{n-3}$
...	...	...	...
n−1	p	1	$p(1+r)^1$
n	p	0	p

or

$$A_n = p \frac{(1+r)^n - 1}{r} \qquad (9.15)$$

Values for $[(1+r)^n - 1]/r$ are given in Table A.7 in the Appendix for various combinations of r and n.

Example 9.14 Mr. Smith pays an insurance company $100/year, which is to accrue interest at a rate of 6 percent/year compounded annually.

(a) Determine the compound maturity value of each payment for 7 years from the first payment, and, using these 7 values, determine the future value of the annuity.
(b) Use Eq. (9.15) to determine the future value of the annuity at the end of 7 years.

Solution

(a) Using Table 9.1 as a guide, we obtain the following compound maturity values:

Payment No.	No. of Periods at 6% Interest Rate	Compound Maturity Value
1	6	$100(1.06)^6 =$ $141.90
2	5	$100(1.06)^5 =$ 133.80
3	4	$100(1.06)^4 =$ 126.20
4	3	$100(1.06)^3 =$ 119.10
5	2	$100(1.06)^2 =$ 112.30
6	1	$100(1.06)^1 =$ 106.00
7	0	100 = 100.00
		Total = $839.30

(b) Using Eq. (9.15), we obtain

$$A_7 = 100 \frac{(1.06)^7 - 1}{.06}$$

Using Table A.7, this becomes

$$A_7 = 100(8.394) = \$839.40$$

[We should note that the 10¢ discrepancy between the answers to the two parts of Example 9.14 is due to roundoff error associated with the seven values in part (a).]

Example 9.15 Mr. Caldwell deposits $75 in his savings account each quarter. He receives interest at the rate of 8 percent/year compounded quarterly. Determine the value of the annuity and the amount of interest which he has received in his account at the end of 3 years.

Solution Since the interest is compounded quarterly, the interest rate must be expressed in terms of the same period; thus the interest rate is equivalent to 2 percent/quarter. Furthermore, the number of conversion periods is the number of quarters in the 3-year period; that is, $n = 3(4) = 12$. Now that the term and the interest rate are in terms of the conversion period, we can use Eq. (9.15) to determine the value of the annuity. Thus

$$A_n = 75 \frac{(1.02)^{12} - 1}{.02}$$

which, by using Table A.7 becomes,

$$A_n = 75(13.412) = \$1,005.90$$

The amount of interest Mr. Caldwell receives is simply the value of the annuity minus the amount of his investment. Thus his investment is $12(75) = \$900$, and, consequently, the interest is $\$1,005.90 - \$900 = \$105.90$.

Fixed Payments Now let us consider the problem of determining how large a deposit must be made at a compound interest rate in order to receive (i.e., take out) a fixed payment each period for a specified length of time. Suppose we wish to receive n equal periodic payments of p dollars from a lump sum S that is invested at an interest rate r per period and for which the interest is compounded each period. Since we wish to know the present value of the annuity, we shall determine the present value for each payment and then sum these values. In Sec. 9.3 we saw that the present value of a single principal invested at a compound interest rate r was given by Eq. (9.12); that is,

$$P = S_n(1 + r)^{-n}$$

where P represents the principal and S_n represents the compound maturity value. Thus, if we wish to determine the principal (or present value of a future payment) that must be invested to yield the payment that we are to receive after n periods, we can employ Eq. (9.12) for this purpose. However, we must use the values appropriate to our problem; that is, the value S_n in Eq. (9.12) will represent the payment value p, and the present values of the respective payments are given in Table 9.2.

Table 9.2

Payment No.	No. of Periods at Interest Rate r	Present Value of Payment
1	1	$p(1+r)^{-1}$
2	2	$p(1+r)^{-2}$
3	3	$p(1+r)^{-3}$
...	...	...
$n-1$	$n-1$	$p(1+r)^{-(n-1)}$
n	n	$p(1+r)^{-n}$

In order to determine the present value of the annuity, we sum the present values of the individual payments given in Table 9.2 and obtain

$$S = p(1+r)^{-1} + p(1+r)^{-2} + p(1+r)^{-3} + \cdots + p(1+r)^{-(n-1)} + p(1+r)^{-n}$$

We note that the present values given in Table 9.2 form a geometric progression for which the first term is $p(1+r)^{-1}$ and the common ratio is $(1+r)^{-1}$; hence S is simply the sum of a geometric progression. Using formula (9.8) for the sum of a geometric progression and the appropriate values for the first term and common ratio, we obtain

$$S = p(1+r)^{-1} \frac{1 - \left[(1+r)^{-1}\right]^n}{1 - (1+r)^{-1}}$$

$$= p \frac{1}{1+r} \frac{1 - (1+r)^{-n}}{r/(1+r)}$$

or

$$S = p \frac{1 - (1+r)^{-n}}{r} \tag{9.16}$$

Since the term of $[1-(1+r)^{-n}]/r$ is used quite often in figuring present

Sec. 9.4 / ANNUITIES

values of annuities, its values for several combinations of r and n are available in table form and are given in Table A.8 in the Appendix. (In the event that the table does not contain the appropriate combination of values for a specific problem, logarithms can be used to determine the value of a term.)

Example 9.16 Determine the present value of each payment of an annuity that pays $4,000/year for 6 years if the interest rate is 6 percent/year and the interest is compounded annually. Determine the present value of the annuity by summing the present values of the six payments and also by using Eq. (9.16).

Solution We are given $p = \$4,000$, $r = .06$, and $n = 6$. The present value of each payment is obtained by substituting these values into Table 9.2, and these new values are given in Table 9.3. Moreover, the present value of the annuity obtained by summing is given as the total in Table 9.3. Using Eq. (9.16) and Table A.8, we obtain

$$S = 4,000 \cdot \frac{1 - (1.06)^{-6}}{.06} = 4,000(4.917) = \$19,668$$

Table 9.3

Payment No.	Payment	$(1.06)^{-n}$ (from Table A.6)	Present Value of Payment
1	$4,000	0.943	$3,772
2	4,000	0.890	3,560
3	4,000	0.840	3,360
4	4,000	0.792	3,168
5	4,000	0.747	2,988
6	4,000	0.705	2,820
			Total $19,668

Example 9.17 Mr. Mason, president of Mason Electronics, wishes to establish a trust fund for his son's college education. He figures that $1,000/quarter should cover his son's entire cost for room, board, books, tuition, and spending money. His son plans to attend college full time, i.e., four quarters per year. The company where Mr. Mason invests his savings will pay 8 percent/year interest compounded quarterly. He wishes to sell a sufficient amount of stock in order to deposit a lump sum to establish the trust that will supply his son with $1,000/quarter for 4 years. Determine the dollar value of the stock that Mr. Mason must sell in order that he may establish the trust fund 3 months before his son begins receiving payments.

Solution We are given $p = \$1,000$. The interest rate is expressed on an annual basis and must be converted to a quarterly basis; thus $r = .08/4 = .02$ per quarter. The conversion period is a quarter; hence we express n in quarters by converting 4 years to 16 quarters; that is, $n = 16$. Substituting these values into Eq. (9.16), we obtain

$$S = 1,000 \cdot \frac{1-(1.02)^{-16}}{.02}$$

From Table A.8, we find the value of $[1-(1.02)^{-16}]/.02$ to be 13.578; thus the present value of the annuity is

$$S = 1,000(13.578) = \$13,578$$

Therefore Mr. Mason must sell $13,578 worth of stock in order to establish the trust fund.

Amortization In the preceding discussions of this section, we have been concerned with determining the present and future values of annuities for a given payment, interest rate, and term. In many situations, the average person is faced with the problem of repaying a debt to a business firm for a credit purchase. If the debt is to be retired by equal periodic payments at a compound interest rate over a specified length of time, the problem becomes one of determining the periodic payment, and the retirement of a debt in this manner is called *amortization*. If we are given the present value of an annuity, which is usually the amount of the debt, we can determine the amount of the payment by solving (9.16) for p. Solving (9.16) for p, we obtain

$$p = S\left[\frac{r}{1-(1+r)^{-n}}\right] \qquad (9.17)$$

[The term in brackets in (9.17) is simply the reciprocal of the values given in Table A.8.]

Example 9.18 Mrs. Ward purchased a new refrigerator on credit from City Hardware for $650. City Hardware will permit her to make monthly payments to retire the debt. However, the interest charge is 1 percent/month on the unpaid balance of her account. What monthly payment should Mrs. Ward make in order to retire the debt in 1 year?

Solution The interest rate is $r = .01$ per month. The term is $n = 12$ months, and the present value of the annuity is $S = \$650$. Substituting into (9.17), we obtain

$$p = 650\left[\frac{.01}{1-(1.01)^{-12}}\right]$$

From Table A.8, we obtain the reciprocal of the term in brackets, and thus Mrs. Ward's monthly payment should be

$$p = 650 \frac{1}{11.255} = \$57.75$$

EXERCISES

1. For each of the following, determine the future value of the annuity and the compound interest if the payment is invested at the indicated rate for the given time period.

(a) $1,000/year, for 10 years, at 6 percent compounded annually
(b) $500/6 months, for 17 years, at 8 percent compounded semiannually
(c) $200/quarter, for 12 years, at 8 percent compounded quarterly
(d) $100/quarter, for 5 years, at 8 percent compounded quarterly
(e) $800/year, for 15 years, at 6 percent compounded annually
(f) $400/6 months, for $10\frac{1}{2}$ years, at 6 percent compounded semiannually

2. For each of the following, determine the present value of the annuity where the payment is received at the indicated rate for the given time period.

(a) $100/quarter, for 3 years, at 8 percent compounded quarterly
(b) $800/year, for 10 years, at 7 percent compounded annually
(c) $500/6 months, for 9 years, at 8 percent compounded semiannually
(d) $200/quarter, for 2 years, at 8 percent compounded quarterly
(e) $600/year, for 6 years, at 6 percent compounded annually
(f) $1,000/year, for 7 years, at 8 percent compounded annually

3. Determine the annual payment for an annuity that is to accumulate $3,000 in 8 years if the payments earn interest at 6 percent compounded annually.

4. Determine the annual payment for an annuity that is to accumulate $10,000 in 20 years if the payments earn interest at 5 percent compounded annually.

5. Determine the annual payment for an annuity that is to accumulate $5,000 in 6 years if the payments earn interest at 7 percent compounded annually.

6. Determine the quarterly payments for an annuity that is to accumulate $3,000 in 6 years if the payments earn interest at 8 percent compounded quarterly.

7. Determine the semiannual payments for an annuity that is to accumulate $4,000 in $7\frac{1}{2}$ years if the payments earn interest at 6 percent compounded semiannually.

8. Ms. Morgan, a stockbroker, assures a client that the stock of a certain company can easily yield 8 percent/year. If all gains and dividends are reinvested annually and the client chooses to invest $1,000/year, what will his investment be worth at the end of (a) 5 years? (b) 10 years? (c) 15 years? (d) 20 years?

9. Mr. Johnson bought a new car and borrowed $3,000 from his bank to pay for it. He wishes to repay the bank at the rate of $150/month. If the bank charges interest at the rate of 12 percent/year compounded monthly, how long will it take Mr. Johnson to pay off the debt?

10. A company can either lease or purchase a copying machine. The cost to lease the machine for 2 years is $12,000, which is to be paid at a rate of $500/month. The purchase price is $10,000. In order to purchase the machine, the $10,000 must be borrowed at an interest rate of 9 percent compounded monthly. The company has agreed to repay the loan at the rate of $300/month. Determine the cheapest way to acquire the machine, assuming that it will have no value at the end of 2 years.

11. A company wants to pay cash for the purchase of a $15,000 machine in 5 years. Two investment alternatives are available for saving the money:

(a) Invest a given amount each year at an annual interest rate of 8 percent compounded annually.

(b) Invest a given amount each quarter at an annual interest rate of 6 percent compounded quarterly.

Determine which alternative will require the smallest yearly outlay of cash.

12. A company borrowed some money at an annual interest rate of 24 percent compounded monthly. If their monthly payments are $200 for a year, how much money did they borrow?

IMPORTANT TERMS AND CONCEPTS

Amortization
Annuity
Arithmetic progression
Common difference
Common ratio
Compound
  amount
  interest
  maturity value
Future value
Geometric progression

Interest
Interest rate
Long-term loan
Maturity value
Present value
Principal
Sequence
Short-term loan
Simple interest
Term

REVIEW PROBLEMS

1. A machine was purchased for $65,000 and will depreciate $5,500 per year. What will the machine be worth after 6 years?

2. Current sales for the XYZ Company are $800,000. Records show that sales have been increasing at a rate of 8 percent/year. If this trend continues, what will sales be in 5 years? What will be the total sales during the 5-year period?

3. A piece of equipment was purchased for $120,000 and will depreciate as follows: $15,000 the first year, $14,000 the second year, $13,000 the third year, and so on. Based on these estimates, what will be the amount of depreciation for the tenth year and what will the equipment be worth after the tenth year?

4. To expand production facilities the Widget Manufacturing Company borrowed $3,500 for 3 years at a rate of 8 percent/year compounded quarterly. Determine the compound maturity value and the amount of interest due at the end of the 3 years.

5. Mr. Smith borrowed $800 from Friendly State Bank 5 years ago and an additional $2,500 today. He will repay $1,000 next year, another $1,500 at the end of 3 years, and a final payment at the end of 5 years. If the loan is given at the rate of 6 percent/year compounded semiannually, determine the amount of the final payment.

6. A steadily growing company made a profit of $50,000 6 years ago and $70,000 this year. What is the annual growth rate compounded annually? What is the total profit made during the 6-year period?

7. A business plans to buy a $50,000 machine in 5 years and has 2 alternatives for saving the money: (1) buy a 5-year note at an interest rate of 9 percent/year compounded every 4 months, and (2) deposit a fixed amount each quarter during the 5 years to accumulate interest at a rate of 12 percent/year compounded quarterly. Determine:

   (a) What principal must be invested for alternative (1) to yield the required $50,000 maturity value
   (b) What amount must be invested each quarter for alternative (2) to yield the required $50,000

8. A $7,500 car was financed at an interest rate of 12 percent/year compounded monthly. If the loan is to be paid by equal monthly payments over a $3\frac{1}{2}$ year period, determine the amount of the monthly payments.

9. The Tick-Tock Clock Company borrowed $5,000 at an interest rate of 6 percent/year compounded semiannually. If the loan is repaid by semiannual payments of $600, how long will it take to pay off the debt?

10. Mr. Kee financed the purchase of his new business at an interest rate of 8 percent compounded quarterly. He is to pay $1,500/quarter for 12 years to retire the loan. Determine the purchase price of the business.

10 DIFFERENTIAL CALCULUS: BASIC METHODOLOGY

10.1 INTRODUCTION

Calculus has obvious potential in business situations because it helps us to represent complex relationships graphically. In many situations where we wish to analyze a function it is helpful to first plot the function and then try to obtain certain information from the graph. Such graphs appear in our daily newspapers to show at a glance what is happening to the Dow Jones Index, weather, air and water pollution levels, economic indicators, and many other indicators. In many cases we are more interested in the relative changes of the function than in the function value. In other words, the shape of the graph is at least as important as the actual data supplied by the graph. Calculus can be used to locate maximum and minimum values of functions, where we seek to maximize profits or revenues and minimize losses or costs. Furthermore, it can be used to convert total cost (or profit or revenue) to marginal cost (or profit or revenue), and in many situations marginal values are of greatest importance to the decision maker.

We study calculus in order to understand the idea of the rate of change for functions; this study is interesting, because it involves new concepts not previously covered in algebra and it has many powerful applications. The term *calculus* as commonly used refers to *infinitesimal calculus*, which encompasses the study of two types of limits: *derivatives* and *integrals*. Thus calculus is divided into two distinct sections called *differential calculus* and *integral calculus*.

We shall approach calculus as we did algebra, by first gaining an understanding of the basic operations and concepts and then applying the concepts to business problems. Since calculus involves the study of two types of limits, we begin by examining the concept of a limit.

10.2 LIMIT OF A FUNCTION

The concept of a *limit*, so fundamental to calculus, is probably not completely new to the reader. In calculus, we are concerned with the *limit of a function*.

Consider a function $f(x)$ and let the independent variable x assume values *near* a given constant a; the function will then assume a corresponding set of values. For example, let $f(x) = x^2 + 2$, and let $a = 4$. Choosing values of x that are near 4, we obtain, for example, $f(3.99) = 17.9201$, $f(4.01) = 18.0801$, $f(3.999) = 17.992001$, $f(4.001) = 18.008001$, $f(3.9999) = 17.99920001$, and $f(4.0001) = 18.00080001$. In this example, the function values corresponding to the x values near 4 are near the value 18; furthermore, the closer the values approach 4, the closer the corresponding function value approaches the value 18. The values of $f(x)$ can be made to differ from 18 by *as little as we please* by taking values of x that are *sufficiently close* to 4. Under these conditions, we say that $f(x)$ *approaches the limit* 18 *as* x *approaches* 4. In the concept of a limit, we are interested, not in the value of $f(x)$ when x is equal to 4, but rather in the values of $f(x)$ when more concisely defined as follows.

Definition 10.1 A function $f(x)$ is said to approach a limit L as x approaches a if the absolute value of the difference between $f(x)$ and L is less than an arbitrary positive number for all values of x that are sufficiently close to a and for which $x \neq a$; this is denoted mathematically as

$$\lim_{x \to a} f(x) = L$$

Example 10.1 Determine the limit of $f(x) = 2x + 4$ as x approaches 1 by determining the function values for six values of x that are within 0.01 of 1 such that three values are less than 1 and three values are greater than 1.

Solution Symbolically the problem is to determine L, where

$$\lim_{x \to 1} (2x + 4) = L$$

Choosing values of x that are symmetrical about 1 and within 0.01, we select 0.99, 0.999, 0.9999, 1.0001, 1.001, and 1.01. Then the corresponding values of $f(x)$ are

$$f(0.99) = 5.98$$
$$f(0.999) = 5.998$$
$$f(0.9999) = 5.9998$$
$$f(1.0001) = 6.0002$$
$$f(1.001) = 6.002$$
$$f.(1.01) = 6.02$$

Thus we see that apparently $f(x)$ approaches 6 as x approaches 1.

In some instances we are concerned with x increasing without bounds. When this is the case, we represent the value that x approaches by the symbol ∞, for "infinity." Thus, if the limit L is a finite number as x increases without bounds, we write

$$\lim_{x \to +\infty} f(x) = L$$

If x were to *decrease* without bounds, we would replace the positive sign by a negative sign and write

$$\lim_{x \to -\infty} f(x) = L$$

Example 10.2 The growth rate of XYZ Corporation is given by the function $f(x) = 1/x^2$, where x is yearly sales. Determine the limit of $f(x)$ as x increases without bounds by determining values of $f(x)$ for increasingly large values of x. What will eventually happen if XYZ Corporation does not do something to change their growth rate?

Solution Symbolically stated, the problem is to determine L where

$$\lim_{x \to +\infty} \frac{1}{x^2} = L$$

For the increasingly large values of x, let us select 100, 1,000, and 10,000; thus we obtain

$$f(100) = 0.0001$$
$$f(1,000) = 0.000001$$
$$f(10,000) = 0.00000001$$

We see that the larger x becomes, the nearer to zero the function value becomes; hence

$$\lim_{x \to +\infty} \frac{1}{x^2} = 0$$

Eventually, the growth rate of XYZ Corporation will become zero; that is, it will cease to grow.

Example 10.3 Determine the limit of $f(x) = 1/x$ as x decreases without bounds by determining values of $f(x)$ for increasingly large negative values of x.

Solution Here the problem is to determine L where

$$\lim_{x \to -\infty} \frac{1}{x} = L$$

Sec. 10.2 / LIMIT OF A FUNCTION

For the increasingly large negative values of x, let us select -100, $-1,000$, and $-10,000$. For these values, we obtain

$$f(-100) = -0.01$$
$$f(-1,000) = -0.001$$
$$f(-10,000) = -0.0001$$

Thus the more negative the value of x becomes, the closer the function approaches zero; hence

$$\lim_{x \to -\infty} \frac{1}{x} = 0$$

In these three examples we have computed several values of $f(x)$ for values of x near a. In practice, however, we do not need to write these values down; instead, we can usually determine the general trend of the function as x approaches a from either direction. If the values that $f(x)$ approaches are finite and equal as x approaches a from the two directions, the limit of the function is the common value that $f(x)$ approaches; otherwise, the limit does not exist.

From Examples 10.1 to 10.3 we have seen that the value of x may either approach a finite value or increase or decrease without bounds while the function approaches a finite limit. However, it is possible for x to do any of the three alternatives and the function not to have a finite limit; when this is the case, we simply say that the limit does not exist because we are interested only in finite values.

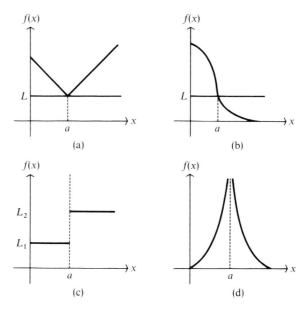

FIGURE 10.1

A better understanding of the limit concept may be obtained from the graphs in Fig. 10.1. Figure 10.1a and b shows functions that have finite limits as x approaches a. Figure 10.1c and d illustrates functions for which no limits exist. However, here the limits do not exist for different reasons: For example, in Fig. 10.1c, the limit would be L_1 as x approaches a from the left and L_2 as x approaches a from the right; unless L_1 and L_2 are equal, there is an inconsistency between the two values, and we say that the limit does not exist. One such function for which the limit does not exist is given by

$$f(x) = \begin{cases} L_1 & x \leq a \\ L_2 & x > a \end{cases}$$

A function represented in this fashion means that $f(x) = L_1$ for $x \leq a$ and $f(x) = L_2$ for $x > a$. Similarly, the function in Fig. 10.1d is such that the limit is unbounded as x approaches a from either direction; thus we say that no limit exists as x approaches a. This type of situation is illustrated by the function $f(x) = 1/(x-a)^2$. However, in Fig. 10.1c and d the limit of the function does exist for any value of x other than a.

Rules for Operating with Limits

The following are the rules for operating with limits. We emphasize that the rules require that each function possess a limit as x approaches the value of concern. Let $f(x)$ and $g(x)$ be any functions of x that possess limits as x approaches a.

Rule 10.1 The limit of the sum of two functions is equal to the sum of their individual limits, provided the individual limits exist:

$$\lim_{x \to a} [f(x) + g(x)] = \lim_{x \to a} f(x) + \lim_{x \to a} g(x) \qquad (10.1)$$

Rule 10.2 The limit of the product of two functions is equal to the product of their individual limits, provided the individual limits exist:

$$\lim_{x \to a} [f(x) g(x)] = \lim_{x \to a} f(x) \lim_{x \to a} g(x) \qquad (10.2)$$

Rules 10.1 and 10.2 can be extended to any finite number of functions.

Rule 10.3 The limit of the quotient of two functions is equal to the quotient of their individual limits, provided the limit of the divisor is not zero and the individual limits exist:

$$\lim_{x \to a} \left[\frac{f(x)}{g(x)} \right] = \frac{\lim_{x \to a} f(x)}{\lim_{x \to a} g(x)} \qquad (10.3)$$

Sec. 10.2 / LIMIT OF A FUNCTION

By applying Rule 10.2 to any number of equal factors, we obtain the following

Rule 10.4 The limit of the nth power of a function is equal to the nth power of the limit of that function, provided the limit exists:

$$\lim_{x \to a} [f(x)]^n = \left[\lim_{x \to a} f(x) \right]^n$$

where n is any positive integral exponent.

Using the result of Rule 10.4, it can also be shown that the following rule holds true:

Rule 10.5 The limit of the nth root of a positive function is equal to the principal nth root of the limit of that function, provided the limit exists:

$$\lim_{x \to a} [f(x)]^{1/n} = \left[\lim_{x \to a} f(x) \right]^{1/n}$$

Example 10.4 Let $f(x) = 2x^2 - 4$ and $g(x) = x + 3$. Determine the limits of the following functions as x approaches 2 by first determining the limits of the individual functions and then applying the above rules for limits of functions:

(a) $f(x) + g(x)$
(b) $f(x) g(x)$
(c) $f(x)/g(x)$
(d) $[f(x)]^3$
(e) $[f(x)]^{1/2}$
(f) $\dfrac{[f(x)]^{1/2} [g(x)]^3}{f(x) + g(x)}$

Solution Determining the limits of the individual functions, we have

$$\lim_{x \to 2} (2x^2 - 4) = 4$$

and

$$\lim_{x \to 2} (x + 3) = 5$$

(a) $\lim\limits_{x \to 2} [f(x) + g(x)] = \lim\limits_{x \to 2} (2x^2 - 4) + \lim\limits_{x \to 2} (x + 3) = 4 + 5 = 9$

(b) $\lim\limits_{x \to 2} [f(x) g(x)] = \lim\limits_{x \to 2} (2x^2 - 4) \lim\limits_{x \to 2} (x + 3) = 4 \cdot 5 = 20$

(c) $\lim\limits_{x \to 2} \dfrac{f(x)}{g(x)} = \dfrac{\lim\limits_{x \to 2} (2x^2 - 4)}{\lim\limits_{x \to 2} (x + 3)} = \dfrac{4}{5}$

(d) $\lim_{x \to 2} [f(x)]^3 = \lim_{x \to 2} (2x^2 - 4)^3 = \left[\lim_{x \to 2} (2x^2 - 4) \right]^3 = (4)^3 = 64$

(e) $\lim_{x \to 2} [f(x)]^{1/2} = \lim_{x \to 2} (2x^2 - 4)^{1/2} = \left[\lim_{x \to 2} (2x^2 - 4) \right]^{1/2} = 4^{1/2} = 2$

(f) $\lim_{x \to 2} \frac{[f(x)]^{1/2}[g(x)]^3}{f(x) + g(x)} = \frac{\lim_{x \to 2} (2x^2 - 4)^{1/2} \lim_{x \to 2} (x + 3)^3}{\lim_{x \to 2} [(2x^2 - 4) + (x + 3)]}$

$= \frac{\left[\lim_{x \to 2} (2x^2 - 4)\right]^{1/2} \left[\lim_{x \to 2} (x + 3)\right]^3}{\lim_{x \to 2} (2x^2 - 4) + \lim_{x \to 2} (x + 3)}$

$= \frac{4^{1/2} \cdot 5^3}{4 + 5} = \frac{250}{9}$

Occasionally we encounter a problem involving the quotient of two functions for which Rule 10.3 does not appear to be applicable. For example, let $f(x) = x^2 - 9$ and $g(x) = x - 3$. If we wish to determine $\lim_{x \to 3} f(x)/g(x)$, we obtain

$$\lim_{x \to 3} f(x) = \lim_{x \to 3} (x^2 - 9) = 0$$

and

$$\lim_{x \to 3} g(x) = \lim_{x \to 3} (x - 3) = 0$$

Thus the divisor has a limit of zero, which nullifies the application of Rule 10.3; however, if we factor the respective functions and divide like terms, we have

$$\frac{f(x)}{g(x)} = \frac{(x - 3)(x + 3)}{x - 3} = x + 3$$

provided $x \neq 3$, and therefore, since we are only interested in values close to 3,

$$\lim_{x \to 3} \frac{f(x)}{g(x)} = \lim_{x \to 3} (x + 3) = 6$$

Continuous Functions In the definition of $\lim_{x \to a} f(x)$, the reader has probably noted that nothing has been mentioned concerning the value of $f(x)$ at $x = a$. The limit depends only on the values of $f(x)$ in the neighborhood of $x = a$, not on the value of $f(x)$ at $x = a$; thus the limit may or may not equal $f(a)$. However, the case where $\lim_{x \to a} f(x) = f(a)$ is of special interest.

Sec. 10.2 / LIMIT OF A FUNCTION

Definition 10.2 A function $f(x)$ is continuous at $x=a$ if the following three conditions are satisfied: (1) $f(a)$ exists, (2) $\lim_{x \to a} f(x)$ exists, and (3) $\lim_{x \to a} f(x) = f(a)$.

Returning to Fig. 10.1, we note that Fig. 10.1a and b illustrate functions that are continuous, while the functions illustrated in c and d are not continuous. A function satisfying Definition 10.2 is said to be *continuous at the point* $x = a$, whereas a function that is continuous at every point in the domain (or an interval) is said to be *continuous in the domain* (or interval). A function that is not continuous at a point $x = a$ is said to be *discontinuous* there. Intuitively, a continuous function is one whose graph constitutes a smooth curve without holes in it.

Example 10.5 Determine whether or not the function $f(x) = 3x^2 + 2x + 1$ is continuous at $x = 1$.

Solution Since $f(1) = 3(1)^2 + 2(1) + 1 = 6$, the function exists at $x = 1$. Also $\lim_{x \to 1}(3x^2 + 2x + 1) = 6$, and hence $\lim_{x \to 1}(3x^2 + 2x + 1) = f(1)$. Therefore, $f(x)$ is continuous at $x = 1$; furthermore, $f(x)$ is continuous for all values of x.

Example 10.6 Determine whether or not the function $f(x) = 1/(x-3)$ is continuous at $x = 3$.

Solution The function value at $x = 3$, $f(3)$, is not defined because division by zero is not permitted. Therefore, the function is not continuous at $x = 3$; however, the function is continuous for all values of x for which $x \neq 3$.

Example 10.7 Determine whether or not the function

$$f(x) = \begin{cases} -x & x < 0 \\ 2 & x = 0 \\ x & x > 0 \end{cases}$$

is continuous at $x = 0$.

Solution Since $f(0) = 2$, we know that $f(0)$ exists. Moreover, as x approaches zero from the left, $f(x)$ approaches zero; likewise, as x approaches zero from the right, $f(x)$ approaches zero. Hence $\lim_{x \to 0} f(x) = 0$. Since $\lim_{x \to 0} f(x) \neq 2$, this means that the function is not continuous at $x = 0$; however, it is continuous for all other values of x.

Example 10.8 Determine whether or not the function

$$f(x) = \begin{cases} \dfrac{1}{x-3} & x \neq 3 \\ 3 & x = 3 \end{cases}$$

is continuous at $x = 3$.

Solution Since $f(3) = 3$, the function is defined at $x = 3$. As x approaches 3 from the left, $1/(x-3)$ decreases without bounds. Therefore, $\lim_{x\to 3} f(x)$ does not exist at $x=3$. Thus the function is not continuous at $x=3$. However, it can be shown that $f(x)$ is continuous for all values of x for which $x \neq 3$.

From the definition of a *limit* and of a *function*, it can be shown that the sums, differences, and products of two continuous functions are themselves continuous functions. Furthermore, the quotient of two continuous functions is continuous at every point for which the denominator does not equal zero. Nearly all of the functions with which we shall be concerned in the remainder of this textbook are continuous [in particular, all *polynomials* are continuous, and all *rational functions* (quotients of polynomials) are continuous, except at points where the denominator equals zero].

EXERCISES

1. Determine each of the following limits:

   (a) $\lim_{x \to 3} (x^2 + 2)$

   (b) $\lim_{x \to 4} \left(\frac{x+4}{x} \right)$

   (c) $\lim_{x \to 1} (3 - x^2)$

   (d) $\lim_{x \to -4} (x^2 + 3x)$

   (e) $\lim_{x \to -6} \left(1 + \frac{4}{x} \right)$

   (f) $\lim_{x \to +\infty} \left(1 + \frac{1}{x} \right)$

   (g) $\lim_{x \to +\infty} \left(\frac{2}{x^2} + 2 \right)$

   (h) $\lim_{x \to +\infty} (1 + 2^{-x})$

   (i) $\lim_{x \to -\infty} (2 + x^{-2})$

   (j) $\lim_{x \to -\infty} \left(1 + \frac{1}{x^3 + 1} \right)$

2. Discuss the behavior of each of the following functions:

   (a) $f(x) = 3 + \frac{1}{x^3}$ (as $x \to 0$)

   (b) $f(x) = \frac{1}{(x-3)^2}$ (as $x \to 3$)

   (c) $f(x) = -4 + \frac{2}{(x+2)^2}$ (as $x \to -2$)

   (d) $f(x) = x^3 + \frac{1}{x^2}$ (as $x \to 0$)

3. Determine each of the following limits:

   (a) $\lim_{x \to 2} \left(1 + \frac{1}{x^2} \right)$

   (b) $\lim_{x \to 2} \left(\frac{x-2}{x+2} \right)$

   (c) $\lim_{x \to 4} \left(x^2 + 1 - \frac{x}{x-2} \right)$

   (d) $\lim_{x \to 0} \left(\frac{x+1}{x+2} \right)$

   (e) $\lim_{x \to -2} \left(3 - \frac{x+2}{x-2} \right)$

   (f) $\lim_{x \to +\infty} \left(\frac{2}{3^x} \right)$

   (g) $\lim_{x \to -1} \left(\frac{x^2 + 1}{x^2 + x + 1} \right)$

   (h) $\lim_{x \to -\infty} (1 + 3^x)$

4. Determine each of the following limits:

   (a) $\lim_{x \to 1} \left(\frac{x^2 - x}{x - 1} \right)$

   (b) $\lim_{x \to 2} \left(\frac{x^2 - 4}{x - 2} \right)$

Sec. 10.3 / DERIVATIVE OF A FUNCTION

(c) $\lim_{x \to 0} \left(\dfrac{x^3 + x^2}{x^2 + 2x} \right)$

(d) $\lim_{x \to 0} \left(\dfrac{x^4 + x^2}{x^3 + 3x^2} \right)$

(e) $\lim_{x \to -3} \left(\dfrac{x^3 - 9x}{x + 3} \right)$

(f) $\lim_{x \to -2} \left(\dfrac{2x^3 + 4x^2}{x + 2} \right)$

(g) $\lim_{x \to 2} \left(\dfrac{x^2 - 5x + 6}{x^2 - x - 2} \right)$

(h) $\lim_{x \to -1} \left(\dfrac{x^2 - 3x - 4}{x^2 - 2x - 3} \right)$

5. Determine the values of x for which each of the following functions is discontinuous:

(a) $f(x) = \dfrac{x - 1}{x + 1}$

(b) $f(x) = \dfrac{2x^2 + 1}{x^2}$

(c) $f(x) = \dfrac{3x}{4 - x^2}$

(d) $f(x) = \dfrac{x}{x^2 + 3x - 4}$

(e) $f(x) = \dfrac{x - 1}{x^2 - 1}$

(f) $f(x) = \dfrac{x^3 + x}{x^2}$

(g) $f(x) = \dfrac{3x + 1}{x(x^2 + 4)}$

(h) $f(x) = \dfrac{x + 2}{x^2 + 4x + 4}$

6. Determine whether or not the following functions are continuous at the given point:

(a) $f(x) = \dfrac{x}{x + 3}$ $(x = 3)$

(b) $f(x) = \dfrac{x^2 + 2x}{x + 2}$ $(x = -2)$

(c) $f(x) = \dfrac{x^2 - x}{x - 1}$ $(x = 1)$

(d) $f(x) = \dfrac{x^3 - 9x}{x + 3}$ $(x = -3)$

(e) $f(x) = \dfrac{x^2 - 5x}{x^3 - 5x^2}$ $(x = 0)$

(f) $f(x) = \dfrac{x^2 - 2x - 3}{x^2 - 3x - 4}$ $(x = 4)$

10.3 DERIVATIVE OF A FUNCTION

The concepts of differential calculus are based on the idea of the *rate of change* of a function; for example, if $f(x)$ is a function of the independent variable x, the rate of change of $f(x)$ is the *ratio* of the *change* in the function value $f(x)$ corresponding to a *change* in the independent variable x. Figure 10.2 illustrates this rate of change for a linear function, where the change in x is represented by Δx, read "delta x," and the change in $f(x)$ is denoted by

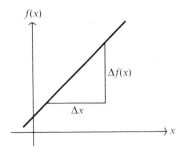

FIGURE 10.2

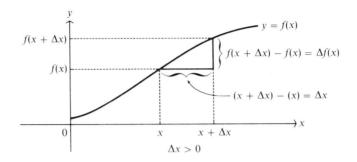

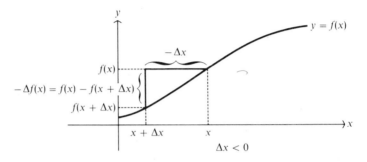

FIGURE 10.3

$\Delta f(x)$. As discussed in Chap. 2, the rate of change, or *slope*, of a straight line is defined as the *ratio of the change in y to the change in x*. Furthermore, the slope of a linear function is constant for all values of x, whereas the slope of nonlinear functions is not constant.

When considering functions in general, we must define what we mean by the slope of a function at a specific value of x. The slope of a function is the *instantaneous rate of change* of the function that is the *limit* of the ratio of the change in the dependent variable to the change in the independent variable. The two graphs in Fig. 10.3 help to visualize this concept. In each graph Δx denotes a small change in x. The expression $x + \Delta x$ should be considered a single real number written as a sum (or difference) of two real numbers. The slope of the hypotenuse of each of the two triangles equals

$$\lim_{\Delta x \to 0} \frac{\Delta f(x)}{\Delta x} = \lim_{\Delta x \to 0} \frac{f(x+\Delta x)-f(x)}{\Delta x}$$

As Δx is taken numerically smaller and smaller, this slope approaches the slope of a line that just touches the graph at the point $(x, f(x))$, as shown in Fig. 10.4. This line is said to be the *tangent line* to the curve at $(x, f(x))$.

Thus the instantaneous rate of change is given by

$$\lim_{\Delta x \to 0} \frac{\Delta f(x)}{\Delta x} = \lim_{\Delta x \to 0} \frac{f(x+\Delta x)-f(x)}{\Delta x}$$

Sec. 10.3 / DERIVATIVE OF A FUNCTION

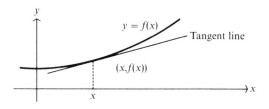

FIGURE 10.4

The terms in this ratio are illustrated in Fig. 10.3. This instantaneous rate of change in $f(x)$ is called the *derivative of f(x)*, denoted by $D_x f(x)$, which is read "the derivative with respect to x of $f(x)$." Two alternative symbols for the derivative of $f(x)$ are $f'(x)$ and df/dx. The choice of which notation to use is generally due to personal preference; however, when we wish merely to indicate the derivative of the function $f(x)$, normally we use $D_x f(x)$ with the appropriate function substituted in place of $f(x)$. On the other hand, when we wish to determine the value of the derivative for a specific value of x, generally we use $f'(x)$ with the value of x substituted in place of x. A possible problem with the df/dx notation is that is appears to be a fraction, which is certainly not the case.

Definition 10.3 The derivative with respect to x of a function $f(x)$ is the limit of the ratio $\Delta f(x)/\Delta x$ as Δx approaches zero and is given by the formula

$$D_x f(x) = \lim_{\Delta x \to 0} \frac{f(x + \Delta x) - f(x)}{\Delta x} \qquad (10.4)$$

whenever the limit exists.

An alternate expression for the derivative of $y = f(x)$ is easily found by making the following substitutions. Let $x = x_1$ and $\Delta x = x_2 - x_1$. Then

$$D_x[f(x_1)] = f'(x_1) = \lim_{(x_2 - x_1) \to 0} \frac{f[x_1 + (x_2 - x_1)] - f(x_1)}{x_2 - x_1}$$

$$= \lim_{x_2 \to x_1} \frac{f(x_2) - f(x_1)}{x_2 - x_1}$$

Hence the derivative at x_1 is the limit of slopes of lines that pass through the points $(x_1, f(x_1))$ and $(x_2, f(x_2))$, where x_1 is fixed and x_2 approaches x_1. As x_2 approaches x_1, the lines passing through these points approach the tangent line at x_1. (See Fig. 10.5.)

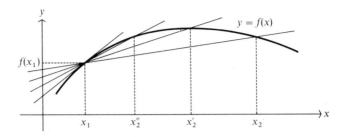

FIGURE 10.5

Before using the definition of the derivative to find $D_x f(x)$ for the particular functions, we need to recall that in functional notation

$$f(x+\Delta x) \neq f(x)+f(\Delta x)$$

Also, in evaluating a function f at a particular x, we insert that value of x for every x in the defining equation of the function. For example, if

$$y=f(x)=\frac{x^3-2x}{x} \qquad x \neq 0$$

then

$$f(1)=\frac{(1)^3-2(1)}{1}=-1$$

$$f(-3)=\frac{(-3)^3-2(-3)}{-3}=\frac{-27+6}{-3}=7$$

$$f(2)=\frac{(2)^3-2(2)}{2}=\frac{8-4}{2}=2$$

$$f(1+1)=\frac{(1+1)^3-2(1+1)}{(1+1)}=\frac{8-4}{2}=2$$

and so

$$f(x+\Delta x)=\frac{(x+\Delta x)^3-2(x+\Delta x)}{(x+\Delta x)}$$

$$=\frac{x^3+3x^2\Delta x+3x\Delta x^2+\Delta x^3-2x-2\Delta x}{x+\Delta x}$$

If

$$y=f(x)=\frac{1}{\sqrt{x}}$$

then

$$f(x+\Delta x)=\frac{1}{\sqrt{x+\Delta x}}$$

Finally, if
$$y = f(x) = e^{3x^4}$$
then
$$f(x+\Delta x) = e^{3(x+\Delta x)^4}$$

In each case we find $f(x+\Delta x)$ by replacing x with $x+\Delta x$ in the defining equation of the given function. We are now prepared to use the definition to find the derivative.

Derivative of Some Special Functions

Now let us verify that Definition 10.3 corresponds with the concept of slope for a linear function. For example, suppose $f(x) = 3x + 2$; from our discussion of the slope of linear functions in Chap. 2, we know that the slope of this function is 3. Using Eq. (10.4) to determine the derivative of the function, we have

$$D_x(3x+2) = \lim_{\Delta x \to 0} \frac{[3(x+\Delta x)+2]-(3x+2)}{\Delta x}$$
$$= \lim_{\Delta x \to 0} \frac{3x + 3(\Delta x) + 2 - 3x - 2}{\Delta x}$$
$$= \lim_{\Delta x \to 0} \frac{3\Delta x}{\Delta x}$$
$$= \lim_{\Delta x \to 0} 3$$
$$= 3$$

Thus, for the given linear function, the slope of the straight line is equal to 3 regardless of the value of x.

Let us consider the general linear function $f(x) = mx + b$ and determine its derivative. Using Eq. (10.4), we have

$$D_x(mx+b) = \lim_{\Delta x \to 0} \frac{[m(x+\Delta x)+b]-(mx+b)}{\Delta x}$$
$$= \lim_{\Delta x \to 0} \frac{mx + m(\Delta x) + b - mx - b}{\Delta x}$$
$$= \lim_{\Delta x \to 0} \frac{m(\Delta x)}{\Delta x}$$
$$= \lim_{\Delta x \to 0} m$$
$$= m$$

Thus the slope of a linear function is the same as the derivative of the function for any value of x.

If we wish to determine the slope of a function at a particular value of x, we simply rely on the functional notation in the usual manner, except that we apply it to the function $f'(x)$; that is, if we wish to determine the slope of a function at $x = c$, we evaluate $f'(c)$.

Example 10.9 Determine the derivative of $f(x) = 2x^2 + 4$, and determine the slope of $f(x)$ at $x = 4$ and $x = 0$.

Solution Using (10.4), we have

$$\begin{aligned}
D_x(2x^2+4) &= \lim_{\Delta x \to 0} \frac{\left[2(x+\Delta x)^2 + 4\right] - (2x^2+4)}{\Delta x} \\
&= \lim_{\Delta x \to 0} \frac{\left\{2\left[x^2 + 2x(\Delta x) + (\Delta x)^2\right] + 4\right\} - (2x^2+4)}{\Delta x} \\
&= \lim_{\Delta x \to 0} \frac{2x^2 + 4x(\Delta x) + 2(\Delta x)^2 + 4 - 2x^2 - 4}{\Delta x} \\
&= \lim_{\Delta x \to 0} \frac{4x(\Delta x) + 2(\Delta x)^2}{\Delta x} \\
&= \lim_{\Delta x \to 0} [4x + 2(\Delta x)] \\
&= 4x
\end{aligned}$$

Thus $D_x f(x) = 4x$, or, alternatively, $f'(x) = 4x$. To determine the slope of $f(x)$ at $x = 4$, we evaluate $f'(4) = 4(4) = 16$; similarly, for the slope of $f(x)$ at $x = 0$, we obtain $f'(0) = 4(0) = 0$.

Example 10.10 The number of salespeople hired by the Cave Manufacturing Company is given by $f(x) = x^3$, where x is the number of sales territories they cover. Determine the derivative of $f(x)$ and the rate of change for the number of salespeople needed when $x = 0$ and $x = 2$.

Solution Using Eq. (10.4), we obtain

$$\begin{aligned}
D_x(x^3) &= \lim_{\Delta x \to 0} \frac{(x+\Delta x)^3 - x^3}{\Delta x} \\
&= \lim_{\Delta x \to 0} \frac{x^3 + 3x^2(\Delta x) + 3x(\Delta x)^2 + (\Delta x)^3 - x^3}{\Delta x} \\
&= \lim_{\Delta x \to 0} \frac{3x^2(\Delta x) + 3x(\Delta x)^2 + (\Delta x)^3}{\Delta x} \\
&= \lim_{\Delta x \to 0} \left[3x^2 + 3x(\Delta x) + (\Delta x)^2\right] \\
&= 3x^2
\end{aligned}$$

Thus $D_x(x^3) = 3x^2$, or, alternatively, $f'(x) = 3x^2$. The rate of change of the number of salespeople when $x = 0$ is determined by evaluating $f'(0) = 3(0)^2 = 0$; likewise, the rate of change when $x = 2$ is $f'(2) = 3(2)^2 = 12$.

Sec. 10.3 / DERIVATIVE OF A FUNCTION

Example 10.11 Determine the derivative of $f(x) = ax^3$.

Solution Applying Eq. (10.4), we obtain

$$D_x(ax^3) = \lim_{\Delta x \to 0} \frac{a(x+\Delta x)^3 - ax^3}{\Delta x}$$

$$= \lim_{\Delta x \to 0} \frac{a\left[x^3 + 3x^2(\Delta x) + 3x(\Delta x)^2 + (\Delta x)^3\right] - ax^3}{\Delta x}$$

$$= \lim_{\Delta x \to 0} \frac{ax^3 + 3ax^2(\Delta x) + 3ax(\Delta x)^2 + a(\Delta x)^3 - ax^3}{\Delta x}$$

$$= \lim_{\Delta x \to 0} \frac{3ax^2(\Delta x) + 3ax(\Delta x)^2 + a(\Delta x)^3}{\Delta x}$$

$$= \lim_{\Delta x \to 0} \left[3ax^2 + 3ax(\Delta x) + a(\Delta x)^2\right]$$

$$= 3ax^2$$

Thus $D_x(ax^3) = 3ax^2$.

Example 10.11 demonstrates that once we determine the derivative of a general type of function such as $f(x) = ax^3$ we can substitute different values for the parameter a in order to obtain the derivative of any function of that particular type. For example, since $D_x(ax^3) = 3ax^2$, we know that for $f(x) = 6x^3$ the derivative is $D_x(6x^3) = 3(6)x^2 = 18x^2$, or, for $f(x) = 15x^3$, the derivative is $D_x(15x^3) = 3(15)x^2 = 45x^2$. This concept is used to determine rules for determining the derivatives of various general types of functions. We shall develop a few simple rules by this method, and the remaining rules will be given without derivation or will be developed by the application of already existing rules.

To determine the derivative for a general function of the type $f(x) = k$, where k is a constant, we use Eq. (10.4), obtaining

$$D_x f(x) = D_x(k) = \lim_{\Delta x \to 0} \frac{k - k}{\Delta x}$$

$$= \lim_{\Delta x \to 0} 0$$

$$= 0$$

Therefore, we have the following rule for the derivative of a constant.

Definition 10.6 If k is a constant, then

$$D_x(k) = 0 \tag{10.5}$$

This rule permits us to differentiate functions that are constant by applying (10.5). Determining the derivative of a function by application of rules such as Rule 10.6 is called *sight* differentiation, or *inspection*, which is the purpose for developing rules for general types of functions.

Example 10.12 Differentiate the following functions:

(a) $f(x) = 8$ \hspace{2cm} (b) $f(x) = -4$
(c) $f(x) = 6$ \hspace{2cm} (d) $f(x) = \frac{1}{3}$

Solution We obtain the results for each part by applying Rule 10.6; that is,

(a) $D_x(8) = 0$ \hspace{2cm} (b) $D_x(-4) = 0$
(c) $D_x(6) = 0$ \hspace{2cm} (d) $D_x(\frac{1}{3}) = 0$

In Example 10.10, we determined the derivative of $f(x) = x^3$. Analogous steps can be employed to determine the derivative of the general function $f(x) = x^n$, where n is a positive integer. Furthermore, the result can be shown to hold true for any value of n; however, we shall not prove the following rule since it is beyond the scope of this text.

Rule 10.7 For any real number n,

$$D_x(x^n) = nx^{n-1} \tag{10.6}$$

Example 10.13 Differentiate each of the following functions:

(a) $f(x) = x^5$ \hspace{2cm} (b) $f(x) = x^{12}$
(c) $f(x) = x^{-1}$ \hspace{2cm} (d) $f(x) = x^{2/3}$
(e) $f(x) = x^{-3/5}$ \hspace{2cm} (f) $f(x) = x$

Solution Applying Rule 10.7, we obtain

(a) $D_x(x^5) = 5x^{5-1} = 5x^4$
(b) $D_x(x^{12}) = 12x^{12-1} = 12x^{11}$
(c) $D_x(x^{-1}) = -1x^{-1-1} = -x^{-2}$
(d) $D_x(x^{2/3}) = \frac{2}{3}x^{2/3-1} = \frac{2}{3}x^{-1/3}$
(e) $D_x(x^{-3/5}) = -\frac{3}{5}x^{-3/5-1} = -\frac{3}{5}x^{-8/5}$
(f) $D_x(x) = 1x^{1-1} = 1x^0 = 1$

Since the derivative of a function at a particular point actually represents the slope of the function *at that point*, we recall from Chap. 2 that a positive slope at a particular point means the function is *increasing* and a negative slope means the function is *decreasing*. If the function of interest is like the one in either Fig. 10.1c or d, it does not have a slope at the point $x = a$. From this observation, we note that the function *must be continuous* at $x = a$ in order for the derivative to exist at $x = a$. However, if the function is continuous at $x = a$, this does *not* necessarily guarantee that the derivative exists at $x = a$. The function in Fig. 10.1a is continuous, but the slope at $x = a$ is undefined because the instantaneous rate of change as x approaches a from the left is not equal to the instantaneous rate of change as x approaches a from the right.

To clarify the last statement, let us examine a specific function that is continuous at $x=0$ but for which the derivative does not exist at $x=0$. Let $f(x)=|x|$, that is, the absolute value of x that is obtained by making the sign of x positive. Another way of representing $f(x)=|x|$ is

$$f(x) = \begin{cases} -x & x<0 \\ x & x \geq 0 \end{cases}$$

If we wish to determine $D_x f(x)$ at $x=0$, we must use Definition 10.3; thus

$$D_x f(x) = \lim_{\Delta x \to 0} \frac{f(x+\Delta x) - f(x)}{\Delta x}$$

However, around $x=0$, the function value $f(x+\Delta x)$ will depend on whether Δx is positive or negative; that is, if Δx is approaching zero from the left (or negative) side, $f(x) = -x$; and if Δx is approaching zero from the right (or positive) side, $f(x) = x$. Thus, if Δx is approaching zero from the left side, denoted by $\Delta x \to 0-$, we have the left limit L such that

$$L = \lim_{\Delta x \to 0-} \frac{-(x+\Delta x)-(-x)}{\Delta x} = \lim_{\Delta x \to 0-} \frac{-\Delta x}{\Delta x} = \lim_{\Delta x \to 0-} -1 = -1$$

However, if Δx is approaching zero from the right side, denoted by $\Delta x \to 0+$, we have the right limit R such that

$$R = \lim_{\Delta x \to 0+} \frac{(x+\Delta x)-x}{\Delta x} = \lim_{\Delta x \to 0+} \frac{\Delta x}{\Delta x} = \lim_{\Delta x \to 0+} 1 = 1$$

Hence the two limits do not have the same value, which means that $D_x f(x)$ does not exist at $x=0$.

Occasionally the functional notation $f(x)$ is replaced by the use of a single variable y, which is called the *dependent* variable where $y=f(x)$. When this occurs, the derivative of $f(x)$ with respect to x is also the derivative of y with respect to x; and this derivative is represented notationally as $D_x y$, dy/dx, or y', all of which are equivalent to $D_x f(x)$ and $f'(x)$. The use of the symbols f, x, and y is not inflexible as long it is understood which variable we are differentiating.

Example 10.14 Determine the derivative of the term on the left side with respect to the variable on the right side for each of the following equations (also, give the different notations for the derivative):

(a) $y = x^4$
(b) $z = t^{-3}$
(c) $g(x) = x^{1/3}$
(d) $u = v^{10}$
(e) $x = y^2$
(f) $y(t) = t^{-5}$

Solution In each case, we use Rule 10.7 to determine the derivative:

(a) $D_x y = y' = \dfrac{dy}{dx} = 4x^3$

(b) $D_t z = z' = \dfrac{dz}{dt} = -3t^{-3-1} = -3t^{-4}$

(c) $D_x g(x) = g'(x) = \dfrac{dg(x)}{dx} = \dfrac{1}{3}x^{1/3-1} = \dfrac{1}{3}x^{-2/3}$

(d) $D_v u = u' = \dfrac{du}{dv} = 10v^{10-1} = 10v^9$

(e) $D_y x = x' = \dfrac{dx}{dy} = 2y^{2-1} = 2y$

(f) $D_t y(t) = y'(t) = \dfrac{dy(t)}{dt} = -5t^{-5-1} = -5t^{-6}$

EXERCISES

In Exercises 1 to 10, determine $D_x f(x)$ from Definition 10.3 and verify your results through use of the appropriate rule.

1. $f(x) = 2$
2. $f(x) = -2$
3. $f(x) = \sqrt{5}$
4. $f(x) = a^2$
5. $f(x) = b^{-3}$
6. $f(x) = a^b$
7. $f(x) = \dfrac{1}{x}$
8. $f(x) = \dfrac{1}{x^2}$
9. $f(x) = x^4$
10. $f(x) = \sqrt{x}$

In Exercises 11 to 18, use the appropriate rule to determine $D_x f(x)$, and determine the slope of $f(x)$ at the given value of x.

11. $f(x) = \tfrac{31}{15}$ $(x = 4)$
12. $f(x) = 5$ $(x = -10)$
13. $f(x) = (\tfrac{1}{2})^3$ $(x = 2)$
14. $f(x) = x^2$ $(x = 1)$
15. $f(x) = x^{-2}$ $(x = -1)$
16. $f(x) = x^5$ $(x = 2)$
17. $f(x) = x^{1/2}$ $(x = 4)$
18. $f(x) = x^{-6}$ $(x = -1)$

In Exercises 19 to 24, determine the derivative of the term on the left side of the equation with respect to the variable on the right side.

19. $y = \dfrac{1}{x^4}$
20. $z = \dfrac{1}{t^{2/3}}$
21. $r = s^{0.2}$
22. $p = q^{7/8}$
23. $h(w) = w^{0.05}$
24. $a = b^3$

Sec. 10.3 / DERIVATIVE OF A FUNCTION

For each of the functions in Exercises 25 and 26, determine whether or not the derivatives exist at the given point.

25. $f(x) = \begin{cases} x^2 - 2 & x > 2 \\ 2 & x = 2 \\ -(2 - x^2) & x < 2 \end{cases}$

26. $f(x) = \begin{cases} x + 2 & x > 4 \\ 6 & x = 4 \\ -x + 10 & x < 4 \end{cases}$

27. A manufacturing company has production costs of 85¢ per unit produced.
   (a) Determine the function that represents the cost of production based on the number of units produced.
   (b) Determine the function that represents the rate of change of cost with respect to production.
   (c) Determine the rate of change of the cost function when production volume is 1,000 and 2,000.
   (d) Graph the cost function as x varies from 0 to 4,000.
   (e) Graph the function that represents the rate of change of cost with respect to production as x varies from 0 to 4,000.

28. A retail store has observed that the number of new customers is 6 times the square of the number of times their window display is changed each month if the number of changes is more than 1 but at most 4, 6 if the number of changes is at most 1, and 96 if the number of changes is greater than 4. Assume part of the display can be changed.
   (a) Determine the function that represents the number of new customers based on the number of times the window display is changed.
   (b) Determine the function that represents the rate of change in the number of new customers with respect to the number of window displays.
   (c) Determine the rate of change in the number of new customers if the number of display changes is 1 and 3.
   (d) Graph the new customer function as x varies from 0 to 8.
   (e) Graph the function that represents the rate of change of new customers with respect to changes in the display as x varies from 0 to 6.

29. The number of errors that a secretary makes is 2 times the square root of the number of pages that she types.
   (a) Determine the function that represents the number of errors based on the number of pages typed.
   (b) Determine the function that represents the rate of change of errors with respect to the number of pages typed.
   (c) Determine the rate of change of errors when 4 and 16 pages are typed.
   (d) Graph the function as x varies from 0 to 25.
   (e) Graph the function that represents the rate of change in the number of errors with respect to the number of pages typed as x varies from 0 to 25.

10.4 RULES OF DIFFERENTIATION

In business most of the functions with which we are concerned are composed of sums and differences of other functions. Since in Sec. 10.3 we developed the rules for determining the derivatives of two very special functions, namely, a constant and a variable raised to a constant power, we now consider the rules that permit us to determine derivatives of functions that are obtained by performing algebraic operations (i.e., addition, subtraction, multiplication, and division) on one or more functions. With these rules at our disposal, we shall have the capability to determine derivatives for a very large group of functions.

Derivative of a Constant Times a Function

Suppose we have a function $h(x)$ that can be represented by a constant k multiplied by another function $f(x)$; that is $h(x) = kf(x)$. In order to determine $D_x h(x)$, we have

$$D_x h(x) = D_x [kf(x)]$$
$$= \lim_{\Delta x \to 0} \frac{kf(x+\Delta x) - kf(x)}{\Delta x}$$
$$= \lim_{\Delta x \to 0} \left[k \frac{f(x+\Delta x) - f(x)}{\Delta x} \right]$$
$$= \lim_{\Delta x \to 0} k \lim_{\Delta x \to 0} \frac{f(x+\Delta x) - f(x)}{\Delta x}$$
$$= k \lim_{\Delta x \to 0} \frac{f(x+\Delta x) - f(x)}{\Delta x}$$
$$= k D_x f(x)$$

Thus we have a rule for determining the derivative of a constant multiplied by a function.

Rule 10.8 The derivative of a constant k multiplied by a function $f(x)$ equals the constant multiplied by the derivative of the function; that is,

$$D_x [kf(x)] = k D_x f(x) \qquad (10.7)$$

Example 10.15 Determine the derivative of each of the following functions:

(a) $f(x) = 6x$ (b) $f(x) = 2x^5$ (c) $f(x) = \frac{1}{2} x^{-4}$
(d) $f(x) = 3x^{-0.5}$ (e) $f(x) = 10x^{-5}$ (f) $f(x) = ax^n$

Solution In each of the cases, first we apply Rule 10.8 and then Rule 10.7 to obtain the answer:

(a) $D_x(6x) = 6 D_x(x) = 6(1) x^{1-1} = 6(1) = 6$

Sec. 10.4 / RULES OF DIFFERENTIATION

(b) $D_x(2x^5) = 2D_x(x^5) = 2(5)x^{5-1} = 10x^4$
(c) $D_x(\frac{1}{2}x^{-4}) = \frac{1}{2}D_x(x^{-4}) = \frac{1}{2}(-4)x^{-4-1} = -2x^{-5}$
(d) $D_x(3x^{-0.5}) = 3D_x(x^{-0.5}) = 3(-0.5)x^{-0.5-1} = -1.5x^{-1.5}$
(e) $D_x(10x^{-5}) = 10D_x(x^{-5}) = 10(-5)x^{-5-1} = -50x^{-6}$
(f) $D_x(ax^n) = aD_x(x^n) = a(n)x^{n-1} = anx^{n-1}$

Derivative of the Sum or Difference of Two Functions If the function $h(x)$ can be expressed as the sum of two functions $f(x)$ and $g(x)$, then we can write $h(x) = f(x) + g(x)$. In order to determine the derivative of $h(x)$, we may write

$$D_x[h(x)] = D_x[f(x) + g(x)]$$
$$= \lim_{\Delta x \to 0} \frac{[f(x+\Delta x) + g(x+\Delta x)] - [f(x) + g(x)]}{\Delta x}$$
$$= \lim_{\Delta x \to 0} \frac{[f(x+\Delta x) - f(x)] + [g(x+\Delta x) - g(x)]}{\Delta x}$$
$$= \lim_{\Delta x \to 0} \frac{f(x+\Delta x) - f(x)}{\Delta x} + \lim_{\Delta x \to 0} \frac{g(x+\Delta x) - g(x)}{\Delta x}$$
$$= D_x f(x) + D_x g(x)$$

Hence we have a rule for determining the derivative of the sum of two functions. If we were to consider the difference of two functions, we would obtain simply the corresponding difference of their respective derivatives.

Rule 10.9 The derivative of the sum (or difference) of two functions equals the sum (or difference) of their respective derivatives; that is,

$$D_x[f(x) + g(x)] = D_x f(x) + D_x g(x) \qquad (10.8)$$
$$D_x[f(x) - g(x)] = D_x f(x) - D_x g(x) \qquad (10.9)$$

Rule 10.9 can be extended to include any finite number of functions that are combined by addition and/or subtraction.

Example 10.16 Determine the derivative of each of the following functions:

(a) $f(x) = 3x^2 + 2x$ (b) $f(x) = 2x^3 - 4x^2$
(c) $f(x) = x^3 + x^2$ (d) $f(x) = 2x - x^5$
(e) $f(x) = 3x + \frac{1}{3}x^3$ (f) $f(x) = 1 + 2x + 3x^2 + x^4 - 3x^5$
(g) $f(x) = 3x^2(3-x)$ (h) $f(x) = x^3(x^2 + 1)$

Solution In parts (a) to (e), first we apply Rule 10.9, then Rule 10.8, and finally Rule 10.7:

(a) $D_x(3x^2 + 2x) = D_x(3x^2) + D_x(2x) = 3D_x(x^2) + 2D_x(x) = 3(2)x + 2$
    $= 6x + 2$

(b) $D_x(2x^3 - 4x^2) = D_x(2x^3) - D_x(4x^2) = 2D_x(x^3) - 4D_x(x^2)$
$= 2(3)x^2 - 4(2)x = 6x^2 - 8x$
(c) $D_x(x^3 + x^2) = D_x(x^3) + D_x(x^2) = 3x^2 + 2x$
(d) $D_x(2x - x^5) = D_x(2x) - D_x(x^5) = 2 - 5x^4$
(e) $D_x(3x + \frac{1}{3}x^3) = D_x(3x) + D_x(\frac{1}{3}x^3) = 3D_x(x) + \frac{1}{3}D_x(x^3)$
$= 3 + \frac{1}{3}(3)x^2 = 3 + x^2$

In part (f), we use the extension of Rule 10.9 to include several terms:

(f) $D_x(1 + 2x + 3x^2 + x^4 - 3x^5) = D_x(1) + D_x(2x) + D_x(3x^2) + D_x(x^4)$
$- D_x(3x^5)$
$= 0 + 2 + 3(2)x + 4x^3 - 3(5)x^4$
$= 2 + 6x + 4x^3 - 15x^4$

To solve parts (g) and (h), we multiply the terms and express the result as a sum or difference:

(g) $D_x[3x^2(3 - x)] = D_x(9x^2 - 3x^3) = D_x(9x^2) - D_x(3x^3) = 18x - 9x^2$
(h) $D_x[x^3(x^2 + 1)] = D_x(x^5 + x^3) = D_x(x^5) + D_x(x^3) = 5x^4 + 3x^2$

(As the reader develops proficiency in applying the rules concerning derivatives, it will not be necessary to write out each step in as much detail as in Example 10.16; however, when omitting intermediate steps, extreme caution must be exerted in the computations.)

Derivative of the Product of Two Functions

In Example 10.16, parts (g) and (h), we were able to determine the derivative of a product by multiplying terms and expressing the result as the appropriate sum or difference. There are many situations in which this is not possible. For example, if $f(x) = (x^2 + 2x)^{1/2}(x^3 - 4)^{1/3}$, we cannot simply multiply the two terms to obtain a sum or difference of two functions; thus we need a *rule for determining the derivative of a product of two functions*. If we let $h(x) = f(x)g(x)$, we could employ Definition 10.3 to obtain the derivative of $h(x)$ in terms of the derivatives of $f(x)$ and $g(x)$; however, since this is rather complicated, we shall state the rule without derivation.

Rule 10.10 The derivative of the product of two functions equals the product of the derivative of the first function and the second function plus the product of the first function and the derivative of the second function; that is,

$$D_x[f(x)g(x)] = D_xf(x)g(x) + f(x)D_xg(x) \qquad (10.10)$$

Example 10.17 Use Rule 10.10 to determine the derivative of each of the following functions:

(a) $f(x) = 3x^2(3 - x)$
(b) $f(x) = x^3(x^2 + 1)$
(c) $f(x) = (x^{3/5} + x^2)(x^{1/2} + x^{-1/2})$
(d) $f(x) = (x^3 + 2x^2 + 1)$
$\times (x^2 - x^{-1})$

Sec. 10.4 / RULES OF DIFFERENTIATION

Solution Note that parts (a) and (b) are identical to parts (g) and (h) of Example 10.16, and would generally be expanded prior to differentiation; hence we may verify our answers from this example with those from Example 10.16. First we shall apply Rule 10.10 to each problem and then apply the appropriate combination of the preceding rules to obtain the derivative.

(a) $D_x[3x^2(3-x)] = D_x(3x^2)(3-x) + 3x^2 D_x(3-x)$
$= 6x(3-x) + 3x^2(-1)$
$= 18x - 6x^2 - 3x^2$
$= 18x - 9x^2$

(b) $D_x[x^3(x^2+1)] = D_x(x^3)(x^2+1) + x^3 D_x(x^2+1)$
$= 3x^2(x^2+1) + x^3(2x)$
$= 3x^4 + 3x^2 + 2x^4$
$= 5x^4 + 3x^2$

(c) $D_x[(x^{3/5} + x^2)(x^{1/2} + x^{-1/2})] = D_x(x^{3/5} + x^2)(x^{1/2} + x^{-1/2})$
$+ (x^{3/5} + x^2) D_x(x^{1/2} + x^{-1/2})$
$= (\frac{3}{5}x^{-2/5} + 2x)(x^{1/2} + x^{-1/2})$
$+ (x^{3/5} + x^2)(\frac{1}{2}x^{-1/2} - \frac{1}{2}x^{-3/2})$

(d) $D_x[(x^3 + 2x^2 + 1)(x^2 - x^{-1})] = D_x(x^3 + 2x^2 + 1)(x^2 - x^{-1})$
$+ (x^3 + 2x^2 + 1) D_x(x^2 - x^{-1})$
$= (3x^2 + 4x)(x^2 - x^{-1})$
$+ (x^3 + 2x^2 + 1)(2x + x^{-2})$

The rule for the derivative of a product can also be extended to include the product of a finite number of functions. The only adjustment that needs to be made is that the derivative of the product of n functions becomes the sum of products, where the first term is the derivative of the first function multiplied by the product of the other $(n-1)$ functions, the second term is the derivative of the second function multiplied by the product of the other $(n-1)$ functions, etc.

Example 10.18 Determine the derivative of each of the following functions:

(a) $f(x) = x^2(x^3 + 1)(x^2 - x)$
(b) $f(x) = x^{1/2}(2 - 3x)(x^2 + 1)$
(c) $f(x) = x^{-2}(x^2 + 3)(x + 5)$

Solution

(a) $D_x[x^2(x^3+1)(x^2-x)] = D_x(x^2)(x^3+1)(x^2-x) + x^2 D_x(x^3+1)(x^2-x)$
$+ x^2(x^3+1) D_x(x^2-x)$
$= 2x(x^3+1)(x^2-x) + x^2(3x^2)(x^2-x)$
$+ x^2(x^3+1)(2x-1)$

(b) $D_x[x^{1/2}(2-3x)(x^2+1)] = D_x(x^{1/2})(2-3x)(x^2+1)$
$+ x^{1/2} D_x(2-3x)(x^2+1)$
$+ x^{1/2}(2-3x) D_x(x^2+1)$
$= \frac{1}{2}x^{-1/2}(2-3x)(x^2+1) + x^{1/2}(-3)(x^2+1)$
$+ x^{1/2}(2-3x)(2x)$

(c) $D_x[x^{-2}(x^2+3)(x+5)] = D_x(x^{-2})(x^2+3)(x+5) + x^{-2}D_x(x^2+3)(x+5)$
$$+ x^{-2}(x^2+3)D_x(x+5)$$
$$= -2x^{-3}(x^2+3)(x+5) + x^{-2}(2x)(x+5)$$
$$+ x^{-2}(x^2+3)(1)$$

Derivative of the Quotient of Two Functions Suppose we have a function $h(x)$ that can be expressed as the quotient of two functions $f(x)$ and $g(x)$; that is, $h(x) = f(x)/g(x)$. Again we could apply Definition 10.3 to the quotient to develop a general formula for determining the derivative of a quotient of two functions; however, since this is rather complicated, we shall state the rule without derivation.

Rule 10.11 The derivative of the quotient of two functions equals the product of the derivative of the numerator and the denominator minus the product of the numerator and the derivative of the denominator, all divided by the square of the denominator; that is,

$$D_x\left[\frac{f(x)}{g(x)}\right] = \frac{D_xf(x)g(x) - f(x)D_xg(x)}{[g(x)]^2} \quad (10.11)$$

Example 10.19 Determine the derivative of each of the following functions:

(a) $f(x) = \dfrac{x+1}{x-1}$ (b) $f(x) = \dfrac{x+1}{x^2+1}$

(c) $f(x) = \dfrac{x^3-x}{x+1}$ (d) $f(x) = \dfrac{1-x+x^2}{1+x+x^2}$

Solution Applying Rule 10.11, we obtain

(a) $D_x\left(\dfrac{x+1}{x-1}\right) = \dfrac{D_x(x+1)(x-1) - (x+1)D_x(x-1)}{(x-1)^2}$

$$= \frac{1(x-1) - (x+1)(1)}{(x-1)^2}$$

$$= -\frac{2}{(x-1)^2}$$

(b) $D_x\left(\dfrac{x+1}{x^2+1}\right) = \dfrac{D_x(x+1)(x^2+1) - (x+1)D_x(x^2+1)}{(x^2+1)^2}$

$$= \frac{1(x^2+1) - (x+1)(2x)}{(x^2+1)^2}$$

$$= \frac{x^2+1-2x^2-2x}{(x^2+1)^2}$$

$$= \frac{-x^2-2x+1}{(x^2+1)^2}$$

Sec. 10.4 / RULES OF DIFFERENTIATION

(c) $D_x\left(\dfrac{x^3-x}{x+1}\right) = \dfrac{D_x(x^3-x)(x+1)-(x^3-x)D_x(x+1)}{(x+1)^2}$

$= \dfrac{(3x^2-1)(x+1)-(x^3-x)(1)}{(x+1)^2}$

$= \dfrac{3x^3+3x^2-x-1-x^3+x}{(x+1)^2}$

$= \dfrac{2x^3+3x^2-1}{(x+1)^2}$

(d) $D_x\left(\dfrac{1-x+x^2}{1+x+x^2}\right) = \dfrac{D_x(1-x+x^2)(1+x+x^2)-(1-x+x^2)D_x(1+x+x^2)}{(1+x+x^2)^2}$

$= \dfrac{(-1+2x)(1+x+x^2)-(1-x+x^2)(1+2x)}{(1+x+x^2)^2}$

$= \dfrac{-1-x-x^2+2x+2x^2+2x^3-1-2x+x+2x^2-x^2-2x^3}{(1+x+x^2)^2}$

$= \dfrac{-2+2x^2}{(1+x+x^2)^2}$

The rules for differentiating sums, differences, products, and quotients can be applied in various ways. For example, when differentiating a quotient of two functions, the numerator could be the product of two functions to which we would apply the rule for differentiating products. In the previous examples we have already seen that the rules for a constant multiplied by a function and for the sum and difference of functions may be applied when differentiating products and quotients. Any of these rules may be applied in combination with the other rules in order to obtain the derivative of a function.

Example 10.20 Determine the derivative of each of the following functions:

(a) $f(x) = \dfrac{(x+6)(2x^2+4)}{x+1}$

(b) $f(x) = \dfrac{x^2+3x+2}{(x+1)(x+2)}$

(c) $f(x) = \dfrac{(x^2+2x)(2x+4)}{(x-3)(x^2+4)}$

Solution

(a) $D_x\left[\dfrac{(x+6)(2x^2+4)}{x+1}\right]$

$= \dfrac{D_x[(x+6)(2x^2+4)](x+1)-(x+6)(2x^2+4)D_x(x+1)}{(x+1)^2}$

$$= \frac{[D_x(x+6)(2x^2+4)+(x+6)D_x(2x^2+4)](x+1)-(x+6)(2x^2+4)(1)}{(x+1)^2}$$

$$= \frac{[(1)(2x^2+4)+(x+6)(4x)](x+1)-(x+6)(2x^2+4)}{(x+1)^2}$$

$$= \frac{(2x^2+4+4x^2+24x)(x+1)-(x+6)(2x^2+4)}{(x+1)^2}$$

$$= \frac{(6x^2+24x+4)(x+1)-(x+6)(2x^2+4)}{(x+1)^2}$$

(b) $D_x\left[\dfrac{x^2+3x+2}{(x+1)(x+2)}\right]$

$$= \frac{D_x(x^2+3x+2)(x+1)(x+2)-(x^2+3x+2)D_x[(x+1)(x+2)]}{[(x+1)(x+2)]^2}$$

$$= \frac{(2x+3)(x+1)(x+2)-(x^2+3x+2)[1(x+2)+(x+1)(1)]}{(x+1)^2(x+2)^2}$$

$$= \frac{(2x+3)(x+1)(x+2)-(x^2+3x+2)(2x+3)}{(x+1)^2(x+2)^2}$$

$$= \frac{(2x+3)[(x+1)(x+2)-(x^2+3x+2)]}{(x+1)^2(x+2)^2}$$

$$= \frac{(2x+3)[(x^2+3x+2)-(x^2+3x+2)]}{(x+1)^2(x+2)^2}$$

$$= 0$$

This result is not at all surprising if we look at the denominator of the quotient. When it is expanded, we obtain the numerator; thus we are determining the derivative of a constant. It would have been simpler to have reduced the quotient to a constant, but these reductions are not always obvious. However, generally it is a good idea to reduce quotients as much as possible.

(c) $D_x\left[\dfrac{(x^2+2x)(2x+4)}{(x-3)(x^2+4)}\right]$

$$= \frac{D_x[(x^2+2x)(2x+4)](x-3)(x^2+4)-(x^2+2x)(2x+4)D_x[(x-3)(x^2+4)]}{[(x-3)(x^2+4)]^2}$$

$$= \frac{[(2x+2)(2x+4)+(x^2+2x)(2)](x-3)(x^2+4)-(x^2+2x)(2x+4)[(1)(x^2+4)+(x-3)(2x)]}{(x-3)^2(x^2+4)^2}$$

Sec. 10.4 / RULES OF DIFFERENTIATION

EXERCISES

In Exercises 1 to 12, determine the derivative of each of the given functions.

1. $f(x) = 3x^4$
2. $f(x) = \frac{1}{2}x^6$
3. $f(x) = 3x + 2x^3$
4. $f(x) = 3x^2 + 5x$
5. $f(x) = 4x^2 - 2x + 10$
6. $f(x) = 5x^4 - 2x^2 + 3x$
7. $y = 6x^2 - 2x^3 + x^4$
8. $y = x^8 + 6x^4 - 2x^{3/2}$
9. $g(t) = 4t^2 + 5t + 2$
10. $g(t) = 7 + 4t^2 - 3t^3$
11. $u = x^4 + 3x^3 - x^2 + \frac{1}{2}x + 2$
12. $s = 3z + 6z^2 - \frac{1}{3}z^3$

In Exercises 13 to 16, determine the value of the derivative of each of the given functions at the given value of the independent variable.

13. $f(x) = x^2 + 2x$ (at $x = 1$)
14. $y = 2x^3 + 4x^2 + 3x + 5$ (at $x = -1$)
15. $u = 2z^4 + 3z^3 + z^2 + 4z + 5$ (at $z = 0$)
16. $g(t) = t^3 - 3t^2 + 3t + 5$ (at $t = 1$)

Determine the derivative of each of the functions in Exercises 17 to 24 by expanding each product before differentiating. Check your results by applying the rule for the product of functions (Rule 10.10).

17. $f(x) = 3(x - 4)(x + 2)$
18. $y = 4x^3(x^2 + x)$
19. $g(t) = 2t^2(t^2 - 3t)$
20. $s = (2w + 1)(w - 4)$
21. $u = z^2(4z - 2)$
22. $h(t) = 3t^4(5t^2 + 2t + 1)$
23. $g(x) = 2x^2(x - 4)(x^2 + x)$
24. $y = 4t(2t^2 + 6)(t^2 - 3)$

Determine the derivative of each of the functions in Exercises 25 to 30 by factoring the numerator and canceling the denominator with the appropriate numerator term before differentiating. Check your results by applying the rule for the quotient of functions (Rule 10.11).

25. $f(x) = \dfrac{x^2 - 2x + 1}{x - 1}$
26. $f(x) = \dfrac{x^2 - 5x + 4}{x - 4}$
27. $y = \dfrac{x^2 + 6x + 9}{x + 3}$
28. $g(x) = \dfrac{2x^2 + 13x + 15}{x + 5}$
29. $h(t) = \dfrac{t^4 - t^2}{t^2}$
30. $u = \dfrac{z^3 + 2z^2 - z}{z^2 + 2z - 1}$

Determine the slope of each of the curves in Exercises 31 to 36 at the given value of the independent variable.

31. $f(x) = \dfrac{x^2+1}{x-1}$ $(x=2)$

32. $f(x) = \dfrac{x^3-x^2}{x-1}$ $(x=0)$

33. $y = \dfrac{3x^2+6x+2}{x^2+2x+1}$ $(x=2)$

34. $u = \dfrac{w^3}{w+3}$ $(w=-2)$

35. $s = \dfrac{z^3-4z^2}{z-4}$ $(z=3)$

36. $g(t) = \dfrac{t^3-t}{t+1}$ $(t=4)$

Determine the values of the independent variable for which the slope equals zero in each of the functions in Exercises 37 to 42.

37. $f(x) = x^2 + 2x + 5$

38. $f(x) = x^3 + 3x^2 - 9x + 7$

39. $g(u) = (u^2+1)(u+2)$

40. $y = (3w^2+4)(2w+5)$

41. $y = \dfrac{x^2+5}{x+2}$

42. $h(t) = \dfrac{t+1}{t^2+3}$

43. The Pure-Rena Pet Food Company has determined empirically the relationship between the production volume of dog food and the associated profit to be given by the expression

$$P = 10v - v^2$$

where P = profits (in thousands of dollars)
v = production volume (in thousands of units of production)

(a) Determine the function that represents the rate of change of profit with respect to production (i.e., determine $D_v P$).
(b) Determine the rate of change of the profit function when production volume is 4 and 5. What is happening to the rate of change of P as v increases from 4 to 5?
(c) Determine the rate of change of the profit function when production volume is 6 and 7. What is happening to the rate of change of P as v increases from 6 to 7?
(d) Graph the profit function.
(e) Graph the function that represents the rate of change of P as v increases from 4 to 7. What happens to the sign of $D_v P$ as v goes from 4 to 7?
(f) What do the previous results indicate about the optimum production volume for maximizing profit?

44. The Imperial Widget Manufacturing Company produces widgets that sell for $20 each; thus the total sales revenue R for widgets is $20x$, where x is the number of units sold; that is, $R = 20x$. Furthermore, it costs $100 to set up the equipment for a production run, and the material and labor costs are proportional to the square of the number of units manufactured with the constant of proportionality being .05; that is, the total production cost C is given by $C = 100 + .05x^2$. If we denote profit by P, then $P = R - C$.

(a) Determine the equation for P as a function of x and graph it for values of x from 0 to 400.
(b) Determine the equation for the rate of change of profit with respect to the number of units sold.

(c) What is the rate of change of profit with respect to the number of units sold when $x = 150$ units? 200 units? 250 units?

(d) What is the total profit when 150 units are sold? 200 units? 250 units?

45. The demand for gadgets is $500x - 5x^2$ where x is the price (in dollars) for one gadget; that is $G = 500x - 5x^2$, where G is the number of gadgets demanded.

(a) Determine the function that represents the rate of change of the number of gadgets demanded with respect to the price.

(b) Determine the rate of change of the demand when the price is $10, $25, and $50.

(c) Graph the demand function as x varies from 0 to 100.

(d) Graph the function that represents the rate of change of demand as the price increases from $0 to $100.

(e) What do the previous results indicate about the optimum price for maximizing demand?

46. A manufacturing company has an agreement with a toll road that their monthly cost of using the toll road will be $.2x^2 + 5x + 30$, where x is the number of the company's trucks that use the road.

(a) Determine the function that represents the rate of change of the cost of using the toll road with respect to the number of trucks.

(b) Determine the rate of change of the cost when the number of trucks is 10, 25, and 50. What is happening to the rate of change of the cost as x increases from 10 to 50?

(c) Graph the cost function for values of x from 0 to 25.

(d) Graph the function that represents the rate of change of the cost as x increases from 0 to 50.

47. A truck driver has determined empirically that his costs are 1¢ times the square of the number of miles driven. He is paid $85/week plus $1.50 per mile driven that week.

(a) Determine the equation for the driver's profit as a function of x and graph it for values of x from 0 to 150.

(b) Determine the equation for the rate of change of profit with respect to the number of miles driven and graph it for values of x from 0 to 150.

(c) What is the rate of change of profit with respect to the number of miles driven when $x = 0$ mi? 25 mi? 75 mi? 100 mi?

(d) What do the previous results indicate about the optimum number of miles to drive to maximize profit?

10.5 COMPOSITE FUNCTIONS

A *composite function* occurs when one function can be considered as a variable in another function; thus if f is a function of g and g is a function of x, then f is also a function of x, and f is said to be a *function of a function*. For example, if $f(g) = 3g^2 - 4g$ and $g(x) = x^2 + 1$, then f can be represented as a function of x by substituting for g. Then, since the function $g(x) = x^2 + 1$ can be viewed as a variable, $f[g(x)] = 3(x^2 + 1)^2 - 4(x^2 + 1)$ is a composite function. (A composite function can also be viewed as a function within a function.)

When composite functions are first encountered, more difficulty is experienced in separating and constructing them than in actually performing the operations necessary to obtain their derivatives; in short, the major problem with composite functions is that of *recognition*.

Example 10.21 Given $f(x)=4x^3$ and $g(x)=2x+3$, determine the composite function $f[g(x)]$.

Solution In any function $f(x)$, x is generally considered as a *position-holder*. This is most evident when we evaluate the function for a specific value of x, say, a. We simply substitute a in place of x and write $f(a)$ or its actual numerical value. Thus, to determine $f[g(x)]$, we replace x in $f(x)$ by $g(x)$, hence obtaining

$$f[g(x)] = 4[g(x)]^3$$
$$= 4(2x+3)^3$$

Example 10.22 Given $f(x)=4x^{1/2}+7x^{3/4}$ and $g(x)=x^2-2$, determine $f[g(x)]$.

Solution We simply replace x in $f(x)$ by $g(x)$ to obtain

$$f[g(x)] = 4[g(x)]^{1/2} + 7[g(x)]^{3/4}$$
$$= 4(x^2-2)^{1/2} + 7(x^2-2)^{3/4}$$

Example 10.23 Given $f[g(x)] = 6(2x+3)^3 - 4(2x+3)^{-1}$, determine $f(x)$ and $g(x)$.

Solution Since $2x+3$ can be viewed as a variable, we let $g(x)=2x+3$; thus $f[g(x)]$ can be written as $f(g) = 6g^3 - 4g^{-1}$. However, we wish to determine $f(x)$, and so we replace g by x, obtaining $f(x) = 6x^3 - 4x^{-1}$. Therefore, we have

$$f(x) = 6x^3 - 4x^{-1}$$
$$g(x) = 2x+3$$

Example 10.24 Given $f[g(x)] = 4(3x^2+2x+10)^{-3}$, determine $f(x)$ and $g(x)$.

Solution In $f[g(x)] = 4(3x^2+2x+10)^{-3}$, we can view $3x^2+2x+10$ as a variable, and hence we have

$$g(x) = 3x^2+2x+10$$

Replacing the expression $3x^2+2x+10$ by x, we obtain

$$f(x) = 4x^{-3}$$

Derivative of a Composite Function

For many composite functions, the derivative with respect to x can be determined by expanding each term and applying the rules of differentiation discussed in Sec. 10.4. However, there are many composite functions for which it is either impossible or inconvient to substitute and simplify. For example, if $f(g) = 5g^{1/2} - 4g^{10}$ and $g(x) = x^2 + 3x + 2$, the composite function of x is $f[g(x)] = 5(x^2 + 3x + 2)^{1/2} - 4(x^2 + 3x + 2)^{10}$. There is no simple expansion of $(x^2 + 3x + 2)^{1/2}$, and, although there is a relatively simple expansion of $(x^2 + 3x + 2)^{10}$, it is extremely inconvenient. Thus it is apparent that we need a rule for determining the *derivative of a composite function*. (The derivation of this rule can be found in almost any textbook on advanced calculus.)

Rule 10.12 If f is a function of g for which the derivative exists and g is a function of x for which the derivative exists, then the derivative of $f[g(x)]$ with respect to x exists and is given by

$$D_x[f(g(x))] = D_g f(g) D_x g(x)$$
$$= f'[g(x)]g'(x)$$
$$= \frac{df}{dg}\frac{dg}{dx} \qquad (10.12)$$

Rule 10.12 is commonly known as the *chain rule* and can be extended to a chain of composite functions. Remember that if we use the last expression of Eq. (10.12), we cannot treat the terms df/dg and dg/dx as fractions, since they each represent the derivative of a function. Let us consider a few examples that involve only two functions.

Example 10.25 Use Rule 10.12 to determine the derivative of $y = (2x+1)^3$.

Solution Since $2x+1$ can be viewed as a variable, we have $g(x) = 2x+1$ and $f(g) = g^3$. In order to apply Rule 10.12, we need to determine $D_g f(g)$ and $D_x g(x)$; thus

$$D_g f(g) = D_g(g^3) = 3g^2$$

Substituting $g(x)$ for g, we obtain

$$D_g f(g) = 3(2x+1)^2$$

Furthermore,

$$D_x g(x) = D_x(2x+1) = 2$$

Applying Rule 10.12, we obtain

$$D_x y = D_x[f(g(x))]$$
$$= D_g f(g) D_x g(x)$$
$$= 3(2x+1)^2 \cdot 2$$
$$= 6(2x+1)^2$$

In this particular case, we could have determined $D_x y$ by first expanding the term and then taking the derivative; that is, expanding, we obtain

$$y = (2x+1)^3$$
$$= (2x)^3 + 3(2x)^2(1) + 3(2x)(1)^2 + 1^3$$
$$= 8x^3 + 12x^2 + 6x + 1$$

and, taking the derivative with respect to x, we have

$$D_x y = 24x^2 + 24x + 6$$
$$= 6(4x^2 + 4x + 1)$$
$$= 6(2x+1)^2$$

Example 10.26 The cost of attending Excell University is given by the equation $y = (2x^2 + 5x + 1)^{3/2}$, where y is the cost and x is the number of hours taken. Determine the function that will represent the rate of change of the cost of attending Excell University.

Solution Since y is a composite function of x, we determine that $g(x) = 2x^2 + 5x + 1$ and $f(g) = g^{3/2}$. The derivatives are

$$D_x g(x) = 4x + 5$$

and

$$D_g f(g) = \tfrac{3}{2} g^{1/2} = \tfrac{3}{2}(2x^2 + 5x + 1)^{1/2}$$

Thus the rate of change of the cost of attending Excell University as a function of the number of hours taken is given by

$$D_x y = D_x [f(g(x))]$$
$$= \tfrac{3}{2}(2x^2 + 5x + 1)^{1/2}(4x + 5)$$

After becoming familiar with the process involved in taking the derivative of a composite function, we begin to do the actual differentiation by viewing $g(x)$ as a variable without explicitly expressing f and g in separate statements. However, it is still necessary that we understand what they are. They can be expressed as part of the statement of Eq. (10.12).

Example 10.27 The cost of manufacturing a batch of parts is given by the function $y = (3x^{1/2} + 2)^2$, where x is the number of parts produced per batch. Determine the function that represents the rate of change of the cost with respect to the number of parts produced per batch.

Solution The rate of change of cost is given by

$$D_x y = D_x (3x^{1/2} + 2)^2$$
$$= D_{3x^{1/2} + 2}(3x^{1/2} + 2)^2 D_x (3x^{1/2} + 2)$$
$$= 2(3x^{1/2} + 2) \cdot \tfrac{3}{2} x^{-1/2}$$
$$= 3x^{-1/2}(3x^{1/2} + 2)$$

Sec. 10.5 / COMPOSITE FUNCTIONS

Example 10.28 Determine $D_x y$ for $y = 2(x^2 + 2x)^2$.

Solution

$$D_x y = D_x\left[2(x^2+2x)^2\right]$$
$$= D_{x^2+2x}\left[2(x^2+2x)^2\right]D_x(x^2+2x)$$
$$= 4(x^2+2x)(2x+2)$$

Example 10.29 Determine $D_x y$ for $y = (x^2+4)^{-5} + 2(x^2+4)^2$.

Solution

$$D_x y = D_x\left[(x^2+4)^{-5} + 2(x^2+4)^2\right]$$
$$= D_{x^2+4}\left[(x^2+4)^{-5} + 2(x^2+4)^2\right]D_x(x^2+4)$$
$$= \left[-5(x^2+4)^{-6} + 4(x^2+4)\right]2x$$
$$= -10x(x^2+4)^{-6} + 8x(x^2+4)$$

As with many other mathematical operations, eventually we become so familiar with taking the derivatives of composite functions that we combine some of the intermediate steps of duplication and simply write down the answer.

Example 10.30 Determine $D_x y$ for $y = 5(3x-4)^4 + 2(3x-4)^2$.

Solution

$$D_x y = D_x\left[5(3x-4)^4 + 2(3x-4)^2\right]$$
$$= \left[20(3x-4)^3 + 4(3x-4)\right]\cdot 3$$
$$= 60(3x-4)^3 + 12(3x-4)$$

Example 10.31 Determine $D_x y$ for

$$y = (x^{1/2} - x^{-1/2})^2 + 3(x^{1/2} - x^{-1/2}) - 4$$

Solution

$$D_x y = D_x\left[(x^{1/2} - x^{-1/2})^2 + 3(x^{1/2} - x^{-1/2}) - 4\right]$$
$$= \left[2(x^{1/2} - x^{-1/2}) + 3\right]\left(\tfrac{1}{2}x^{-1/2} + \tfrac{1}{2}x^{-3/2}\right)$$

Rule 10.12 can be extended to a chain of composite functions. For a chain of three functions such as $f[g(h(x))]$, Eq. (10.12) becomes

$$D_x[f(g(h(x)))] = D_g f \cdot D_h g \cdot D_x h$$
$$= \frac{df}{dg}\frac{dg}{dh}\frac{dh}{dx} \qquad (10.13)$$

Thus, (10.31) indicates that when applying the chain rule, we differentiate from the outside in; it may also be viewed as differentiating the largest possible quantity first, then the second largest, and so on. Again, remember that we cannot treat the last three terms in (10.13) as fractions.

Example 10.32 Use (10.13) to determine $D_x y$ for

$$y = \left[(x^2 + 3x)^3\right]^{1/2}$$

Solution We note that x is the variable for $h(x) = x^2 + 3x$. We can view $h(x) = x^2 + 3x$ as the variable for $g[h(x)] = (x^2 + 3x)^3$; and, of course, we can view $g[h(x)] = (x^2 + 3x)^3$ as the variable for $f[g(h(x))] = [(x^2 + 3x)^3]^{1/2}$. Thus, applying (10.13), we have

$$D_x \left\{ \left[(x^2 + 3x)^3\right]^{1/2} \right\} = \tfrac{1}{2}\left[(x^2 + 3x)^3\right]^{-1/2} 3(x^2 + 3x)^2 (2x + 3)$$

EXERCISES

For the pairs of functions in Exercises 1 to 10, determine $f[g(x)]$ and $g[f(x)]$.

1. $f(x) = 2x$
   $g(x) = x^2$

2. $f(x) = 3x^2$
   $g(x) = x$

3. $f(x) = x$
   $g(x) = x$

4. $f(x) = x^2 - 2x + 1$
   $g(x) = x + 2$

5. $f(x) = x^2 - 5$
   $g(x) = 2x + 7$

6. $f(x) = 3x^2 - 4x + 6$
   $g(x) = 2x^2 + 3x + 2$

7. $f(x) = 2x^{1/2} + x$
   $g(x) = x + 3$

8. $f(x) = 5^x$
   $g(x) = 4x + 5$

9. $f(x) = \log_{10} x$
   $g(x) = 2x + 3$

10. $f(x) = 7^x + \log_{10} x$
    $g(x) = x^2 + 2x + 5$

For the functions in Exercises 11 to 14 determine $f[g(h(x))]$.

11. $f(x) = x$
    $g(x) = x + 2$
    $h(x) = 3x + 1$

12. $f(x) = x^2 + 3$
    $g(x) = 2x - 1$
    $h(x) = 3x^2$

13. $f(x) = \log_{10}(x + 4)$
    $g(x) = \sqrt{x^2 + 2x}$
    $h(x) = 3x$

14. $f(x) = e^{3x}$
    $g(x) = x^3 - 2$
    $h(x) = \sqrt{x} + 4$

For each of the composite functions $f[g(x)]$ in Exercises 15 to 24, determine the component parts of $f(x)$ and $g(x)$.

Sec. 10.5 / COMPOSITE FUNCTIONS

15. $(3x+1)^3$
16. $(2x^3-7)^4$
17. $(2x+6)^8$
18. $(12x+3x^2)^{12}$
19. $(10x^2-3x+2)^4$
20. $1/(2x+5)^3$
21. $2/(7x^2+3)^4$
22. $(3x^2+2x)^{-4}$
23. $\log_4(x^2+3x-1)$
24. e^{2x^2+7x+3}

For each of the composite functions $f[g(h(x))]$ in Exercises 25 to 30, determine the component parts $f(x)$, $g(x)$, and $h(x)$.

25. $[(x+2)^3]^{-2}$
26. $\sqrt[5]{(3x+7)^2}$
27. $[(x^2-4)^{2/3}]^4$
28. $e^{(x+4)^3}$
29. $\log_e(x^2-4)^3$
30. $e^{(4x+2)^2} - \log_5(4x+2)^2$

Determine the derivative with respect to x for each of the functions in Exercises 31 to 50.

31. $(4x+3)^2$
32. $(6x-7)^4$
33. $(5x+7)^{-3}$
34. $(3x^2+4)^5$
35. $(3x^2+4x-2)^4$
36. $(6x^2-7x+3)^{-5}$
37. $(3x+4)^{-2}$
38. $(2x-5)^{1/3}$
39. $\sqrt[3]{2x^2+3x+1}$
40. $\sqrt[4]{x^2+4}$
41. $3(x^2+4x)^3+(x-5)^4$
42. $5(2x+4)^{-3}-3(x^2+4x+2)^2$
43. $(x^2+3x)^3(4x+1)^2$
44. $(3x+6)^5(2x-3)^4$
45. $(x^2-4)^3/(2x+3)^2$
46. $(2x^2+3x+1)^2/(x^2+3)^3$
47. $(x^2-4)^3(2x+3)^{-2}$
48. $(2x^2+3x+1)^2(x^2+3)^{-3}$
49. $\sqrt{(10x^2+4x)^3}$
50. $[2(x^2+4)^3-3(x^2+4)^2]^4$

51. A salesman sells parts for $20 each and has expenses of $50/week. The salesman must pay the manufacturer $\frac{1}{2}$ of his receipts after deducting his expenses.

   (a) Determine the function, $f(x)$, that represents the salesman's receipts less his expenses.
   (b) Determine the function, $g(x)$, that represents the amount paid to the manufacturer, where x represents his receipts from sales minus expenses.
   (c) Determine $g(f(x))$.
   (d) Determine the amount of money the salesman will make after he pays the manufacturer if he sells 100 units one week.

52. A company has a policy to pay 40 percent of their yearly profit to stockholders. They make a profit of $100x - 15,000$, where x is the number of boxes of gadgets sold.

   (a) Determine the function, $f(x)$, to represent the company's yearly profit.
   (b) Determine the function, $g(x)$, to represent the stockholders' share given a yearly profit of x.

(c) Determine the function to represent the stockholders' share given the number of boxes of gadgets sold.

(d) Show that the equation from part (c) is correct by determining the stockholders' share of the profit by using the equations from parts (a) and (b) and then by using the equation from part (c) when 500 boxes are sold.

53. A machine works 90 percent of the time that it is needed. The machine is needed 5 hr for every unit of product A that is produced.

(a) Determine the function, $f(x)$, which represents the number of hours that the machine is needed.

(b) Determine the function, $g(x)$, which represents the number of hours the machine works given the number of hours x that it is needed.

(c) Determine the function to represent the number of hours the machine works given the number of units of product A that are to be produced.

10.6 HIGHER-ORDER DERIVATIVES

We have shown that the derivative of a function, when evaluated at a particular point, say, $x = a$, is the slope of the function at $x = a$; more precisely, it is the *instantaneous rate of change of the function at* $x = a$. Furthermore, the derivative of a function is also a function. Thus, if we are interested in determining the instantaneous rate of change of the slope of a function at $x = a$, we could simply determine the derivative of the function that yields the slope and evaluate it at $x = a$, which would involve differentiating a function that is already a derivative. Fortunately, since each derivative of a function is also a function, we can continue to determine derivatives as long as we can differentiate. The derivatives obtained by differentiating functions that are already derivatives are called *higher-order derivatives*, and each higher-order derivative gives information concerning previous derivatives. (In Chap. 12 we shall apply the concept of higher-order derivatives.)

Several symbols are used to denote the derivative of a function, namely, $D_x f(x), f'(x), dy/dx,$ and y'. If we wished to determine the derivative of $D_x f(x)$, we would simply differentiate the function given by $D_x f(x)$ in the usual manner and denote it as $D_x[D_x f(x)]$. However, for notational convenience, we denote this as $D_x^2 f(x)$, where the superscript 2 *does not* represent an exponent but rather the number of consecutive derivatives we have taken. Similarly, if we wished to determine the derivative of $D_x^2 f(x)$, we would simply differentiate the function for $D_x^2 f(x)$ in the usual manner and denote it by $D_x[D_x^2 f(x)]$, or, more simply by $D_x^3 f(x)$. This notation can be extended for as many consecutive derivatives as necessary, and, in general, the nth such derivative is denoted by $D_x^n f(x)$.

Extensions of the other symbols can also be used to represent the higher-order derivatives, and a few of these are given in Table 10.1. It should be noted that $f^{(n)}(x)$ represents the nth *derivative* of $f(x)$ and not $f(x)$ raised to the nth power, which we denote by $[f(x)]^n$. However, when referring to other textbooks, the reader should be careful to verify what notation is being used. For example, authors use $f^n(x)$ to represent the function raised to the

Table 10.1

Function	First Derivative	Second Derivative	Third Derivative	nth Derivative
$y = f(x)$	$D_x f(x)$	$D_x^2 f(x)$	$D_x^3 f(x)$	$\ldots D_x^n f(x)$
$y = f(x)$	$D_x y$	$D_x^2 y$	$D_x^3 y$	$\ldots D_x^n y$
$y = f(x)$	dy/dx	d^2y/dx^2	d^3y/dx^3	$\ldots d^n y/dx^n$
$y = f(x)$	$f'(x)$	$f''(x)$	$f'''(x)$	$\ldots f^{(n)}(x)$
$y = f(x)$	$f'(x)$	$f^{(2)}(x)$	$f^{(3)}(x)$	$\ldots f^{(n)}(x)$
$y = f(x)$	df/dx	d^2f/dx^2	d^3f/dx^3	$\ldots d^n f/dx^n$

nth power, which can be easily confused with $f^{(n)}(x)$ because they are similar in appearance. In this textbook we shall adhere to the $D_x^n f(x)$ notation, except when evaluating the derivatives at $x = a$; then we shall use $f^{(n)}(a)$.

Suppose we wish to determine the first four consecutive derivatives of $f(x) = x^5 + 3x^4 + 4x^3 - 10x^2 + x - 1$. We simply obtain the first derivative in the usual manner; next we determine the second derivative by differentiating the function that represents the first derivative; etc. Proceeding until we have obtained the fourth derivative of $f(x)$, we obtain the following derivatives:

$$D_x f(x) = 5x^4 + 12x^3 + 12x^2 - 20x + 1$$

$$D_x^2 f(x) = 20x^3 + 36x^2 + 24x - 20$$

$$D_x^3 f(x) = 60x^2 + 72x + 24$$

$$D_x^4 f(x) = 120x + 72$$

Evaluating the respective derivatives at $x = 1$, we obtain the following values:

$$f'(1) = 10$$
$$f^{(2)}(1) = 60$$
$$f^{(3)}(1) = 156$$
$$f^{(4)}(1) = 192$$

Example 10.33 Determine the first four derivatives of $f(x) = x^3 + 3x^2 - 4x + 7$, and determine the values of the function and the respective derivatives at $x = 0$ and $x = 2$.

Solution

$$D_x f(x) = 3x^2 + 6x - 4$$
$$D_x^2 f(x) = 6x + 6$$
$$D_x^3 f(x) = 6$$
$$D_x^4 f(x) = 0$$

The function values are as follows:

$$f(0) = 7 \quad f(2) = 19$$
$$f'(0) = -4 \quad f'(2) = 20$$
$$f^{(2)}(0) = 6 \quad f^{(2)}(2) = 18$$
$$f^{(3)}(0) = 6 \quad f^{(3)}(2) = 6$$
$$f^{(4)}(0) = 0 \quad f^{(4)}(2) = 0$$

Example 10.34 Determine the first five derivatives of $f(x) = (2x-3)^4$, and determine the values of the function and the respective derivatives at $x=1$, $x=1\frac{1}{2}$, and $x=2$.

Solution

$$D_x f(x) = 4(2x-3)^3 \cdot 2 = 8(2x-3)^3$$
$$D_x^2 f(x) = 24(2x-3)^2 \cdot 2 = 48(2x-3)^2$$
$$D_x^3 f(x) = 96(2x-3) \cdot 2 = 192(2x-3)$$
$$D_x^4 f(x) = 384$$
$$D_x^5 f(x) = 0$$

The respective function values are as follows:

$$f(1) = 1 \quad f(\tfrac{3}{2}) = 0 \quad f(2) = 1$$
$$f'(1) = -8 \quad f'(\tfrac{3}{2}) = 0 \quad f'(2) = 8$$
$$f^{(2)}(1) = 48 \quad f^{(2)}(\tfrac{3}{2}) = 0 \quad f^{(2)}(2) = 48$$
$$f^{(3)}(1) = -192 \quad f^{(3)}(\tfrac{3}{2}) = 0 \quad f^{(3)}(2) = 192$$
$$f^{(4)}(1) = 384 \quad f^{(4)}(\tfrac{3}{2}) = 384 \quad f^{(4)}(2) = 384$$
$$f^{(5)}(1) = 0 \quad f^{(5)}(\tfrac{3}{2}) = 0 \quad f^{(5)}(2) = 0$$

Example 10.35 Determine the first three derivatives of $f(x) = (x^3+2)^2$, and determine the value of the function and the derivatives at $x=1$.

Solution

$$D_x f(x) = 2(x^3+2)(3x^2) = 6x^2(x^3+2)$$
$$D_x^2 f(x) = 12x(x^3+2) + 6x^2(3x^2)$$
$$= 30x^4 + 24x$$
$$D_x^3 f(x) = 120x^3 + 24$$

The respective function values are

$$f(1) = 9$$
$$f'(1) = 18$$
$$f^{(2)}(1) = 54$$
$$f^{(3)}(1) = 144$$

EXERCISES

For each of the functions in Exercises 1 to 8, determine all of the successive derivatives until all remaining derivatives equal zero and evaluate the derivatives for the given value.

1. $f(x) = x^2 + 3x - 4$ $(x = -2)$
2. $f(x) = 4x - x^3$ $(x = 5)$
3. $f(x) = x^4 + 5x^3 - 7x^2 + 3x - 10$ $(x = 1)$
4. $f(x) = (x^2 + 4)(x^3 - 5)$ $(x = 1)$
5. $f(x) = x(x-1)^3$ $(x = 1)$
6. $s(t) = t^3(t+2)^2$ $(t = 0)$
7. $z(t) = t(t+1)(t+2)$ $(t = -1)$
8. $g(z) = (z+1)^2(z+2)^2$ $(z = -1)$

For each of the functions in Exercises 9 to 16, determine the first and second derivatives.

9. $\dfrac{1}{\sqrt{x}}$

10. $\dfrac{x-1}{x+1}$

11. $\dfrac{(x+2)^2}{x}$

12. $\dfrac{x^2}{x+1}$

13. $\sqrt{3x - x^3}$

14. $(x^2 + 4)^3$

15. $(x^2 + 3x + 2)^2$

16. $(x^4 - x^2 + 1)^3$

17. The molding department can process $5x^4 + x^3 + 3x^2 - 5$ boxes of screws per hour, where x is the number of machines working.

   (a) Determine the function that represents the instantaneous rate of change of the number of boxes of screws processed with respect to the number of machines working.
   (b) Determine the rate of change of the number of boxes produced when $x = 1$ machine, 3 machines, and 5 machines.
   (c) Determine the function that represents the rate of change of the slope of the function in part (a).

(d) Determine the instantaneous rate of change of the slope when $x=1$ machine, 3 machines, and 5 machines.

18. A manufacturing process requires $(2x^2-3)^2$ hr, where x is the number of units made per run.

(a) Determine the function that represents the rate of change of the hours required with respect to the number of units to be processed in a run.
(b) Determine the rate of change of the number of hours required when $x=10$ units, 12 units, 16 units.
(c) Determine the function that represents the rate of change of the slope of the function in part (a).
(d) Determine the instantaneous rate of change of the slope when $x=10$ units, 12 units, 16 units.
(e) Determine the number of hours required to make 10 units, 12 units, 16 units.

19. A bottling company can fill $(x+1)^3(x^2+5)^2$ bottles per hour, where x is the number of machines used that hour.

(a) Determine the function that represents the rate of change of the number of bottles filled with respect to the number of machines used.
(b) Determine the rate of change of the number of bottles when $x=1$ machine, 3 machines, 5 machines.
(c) Determine the function that represents the rate of change of the slope of the function that represents the rate of change of the number of bottles filled.
(d) Determine the instantaneous rate of change of the slope when $x=1$ machine, 3 machines.
(e) Determine the number of bottles filled when $x=1$ machine, 3 machines, 5 machines.

20. A painter has rented a booth at a flea market for $\$25+\sqrt{x+5}$, where x is the number of paintings he sells.

(a) Determine the rental fee if the number of paintings sold is 0, 11, 20.
(b) Determine the function that represents the rate of change of the rental fee with respect to the number of paintings sold.
(c) Determine the rate of change of the rental fee when the number of paintings sold is 0, 11, 20.
(d) Determine the function that represents the rate of change of the slope of the function that represents the rate of change of the rental fee.
(e) Determine the instantaneous rate of change of the slope when $x=0$ sales, 11 sales, 20 sales.

IMPORTANT TERMS AND CONCEPTS

Calculus
Chain rule
Continuous
   at a point
   function
   on the domain

Composite function
Derivative
Derivative of a
   composite function
   constant
   difference

function
product
quotient
sum
Differential calculus
Discontinuous function
Function
Infinity

Instantaneous rate of change
Integral
Integral calculus
Limit of a function
Limits
Rate of change
Slope

REVIEW PROBLEMS

1. Determine each of the following limits:

(a) $\lim_{x \to 4} \left(1 - \dfrac{x+2}{x-2}\right)$

(b) $\lim_{x \to 1} \left(x + \dfrac{x}{x+1}\right)$

(c) $\lim_{x \to 2} \dfrac{x^2 - 4x + 4}{x^2 - 2x}$

(d) $\lim_{x \to -3} \dfrac{x^2 - 9}{-(3x+9)}$

(e) $\lim_{x \to 1} \dfrac{x^2 + 4x - 5}{x^2 - 1}$

(f) $\lim_{x \to 0} \dfrac{x^3 - 3x^2}{x^2 + 4x}$

2. Determine whether or not the following functions are continuous at the given point:

(a) $f(x) = \dfrac{1}{x-1}$ $(x = -1)$

(b) $f(x) = \dfrac{x^2 - 2x}{x}$ $(x = 0)$

(c) $f(x) = \dfrac{x^2 - 4}{x^2 - x - 6}$ $(x = 3)$

(d) $f(x) = \dfrac{x^3 - 4x}{x^3 - 16x}$ $(x = 4)$

(e) $f(x) = \dfrac{x^4 - 1}{x^2 - 1}$ $(x = 1)$

In Problems 3 to 7, use the appropriate rule to determine $D_x f(x)$, and determine the slope of $f(x)$ at the given value of x.

3. $f(x) = 4$ $(x = 4)$

4. $f(x) = x^{-3}$ $(x = 2)$

5. $f(x) = \dfrac{x^3}{4} + x^2 - \dfrac{x}{2} + 1$ $(x = -2)$

6. $f(x) = x^3(x^2 - 1)(x^{-3} + x^{-1})$ $(x = 1)$

7. $f(x) = \dfrac{x-2}{x+3}$ $(x = 2)$

8. For the following composite functions $f[g(x)]$, determine the component parts of $f(x)$ and $g(x)$:

(a) $3(4x-9)^4$

(b) $\dfrac{-3}{x^2-x+1}$

(c) $e^{-(2x+4)}$

(d) $\dfrac{3x^3-5}{2}$

9. Determine the derivative with respect to x for each of the following functions:

(a) $(7x+2)^9$

(b) $(3x^2-5x+2)^3$

(c) $\sqrt{x^4+3x^3-x+1}$

(d) $(x^2)(4x^3-2x)^3$

(e) $(x^2+1)^2(x^3-2x+2)^2$

10. For each of the following functions, determine all of the successive derivatives until all remaining derivatives equal zero:

(a) $f(x)=x^4-3x^3+\dfrac{x^2}{2}-1$

(b) $f(x)=(x^2-2)^2(x+5)$

(c) $f(x)=x(x^2+1)(x-2)^2$

(d) $f(x)=(4x^3-5x^2)(2x^2+5)$

11. For the following functions, determine the first and second derivatives:

(a) $f(x)=\dfrac{x}{x^2+2x-1}$

(b) $f(x)=\dfrac{2x^2+x-5}{x-2}$

(c) $f(x)=(x^6-4x^4)^3$

(d) $f(x)=x^4(x^3-2)(x^2+x)^2$

# DIFFERENTIAL CALCULUS: ADVANCED METHODOLOGY

11.1 INTRODUCTION

In many business situations the function for representing a variable of interest is either exponential or logarithmic. Consequently, we may need to determine the rate of change of the function using differential calculus. For example, the profits of a firm may be increasing according to some exponential function and we wish to know the rate of increase of the profit as the number of units sold increases. Or the sales of a particular division may be changing according to some logarithmic function and we wish to know how the sales are increasing as the advertising budget increases. In determining the derivatives of sums, differences, products, quotients, and composite functions, we used the rule for determining the derivative of a variable to a constant power, namely Rule 10.7. In other words, we considered only those functions that were sums, differences, products, quotients, powers, and roots of polynomial functions. Such functions are commonly known as *algebraic functions*.

A class of nonalgebraic functions that is of concern to us is the class of *transcendental functions*, which includes mainly exponential, logarithmic, and trigonometric functions. Since trigonometric functions are rarely important in business applications, we shall only concern ourselves here with exponential and logarithmic functions.

11.2 INVERSE FUNCTIONS

The functions considered in Chap. 10 were of the form $y = f(x)$, where y is the dependent variable and x is the independent variable. If the function is such that a different value of x will always yield a different value of y, then the function will be such that for each element in the range there is one and only one element in the domain. Of course, since $f(x)$ is a function, the converse is also true; that is, for each element in the domain, there is one and only one element in the range. This correspondence between the elements in the domain and the range is called a one-to-one correspondence. Thus, for any element in the range, the value in the domain is uniquely determined. When a function $f(x)$ has a one-to-one correspondence between the elements, it will have an inverse function $x = f^{-1}(y)$ (read "x is an inverse function of y"). However, to be consistent with x representing the independent variable and y the dependent variable, we denote the inverse function by $y = f^{-1}(x)$. (*Warning:* f^{-1} does not mean $1/f$.)

For a nonnumerical example, consider the correspondence between husbands and wives in a monogamous society. Each husband has a unique wife, and each wife has a unique husband. Thus the rule that identifies the wife of a given husband can be inverted (so to speak) to determine the husband of a given wife. On the other hand, the correspondence between the set of all fathers to the set of all sons is not a one-to-one correspondence, because a father may have more than one son, while each son has a unique father.

In simple cases, the inverse of the function $y = f(x)$ can be obtained by solving for x in terms of y and then interchanging the roles of x and y. For example, the inverse of the function $y = 5x + 7$ is found by solving for x in terms of y to obtain $x = (y-7)/5$ and interchanging the roles of x and y to obtain $y = (x-7)/5$. Thus, for the function $f(x) = 5x + 7$, the inverse function is $f^{-1}(x) = (x-7)/5$.

Two important functions are the exponential and logarithmic functions $f(x) = b^x$ and $g(x) = \log_b x$. Each of these functions do in fact have a one-to-one correspondence between the elements in the domain and the range. Actually $g(x)$ is the inverse function of $f(x)$. You should verify this through the steps described above for determining the inverse. Also verify that $f(x)$ is the inverse function of $g(x)$.

It is important to note that the inverse relationship obtained by solving for x in terms of y must also satisfy the requirements of a function. One of the reasons why the square and square-root functions are important is that they provide an example where simply solving $y = x^2$ for x does not yield a function; that is, we obtain $x = \pm\sqrt{y}$, and thus for any positive value in the domain, there are *two* values in the range. For example, if $y = 16$, $x = \pm\sqrt{16} = \pm 4$; hence there are two values $+4$ and -4 in the range for one value in the domain. However, if we limit the domain of $y = x^2$ to nonnegative values of x, the solution of x yields the inverse function $x = \sqrt{y}$; and if we limit the domain to nonpositive values of x, the solution yields the inverse function $x = -\sqrt{y}$.

Since there are some functions for which it is very difficult (and sometimes impossible) to solve for the inverse function, it would be most convenient to have a rule that provides the derivative of the inverse function without requiring the solution for the inverse function. Fortunately such a rule is available and is stated below without derivation.

Rule 11.1 If $y = f(x)$ is a single-valued function on the domain

$$D = \{x \mid a \leqslant x \leqslant b\}$$

Sec. 11.2 / INVERSE FUNCTIONS

that has a derivative for each value in D such that the derivative does not equal zero, then the inverse function $x = g(y)$ exists and has a derivative such that

$$D_y x = \frac{1}{D_x y} \qquad (11.1)$$

We should note that in Rule 11.1 we do not interchange the roles of x and y, but rather, we simply differentiate with respect to the variable in the function. Also, Rule 11.1 is applicable only when the function involved has a one-to-one correspondence between its domain and range. When dealing with functions that do not possess this one-to-one correspondence, we may be able to partition the curve into separate functions possessing a one-to-one correspondence, each with a restricted domain. To each of these, Rule 11.1 can again apply.

Example 11.1 For the function $y = 6 + 3x$, determine the inverse function $y = g(x)$ and its derivative. Also, use Rule 11.1 to determine the derivative of the inverse function.

Solution Solving $y = 6 + 3x$ for x, we obtain

$$x = \tfrac{1}{3} y - 2$$

Determining $D_y x$ directly, we have $D_y x = \tfrac{1}{3}$. Using Rule 11.1 to determine $D_y x$, we have

$$D_y x = \frac{1}{D_x y} = \frac{1}{D_x(6 + 3x)} = \frac{1}{3}$$

Example 11.2 For the function $y = x^3 + x$, determine $D_x y$ and $D_y x$.

Solution Applying the rules of Chap. 10, we obtain

$$D_x y = D_x(x^3 + x) = 3x^2 + 1$$

Applying Rule 11.1, we obtain

$$D_y x = \frac{1}{D_x y} = \frac{1}{3x^2 + 1}$$

Example 11.3 For the function $y = 3x^2 + 2x + 4$ defined on the domain $D = \{x \mid x > -\tfrac{1}{3}\}$, determine $D_x y$ and $D_y x$.

Solution Using the rules of Chap. 10, we obtain

$$D_x y = D_x(3x^2 + 2x + 4) = 6x + 2$$

Using Rule 11.1, we obtain

$$D_y x = \frac{1}{D_x y} = \frac{1}{6x+2}$$

All the functions that we have considered are of such form that either y is expressed in terms of x or x is expressed in terms of y; that is, the functions are of a form that *explicitly* states that one variable is a function of the other. For example, the function $y = x^3 + x$ explicitly defines y as a function of x. These functions are called *explicit functions*. With functions in this form, generally there is no difficulty in determining the derivative of the function or its inverse because we can determine the derivatives directly through the rules of Chap. 10 or indirectly through Rule 11.1.

A function can also be stated *implicitly*; that is, the function is stated such that it is not apparent which variable is the dependent variable. Such functions are called *implicit functions*. For example, the equation $3x - 6y = 12$ implicitly defines x in terms of y; thus, if y is permitted to assume a given value, then x is implicitly defined by the equation. Conversely, the equation also implicitly defines y in terms of x. Furthermore, the equation can be solved such that y is explicitly defined in terms of x; that is, $y = \frac{1}{2}x + (-2)$. It can also be solved such that x is explicitly defined in terms of y; that is, $x = 2y + 4$.

However, there are many implicit functions that cannot conveniently be converted to explicit functions of the form $y = f(x)$ or $x = g(y)$. For example, the implicit function $y^3 + 3x^2y + x^3 = 25$ cannot conveniently be converted into an explicit function by solving for either x or y. Thus we cannot apply Rule 11.1 to determine the derivative of x or y. Fortunately there is a method of differentiation that permits us to determine the derivative of y with respect to x or the derivative of x with respect to y. (Moreover, when one of these derivatives is obtained, the other can be obtained by applying Rule 11.1.) The method is known as *implicit differentiation* because it can be applied to an *implicit function*. It involves differentiating each term in the function with respect to the variable of interest and applying the chain rule where necessary.

To determine $D_x y$ for the previous *implicit* function, we simply differentiate each term with respect to x as follows:

$$D_x(y^3) + D_x(3x^2y) + D_x(x^3) = D_x(25)$$

Remembering that $D_x y$ is a variable, treating y as a function of x, and applying the chain rule, we obtain

$$3y^2 D_x y + (6xy D_x x + 3x^2 D_x y) + 3x^2 D_x x = 0$$

Finally, replacing $D_x x$ by 1 and solving for $D_x y$, we have

$$D_x y (3y^2 + 3x^2) = -3x^2 - 6xy$$

Sec. 11.2 / INVERSE FUNCTIONS

or

$$D_x y = -\frac{3x^2 + 6xy}{3y^2 + 3x^2}$$

$$= -\frac{x^2 + 2xy}{y^2 + x^2}$$

When using implicit differentiation, it is quite common for the resulting derivative to contain both variables, but this rarely causes any difficulty with further use of the derivative and can be corrected where it is possible to solve for either variable in terms of the other.

To determine $D_y x$ for the previous *implicit* function, we differentiate each term of the equation with respect to y as follows:

$$D_y(y^3) + D_y(3x^2 y) + D_y(x^3) = D_y(25)$$

Treating x as a function of y and applying the chain rule, we obtain

$$3y^2 + (6x D_y x\, y + 3x^2) + 3x^2 D_y x = 0$$

Gathering terms and solving for $D_y x$, we obtain

$$D_y x (6xy + 3x^2) = -3y^2 - 3x^2$$

or

$$D_y x = -\frac{3y^2 + 3x^2}{3x^2 + 6xy}$$

$$= -\frac{y^2 + x^2}{x^2 + 2xy}$$

We could have applied Rule 11.1 to obtain $D_y x$ because we have already determined $D_x y$. Using Rule 11.1 and the result for $D_x y$, we obtain

$$D_y x = \frac{1}{D_x y} = \frac{1}{-(x^2 + 2xy)/(y^2 + x^2)} = -\frac{y^2 + x^2}{x^2 + 2xy}$$

Example 11.4 Use the method of implicit differentiation to determine $D_x y$ and $D_y x$ for

$$x^2 + xy + y^2 = 16$$

Solution Differentiating each term with respect to x and applying the chain rule where necessary, we obtain

$$D_x(x^2) + D_x(xy) + D_x(y^2) = D_x(16)$$

or

$$2x + (y + xD_xy) + 2yD_xy = 0$$

Solving for D_xy, we have

$$D_xy(x+2y) = -2x - y \quad \text{or} \quad D_xy = -\frac{2x+y}{x+2y}$$

To determine D_yx by implicit differentiation, we differentiate each term with respect to y and apply the chain rule where necessary to obtain

$$D_y(x^2) + D_y(xy) + D_y(y^2) = D_y(16)$$

or

$$2xD_yx + (D_yxy + x) + 2y = 0$$

Solving for D_yx, we obtain

$$D_yx(2x+y) = -x - 2y$$

or

$$D_yx = -\frac{x+2y}{2x+y}$$

We can verify D_yx by applying Rule 11.1 to D_xy; thus we have

$$D_yx = \frac{1}{-(2x+y)/(x+2y)} = -\frac{x+2y}{2x+y}$$

EXERCISES

Determine the inverse of each of the functions in Exercises 1 to 6.

1. $y = 3x + 5$

2. $y = \frac{x+1}{x}$

3. $y = \frac{2x+4}{3x}$

4. $y = \frac{x-5}{2x}$

5. $y = \frac{1}{x}$

6. $y = \frac{1}{3x+5}$

For each of the equations in Exercises 7 to 18, determine D_xy and D_yx.

7. $y = 3x + 5$

8. $y = \frac{x+1}{x}$

9. $y = \frac{2x+4}{3x}$

10. $y = \frac{x-5}{2x}$

11. $y = \frac{1}{x}$

12. $y = \frac{1}{3x+5}$

13. $y^2 + 3y = 4x^2 + 2x + 1$ **14.** $xy - y^2 = 2x^2y - 3xy^2$

15. $3x^2y^3 + 6xy = 4x^3y^2 + 2x$ **16.** $x^2y = (x^2y + y^2x)(xy + y^2)$

17. $3xy^2 = 5x^2 - 6y^2 + 4x$ **18.** $3xy = \dfrac{1}{x^2 + y^3}$

Determine the slope of each of the curves in Exercises 19 to 22 at the given point.

19. $x^2 + xy + y^2 = 12$ $(2, 2)$

20. $xy + 2x^2 = 9$ $(3, -3)$

21. $y^3x + 3x^2y = 14$ $(1, 2)$

22. $y^2 + 3y = 4x^2 + 11x + 1$ $(-3, 1)$

23. A company has weekly profits of $20x - 50$, where x is the number of widgets sold in a week.

  (a) Let Y equal weekly profits and determine the equation to compute the weekly profits given the number of units sold.
  (b) Determine the function that represents the rate of change of profits with respect to the number of units sold.
  (c) Determine the function that represents the rate of change of the number of widgets sold with respect to weekly profits.

24. A store has determined empirically that the demand of product A plus the demand of product B equals 10 less than the product of the demand of A and the demand of B.

  (a) Determine the equation to represent this relationship.
  (b) Determine the function that represents the rate of change of the demand of product A with respect to product B.
  (c) What is the rate of change of the demand of product A when the demand for product B is 0? 5? 10?
  (d) Determine the function that represents the rate of change of the demand of product B with respect to product A.
  (e) What is the rate of change of the demand of product B when the demand for product A is 0? 3? 6?

25. When 4.6 is added to the time required to repair defective items multiplied by one-fifth of the number of items produced it is equal to the square of the number of items produced minus the square of the time required to repair the defective items.

  (a) Determine the equation to represent this relationship.
  (b) Determine the slope of the equation in part (a) at the point (4,3).
  (c) Determine $D_y x$.

11.3 EXPONENTIAL AND LOGARITHMIC FUNCTIONS

Exponential Functions

In Sec. 3.5, we demonstrated how an exponential function with base b could be converted into one with base e, and the relationship obtained was

$$b^x = e^{(\ln b)x} \qquad (11.2)$$

We also mentioned that one of the reasons for converting an exponential function to one with base e was because certain results in calculus are simpler for exponential functions with base e.

If we wish to determine the derivative of $f(x) = e^x$, we may apply Definition 10.3 and obtain

$$D_x e^x = \lim_{\Delta x \to 0} \frac{e^{x+\Delta x} - e^x}{\Delta x}$$

$$= \lim_{\Delta x \to 0} e^x \left(\frac{e^{\Delta x} - 1}{\Delta x} \right)$$

$$= e^x \lim_{\Delta x \to 0} \frac{e^{\Delta x} - 1}{\Delta x}$$

However, it can be shown that

$$\lim_{\Delta x \to 0} \frac{e^{\Delta x} - 1}{\Delta x} = 1$$

Thus we have a rule for differentiating $f(x) = e^x$.

Rule 11.2

$$D_x e^x = e^x \tag{11.3}$$

At first glance, Rule 11.2 does not appear to be very important because there is only one function of the form $f(x) = e^x$; however, this is not the case. It has the peculiar property that the derivative is the same as the original function. Also, we can utilize Rule 11.1 in conjunction with the chain rule and determine the derivative of a rather large class of functions. For example, suppose $f(x) = e^{5x}$. Treating this as a composite function for which the variable is $g(x) = 5x$, we can write $f(g) = e^g$. Thus, employing Rule 11.2, we determine the derivative *with respect to g* to be

$$D_g e^g = e^g$$

Furthermore, applying the chain rule, we obtain the derivative with respect to x of $f(x)$ to be

$$D_x e^{5x} = e^{5x} \cdot 5 = 5 e^{5x}$$

Similarly, if $f(x) = e^{g(x)}$ and the derivative with respect to x of $g(x)$ exists, we apply Rule 11.2 and the chain rule to obtain

$$D_x e^{g(x)} = e^{g(x)} D_x g(x)$$

Sec. 11.3 / EXPONENTIAL AND LOGARITHMIC FUNCTIONS

Thus we have a rule for determining the derivative of a function of the form $f(x) = e^{g(x)}$.

Rule 11.3 If $D_x g(x)$ exists, then

$$D_x e^{g(x)} = e^{g(x)} D_x g(x) \qquad (11.4)$$

Example 11.5 Determine the derivative with respect to x for each of the following functions:

(a) e^{7x}
(b) e^{x^3}
(c) $e^{x^2 + 3x + 1}$
(d) $e^{6x^3 + 4x^{1/2}}$

Solution Employing Rule 11.3 and the rules of differentiation from Chap. 10, we obtain

(a) $D_x e^{7x} = e^{7x} \cdot 7 = 7 e^{7x}$
(b) $D_x e^{x^3} = e^{x^3} 3x^2 = 3x^2 e^{x^3}$
(c) $D_x e^{x^2 + 3x + 1} = e^{x^2 + 3x + 1}(2x + 3) = (2x + 3) e^{x^2 + 3x + 1}$
(d) $D_x e^{6x^3 + 4x^{1/2}} = e^{6x^3 + 4x^{1/2}}(18x^2 + 2x^{-1/2}) = (18x^2 + 2x^{-1/2}) e^{6x^3 + 4x^{1/2}}$

On several occasions we have stated that the advantage of exponential functions with base e is that they yield relatively simple results. Rules 11.2 and 11.3 verify the simplicity of the results for exponential functions with base e. Let us now consider the derivative of an exponential function for which the base is b. To determine $D_x b^x$, we use Eq. (11.2) and Rule 11.3 to obtain

$$D_x b^x = D_x e^{x \ln b} = e^{x \ln b} \ln b$$

However, since $b^x = e^{x \ln b}$, we make this substitution on the right side and obtain the following rule.

Rule 11.4 If b is a positive number that does not equal 1, then

$$D_x b^x = b^x \ln b \qquad (11.5)$$

Example 11.6 Determine the derivative with respect to x for each of the following functions. Use Table A.9 in the Appendix to determine the numerical values for the natural logarithms.

(a) 6^x
(b) 10^x
(c) 2.51^x
(d) 11^x

Solution

(a) $D_x 6^x = 6^x \ln 6 = 1.79176 \cdot 6^x$
(b) $D_x 10^x = 10^x \ln 10 = 2.30259 \cdot 10^x$
(c) $D_x 2.51^x = 2.51^x \ln 2.51 = 0.92028 \cdot 2.51^x$
(d) $D_x 11^x = 11^x \ln 11 = 2.39790 \cdot 11^x$

From Example 11.6 it is apparent that the advantage of converting exponential functions to those with base e is that we shall not need to determine the natural logarithm of the base. Furthermore, there exist tables of e^x for various values of x.

To determine the derivative of a composite function of the form $b^{g(x)}$, we simply employ Rule 11.4 and the chain rule, which yields a rule for differentiating such a function.

Rule 11.5 If $D_x g(x)$ exists, then

$$D_x b^{g(x)} = b^{g(x)} \ln b \, D_x g(x) \qquad (11.6)$$

Example 11.7 Determine the derivative with respect to x for each of the following functions:

(a) 10^{x^2}
(b) 5^{3x^2+2x}
(c) $3^{x^{1/2}}$
(d) $6^{3e^{x^2}}$

Solution

(a) $D_x 10^{x^2} = 10^{x^2} \ln 10 \cdot 2x = 4.60518 x \, 10^{x^2}$
(b) $D_x 5^{3x^2+2x} = 5^{3x^2+2x} \ln 5 \cdot (6x+2) = 1.60944(6x+2) 5^{3x^2+2x}$
(c) $D_x 3^{x^{1/2}} = 3^{x^{1/2}} \ln 3 \cdot \frac{1}{2} x^{-1/2} = 0.54931 x^{-1/2} 3^{x^{1/2}}$
(d) $D_x 6^{3e^{x^2}} = 6^{3e^{x^2}} \ln 6 \cdot (3e^{x^2} 2x) = 10.75056 x 6^{3e^{x^2}} e^{x^2}$

Example 11.8 Determine the first four derivatives of $f(x) = e^x$.

Solution

$$D_x f(x) = e^x \qquad D_x^2 f(x) = e^x$$
$$D_x^3 f(x) = e^x \qquad D_x^4 f(x) = e^x$$

Example 11.9 Determine the first three derivatives of $f(x) = e^{x^2}$.

Solution

$$D_x f(x) = 2x e^{x^2}$$
$$D_x^2 f(x) = 2 e^{x^2} + 4x^2 e^{x^2} = (2 + 4x^2) e^{x^2}$$
$$D_x^3 f(x) = 8x e^{x^2} + (2 + 4x^2) 2x e^{x^2} = (12x + 8x^3) e^{x^2}$$

Logarithmic Functions Having determined the rules for differentiating exponential functions, let us turn our attention to developing rules for differentiating logarithmic functions. Suppose we wish to differentiate $y = \ln x$ with respect to x. Thus we wish to determine $D_x y$, but we do not yet have a rule for logarithmic

Sec. 11.3 / EXPONENTIAL AND LOGARITHMIC FUNCTIONS

functions. However, we do have a sufficient set of rules to permit us to determine its derivative; and it is simply a matter of utilizing the tools that are available to us.

In order to differentiate the function $y = \ln x$, let us solve for x and express the function in terms of y; that is, let us determine the inverse function of y. Using the relationship between logarithms and exponents, we convert the logarithmic function

$$y = \ln x$$

to its inverse

$$x = e^y$$

Applying Rule 11.2, we determine the derivative with respect to y of x to be

$$D_y x = e^y = x$$

However, since we really want $D_x y$ but have $D_y x$, we employ Rule 11.1 to obtain

$$D_x y = \frac{1}{D_y x} = \frac{1}{x}$$

when $y = \ln x$. Thus we have the following rule for differentiating *natural* logarithm functions.

Rule 11.6 For positive x,

$$D_x \ln x = \frac{1}{x} \tag{11.7}$$

To determine the derivative for the more general function $y = \ln f(x)$, we jointly apply Rule 11.6 and the chain rule to obtain the following rule.

Rule 11.7 For $f(x) > 0$,

$$D_x [\ln f(x)] = \frac{1}{f(x)} D_x f(x) \tag{11.8}$$

Example 11.10 For positive values of x, determine the derivative with respect to x for each of the following functions:

(a) $\ln x^2$ \hspace{2cm} (b) $\ln(x^2 + 3x + 4)$

(c) $\ln x^{1/2}$ \hspace{2cm} (d) $\ln \dfrac{2x+1}{3x+2}$

Solution

(a) $D_x(\ln x^2) = \dfrac{1}{x^2}(2x) = \dfrac{2}{x}$

(b) $D_x[\ln(x^2+3x+4)] = \dfrac{1}{x^2+3x+4}(2x+3) = \dfrac{2x+3}{x^2+3x+4}$

(c) $D_x(\ln x^{1/2}) = \dfrac{1}{x^{1/2}}\left(\dfrac{1}{2}x^{-1/2}\right) = \dfrac{1}{2x}$

(d) $D_x\left(\ln\dfrac{2x+1}{3x+2}\right) = \dfrac{1}{(2x+1)/(3x+2)} \dfrac{[2(3x+2)-(2x+1)3]}{(3x+2)^2}$

$= \dfrac{6x+4-6x-3}{(2x+1)(3x+2)}$

$= \dfrac{1}{(2x+1)(3x+2)}$

Alternative ways of determining the derivatives for parts (a), (c), and (d) involve the use of rules for logarithms of powers and quotients, i.e., the rules given in Eqs. (3.12) and (3.13), respectively. Applying these rules, we differentiate in the following manner:

(a) $D_x(\ln x^2) = D_x(2\ln x) = 2D_x(\ln x) = \dfrac{2}{x}$

(c) $D_x(\ln x^{1/2}) = D_x\left(\dfrac{1}{2}\ln x\right) = \dfrac{1}{2}D_x(\ln x) = \dfrac{1}{2x}$

(d) $D_x\left(\ln\dfrac{2x+1}{3x+2}\right) = D_x[\ln(2x+1) - \ln(3x+2)]$

$= D_x[\ln(2x+1)] - D_x[\ln(3x+2)]$

$= \dfrac{1}{2x+1}\cdot 2 - \dfrac{1}{3x+2}\cdot 3$

$= \dfrac{1}{(2x+1)(3x+2)}$

Example 11.10 illustrates the advantage of using the rules for logarithms; usually it is more convenient to apply the rules given in Eqs. (3.11) to (3.13) before differentiating.

As in the case for exponential functions, first we have determined the derivatives for logarithmic functions with base e. Now we consider differentiating logarithmic functions with base b. To determine $D_x y$ for $y = \log_b x$, we convert the function to its inverse

$$x = b^y$$

Sec. 11.3 / EXPONENTIAL AND LOGARITHMIC FUNCTIONS

Using Rule 11.4, we obtain $D_y x = b^y \ln b = x \ln b$. Applying the rule for inverse functions, we obtain

$$D_x y = \frac{1}{D_y x} = \frac{1}{x \ln b}$$

Thus we obtain the following rule for differentiating $\log_b x$.

Rule 11.8 For positive x,

$$D_x(\log_b x) = \frac{1}{x \ln b} \qquad (11.9)$$

Example 11.11 Determine derivatives for the following functions:

(a) $\log_{10} x$ (b) $\log_2 x$
(c) $\log_5 x$ (d) $\log_7 x$

Solution

(a) $D_x \log_{10} x = \frac{1}{x \ln 10} = \frac{1}{2.30259 x}$ (b) $D_x \log_2 x = \frac{1}{x \ln 2} = \frac{1}{0.69315 x}$

(c) $D_x \log_5 x = \frac{1}{x \ln 5} = \frac{1}{1.60944 x}$ (d) $D_x \log_7 x = \frac{1}{x \ln 7} = \frac{1}{1.94591 x}$

By using Rule 11.8 and the chain rule, we obtain the rule for differentiating $\log_b f(x)$.

Rule 11.9 For $f(x) > 0$,

$$D_x[\log_b f(x)] = \frac{1}{f(x) \ln b} D_x f(x) \qquad (11.10)$$

Example 11.12 For positive x, determine the derivative with respect to x for each of the following functions:

(a) $\log_{10} x^2$ (b) $\log_5 x^4$
(c) $\log_3 x^2/(x+1)$ (d) $\log_{11}(x+2)(x^2+1)$

Solution

(a) $D_x(\log_{10} x^2) = D_x(2 \log_{10} x) = \dfrac{2}{x \ln 10}$

(b) $D_x(\log_5 x^4) = D_x(4 \log_5 x) = \dfrac{4}{x \ln 5}$

(c) $D_x\left(\log_3 \dfrac{x^2}{x+1}\right)$ $= D_x[2\log_3 x - \log_3(x+1)]$

$$= \dfrac{2}{x\ln 3} - \dfrac{1}{(x+1)\ln 3}$$

$$= \dfrac{2x+2-x}{x(x+1)\ln 3}$$

$$= \dfrac{x+2}{x(x+1)\ln 3}$$

(d) $D_x\{\log_{11}[(x+2)(x^2+1)]\} = D_x[\log_{11}(x+2) + \log_{11}(x^2+1)]$

$$= \dfrac{1}{(x+2)\ln 11} + \dfrac{1}{(x^2+1)\ln 11}2x$$

$$= \dfrac{x^2+1+2x^2+4x}{(x+2)(x^2+1)\ln 11}$$

$$= \dfrac{3x^2+4x+1}{(x+2)(x^2+1)\ln 11}$$

$$= \dfrac{(3x+1)(x+1)}{(x+2)(x^2+1)\ln 11}$$

Example 11.13 Determine the first four derivatives of $f(x) = \ln x$.

Solution

$$D_x f(x) = \dfrac{1}{x} = x^{-1} \qquad D_x^2 f(x) = -x^{-2}$$

$$D_x^3 f(x) = 2x^{-3} \qquad D_x^4 f(x) = -6x^{-4}$$

Example 11.14 Determine the first and second derivatives of

$$f(x) = \ln \dfrac{x^2+2}{x^3+4x}$$

Solution

$$f(x) = \ln \dfrac{x^2+2}{x^3+4x} = \ln(x^2+2) - \ln(x^3+4x)$$

$$D_x f(x) = \dfrac{2x}{x^2+2} - \dfrac{3x^2+4}{x^3+4x}$$

$$D_x^2 f(x) = \dfrac{2(x^2+2) - 2x(2x)}{(x^2+2)^2} - \dfrac{6x(x^3+4x) - (3x^2+4)(3x^2+4)}{(x^3+4x)^2}$$

$$= \dfrac{-2x^2+4}{(x^2+2)^2} + \dfrac{3x^4+16}{(x^3+4x)^2}$$

EXERCISES

Determine the derivative with respect to x for each of the functions in Exercises 1 to 28.

1. e^{2x}
2. $e^{x/2}$
3. e^{x^2}
4. e^{3x^2}
5. e^{2x^2+3x-1}
6. $e^{(2x+1)/(3x+2)}$
7. 4^x
8. 5^x
9. 10^{3x}
10. 6^{6x^2}
11. 8^{2x^2+x+5}
12. $56^{2x^3-3x^2}$
13. $\ln 3x$
14. $\ln(6x^2+4x+1)$
15. $\ln(3x^2+4)^{1/2}$
16. $\ln[(4x+10)(x^2+5x+10)]$
17. $\log_5 x$
18. $\log_2 x$
19. $\log_3(3x^2+2x+7)$
20. $\log_{10}(3-4x+7x^3)$
21. $\log_5[(x+6)/(x^2+4)]$
22. $\log_9[(x^2+5x)(x^3+1)]$
23. $\log_{11}[(x+4)^3(x^2+3)^{1/2}]$
24. $\log_6[(x+3)^{11}/(x^4+2)^{1/2}]$
25. $\log_{10}(\ln 3x)$
26. $\ln[\log_{10}(3x^2+8)^{3/2}]$
27. $\{\log_{10}[\ln(x^2+6)^5]^3\}^{1/2}$
28. $\log_3\{\log_2[(x+4)^{1/2}(x+3)^{1/2}/(x^2+2)]\}$

Determine the first and second derivatives of each function in Exercises 29 to 38.

29. e^{2x^2}
30. $\ln[x^2/(2x+1)]$
31. e^{x^3}
32. 10^{x^2+2x}
33. $\log_5(x^2+1)^2$
34. $[\log_3(x^2+1)]^2$
35. $\log_{10} x + x \ln x$
36. $e^x - \ln(x^2+1)$
37. $(x^2+4)^3 - \ln\sqrt{x^2+6}$
38. $3^{x^3} - \ln\sqrt{x^3}$

39. An insurance office has an increase of e^{x^2+4x} customers per year, where x is the number of customers, in 1,000's, the previous year.

   (a) Determine the function that represents the rate of change of the number of customers.
   (b) Determine the rate of increase when $x=0$, 2, and 5. (Leave answers in terms of e.)
   (c) Determine the function that represents the rate of change of the slope of the function in part (a).

40. The increase in the selling price of a product is $\ln[(x^3-5)/(x^2+1)]$, where x represents the amount of a certain ingredient used in its manufacture.

(a) Determine the function that represents the rate of change in the increase in the price.
(b) Determine the function that represents the rate of change of the slope of the function in part (a).

41. The increase in the number of customers entering a store has been increasing by $0.3\ln(x^3+4x^2+5)$ customers per month, where x is the number of customers, in 1,000's, the previous month.

(a) Determine the function that represents the rate of change of the increase in the number of customers.
(b) Determine the rate of change in the number of customers when $x=2$, 4, and 6.

11.4 LOGARITHMIC DIFFERENTIATION

We have considered rules for differentiating k (a constant), x^n (a variable to a constant power), and b^x (a constant to a variable power). We have also considered rules for differentiating sums, differences, products, quotients, and composite functions involving these types of functions. One type of function that fits into this general category is that of a variable to a variable power. In order to determine the derivative of such a function, we shall make use of a process called *logarithmic differentiation*, which is simply an application of the various rules that we have already considered.

Suppose we have a function of the form $y=f(x)^{g(x)}$. If $f(x)>0$ and we take the natural logarithm of both sides, we obtain

$$\ln y = \ln f(x)^{g(x)} = g(x)\ln f(x) \tag{11.11}$$

We could use the logarithm to any base b; however, when differentiating, the results would be more cumbersome. (Recall that this is the primary reason for using base e for exponential functions and logarithms.) Using implicit differentiation for the left and right sides of (11.11), we obtain

$$D_x \ln y = D_x\big[\, g(x)\ln f(x)\,\big]$$

or

$$\frac{1}{y}D_x y = D_x g(x)\ln f(x) + g(x)\frac{1}{f(x)}D_x f(x)$$

Since we wish to determine $D_x y$, we need solve only the last equation for $D_x y$, and, doing so, we have

$$D_x y = y\left[D_x g(x)\ln f(x) + \frac{g(x)}{f(x)}D_x f(x)\right]$$

Substituting $f(x)^{g(x)}$ in place of y, we obtain

$$D_x\big[f(x)^{g(x)}\big] = f(x)^{g(x)}\left[D_x g(x)\ln f(x) + \frac{g(x)}{f(x)}D_x f(x)\right] \tag{11.12}$$

Sec. 11.4 / LOGARITHMIC DIFFERENTIATION

Equation (11.12) *could* now be used to determine the derivative with respect to x of any function of the form $y = f(x)^{g(x)}$. However, the reader is strongly advised to use the *process* of logarithmic differentiation instead of the resulting formula. This advice is given for several reasons: The formula is too complex to memorize, and it is usually easier to differentiate a function using the logarithmic differentiation process than to substitute into Eq. (11.12).

Example 11.15 Use the process of logarithmic differentiation to determine $D_x y$ for $y = (2x^3)^{5x+4}$.

Solution Taking the natural logarithm of both sides, we have

$$\ln y = (5x+4)\ln(2x^3)$$

Differentiating the implicit function with respect to x, we obtain

$$\frac{1}{y} D_x y = 5\ln 2x^3 + (5x+4)\frac{1}{2x^3} 6x^2$$

Simplifying, solving for $D_x y$, and substituting the function value of y, we obtain

$$D_x y = (2x^3)^{5x+4}\left[5\ln 2x^3 + \frac{3(5x+4)}{x} \right]$$

Example 11.16 Determine $D_x y$ for $y = x^x$.

Solution

$$\ln y = x \ln x$$

$$\frac{1}{y} D_x y = \ln x + x \frac{1}{x}$$

$$D_x y = x^x (\ln x + 1)$$

Example 11.17 Determine $D_x y$ for $y = (3x^4 + 2x)^{3x^2 + 6x^{-2}}$.

Solution

$$\ln y = (3x^2 + 6x^{-2})\ln(3x^4 + 2x)$$

$$\frac{1}{y} D_x y = (6x - 12x^{-3})\ln(3x^4 + 2x) + (3x^2 + 6x^{-2})\frac{1}{3x^4 + 2x}(12x^3 + 2)$$

$$D_x y = (3x^4 + 2x)^{3x^2 + 6x^{-2}}\left[(6x - 12x^{-3})\ln(3x^4 + 2x) + \frac{(3x^2 + 6x^{-2})(12x^3 + 2)}{3x^4 + 2x} \right]$$

EXERCISES

In each of the functions in Exercises 1 to 18, determine the derivative with respect to x.

1. x^{x+1} **2.** $(x+1)^x$

3. $(3x-2)^{2x}$ 	 4. $(10x+3)^x$

5. x^{x^2+4x+1} 	 6. $(5x+4)^{5x+4}$

7. $(3x^2+6x+1)^{2x+3}$ 	 8. $(2x+3x^2)^{3x+2}(2x+3)$

9. $(e^{3x^2})^{2x}$ 	 10. $[\ln(x^2+2x)]^{x+2}$

11. $\ln(x^2+2x)^{x+2}$ 	 12. $[(x+2)/(x+5)]^x$

13. $b^x x^b$ 	 14. $[\log_3(x+2)]^{x+1}$

15. $x^x + x^4$ 	 16. $(x+1)^x + (2x+3)^3$

17. $(x+2)^x/(x+3)$ 	 18. $[(2x+5)^{2x+1}]^3 + (3x^2+4)^x$

In each of the functions in Exercises 19 to 34, determine the derivative by the rules that are normally applicable to the respective function and verify the result through use of logarithmic differentiation.

19. e^5 	 20. 6^4

21. x^7 	 22. x^n

23. $(x^2+3x+2)^{-1/2}$ 	 24. $(6x+5)^4(x+3)^{-2}$

25. $(3x+2)^2/(x-7)^3$ 	 26. $(3x^2+x-1)^4(x^2+2x)^3/(x+3)^2$

27. e^x 	 28. e^{4x+6}

29. e^{x^2+3x+2} 	 30. $e^{\ln x}$

31. b^x 	 32. 5^x

33. 10^{4x+3} 	 34. 9^{x^2+3x+7}

11.5 PARTIAL DERIVATIVES

In Secs. 11.1 to 11.4, we have dealt with functions of one independent variable. In this section, we shall consider functions of several variables; such functions are called *multivariate functions*. A function $f(x,y)$ of two variables is a correspondence that associates with each pair of possible values of the independent variables (x,y) one and only one value of the dependent variable $f(x,y)$. This concept can be extended to include as many independent variables as desired, but, regardless of the number of independent variables, there is only one *dependent* variable.

When working with functions of several variables, we can manipulate them in the same way as we do with functions of one variable; that is, we can construct sums, differences, products, quotients, and composite functions in the same manner. However, an increase in the number of variables tends to increase the complexity of the problem, which seems to make performing the operations more difficult and increases the possibilities for confusion. Nevertheless, the algebraic operations are the same as for the single-variable function, and the process of differentiating multivariate functions is practically the same as for one-variable (univariate) functions. Here is an example of the algebraic operations for a two-variable function.

Example 11.18 If $f(x,y) = x^2 + 3xy + y^2$ and $g(x,y) = 2x^2 + 4xy$, determine the function $h(x,y)$ obtained by each of the following algebraic operations:

(a) $f + g$ (b) $f - g$
(c) fg (d) f/g

Solution

(a) $h(x, y) = f(x,y) + g(x,y)$
$= x^2 + 3xy + y^2 + 2x^2 + 4xy$
$= 3x^2 + 7xy + y^2$

(b) $h(x, y) = f(x,y) - g(x,y)$
$= x^2 + 3xy + y^2 - (2x^2 + 4xy)$
$= -x^2 - xy + y^2$

(c) $h(x, y) = f(x,y)g(x,y)$
$= (x^2 + 3xy + y^2)(2x^2 + 4xy)$

(d) $h(x,y) = f(x,y)/g(x,y)$
$= \dfrac{x^2 + 3xy + y^2}{2x^i + 4xy}$

First-Order Partial Derivatives

In a function of two variables, say, $u = f(x,y)$, each of the independent variables can be varied *independently* of the other. In changing x for a fixed y, actually we are determining changes in u that are caused by the corresponding changes in x. Essentially we have reduced the two-variable function $u = f(x,y)$ to a one-variable function, say, $z = g(x)$ because y is fixed and consequently is being treated as a constant. Thus we may refer to the derivative of $f(x,y)$ with respect to x and mean essentially the same as the derivative of $g(x)$ with respect to x; the only difference is that the actual function for $g(x)$ can change for different values of y. Nevertheless, the derivative of $f(x,y)$ with respect to x refers to the instantaneous rate of change of the function for a constant value of y.

Derivatives of multivariate functions are called *partial derivatives* because the derivatives are taken with respect to a single variable while the remaining variables are temporarily held constant. If we represent the function by $u = f(x,y)$, the derivative of the function with respect to x is generally denoted by one of the following symbols:

$$\frac{\partial f}{\partial x} \qquad \frac{\partial f(x,y)}{\partial x} \qquad f_x \qquad f_x(x,y) \qquad \frac{\partial u}{\partial x} \qquad u_x$$

You may have already noticed that we have replaced D_x by the symbol $\partial/\partial x$ in order to denote that we are determining *partial* derivatives instead of derivatives of one-variable functions. Since we are temporarily holding constant the variables other than the one of interest, the definition of partial derivatives should not be completely unfamiliar.

Definition 11.1 The partial derivative of $f(x,y)$ with respect to x is

$$\frac{\partial f}{\partial x} = \lim_{\Delta x \to 0} \frac{f(x+\Delta x, y) - f(x,y)}{\Delta x}$$

and the partial derivative of $f(x,y)$ with respect to y is

$$\frac{\partial f}{\partial y} = \lim_{\Delta y \to 0} \frac{f(x, y+\Delta y) - f(x,y)}{\Delta y}$$

provided the limits exist.

We could employ Definition 11.1 to determine the partial derivatives of any particular two-variable function; however, this is not necessary. Since we are temporarily holding one variable constant, we can differentiate with respect to the remaining variable as though the function were a one-variable function. Thus, to determine the partial derivatives, we can simply apply the rules of differentiation that were developed for the one-variable case; however, the result may still be a function of the variable that was being held constant. For example, if $f(x,y) = 6xy + 5y + 3x$, the partial derivative with respect to x is obtained by considering y to be constant and differentiating with respect to x; thus $\partial f/\partial x = 6y + 0 + 3 = 6y + 3$. Similarly, the partial derivative with respect to y is obtained by considering x to be constant and differentiating with respect to y; differentiating, we obtain $\partial f/\partial y = 6x + 5 + 0 = 6x + 5$.

We note that a two-variable function has two first-order partial derivatives $\partial f/\partial x$ and $\partial f/\partial y$, whereas a one-variable function has only one first-order derivative. Similarly, a function of n variables will have n first-order partial derivatives.

Example 11.19 Determine the two first-order partial derivatives for $f(x,y) = 6x^2y + 3xy + y^2$.

Solution

$$\frac{\partial f}{\partial x} = 12xy + 3y + 0 = 12xy + 3y$$

$$\frac{\partial f}{\partial y} = 6x^2 + 3x + 2y$$

Example 11.20 Determine the two first-order partial derivatives for $f(x,y) = (x^2y + xy^2)^2$.

Solution

$$\frac{\partial f}{\partial x} = 2(x^2y + xy^2)(2xy + y^2)$$

$$\frac{\partial f}{\partial y} = 2(x^2y + xy^2)(x^2 + 2xy)$$

Example 11.21 Determine all first-order partial derivatives for $f(x,y) = e^{x^3+y^2}$.

Solution

$$\frac{\partial f}{\partial x} = 3x^2 e^{x^3+y^2}$$

$$\frac{\partial f}{\partial y} = 2y e^{x^3+y^2}$$

For multivariate functions of three or more variables, we determine the partial derivatives with respect to each variable by temporarily holding all other variables constant and differentiating in the usual manner.

Example 11.22 Determine all first-order partial derivatives for $f(x,y,z) = 6x^2y + 4y^2z + 3xyz$.

Solution

$$\frac{\partial f}{\partial x} = 12xy + 0 + 3yz = 12xy + 3yz$$

$$\frac{\partial f}{\partial y} = 6x^2 + 8yz + 3xz$$

$$\frac{\partial f}{\partial z} = 0 + 4y^2 + 3xy = 4y^2 + 3xy$$

Example 11.23 Determine all first-order partial derivatives for $f(x,y,z,w) = e^{x^2y} + 2xzw + 4w^2z^2$.

Solution

$$\frac{\partial f}{\partial x} = 2xy e^{x^2y} + 2zw + 0 = 2xy e^{x^2y} + 2zw$$

$$\frac{\partial f}{\partial y} = x^2 e^{x^2y} + 0 + 0 = x^2 e^{x^2y}$$

$$\frac{\partial f}{\partial z} = 0 + 2xw + 8w^2z = 2xw + 8w^2z$$

$$\frac{\partial f}{\partial w} = 0 + 2xz + 8wz^2 = 2xz + 8wz^2$$

Higher-Order Partial Derivatives

When considering higher-order derivatives for multivariate functions, we encounter a considerable increase in the number of partial derivatives of a particular order. For example, a two-variable function has two first-order partial derivatives. Consequently, to determine all second-order derivatives, we must differentiate *each* first-order partial derivative with respect to each variable, which yields four second-order partial derivatives, as shown in Fig. 11.1. Furthermore, if we were to determine the third-order partial derivatives, we would differentiate each of the second-order partial derivatives with respect to each of the two variables, which would yield eight third-order

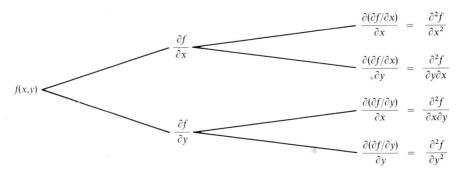

FIGURE 11.1

partial derivatives. Thus, for a two-variable function, there are 2^n nth-order partial derivatives. In general, for a k-variable function, there are k^n nth-order partial derivatives. (In this problem involving multivariate functions, determining all of those higher-order derivatives for a one-variable function was not really as bad as was first imagined, at least relatively speaking. Fortunately, we shall not consider any derivatives beyond second-order partial derivatives.)

The process of determining higher-order partial derivatives is essentially the same as determining higher-order derivatives for one-variable functions. That is, we simply determine the partial derivative of each function that represents a partial derivative of the preceding order; also, we must do this for *each* variable. Notationally this is demonstrated in Fig. 11.1. For higher-order partial derivatives the notation used is such that the superscript in the notation $\partial^2 f/\partial x\, \partial y$ represents the *order* of the partial derivative. The lower portion of the symbol (that appears to be the denominator but is not because the term is not a fraction) represents the order in which the derivatives were taken. The variable on the *right* represents the *first* variable with respect to which the first-order partial derivative was taken. The variable on the *left* represents the *last* variable of differentiation.

Thus $\partial^2 f/\partial x\, \partial y$ represents the function that is determined by first differentiating $f(x,y)$ with respect to y and then differentiating the resulting first-order partial derivative $\partial f/\partial y$ with respect to x. Conversely, $\partial^2 f/\partial y\, \partial x$ represents the second-order partial derivative obtained by first differentiating $f(x,y)$ with respect to x and then differentiating $\partial f/\partial x$ with respect to y. The second-order partial derivatives denoted by $\partial^2 f/\partial x^2$ and $\partial^2 f/\partial y^2$ are obtained by differentiating both the function and the resulting first-order partial derivative with respect to x and y, respectively.

Example 11.24 Determine all first- and second-order partial derivatives for $f(x,y) = x^3 y^2 + 3x^2 y + 2y^3$.

Solution The first-order partial derivatives are

$$\frac{\partial f}{\partial x} = 3x^2 y^2 + 6xy$$

Sec. 11.5 / PARTIAL DERIVATIVES

and

$$\frac{\partial f}{\partial y} = 2x^3y + 3x^2 + 6y^2$$

The second-order partial derivatives are

$$\frac{\partial^2 f}{\partial x^2} = \frac{\partial}{\partial x}\left(\frac{\partial f}{\partial x}\right) = 6xy^2 + 6y$$

$$\frac{\partial^2 f}{\partial y \partial x} = \frac{\partial}{\partial y}\left(\frac{\partial f}{\partial x}\right) = 6x^2y + 6x$$

$$\frac{\partial^2 f}{\partial x \partial y} = \frac{\partial}{\partial x}\left(\frac{\partial f}{\partial y}\right) = 6x^2y + 6x + 0 = 6x^2y + 6x$$

$$\frac{\partial^2 f}{\partial y^2} = \frac{\partial}{\partial y}\left(\frac{\partial f}{\partial y}\right) = 2x^3 + 0 + 12y = 2x^3 + 12y$$

Example 11.25 Determine all first- and second-order partial derivatives for $f(x,y) = (2xy + x^2y)^2$.

Solution The first-order partial derivatives are

$$\frac{\partial f}{\partial x} = 2(2xy + x^2y)(2y + 2xy)$$
$$= 4xy^2(2 + x)(1 + x)$$
$$= 4y^2(2x + 3x^2 + x^3)$$

and

$$\frac{\partial f}{\partial y} = 2(2xy + x^2y)(2x + x^2)$$
$$= 2y(2x + x^2)(2x + x^2)$$
$$= 2y(2x + x^2)^2$$

The second-order partial derivatives are

$$\frac{\partial^2 f}{\partial x^2} = 4y^2(2 + 6x + 3x^2)$$

$$\frac{\partial^2 f}{\partial y \partial x} = 8y(2x + 3x^2 + x^3)$$

$$\frac{\partial^2 f}{\partial x \partial y} = 2y2(2x + x^2)(2 + 2x) = 8y(2x + 3x^2 + x^3)$$

$$\frac{\partial^2 f}{\partial y^2} = 2(2x + x^2)^2$$

Example 11.26 Determine all first- and second-order partial derivatives for $f(x,y) = e^{xy}$.

Solution The first-order partial derivatives are

$$\frac{\partial f}{\partial x} = ye^{xy}$$

and

$$\frac{\partial f}{\partial y} = xe^{xy}$$

The second-order partial derivatives are

$$\frac{\partial^2 f}{\partial x^2} = y^2 e^{xy}$$

$$\frac{\partial^2 f}{\partial y\,\partial x} = e^{xy} + xye^{xy} = (1+xy)e^{xy}$$

$$\frac{\partial^2 f}{\partial x\,\partial y} = e^{xy} + xye^{xy} = (1+xy)e^{xy}$$

$$\frac{\partial^2 f}{\partial y^2} = x^2 e^{xy}$$

Example 11.27 Determine all first- and second-order partial derivatives for $f(x,y,z) = x^2 + y^2 + z^2 + 2xyz$.

Solution The three first-order partial derivatives are

$$\frac{\partial f}{\partial x} = 2x + 2yz$$

$$\frac{\partial f}{\partial y} = 2y + 2xz$$

$$\frac{\partial f}{\partial z} = 2z + 2xy$$

The nine second-order partial derivatives are

$$\frac{\partial^2 f}{\partial x^2} = 2 \qquad \frac{\partial^2 f}{\partial x\,\partial y} = 2z \qquad \frac{\partial^2 f}{\partial x\,\partial z} = 2y$$

$$\frac{\partial^2 f}{\partial y^2} = 2 \qquad \frac{\partial^2 f}{\partial y\,\partial x} = 2z \qquad \frac{\partial^2 f}{\partial y\,\partial z} = 2x$$

$$\frac{\partial^2 f}{\partial z^2} = 2 \qquad \frac{\partial^2 f}{\partial z\,\partial x} = 2y \qquad \frac{\partial^2 f}{\partial z\,\partial y} = 2x$$

The partial derivatives $\partial^2 f/\partial x\,\partial y$ and $\partial^2 f/\partial y\,\partial x$ are called *cross partial derivatives* and are equal if their respective functions are continuous in some region, which means simply that the order of differentiation is immaterial if

Sec. 11.5 / PARTIAL DERIVATIVES

the *continuity condition* is satisfied. (This continuity condition is satisfied for nearly all functions that we shall consider in this book.) In each of Examples 11.24 to 11.27 the reader should verify that the corresponding cross partial derivatives are, in fact, equal.

EXERCISES

1. Given $f(x,y)=2x^2+4xy+2x+y^2$ and $g(x,y)=x^2-xy+3y-x+2y^2$, determine
   (a) $f+g$ (b) $f-g$
   (c) fg (d) f/g

2. Given $f(x,y)=e^{x^2+2xy+y^2}$ and $g(x,y)=e^{x^2+y^2}$, determine
   (a) $f+g$ (b) $f-g$
   (c) fg (d) f/g

In Exercises 3 to 16, determine all first-order partial derivatives.

3. $f(x,y)=2x+5$
4. $f(x,y)=6x-3y$
5. $f(x,y)=x^2+2xy$
6. $f(x,y)=y^2-3xy+4x^2$
7. $f(x,y)=y^2+6xy+x^2-4x+3y$
8. $f(x,y)=4x^2+5x^{-6}+2y^3$
9. $f(x,y)=e^{x^2y}$
10. $f(x,y)=e^{4xy+x^2}$
11. $f(x,y)=(3x^2+2xy)^3$
12. $f(x,y)=11^{6xy}$
13. $f(x,y)=4^{x^2-y^2}$
14. $f(x,y)=\ln(xy+2)$
15. $f(x,y,z)=xy/z$
16. $f(x,y,z)=x^2+\ln(y^2z^2)$

17. Using f and g as given in Exercise 1, verify that $\partial(f+g)/\partial x = \partial f/\partial x + \partial g/\partial x$, and $\partial(f+g)/\partial y = \partial f/\partial y + \partial g/\partial y$.

18. Using f and g as given in Exercise 2, verify that $\partial(fg)/\partial x = (\partial f/\partial x)g + f(\partial g/\partial x)$, and $\partial(fg)/\partial y = (\partial f/\partial y)g + f(\partial g/\partial y)$.

In each of the functions in Exercises 19 to 28, determine all first- and second-order partial derivatives and verify that the respective cross partial derivatives are equal.

19. $f(x,y)=x^2+3y^2$
20. $f(x,y)=x^2+2y^2+3x+4y+1$
21. $f(x,y)=xy+2x^2-3y^2$
22. $f(x,y)=e^{3x-2y}$
23. $f(x,y)=2x-4y+3x/y$
24. $f(x,y)=(x+1)\ln(y+1)$
25. $f(x,y)=2x^2-y+ye^x$
26. $f(x,y)=\ln(xy)$
27. $f(x,y)=\ln x^y+4y^3x$
28. $f(x,y,z)=4xyz+2xz$

29. For $f(x,y)=2x^2+3y^2$ at $(2,2)$, determine the instantaneous rate of change for (a) x and (b) y.

30. For $f(x,y)=4xy$ at $(2,1)$, determine the instantaneous rate of change for (a) x and (b) y.

31. For $f(x,y)=2x^2+3xy-y^2$ at $(2,3)$, determine the instantaneous rate of change for (a) x and (b) y.

32. Product A has a contribution margin of \$5 per unit, and product B has a contribution margin of \$6 per unit. Fixed costs are \$250/month.

   (a) Determine the function that represents monthly profits. (Profit is equal to the sum of the contribution margin of each product minus fixed costs.)
   (b) Determine the function that represents the instantaneous rate of change for the contribution margin of product A.
   (c) Determine the function that represents the instantaneous rate of change for the contribution margin of product B.
   (d) Determine the instantaneous rate of change for product A and product B at the point $(50,40)$.

33. The number of minutes required in department A to produce products X and Y each day is equal to the number of units of product X produced squared plus the number of units of product Y produced cubed minus 4 times the number of units of product X produced multiplied by the number of units of product Y produced.

   (a) Determine the function that represents the number of minutes required to process products X and Y in department A.
   (b) Determine the function that represents the instantaneous rate of change for product X.
   (c) Determine the function that represents the instantaneous rate of change for product Y.
   (d) Determine the instantaneous rate of change for product X and product Y at the point $(6,5)$.
   (e) Determine all of the second-order partial derivatives for the function in part (a).

34. The increase in the demand for widgets is equal to the increase in the amount, in thousands of dollars, of advertising times the natural logarithm of the number of salespeople.

   (a) Determine the function that represents the increase in the demand for widgets.
   (b) Determine the function that represents the instantaneous rate of change for the amount, in thousands of dollars, spent on advertising.
   (c) Determine the function that represents the instantaneous rate of change for the number of salespeople.
   (d) Determine all of the second-order partial derivatives for the function in part (a).

IMPORTANT TERMS AND CONCEPTS	Algebraic function Cross partial derivatives Derivative of derivatives	exponential function implicit function inverse function logarithmic function

multivariate function
partial derivatives
Explicit function
Higher-order derivatives
Higher-order partial derivatives
Implicit differentiation
Implicit function

Inverse function
Logarithmic differentiation
Multivariate function
Partial derivative
Transcendental function
Univariate function

REVIEW PROBLEMS

1. For the following equations, determine $d_x y$ and $d_y x$:

   (a) $y = 2x - 5$

   (b) $y = \dfrac{3x^2 + 1}{3x}$

   (c) $y = 4x^2 + 2x + 1$

   (d) $x = \dfrac{4y + 6}{3y}$

   (e) $4x^2 + 2xy = 3y^2$

   (f) $2xy + 3x^2 y^2 = 4x^2 y^3 - 2x^2$

2. Determine the slope of the following curves at the given point:

   (a) $x^2 - 2y^2 = 4$ $(2, 0)$
   (b) $2x^2 = 8 - 2xy$ $(-4, 3)$
   (c) $2x^2 y = 3y^2 - xy + 13y$ $(2, -1)$

In each of the functions in Problems 3 to 10, determine the derivative with respect to x.

3. $e^{(3x+1)(2x-1)}$

4. $\ln \dfrac{x^2 + 2}{x - 2}$

5. $4^{2x^2 + x - 1}$

6. $\log_8 (3x^3 + 2x^2 - 5x + 2)$

7. $4^{x^3 - 2x} + \ln[(x + 2)(x - 1)]$

8. $x^{2x - 4}$

9. $(x^2 + 3x - 1)^{x - 5}$

10. $(x^2 + 2)^x + (x + 2)^{x^2 + 2}$

In each of the functions in Problems 11 to 13, determine all first- and second-order partial derivatives and verify that the respective cross partial derivatives are equal.

11. $f(x, y) = 4x^3 + 2x^2 y + 3y^2$

12. $f(x, y) = 2 \ln(4x^2 - 3xy + 5y^2)$

13. $f(x, y) = 4^{xy} + e^{x^2 + y^2}$

DIFFERENTIAL CALCULUS: APPLICATIONS

12.1 INTRODUCTION

The applications of differential calculus to business situations are numerous. For example, we may wish to determine the number of units to produce that will yield the maximum profit for our company, or we may wish to find the amount of fuel that will meet our requirements and minimize our fuel costs. A banker may wish to find the marginal cost of servicing an additional account, or a rancher may need to find the shape of pasture that will cost the least to fence. So far, we have considered the meaning and mechanics associated with determining derivatives of various functions. In this chapter, we utilize concepts that were presented in the preceding chapters. Specifically we utilize differential calculus in graphing functions, and determining optimal function values, marginal rates of change, extrema for multivariate functions, and extreme values for functions that are subject to certain constraints. The concepts presented in this chapter can be applied to numerous business situations because at the essence of decision making is the concept of optimizing certain variables, e.g., profit, cost, revenue, and income.

12.2 MAXIMA AND MINIMA

In Chaps. 5 and 6 we considered the problem of determining the optimum (i.e., maximum or minimum) value of a *linear* objective function that was subject to certain *linear* constraints. Now we shall utilize differential calculus to determine the optimum value of a function that is not necessarily linear. Furthermore, in this section we shall consider determining the optimum value of functions that are not subject to any other constraints; in Sec. 12.6 we shall consider optimizing functions that are constrained.

When first considering the derivative of a function, we noted that the derivative evaluated at a point, say $x = a$, is the instantaneous rate of change, or slope, of the function at $x = a$. We noted that a positive slope indicates that as the independent variable increases, the function increases and a negative slope indicates that the function decreases; moreover, if the slope equals zero, the function neither increases nor decreases at that particular point.

The last item becomes of considerable importance when graphing a specific function. Thus, when the slope of a function equals zero, we know that the function is behaving in one of four different ways, shown in Fig. 12.1, where at each given point the slope equals zero. However, the difference among the four points is the behavior of the curve on either side of the points. For example, at $x = a$, the curve has zero slope but it decreases on both sides of $x = a$; thus, if we proceed from the left to the right in the region around $x = a$, the curve decreases, except at $x = a$, where it is stalled in its downward trend. To the left of $x = b$, the curve decreases; at $x = b$, the curve is stalled in its decrease; and to the right of $x = b$, the curve increases. Similarly, around $x = c$, the curve increases with a hesitation and then it continues to increase. Likewise around $x = d$, the curve increases to the left of $x = d$; it stalls at $x = d$, and then decreases to the right of $x = d$.

At each of the four points in Fig. 12.1, the slope of the curve equals zero; that is, it has temporarily stalled from its preceding trend. Those x values at which a curve stalls are called *critical values*; those points (x,y) at which a curve stalls and then continues the same trend are called *horizontal inflection points*, for example, points a and c in Fig. 12.1. Those critical values at which a curve stalls and changes direction are called *local maxima* if the trend is from increasing to decreasing, such as at $x = d$, or *local minima* if the trend is from decreasing to increasing, such as at $x = b$. The values are *local* maxima or minima because they represent the maximum or minimum within a local neighborhood of the point on the curve. From Fig. 12.1 it is apparent that points b and d are not the absolute minimum or maximum for the curve because there are values on the curve to the right of $x = d$ that are less than the function value at $x = b$ and to the left of $x = a$ that are greater than the function value at $x = d$.

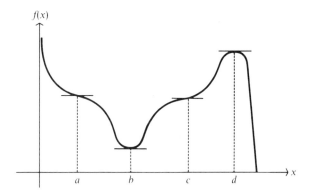

FIGURE 12.1

As seen in Fig. 12.1, horizontal inflection points, local maxima, and local minima have a common characteristic; that is, they each occur at a point on the curve where the slope equals zero. In the terminology of differential calculus, they occur at those critical points for which the first derivative equals zero. Thus, in order to locate the values of the independent variable corresponding to such points, we simply need to solve for those values that render the first derivative equal to zero. For example, to determine the points for which the curve of $f(x) = x^3 - 6x^2 + 9x + 4$ has zero slope, we determine $D_x f(x) = 3x^2 - 12x + 9$, set the derivative equal to zero, and determine the roots of the equation. That is, setting the derivative equal to zero, factoring, and solving for the x values, we have

$$3x^2 - 12x + 9 = 0$$
$$3(x^2 - 4x + 3) = 0$$
$$3(x - 3)(x - 1) = 0$$

where $x = 3$ and $x = 1$. Thus, at the critical values $x = 1$ and $x = 3$, we have either a horizontal inflection point, a local maximum, or a local minimum.

In order to determine which of the three types of points we have at $x = 1$, we can compare the value of $f(1)$ with values of the function for x values that are on each side of $x = 1$ and are relatively close to $x = 1$. By the expression *relatively close*, generally we mean that the value must be between the critical point being checked and the next consecutive critical point, that is, for $x = 1$, anywhere to the left of 1 and between 1 and 3. Thus, evaluating $f(x)$ at $x = 0$ and $x = 2$, we have $f(0) = 4$ and $f(2) = 6$. Comparing these values with $f(1) = 8$, we note that both values are less than 8, which means that the curve has a local maximum at $x = 1$. Likewise, we compare $f(3) = 4$ with $f(2) = 6$ and $f(4) = 8$ and note that $f(3)$ is less than either value, indicating that the curve has a local minimum at $x = 3$.

Utilizing the information obtained in checking for horizontal inflection points, local maxima, and local minima, we can sketch the curve of the function $f(x) = x^3 - 6x^2 + 9x + 4$. We have determined five function values, and we know that the function has a local maximum at $x = 1$ and a local

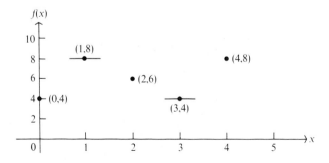

FIGURE 12.2

Sec. 12.2 / MAXIMA AND MINIMA

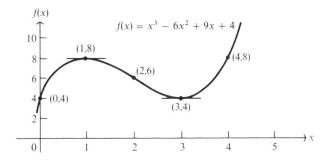

FIGURE 12.3

minimum at $x=3$. Furthermore, we know that it does not possess any other horizontal inflection points or local maxima or minima. Figure 12.2 presents graphically the basic information that we have determined, and Fig. 12.3 presents a rough sketch of the curve that is obtained by connecting the points in Fig. 12.2 with smooth curves. Thus we can utilize differential calculus in determining the horizontal inflection points and the local maxima and minima for a specific function. This information can then be used in sketching the general pattern of the curve given by the graph of the function.

Example 12.1 Determine the horizontal inflection points and local maxima and minima for $f(x) = x^2 - 5x + 6$ and sketch the related graph.

Solution The first derivative is

$$D_x f(x) = 2x - 5$$

Setting this equal to zero and solving for x, we obtain

$$2x - 5 = 0$$

or

$$x = \tfrac{5}{2}$$

Thus we have only one critical value.

Choosing the values $x=2$ and $x=3$ on each side of the critical value and evaluating $f(x)$, we have $f(2)=0$, $f(\tfrac{5}{2}) = -\tfrac{1}{4}$, and $f(3)=0$. These values are such that $f(\tfrac{5}{2})$ is less than either of the adjacent values; hence the critical value $x = \tfrac{5}{2}$ represents the x value for which $f(x)$ has a local minimum. In Fig. 12.4 the previous three points are plotted and connected by a smooth curve to give a sketch of the related graph of $f(x) = x^2 - 5x + 6$.

Since the function in Example 12.1 is a quadratic function, we could employ the information from Chap. 2 and achieve the same results; however, for more complicated functions, generally we shall rely on differential calculus to obtain the desired information.

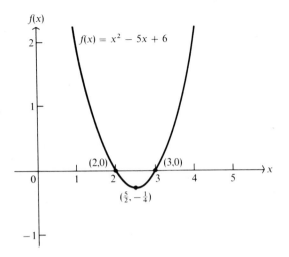

FIGURE 12.4

Example 12.2 Use differential calculus to sketch the graph of $f(x) = x^5 - 15x^3$.

Solution To determine the critical values of x, we determine the derivative of $f(x)$:

$$D_x f(x) = 5x^4 - 45x^2$$

Setting the derivative equal to zero and solving, we have

$$5x^4 - 45x^2 = 0$$
$$5x^2(x^2 - 9) = 0$$
$$5x^2(x+3)(x-3) = 0$$

which yields the critical values

$$x = -3, \quad x = 0, \quad \text{and} \quad x = 3$$

Determining the value of the function at each critical value and two adjacent x values, we obtain

$$\begin{array}{lll} f(-4) = -64 & f(-1) = 14 & f(2) = -88 \\ f(-3) = 162 & f(0) = 0 & f(3) = -162 \\ f(-2) = 88 & f(1) = -14 & f(4) = 64 \end{array}$$

Thus we see that the curve has a local maximum at $x = -3$, a horizontal inflection point at $x = 0$, and a local minimum at $x = 3$. The nine points are plotted and a sketch of the curve is presented in Fig. 12.5.

Sec. 12.2 / MAXIMA AND MINIMA

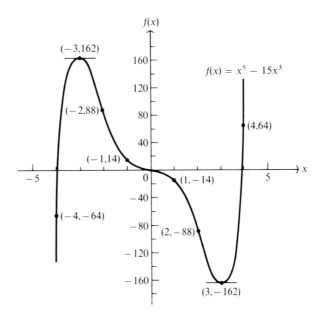

FIGURE 12.5

The First and Second Derivative Tests

Examples 12.1 and 12.2 show that the major contribution that differential calculus makes to the graphing of functions is the indication of regions that require further investigation of the shape of the curve. That is, it tells us where the directional changes of the function are occurring, and hence we can concentrate on evaluating the function around these critical values.

In many situations, it is rather awkward and tedious to evaluate a function; occasionally it is easier to evaluate the first and/or second derivative. Fortunately there are tests that utilize the first and second derivatives for determining whether or not the function has attained a local maximum or local minimum at the critical value. Two of these tests are stated below without proof.

Rule 12.1 If $f(x)$ is a continuous function for which the first derivative exists and x^* is the only critical value in the interval $a \leqslant x \leqslant b$, then
(a) $f(x^*)$ is a local maximum if $f'(a) > 0$ and $f'(b) < 0$
(b) $f(x^*)$ is a local minimum if $f'(a) < 0$ and $f'(b) > 0$.
Otherwise, $f(x^*)$ is not an extremum.

This rule is commonly known as the *first derivative test* because it uses only the values of the first derivative to determine whether or not the critical value yields a local maximum, a local minimum, or a horizontal inflection

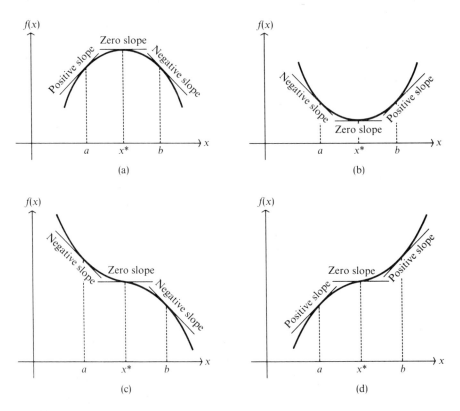

FIGURE 12.6

point. It is simply a different way of stating the method used in the previous examples because it evaluates first derivatives. A graphic analogy of Rule 12.1 is given in Fig. 12.6, where the four parts are analogous to the four situations in Fig. 12.1. If we keep in mind the general relationships given in Fig. 12.6, we should not have any difficulty with the first derivative test. Figure 12.6a corresponds to part (a) of Rule 12.1; Fig. 12.6b corresponds to part (b); and Fig. 12.6c and d correspond to the situation where x^* is not an extremum but rather a horizontal inflection point.

Rule 12.2 If the first and second derivatives of $f(x)$ exist for every value of x in the interval $a \leqslant x \leqslant b$ and $f'(x^*) = 0$, then
 (a) $f(x^*)$ is a local maximum if $f''(x^*) < 0$.
 (b) $f(x^*)$ is a local minimum if $f''(x^*) > 0$.
 (c) The test fails if $f''(x^*) = 0$.

This rule is commonly known as the *second derivative test* because it uses the value of the second derivative to determine whether a critical value

Sec. 12.2 / MAXIMA AND MINIMA

produces a local maximum or a local minimum. However, the test does not yield conclusive information if the second derivative equals zero. Let us apply Rules 12.1 and 12.2 to the functions in Examples 12.1 and 12.2.

Example 12.3 Apply Rules 12.1 and 12.2 to determine the inflection points, local maxima, and local minima for $f(x) = x^2 - 5x + 6$.

Solution Determining the critical values by use of differential calculus, we have

$$D_x f(x) = 2x - 5 = 0$$

or $x = \frac{5}{2}$ is the only critical value. Applying the first derivative test (Rule 12.1) for $a = 2$ and $b = 3$, we have

$$f'(2) = 2(2) - 5 = -1$$

and

$$f'(3) = 2(3) - 5 = 1$$

Thus $f'(a) < 0$ and $f'(b) > 0$, which means that $f(x) = x^2 - 5x + 6$ attains a local minimum at $x = \frac{5}{2}$.

Applying the second derivative test (Rule 12.2), we have

$$D_x^2 f(x) = 2$$

Thus, for the critical value $x^* = \frac{5}{2}$, we have $f''(\frac{5}{2}) = 2 > 0$; therefore, $f(x) = x^2 - 5x + 6$ has a local minimum at $x = \frac{5}{2}$.

Example 12.4 Use the first and second derivative tests to determine the horizontal inflection points, local maxima, and local minima for $f(x) = x^5 - 15x^3$.

Solution Determining the critical values, we have

$$\begin{aligned} D_x f(x) &= 5x^4 - 45x^2 &= 0 \\ &= 5x^2(x^2 - 9) &= 0 \\ &= 5x^2(x+3)(x-3) &= 0 \end{aligned}$$

which yields the critical values $x = -3$, $x = 0$, and $x = 3$.

Applying the first derivative test for each of the critical values, we choose a value from each side of the critical values and determine the values of $f'(x) = 5x^4 - 45x^2$ as follows:

$x^* = -3$	$x^* = 0$	$x^* = 3$
$f'(-4) = 560$	$f'(-1) = -40$	$f'(2) = -100$
$f'(-2) = -100$	$f'(1) = -40$	$f'(4) = 560$

Thus we see that $f(-3)$ is a local maximum, $f(3)$ is a local minimum, and $f(0)$ is not an extremum. Hence a horizontal inflection point occurs when $x^* = 0$ because one of the three possibilities must occur and two of them have been eliminated.

Applying the second derivative test, we have

$$D_x^2 f(x) = 20x^3 - 90x$$

which yields

$$f''(-3) = -270$$
$$f''(0) = 0$$
$$f''(3) = 270$$

Thus $f''(-3)$ is negative, which indicates that $f(-3)$ is a local maximum, and $f''(3)$ is positive, which indicates that $f(3)$ is a local minimum. However, $f''(0)$ is zero, which tells us that the test fails. Hence we should use the first derivative test to determine that a horizontal inflection point occurs at $x^* = 0$.

The choice of which of the three methods to use for determining the optimum values is dependent upon the complexity of the particular function. The second derivative test is generally the easiest to use because it involves determining only one function value per critical value; however, for those situations where this test is very difficult to evaluate, it may be better to use one of the other methods. Also, when the second derivative equals zero at x^*, the test fails and further evaluation is necessary using another test. The first derivative test is generally better than the method of evaluating the original function because it involves determining only two function values for each critical value whereas the latter method requires us to determine three function values per critical value. Nevertheless, if the function for the first derivative is difficult to evaluate, it may be best to use the method of evaluating the original function at each critical value and two adjacent values. In any event, we are required to determine the critical values.

So far, we have considered the application of differential calculus only as an aid in graphing functions, but there exist many situations in which we wish to determine the optimum value of a function regardless of whether or not we wish to graph it. However, for these situations, we need to remember that we determine *local* maxima and minima when using differential calculus. A local maximum or minimum may or may not be the absolute maximum or minimum over the entire domain of the function. The absolute maximum or minimum may occur at a critical point, an end point, or a point where the first derivative does not exist.

Example 12.5 Estatic Plastics, Inc., has determined that the cost function for producing a batch of a particular type of unit is given by $C(x) = x^2 - 14x + 100$, where $C(x)$ is the cost in dollars and x is hundreds of units per batch. Their equipment is such that they cannot produce more than 1,500

Sec. 12.2 / MAXIMA AND MINIMA

units in a single batch. Determine the batch size that will yield a minimum cost.

Solution First we determine the critical values by obtaining the first derivative and setting it equal to zero:

$$C(x) = x^2 - 14x + 100$$
$$D_x C(x) = 2x - 14$$
$$2x - 14 = 0$$
$$x = 7$$

Having determined the critical value, now we need to determine the second derivative and evaluate it at $x=7$; thus

$$D_x^2 C(x) = 2$$

and

$$C''(7) = 2 > 0$$

Therefore, the cost function has a *local* minimum at $x = 7$.

The practical values for the batch size range from 0 to 1,500; thus the local minimum cost occurs when the batch size is 700 units, which yields a cost per batch of

$$C(7) = (7)^2 - 14(7) + 100 = \$51$$

To determine the absolute minimum, we check the values at the two ends of the range, that is, 0 and 1,500, obtaining

$$C(0) = (0)^2 - 14(0) + 100 = \$100$$

and

$$C(15) = (15)^2 - 14(15) + 100 = \$115$$

Hence the minimum cost per batch is $51 and is achieved when 700 units per batch are made.

Example 12.6 The Rocking W Cattle Ranch wishes to pasture some cattle on a field of winter wheat. They have available 1,500 ft of wire for building a fence. Due to the high cost, they do not wish to buy any more wire; however, they do wish to fence a rectangular area as large as possible with the available wire. Determine the dimensions of the rectangular field that will produce the maximum possible area. Determine the maximum area of the field.

Solution The area of a rectangle is obtained by multiplying the length by the width. Furthermore, since there is available only 1,500 ft of wire, the perimeter of the rectangular field must be 1,500 ft. Thus if we let L be the length and W be the width, the perimeter is given by $2L+2W=1,500$, or the sum of one length and one width is $L+W=750$. Moreover, we note that the length is related to the width by $L=750-W$. Letting x represent the width, the length is equal to $750-x$. Computing the area of the field, we have $A(x)=x(750-x)$. Using differentiation to determine the value of x that will maximize $A(x)$, we determine the first derivative to be

$$D_x A(x) = 750 - 2x$$

Setting this derivative equal to zero and solving for x, we get $750-2x=0$, or $x=375$. Determining the second derivative, we obtain

$$D_x^2 A(x) = -2 < 0$$

which is less than zero; thus $A''(375)=-2$, indicating a maximum area at $x=375$. Therefore, the width of the rectangular field is $x=375$ ft and the length is $750-x$ or 375 ft. The maximum area of the field is given by $A(x)=x(750-x)=(375)(375)=140,625$ sq. ft.

Example 12.7 The Johnson Tool Company sells oil-well-drilling bits for $300 each, and the cost of making x bits per month is given by $C(x)=100+10x+5x^2$. Determine the number of bits that they should make per month and the associated profit.

Solution First we must determine the profit function, which can be obtained by subtracting the total cost from the total sales (or revenue). Thus the total sales in dollars is $S(x)=300x$, and the total cost is $C(x)=100+10x+5x^2$. The profit function is given by

$$P(x) = S(x) - C(x)$$
$$= 300x - 100 - 10x - 5x^2$$
$$= -100 + 290x - 5x^2$$

Determining the critical values of the profit function, we have

$$D_x P(x) = 290 - 10x$$
$$290 - 10x = 0$$
$$x = 29$$

Determining the second derivative, we have

$$D_x^2 P(x) = -10$$

Evaluating $P''(29)$, we have $P''(29)=-10$, which indicates that $P(x)$ has a local maximum at $x=29$. Since $P(x)$ is decreasing everywhere on both sides of $x=29$, the absolute maximum is also attained at $x=29$. The maximum profit is

$$P(29) = -100 + 290(29) - 5(29)^2 = 4,105$$

Thus the Johnson Tool Company can make a maximum profit of $4,105 per month if they make 29 bits per month.

EXERCISES

In Exercises 1 to 6, use differential calculus to determine the critical values. Evaluate the original function for the critical values and two adjacent values to determine the nature of the function at each critical value. Roughly sketch each function from the available information.

1. $f(x) = 2x^2 + 6x - 8$
2. $f(x) = x^2 - 16$
3. $f(x) = x^3 - 6x^2 - 15x + 4$
4. $f(x) = x^3 - 3x$
5. $f(x) = (x^2 - 4)^2$
6. $f(x) = (x^2 - 5x + 6)^3$

In Exercises 7 to 12, apply the first derivative test to determine the nature of the function at its critical values. Roughly sketch each function from the available information.

7. $f(x) = x^2 + 4x - 5$
8. $f(x) = x^2 - 16$
9. $f(x) = x^4 - 8x^2 + 16$
10. $f(x) = e^{x^2 + 2x + 1}$
11. $f(x) = \ln(x^2 + 4x + 5)$
12. $g(t) = t^3 - 3t^2 + 3t - 1$

In Exercises 13 to 22, apply the second derivative test to determine the nature of the function at its critical values. If the test fails, apply the first derivative test. Roughly sketch each function from the available information.

13. $f(x) = x^2 - 6x + 4$
14. $f(x) = 3 + 8x - x^2$
15. $f(x) = x^5$
16. $f(x) = 6x^8 - 32x^6$
17. $f(x) = x^6$
18. $f(x) = 4 - x^4$
19. $h(s) = \ln(s^2 + 2s + 6)$
20. $w(y) = e^{y^2 + 6y + 5}$
21. $g(w) = -3w^3 + w^2 + 7w + 15$
22. $y(z) = 9^{z^2 + 7z + 10}$

23. The profit in dollars for the manufacturing and sale of x units is given by
$$P(x) = 50x - 0.002x^2$$
Determine the number of units that will maximize profit and the associated profit.

24. The cost in dollars for the manufacture of x units is given by
$$C(x) = x^2 - 1{,}000x + 400{,}000$$
Determine the number of units that will minimize cost and the associated cost.

25. The total cost in dollars of producing x units is given by
$$C(x) = 0.04x^2 + 8x + 80$$
The selling price of a unit is $16. Determine

(a) The profit as a function of x
(b) The number of units that will maximize profit
(c) The maximum profit

26. The Dixie Transport Company has determined that the cost per thousand miles for operating a certain type of truck is given by $C(x) = 1,000,000/(100x - x^2)$, where x is thousands of miles driven. Company policy states that no truck will be sold prior to accumulating 20,000 mi and no truck will be retained beyond 90,000 mi. Determine the optimum number of miles that a truck should be driven before disposing of it and the associated cost per thousand miles.

27. The Ajax Metal Company manufactures steel vaults for banks. The cost of manufacturing a batch of x vaults is given by $C(x) = x^3 - 36x^2 - 156x + 20,000$. Determine the optimum number of vaults to manufacture in a batch and the cost of manufacturing the batch.

28. The Leisure-Time Manufacturing Company has determined that the function that represents manufacturing cost for dune buggies is given by $C(x) = x^2 - 250x + 20,000$, where x is the number of buggies produced. Their revenue function was determined to be $R(x) = 650x - x^2$. Determine their profit function, the optimum number of dune buggies to maximize profits, and the associated profit.

29. The Puppy Dog Food Company sells a case of dog food for $13. The cost of manufacturing and shipping an order of dog food is represented by the function $C(x) = x^2 - 3x + 5$, where x is the number of cases shipped. Determine their profit function, the optimum number of cases to ship at a time to maximize profits, and the associated profit.

30. The time required to make a batch of screws is given by $T(x) = x^4 - 2x^3 - 8x^2 - 24x + 500$, where x is the number of screws, in thousands, produced in a batch. Determine the optimum number of screws to manufacture in a batch to minimize manufacturing time.

31. The Silver Printing Company purchases its paper at the cost represented by the function $C(x) = 25 - 2x + 0.2x^2$, where x is the number of boxes purchased at one time. Determine the optimum number of boxes of paper to order at one time to minimize costs and the cost of purchasing the order.

32. A manufacturing company has determined that the percentage of good parts manufactured is represented by the function $G(x) = 0.75 + 0.02x - 0.001x^2$, where x is the number of days between routine maintenance of their machines. Determine the optimum number of days between routine maintenance and the percentage of good parts manufactured.

33. The Southwest Metal Company wishes to manufacture tin boxes from a 10-in. by 10-in. piece of tin by cutting squares from each corner and folding up the sides. Determine the size of the square that should be cut from each corner in order to produce a box with maximum volume. Determine the maximum volume.

34. Green Acres Farm grows wheat and has harvested 25,000 bu this season. When the wheat is harvested, the price per bushel is $3.50. The cost of storing the wheat is $15 + 0.0003x$ per day, where x represents the number of bushels stored. The price per bushel is expected to increase at the rate of $0.02t - 0.002t^2$, where t is time in weeks. Determine how long they should hold the wheat before selling in order to obtain maximum profit.

35. International Travel Agency is offering a plane tour of Europe and the Middle East. Air East has agreed to charter a DC-10 to the travel agency if they are guaranteed at least 60 passengers. The air fare is to be $1,000 per person if 60 people go and it will decrease by $2 per person for every person above the minimum who goes. A maximum of 300 spaces are available.

(a) Determine the number of tourists that will give Air East the maximum revenue.
(b) Will they be able to get maximum revenue?
(c) Determine the revenue if all available spaces are occupied.

36. The Southwest Laundry Company manufactures automatic coin-operated washers. The total cost of manufacturing x washers is $2x^2 - 270x + 200$ dollars, and the selling price for each unit is $525 - 3x$ dollars. Determine the number of washers they should produce to maximize profit.

37. The Jurgens Equipment Company determines that its profit is related to the number of products it sells, x, by $P(x) = 720 + 352 - 4x^2$. Determine the optimum number of products Jurgens should sell in order to maximize profit.

12.3 MARGINAL ANALYSIS

In the preceding section, we were concerned with determining the minimum and/or maximum of certain functions. Specifically, we considered cost, revenue, and profit functions; however, those functions were for total cost, total revenue, and total profit. Another interpretation that is of considerable interest in business decision making is that the derivative of an aggregate function, such as total cost, total revenue, or total profit, represents the marginal function, i.e., marginal cost, marginal revenue, or marginal profit. This appears reasonable because *marginal* is defined as the *change in the total figure that results from a small increase in the independent variable*. For example, marginal revenue is the additional revenue generated from the sale of an additional item. Thus the comparison of marginal versus aggregate is compatible with the comparison of derivative versus original function because the derivative of a function indicates the instantaneous rate of change of the function.

In many business situations, the information about marginal values is more important and useful than information about total values. For example, generally management is concerned with questions such as whether or not to produce a larger or smaller batch of units, to increase or decrease the number of machines for production, to increase or decrease the manpower for a certain task, or to increase or decrease advertising expenditures. If a company wished to maximize total profit, they would be willing to spend more on advertising only if the return would be at least sufficient to offset the expenditure; similarly, if they wished to minimize total cost, they would be willing to add another machine if the reduction in production cost would be sufficient to offset the cost of using the additional machine. The analysis of problems that consider the effect of a decision on the total function is termed *marginal analysis*.

For those situations where it is of interest to determine the marginal cost for a specific value of the independent variable and for which the total cost

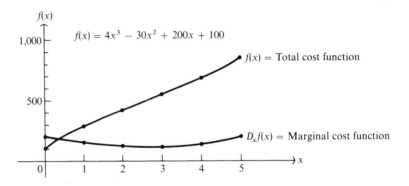

FIGURE 12.7

function is known, the analysis is performed by differentiating the total cost function to obtain the marginal cost function. For example, if the total cost function is given by $f(x) = 4x^3 - 30x^2 + 200x + 100$, the marginal cost function is given by $D_x f(x) = 12x^2 - 60x + 200$. (The graphic relationship for these two functions is illustrated in Fig. 12.7.) Furthermore, if we wish to determine the value of x that renders the minimum *marginal* cost, we simply differentiate the marginal cost function (i.e., the second derivative of the total cost function) to determine the critical values and then proceed in the usual manner. The same procedure applies for marginal revenue and marginal profit.

Example 12.8 Total cost in dollars for an assembly operation is given by $C(x) = x^3 - 21x^2 + 360x + 3{,}025$, where x represents the number of units made. Determine

(a) The marginal cost of the tenth unit
(b) The number of units for which the marginal cost is a minimum
(c) The minimum marginal cost
(d) The total cost and average cost for the number of units that minimizes the marginal cost
(e) The total cost and average cost for 10 units

Solution The marginal cost function $c(x)$ is the first derivative of the total cost function; that is,

$$c(x) = D_x C(x) = 3x^2 - 42x + 360$$

(a) The marginal cost of the tenth unit is

$$c(10) = 3(10)^2 - 42(10) + 360 = \$240$$

(b) The first derivative of the marginal cost function (i.e., the second derivative of the total cost function) is

$$D_x c(x) = D_x^2 C(x) = 6x - 42$$

Setting this equal to zero and solving, we obtain the critical value

$$6x - 42 = 0$$

or

$$x = 7$$

Applying the second derivative test, we have $D_x^2 c(x) = D_x^3 C(x) = 6$; thus $c''(7) = 6 > 0$, which means that $c(x)$ attains a minimum at $x = 7$.

(c) The minimum marginal cost is

$$c(7) = 3(7)^2 - 42(7) + 360 = \$213$$

(d) The total cost for $x = 7$ is

$$C(7) = (7)^3 - 21(7)^2 + 360(7) + 3{,}025 = \$4{,}859$$

The average cost for $x = 7$ is

$$\frac{\$4{,}859}{7} = \$694.14/\text{umit}$$

(e) The total cost for $x = 10$ is

$$C(10) = (10)^3 - 21(10)^2 + 360(10) + 3{,}025 = \$5{,}525$$

The average cost for $x = 10$ is

$$\frac{\$5{,}525}{10} = \$552.50/\text{unit}$$

From Example 12.8 we see that the lowest marginal cost does not necessarily give the lowest average cost. It can also be shown that the lowest total cost does not necessarily yield the lowest average cost.

Example 12.9 The total profit function in hundreds of dollars for an executive training course is given by $P(x) = -x^3 + 30x^2 + 600x - 1{,}000$, where x represents the dollars in thousands spent on advertising. Determine

(a) The marginal profit function and the marginal profit for spending \$9,000 and \$11,000 on advertising
(b) The advertising expenditure that yields a maximum marginal profit and the associated maximum marginal profit
(c) The average profit per thousand dollars spent on advertising for the advertising expenditure obtained in part (b)

Solution

(a) The marginal profit function is $p(x) = D_x P(x) = -3x^2 + 60x + 600$.

$$p(9) = -3(9)^2 + 60(9) + 600 = 897 \quad \text{or} \quad \$89{,}700$$
$$p(11) = -3(11)^2 + 60(11) + 600 = 897 \quad \text{or} \quad \$89{,}700$$

(b) The first derivative of the marginal profit function is $D_x p(x) = -6x + 60$, which, when set equal to zero, yields $x = 10$. Applying the second derivative test, we have

$$D_x^2 p(x) = -6$$

Thus $p''(10) = -6 < 0$, and therefore the marginal profit function is a maximum at $x = 10$. The associated marginal profit is

$$p(10) = -3(10)^2 + 60(10) + 600 = 900$$

or

$$p(10) = \$90,000$$

(c) The average profit per thousand dollars spent is

$$\frac{P(10)}{10} = \frac{1}{10}\left[-(10)^3 + 30(10)^2 + 600(10) - 1,000 \right] = \frac{7,000}{10} = 700$$

or $70,000/$1,000 spent on advertising.

In Example 12.7, we considered the total profit function as a difference between the sales (or revenue) function $S(x)$ and the cost function $C(x)$; that is,

$$P(x) = S(x) - C(x)$$

To determine the x value that yields maximum profit, we applied an appropriate test to $P(x)$. As an alternative method for solving for the maximum profit, we note that

$$P'(x) = S'(x) - C'(x)$$

In terms of marginal analysis, this simply means that marginal profit equals marginal sales (or revenue) minus marginal cost. Furthermore, to obtain the critical values, we have

$$S'(x) - C'(x) = 0$$

which is satisfied when

$$S'(x) = C'(x)$$

Again, in terms of marginal analysis, this says that maximum profit occurs only when marginal sales (or revenue) equal marginal cost. When this occurs, we say that we have satisfied the *optimality condition of marginal analysis*.

Sec. 12.3 / MARGINAL ANALYSIS

Furthermore, applying the second derivative test, the profit will be a maximum if the second derivative of $P(x)$ is negative at the critical values. Thus

$$P''(x) = S''(x) - C''(x) < 0$$

or

$$S''(x) < C''(x)$$

which means that the marginal sales (or revenue) function is increasing at a slower rate than the marginal cost function. In summary, the profit will attain its local maxima at those points for which the marginal sales equals marginal cost and for which the rate of increase of sales is less than the rate of increase of costs.

Example 12.10 Use the information given in Example 12.7 and the marginal analysis approach to determine the number of drilling bits that will maximize profit.

Solution Essentially we are given the sales and cost functions

$$S(x) = 300x$$

and

$$C(x) = 100 + 10x + 5x^2$$

respectively. Thus the marginal sales function is

$$S'(x) = 300$$

and the marginal cost function is

$$C'(x) = 10 + 10x$$

The maximum will occur when

$$S'(x) = C'(x)$$

or

$$300 = 10 + 10x$$

Solving, $x = 29$ is the number of bits that gives marginal sales equal to marginal cost.

To ensure that we have a maximum instead of an inflection point or a minimum, we compare the rates of change of the two marginal functions; thus $S''(x) = 0$ and $C''(x) = 10$. Therefore, for any value of x, the rate of change of sales is less than the corresponding rate of change of costs.

Specifically, this is satisfied at $x=29$, and hence a local maximum profit is attained when the Johnson Tool Company produces 29 bits per month.

EXERCISES

1. The total cost of making x units of a colored television set is given by

$$C(x) = 0.0002x^3 - 0.3x^2 + 50x + 100{,}000$$

Determine

   (a) The marginal cost of producing the one thousandth unit
   (b) The total cost of producing 1,000 units

2. If the cost in dollars of making x electric frying pans is given by

$$C(x) = 0.0009x^3 - 0.54x^2 + 108x + 2{,}000$$

determine the marginal cost function, and plot the marginal cost function for the following x values: 100, 150, 200, 250, and 300. Use differential calculus to determine the number of units that yield a minimum marginal cost.

3. The total cost for manufacturing x units of a certain type of radio equipment for military aircraft is given by the function

$$C(x) = 50{,}000 + 2{,}000x - 15x^2 + x^3$$

Determine

   (a) The marginal cost function
   (b) The marginal cost of the tenth unit
   (c) The total cost of manufacturing 10 units
   (d) The minimum marginal cost
   (e) The total cost for producing the number of units that minimizes marginal cost

4. The total profit function in hundreds of dollars for a building contractor is given by the function

$$P(x) = 400x - 40x^2 + 4{,}444e^{-x}$$

where x represents the number of four-unit apartment houses constructed. Determine

   (a) The marginal profit function
   (b) The marginal profit of the fifth apartment house
   (c) The total profit for five apartment houses

5. The Excellent Encyclopedia Company is considering increasing their number of salespeople from the current level of 225. They have established that their profit function in hundreds of dollars is $P(x) = 60x^{3/2} - 200x$, where x represents the number of salespeople.

Sec. 12.3 / MARGINAL ANALYSIS

(a) Determine the total profit for the current staff of salespeople.
(b) Determine the marginal profit for the current staff of salespeople.

6. The manufacturer of Excello Widgets has determined that the cost of manufacturing x widgets is given by

$$C(x) = 2x^2 - 10x + 100$$

and the sales function is given by

$$S(x) = x^2 + 40x - 200$$

(a) Determine the profit function and use calculus to determine the optimal number of units and the associated profit.
(b) Verify the results of part (a) by applying the optimality condition of marginal analysis.

7. Dewey Bayfield is a candidate for the United States Senate. His campaign manager has told him that the total number of votes received (in thousands) is related to campaign expenditures x (in hundreds of dollars) by

$$V(x) = 100 + 4x + 2\sqrt{x}$$

Determine the marginal number of votes gained when campaign expenditure is

(a) $x = 100$ (i.e., $10,000)
(b) $x = 900$ (i.e., $90,000)

8. The Freehoff Trailer Company has determined that the total cost (in hundreds of dollars) of making x trailers is given by

$$C(x) = 12x - 0.06x^2 + 0.0001x^3$$

Determine

(a) The marginal cost function
(b) The marginal cost of the three-hundredth trailer
(c) The total cost of manufacturing 200 trailers
(d) The minimum marginal cost
(e) The total cost of making the number of trailers that minimizes marginal cost

9. Weekly profit from using x machines is given by $P(x) = 300x - (21x^2 - x^3)$. The company is currently using 10 machines and has a chance to sell any of the machines they do not need. If their objective is to minimize marginal costs, determine whether or not they should sell any machines. If they should sell some machines, how many should they sell? Determine the weekly profits when using the number of machines that will minimize marginal costs.

10. A salesman's monthly expenses are given by the function

$$C(x) = -(1/3)x^3 + 20x^2 - 15x - 1{,}750$$

where x is the number of miles traveled in hundreds of miles. He travels between 1,200 and 2,500 miles per month. Determine

(a) The marginal cost function
(b) The marginal cost of the twelve-hundredth mile
(c) The total cost of traveling 1200 mi
(d) The number of miles to travel to minimize marginal costs
(e) The minimum marginal cost
(f) The total cost of traveling the number of miles that minimizes marginal cost

11. The cost of running a train from Los Angeles to New York is given by

$$C(x) = x^3 - 45x^2 + 750x + 6,500$$

where x is the number of box cars. Determine

(a) The marginal cost function
(b) The number of box cars per train to minimize marginal cost
(c) The minimum marginal cost
(d) The total cost for the train having the number of box cars that minimizes marginal cost

12. The XYZ Flour Mill can make a batch of flour of x lb, where x is in 100's, at a cost given by

$$C(x) = 3x^2 - 6x - 2$$

and the revenue function is given by

$$R(x) = x^2 + 10x + 20$$

(a) Determine the profit function and use calculus to determine the optimal number of units and the associated profit.
(b) Verify the results of part (a) by applying the optimality condition of marginal analysis.

12.4 EXTREMA FOR BIVARIATE FUNCTIONS

In Sec. 12.2 we considered the application of differential calculus to determine the local maxima and/or local minima for a one-variable function. We shall now consider the corresponding problem for a two-variable (bivariate) function.

If the function $f(x,y)$ is a continuous function with continuous partial derivatives $\partial f/\partial x$ and $\partial f/\partial y$, the function can be represented by a smooth unbroken surface whose equation is $u = f(x,y)$. The function $f(x,y)$ has a *local maximum* at a point (a,b) if the value of $f(a,b)$ is greater than the values of $f(x,y)$ at all points in the neighborhood of (a,b). The *local minimum* can be defined in the same manner by replacing "greater" with "less." Both maxima and minima are included under the term *extreme values*. The geometric illustration of a local maximum is given in Fig. 12.8, with the maximum occurring at the point (a,b,c), where $c = f(a,b)$.

Sec. 12.4 / EXTREMA FOR BIVARIATE FUNCTIONS

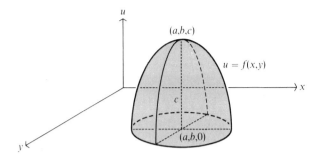

FIGURE 12.8

If $f(x,y)$ and its first derivatives are continuous in a region that includes the point (a,b) a *necessary condition* for $f(a,b)$ to be an extreme (maximum or minimum) value of the function $f(x,y)$ is that

$$\frac{\partial f}{\partial x} = 0 \quad \text{and} \quad \frac{\partial f}{\partial y} = 0$$

where the partial derivatives are evaluated at (a,b). If we denote $\partial f/\partial x$ by $f_x(x,y)$ and $\partial f/\partial y$ by $f_y(x,y)$, the necessary condition is that

$$f_x(a,b) = 0 \quad \text{and} \quad f_y(a,b) = 0$$

It must be understood that this necessary condition does not ensure that we have an extreme value at (a,b), which is similar to the second derivative test of a one-variable function for which the first derivative must equal zero at a critical value before it is possible to have a maximum or minimum. We need a test for the bivariate case to enable us to distinguish between a maximum and a minimum. Such a test is now given, and it provides *sufficient conditions* for the existence of an extreme value.

Extending the above notation for partial derivatives to the second-order partial derivatives, we denote $\partial^2 f/\partial x^2$ by $f_{xx}(x,y)$ and $\partial^2 f/\partial y^2$ by $f_{yy}(x,y)$; for the cross partial derivatives we denote $\partial^2 f/\partial x \partial y$ by $f_{xy}(x,y)$ and $\partial^2 f/\partial y \partial x$ by $f_{yx}(x,y)$. However, for the types of problems that are of concern to us, we shall have $\partial^2 f/\partial x \partial y = \partial^2 f/\partial y \partial x$ because they will satisfy the required condition given in Sec. 11.5.

If at the point (a,b)

$$f_x(a,b) = 0 \quad \text{and} \quad f_y(a,b) = 0 \qquad (12.1)$$

and if

$$D(a,b) = f_{xx}(a,b) f_{yy}(a,b) - \left[f_{xy}(a,b) \right]^2 > 0$$

then $f(x,y)$ has a *maximum* value at (a,b) if

$$f_{xx}(a,b)<0$$

or a *minimum* value if

$$f_{xx}(a,b)>0$$

If condition (12.1) holds and

$$D(a,b)<0$$

then $f(a,b)$ is neither a maximum nor a minimum. If

$$D(a,b)=0$$

the test fails to give any information.

The previous test can be stated in the following form:

Extreme Value Test for Bivariate Functions Given a point (a,b) that satisfies (12.1):

1. If $D(a,b)>0$ and $f_{xx}(a,b)<0$, then $f(a,b)$ is a maximum.
2. If $D(a,b)>0$ and $f_{xx}(a,b)>0$, then $f(a,b)$ is a minimum.
3. If $D(a,b)<0$, then $f(a,b)$ is not an extreme value.
4. If $D(a,b)=0$, the test fails.

When a point satisfies (12.1) and extreme value test condition 3, it is called a *saddle point*. This name evolves from the analogy of the function at this point and the characteristics of a saddle; that is, from the point in question, the function increases in both directions along one axis and decreases in both directions along the other axis. (A graphic illustration of such a point is given in Fig. 12.9.)

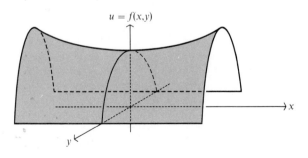

FIGURE 12.9

Sec. 12.4 / EXTREMA FOR BIVARIATE FUNCTIONS

Example 12.11 Examine the function $f(x,y) = x^2 + y^2 + xy$ for extreme values.

Solution To determine the points for which condition (12.1) is satisfied, we determine the first derivatives and set them equal to zero:

$$\frac{\partial f}{\partial x} = 2x + y$$

$$\frac{\partial f}{\partial y} = 2y + x$$

Rearranging the two equations and setting them equal to zero, we have

$$2x + y = 0$$
$$x + 2y = 0$$

Simultaneously solving this system of equations, we determine all points that are candidates for extreme values. The system has only one point that satisfies it, namely, $(0,0)$. Thus $(0,0)$ is the only point that satisfies condition (12.1).

Determining the second-order partial derivatives, we obtain $\partial^2 f/\partial x^2 = 2$, $\partial^2 f/\partial y^2 = 2$, and $\partial^2 f/\partial x \partial y = 1$; therefore,

$$D(0,0) = f_{xx}(0,0) f_{yy}(0,0) - [f_{xy}(0,0)]^2$$
$$= 2 \cdot 2 - (1)^2$$
$$= 3$$

Furthermore, $f_{xx}(0,0) = 2 > 0$, which means that extreme value test condition 2 is satisfied; hence $f(x,y) = x^2 + y^2 + xy$ is at its minimum value at the point $(0,0)$ and the minimum value is $f(0,0) = 0$.

Example 12.12 Examine the function $f(xy) = x^3 - 3x - y^2$ for extreme values.

Solution First we determine the points that satisfy condition (12.1); thus we obtain the first-order partial derivatives

$$\frac{\partial f}{\partial x} = 3x^2 - 3$$

$$\frac{\partial f}{\partial y} = -2y$$

Setting the two partial derivatives equal to zero and solving simultaneously, we have

$$3x^2 - 3 = 0$$
$$-2y = 0$$

The first equation yields $x = -1$ and $x = 1$, while the second equation yields $y = 0$; thus the two points $(-1, 0)$ and $(1, 0)$ are candidates for rendering any extreme values for $f(x, y)$.

Next we determine the status of $(-1, 0)$. The second-order partial derivatives are $\partial^2 f/\partial x^2 = 6x$, $\partial^2 f/\partial y^2 = -2$, and $\partial^2 f/\partial x \partial y = 0$; thus

$$D(-1, 0) = [6(-1)](-2) - (0)^2 = 12 > 0 \quad \text{and} \quad f_{xx}(-1, 0) = -6 < 0$$

Therefore, extreme value test condition 1 is satisfied, and hence $f(-1, 0)$ is a maximum. Checking the status $(1, 0)$, we determine

$$D(1, 0) = [6(1)](-2) - (0)^2 = -12 < 0$$

Thus extreme value test condition 3 is satisfied, and $f(1, 0)$ is *not* an extreme value. Hence the point $(1, 0)$ is a saddle point of $f(x, y)$.

EXERCISES

In Exercises 1 to 12, examine the given function for extreme values.

1. $f(x, y) = 4x^2 + y^2 - 4y$
2. $f(x, y) = 4x^2 + y^2 + 8x - 2y - 1$
3. $f(x, y) = xy$
4. $f(x, y) = x^2 + y^3 - 3y$
5. $f(x, y) = x^2 + xy + y^2 - 6x - 2y + 5$
6. $f(x, y) = x^3 + x^2 + y^2 + xy + 10$
7. $f(x, y) = x^3 + y^3 - 3xy$
8. $f(x, y) = x^2 y - 4x^2 + y^3$
9. $f(x, y) = y^2 - x^2$
10. $f(x, y) = y^3 - 3y - x^2$
11. $f(x, y) = 3x^3 - 5x^2 + x - 5y^3 + 6y^2 + 3y - 10$
12. $f(x, y) = \ln x^6 + x^2 + y^2 + 4xy + 10$

13. The labor cost of a certain construction project depends upon the number of skilled workers x and the number of unskilled workers y. The labor cost function has been determined to be

$$C(x, y) = 90{,}000 + 60x^2 - 240xy + 10y^3$$

(a) Examine $C(x, y)$ for extreme values.
(b) Determine the numbers of skilled and unskilled workers that yield the minimum labor cost.
(c) Determine the minimum labor cost.

Sec. 12.4 / EXTREMA FOR BIVARIATE FUNCTIONS

14. The Resthaven Funeral Home determines that the profit per month in hundreds of dollars is approximated by

$$P(x,y) = 2{,}800 + 8x + 10y - x^2 + 6xy - 10y^2$$

where x is salary expense and y is advertising expense, both in thousands of dollars.

(a) Examine $P(x,y)$ for extreme values.
(b) Determine the salary and advertising expenses that yield maximum profit.
(c) Determine the maximum profit.

15. The Peyton Manufacturing Company can approximate their monthly sales S in thousands of dollars by the function

$$S(x,y) = 100x + 10y + 6xy$$

and they can approximate their monthly cost C in thousands of dollars by the function

$$C(x,y) = 16x + 2y + 9x^2 + 2xy + y^2$$

where x is inventory value in thousands of dollars and y is floor space in thousands of square feet.

(a) Determine the profit function $P(x,y)$.
(b) Examine the profit function for extreme values.
(b) Determine the inventory value and floor space that yield maximum profit.
(d) Determine the maximum profit.

16. The Green Acres Cattle Ranch fattens steers to sell to meat-processing plants. The total profit per head in dollars can be approximated by

$$P(x,y) = 2.4x + 3y + 0.5xy - 0.5x^2 - 0.3y^2$$

where x represents the amount of especially mixed feed in hundreds of pounds and y represents the number of bales of hay required to fatten one steer.

(a) Examine the profit function for extreme values.
(b) Determine the amounts of feed and hay that yield maximum profit.
(c) Determine the maximum profit.

17. The cost of making 1 unit of product X is given by the function

$$C(x,y) = 3 + 4x^3 - 10xy + 6y^2$$

where x is the dollar cost of materials and y is the dollar cost of labor.

(a) Examine $C(x,y)$ for extreme values.
(b) Determine the amount of materials and labor to use in each unit of product X to minimize costs.
(c) Determine the minimum cost for a unit of product X.

18. The daily cost of Gizmo Manufacturing Company is given by

$$C(x,y) = x^3 + 13x^2 - 20x + y^3 + 15y^2 - 25y + 85$$

where x is the number of widgets produced in thousands and y is the number of gadgets produced in thousands.

(a) Determine the number of widgets and gadgets to produce in order to minimize costs by examining $C(x,y)$ for extreme values.
(b) Determine the minimum cost.

19. The profit for making and selling 5,000 boxes of screws is given by

$$P(x,y) = 0.5x^2 + 0.9xy - y^2 + 10x + 15y$$

where x is the time, in hours, required to make the screws and y is the number of boxes of screws made per batch.

(a) Examine $P(x,y)$ for extreme values.
(b) Determine the number of boxes per batch and the time required to make the screws that will maximize the total profit.
(c) Determine the maximum profit.
(d) Determine the profit per box of screws.

20. The XYZ Corporation has weekly revenue R in hundreds of dollars that can be represented by the function

$$R(x,y) = 50x + 60y + 10xy$$

and weekly cost C in hundreds of dollars that can be represented by the function

$$C(x,y) = 4x^2 + 5y^2 + 5x + 8y + 2xy$$

where x is the number of vending machines, in hundreds, filled per week and y is the number of people employed by XYZ Corporation.

(a) Determine the profit function $P(x,y)$.
(b) Examine the profit function for extreme values.
(c) Determine the number of vending machines and employees that yield maximum profit.
(d) Determine the maximum profit.

12.5 CONSTRAINED OPTIMA

In Chaps. 5 and 6, we considered the problem of determining the optimum value of an objective function that was subject to certain constraints. In Sec. 12.2, we considered ways of determining optimum values of objective functions that were not constrained. However, there is one very major difference between the types of functions considered: In Chaps. 5 and 6 the functions that were constrained were linear functions and, furthermore, the constraints were also linear, while in this chapter the functions considered were neither constrained nor limited to linear functions. Actually, nearly all functions to which we applied differential calculus to determine the optimum were nonlinear functions.

There are many situations in which the problem of minimizing production cost is constrained by certain demand requirements that must be satis-

Sec. 12.5 / CONSTRAINED OPTIMA

fied; similarly, the problem of maximizing profit is quite often constrained by limitations on working capital or manpower; likewise, the maximization of sales is generally constrained by production capacity and advertising budget. Such problems of determining the optimum values of nonlinear objective functions that are subject to nonlinear constraints can be solved by differential calculus. Actually the objective function and/or the constraints can be linear and calculus can still be utilized; however, if both are linear, then linear programming is usually the easier method to employ.

The problem of determining extreme values of a function $f(x,y)$ that is subject to a constraint can be stated as that of finding the extreme values of $f(x,y)$ subject to the constraint $g(x,y)=0$. One method of solving for the extreme values of a constrained optima problem is to solve the constraining equation for one variable in terms of the remaining variable and substitute the result into $f(x,y)$. This reduces the objective function $f(x,y)$ to a function of one variable, which can be solved by the methods given in Sec. 12.2; this procedure is called the *method of substitution*.

Example 12.13 Determine the extreme values of $f(x,y)=x^2+y^2+xy$ subject to $x+y=10$.

Solution Solving $x+y=10$ for y in terms of x and substituting into $f(x,y)=x^2+y^2+xy$, we obtain $y=10-x$ and

$$\begin{aligned} f(x,y) &= f(x,10-x) \\ &= x^2+(10-x)^2+x(10-x) \\ &= x^2+100-20x+x^2+10x-x^2 \\ &= x^2-10x+100 \end{aligned}$$

Thus, with the constraint substituted into $f(x,y)$, the function is dependent only on x, and hence we need simply to determine the optimum of $h(x) = x^2 - 10x + 100$.

Applying the second derivative test from Sec. 12.2, we determine the first derivative

$$D_x h(x) = 2x - 10$$

Setting it equal to zero and solving for x, we obtain

$$2x - 10 = 0$$

or

$$x = 5$$

The second derivative of $h(x)$ is $D_x^2 h(x) = 2$; thus $h''(5) = 2 > 0$, which means that $h(x)$ is a minimum at $x=5$.

Returning to the original two-variable problem, we determine $y = 10 - x = 10 - 5 = 5$. Hence, when the function $f(x,y) = x^2 + y^2 + xy$ is constrained by

$x+y=10$, the minimum occurs at the point $(5,5)$ and is computed in the usual manner to be

$$f(5,5) = (5)^2 + (5)^2 + (5)(5) = 75$$

In Example 12.11 the minimum of the function $f(x,y) = x^2 + y^2 + xy$ was determined to occur at the point $(0,0)$ and was found to be $f(0,0) = 0$. Thus we see that when the function is constrained by $x + y = 10$, the minimum value is considerably increased, which is the penalty paid to satisfy the constraint.

Unfortunately the method of substitution is not widely applicable because in nonlinear equations generally it is difficult to solve for one variable in terms of the remaining variable. For example, the method of substitution quickly encounters difficulty in a constraint such as $3x^2 - 4xy + y^3 + 4x = 0$. Fortunately an alternative method is available: It is called the *method of lagrangian multipliers*, named after the famous mathematician Comte Joseph Louis Lagrange, who discovered it during the eighteenth century.

The method of lagrangian multipliers is simply a technique for expressing the original function $f(x,y)$ and the constraint $g(x,y) = 0$ together as a single function. Since $g(x,y) = 0$, we can let λ (lambda) be a third variable, called the *lagrangian multiplier*, and form the new three-variable function

$$F(x,y,\lambda) = f(x,y) + \lambda g(x,y)$$

We see that the value of $F(x,y,\lambda)$ is the same as the value of $f(x,y)$ for any set of values of x,y and λ because $g(x,y) = 0$; thus we have not altered the function value. Furthermore, taking the first-order partial derivatives with respect to each of the three variables, we obtain

$$\frac{\partial F}{\partial x} = \frac{\partial f}{\partial x} + \frac{\lambda \partial g}{\partial x}, \quad \frac{\partial F}{\partial y} = \frac{\partial f}{\partial y} + \frac{\lambda \partial g}{\partial y}, \quad \text{and} \quad \frac{\partial F}{\partial \lambda} = g(x,y)$$

If we extend to our three-variable function the *necessary* conditions for a function to attain an extreme value, we have

$$\frac{\partial F}{\partial x} = 0, \quad \frac{\partial F}{\partial y} = 0, \quad \text{and} \quad \frac{\partial F}{\partial \lambda} = 0$$

However, expressing these conditions in terms of the original objective function and the constraint, we have

$$\frac{\partial f}{\partial x} + \lambda \frac{\partial g}{\partial x} = 0, \quad \frac{\partial f}{\partial y} + \lambda \frac{\partial g}{\partial y} = 0, \quad \text{and} \quad g(x,y) = 0$$

Thus, in order for the function $f(x,y,\lambda)$ to have an extreme value at the point (a,b,α), it is necessary that the following conditions be satisfied:

$$f_x(a,b) + \alpha g_x(a,b) = 0, \quad f_y(a,b) + \alpha g_y(a,b) = 0, \quad \text{and} \quad g(a,b) = 0$$

(12.2)

In order to determine the critical values of x, y, and λ, we would solve simultaneously the three equations of (12.2) to determine the values a, b, and α.

Necessary conditions for $F(x,y,\lambda)$ to assume an extreme value are given in (12.2), and these conditions are very similar to those for bivariate functions. However, when $F(x,y,\lambda)$ assumes an extreme value, $f(x,y)$ will assume an extreme value subject to $g(x,y)=0$. Furthermore, the extreme *value* of $F(x,y,\lambda)$ is not *directly* dependent upon the value of λ because $g(x,y)=0$, which leads us to the consideration of sufficient conditions for $F(x,y,\lambda)$ to assume an extreme value.

We note that $\partial F/\partial \lambda$ is the constraint itself, and so it is unnecessary to differentiate F with respect to λ. However, it is necessary that we use this constraint in solving the system of equations given in (12.2), which means that the critical values are influenced by the value of λ because it is one of the variables in the system of equations that represents the necessary conditions. Furthermore, the sufficient conditions that are given below are directly influenced by the value of λ.

Extreme Value Test for Constrained Bivariate Functions Given $F(x,y,\lambda)=f(x,y)+\lambda g(x,y)$, where $g(x,y)=0$, and $D(x,y,\lambda) = F_{xx}(x,y,\lambda)F_{yy}(x,y,\lambda)-[F_{xy}(x,y,\lambda)]^2$; then the critical point (a,b,α), which satisfies the equations of (12.2), is such that

1. If $D(a,b,\alpha)>0$ and $F_{xx}(a,b,\lambda)<0$, then $f(a,b)$ is a maximum.
2. If $D(a,b,\alpha)>0$ and $F_{xx}(a,b,\lambda)>0$, then $f(a,b)$ is a minimum.
3. If $D(a,b,\alpha)\leq 0$, the test fails.

We note that condition 3 of the previous test is essentially a combination of conditions 3 and 4 of the unconstrained extreme value test. Thus, if condition 3 is encountered, we must use a different approach similar to the first derivative test for a one-variable function; that is, for a critical point (a,b,λ), we simply determine the values of $f(a,b)$, $f(a+\Delta x, b-\Delta y)$, and $f(a-\Delta x, b+\Delta y)$, where both points $(a+\Delta x, b-\Delta y)$ and $(a-\Delta x, b+\Delta y)$ satisfy $g(x,y)=0$. If

$$f(a,b)>f(a+\Delta x, b-\Delta y)$$

and

$$f(a,b)>f(a-\Delta x, b+\Delta y)$$

$f(a,b)$ is a constrained maximum. If

$$f(a,b)<f(a+\Delta x, b-\Delta y)$$

and

$$f(a,b) < f(a-\Delta x, b+\Delta y)$$

$f(a,b)$ is a constrained minimum.

Example 12.14 Determine the extreme values $f(x,y) = x^2 + y^2 + xy$ subject to $x+y=10$.

Solution From the constraint, we determine $g(x,y) = x+y-10 = 0$. Thus, using a lagrangian multiplier, we have

$$F(x,y,\lambda) = x^2 + y^2 + xy + \lambda(x+y-10)$$

Determining the first-order partial derivatives of $f(x,y,\lambda)$ and setting them equal to zero, we obtain

$$\frac{\partial F}{\partial x} = 2x + y + \lambda = 0$$

$$\frac{\partial F}{\partial y} = 2y + x + \lambda = 0$$

$$\frac{\partial F}{\partial \lambda} = x + y - 10 = 0$$

Subtracting the second equation from the first, we eliminate λ and combine the resulting equation with the third equation to have two equations in two unknowns; that is,

$$x - y = 0$$
$$x + y = 10$$

Solving this reduced system for x and y, we have $x=5$ and $y=5$; substituting these values into the equation for $\partial F/\partial x = 0$ (or $\partial F/\partial y = 0$), we obtain $\lambda = -15$. Thus the only critical point (x,y,λ) is $(5,5,-15)$.

Having determined a critical point, next we examine the values $D(x,y,\lambda)$. Obtaining all second-order partial derivatives of $F(x,y,\lambda)$ with respect to x and y, we have

$$\frac{\partial^2 F}{\partial x^2} = 2, \quad \frac{\partial^2 F}{\partial y^2} = 2, \quad \text{and} \quad \frac{\partial^2 F}{\partial x \partial y} = 1$$

Thus regardless of the specific values of the critical point, we have

$$D(x,y,\lambda) = (2)(2) - (1)^2 = 3 > 0$$

which means that $D(5,5,-15) = 3 > 0$. Furthermore, $\partial^2 F/\partial x^2 = 2 > 0$, which means that $F_{xx}(5,5-15) = 2 > 0$ and condition 2 of the extreme value test is satisfied. Hence $f(x,y) = x^2 + y^2 + xy$ assumes a minimum value of

$$f(5,5) = (5)^2 + (5)^2 + (5)(5) = 75$$

Sec. 12.5 / CONSTRAINED OPTIMA

We should note that Example 12.14 presents the same problem as Example 12.13; however, the method of solution utilizes lagrangian multipliers.

Example 12.15 Determine the extreme values of $f(x,y) = x+y$ subject to $x^2 + y^2 = 2$.

Solution Using a lagrangian multiplier, we have

$$F(x,y,\lambda) = x + y + \lambda(x^2 + y^2 - 2)$$

Determining the critical values, we have

$$\frac{\partial F}{\partial x} = 1 + 2\lambda x = 0$$

$$\frac{\partial F}{\partial y} = 1 + 2\lambda y = 0$$

$$\frac{\partial F}{\partial \lambda} = x^2 + y^2 - 2 = 0$$

Solving the system of equations, we obtain from the first two equations $\lambda = -1/2x$ and $\lambda = -1/2y$, which yields $x = y$. Substituting into the third equation, we have $x^2 + x^2 - 2 = 0$, or $x^2 = 1$, which yields $x = -1$ and $x = 1$ as solutions. Thus the critical points are $(-1, -1, \frac{1}{2})$ and $(1, 1, -\frac{1}{2})$.

Determining the values of $D(x,y,\lambda)$ for the two critical points, we determine the required second-order partial derivatives and substitute the values for each point:

$$\frac{\partial^2 F}{\partial x^2} = 2\lambda \qquad \frac{\partial^2 F}{\partial y^2} = 2\lambda \qquad \frac{\partial^2 F}{\partial x \partial y} = 0$$

Thus $D(x,y,\lambda) = (2\lambda)(2\lambda) - (0)^2 = 4\lambda^2$, which is positive for any nonzero real value of λ. For the critical point $(-1, -1, \frac{1}{2})$, $D(-1, -1, \frac{1}{2}) = 1$, and $F_{xx}(-1, -1, \frac{1}{2}) = 1 > 0$; therefore, $f(-1, -1) = -2$ is a minimum. Similarly, $D(1, 1, -\frac{1}{2}) = 1$, and $F_{xx}(1, 1, -\frac{1}{2}) = -1 < 0$; hence $f(1, 1) = 2$ is a maximum.

In the event a two-constraint problem is encountered, we may employ a process that uses two lagrangian multipliers, the mechanics of which are quite similar to those for the one-constraint problem.

EXERCISES

Determine the extreme values of each of the functions in Exercises 1 to 6 subject to the given constraints:

1. $f(x,y) = 4x^2 + 3y^2 - xy$ (if $x + 2y = 21$)
2. $f(x,y) = x^2 + y^2 - xy$ (if $x - y = 8$)

3. $f(x,y) = 10x^2 + y^2$ (if $x - y = 22$)

4. $f(x,y) = 3x + 2y$ (if $x^2 + y^2 = 125$)

5. $f(x,y) = 2x^2 + y^2 - xy$ (if $x + y = 8$)

6. $f(x,y) = 6x^2 + 5y^2 - xy$ (if $2x + y = 24$)

7. The Hayden Manufacturing Company has determined that the number of units of production, as a function of the number of tons of raw material x and the number of man-hours y in hundreds of hours, is given by

$$P(x,y) = -4x^2 + 5xy + y^2$$

Determine the amounts of raw material and labor that maximize production if $3x + 2y = 74$. Also, determine the maximum number of units of production.

8. The Fritter-Lee Potato Chip Company uses an automatic packaging machine that has two critical parts. The number of breakdowns of the machine, as a function of the number of replacements x and y of the parts A and B, respectively, is given by

$$f(x,y) = x^2 + 3y^2 + 2xy - 11x + 15$$

Determine the number of replacements that should be made for each part in order to minimize the number of breakdowns if for every replacement of part B we must make two replacements of part A. Also determine the minimum number of breakdowns.

9. The cost of repairs for the machine in Exercise 8 is a function of the numbers x and y of inspections per week of the two parts A and B, respectively. The cost function is given by

$$f(x,y) = 2x^2 + 5xy + 4y^2 - 40y + 100$$

If the total number of inspections is 20, determine the number of inspections per week that should be made for each part in order to minimize repair costs.

10. For the Resthaven Funeral Home of Exercise 14, Sec. 12.5, determine the salary and advertising expenses that yield maximum profit if the total amount spent on salary and advertising must equal $55,000. Also determine the maximum profit.

11. For the XYZ Corporation of Exercise 20, Sec. 12.5, determine the number of vending machines and employees that will yield maximum profit if an employee can only service 100 vending machines per week. Also determine the maximum profit.

12. The number of man-hours required per day in department A is given by

$$T(x,y) = 5x^2 + 4y^2 - 2x - y + 2xy$$

where x is the number of tons of raw material A used and y is the number of tons of raw material B used. Determine the amount of raw materials A and B that minimizes time if $x + 3y = 50$. Also determine the minimum amount of time required per day.

13. The weekly demand for widgets in units, as a function of the material input x in dollars and the selling price y, is given by

$$f(x,y) = -y^2 + 25y - 2x^2 + 5x + xy$$

Determine the amount of material input and the selling price that maximizes demand if the price is three times the amount of material. Also determine the maximum demand of widgets.

IMPORTANT TERMS AND CONCEPTS

Constrained optima
Critical values
Extreme value tests
  bivariate functions
  constrained bivariate functions
Extreme values
First derivative test
Inflection point
Lagrangian multiplier
Local maxima
Local minima
Marginal analysis

Methods of substitution
Necessary conditions for extreme values
Optimality condition of marginal analysis
Optimum
Roots
Saddle point
Second derivative test
Sufficient conditions for extreme values

REVIEW PROBLEMS

1. The Slick Company manufactures ice skates. The cost of manufacturing a batch of x pairs of ice skates is given by $C(x) = 2x^3 - 223x^2 - 300x + 435{,}000$. Determine the optimum number of pairs of ice skates to manufacture in a batch and the cost of manufacturing the batch.

2. The function that represents the cost of manufacturing color televisions is $C(x) = x^2 + 22x + 4{,}000$, where x is the number of televisions produced per week. Their revenue function was determined to be $R(x) = 460x - 2x^2$. Determine their profit function, the optimum number of televisions to maximize weekly profits, and their associated profits.

3. The cost of manufacturing widgets is given by $C(x) = x^3/3 - 18x^2 + 200x + 750$, where x represents the number of widgets produced. Determine

  (a) The marginal cost function
  (b) The marginal cost of the fifteenth widget
  (c) The total cost and average cost for 15 widgets
  (d) The number of widgets for which the marginal cost is a minimum
  (e) The minimum marginal cost

(f) The total cost and average cost for the number of widgets that minimizes the marginal cost

4. For the profit function determined in Problem 2 determine
   (a) The marginal profit function
   (b) The marginal profit of the seventy-third television
   (c) The marginal profit of the forty-fifth television
   (d) The total profit for 45 televisions

5. The sales manager of the Keystone Kettle Company has forecast that daily demand for the new copper kettle will be 10 to 12 units. The cost of manufacturing the new kettle is given by $C(x) = x^3 - 17x^2 - 5x + 1,000$, where x is the number of kettles produced.
   (a) Is there a local maxima and/or minima between 10 and 12?
   (b) Determine the optimum daily production rate, accurate to the nearest integer.
   (c) If they sell all of the kettles for $30 each, what is the total daily profit?

6. The Armstrong Leather Company has determined that the weekly profit for its belts is given by $P(x) = x^3 - 20x^2 + 100x - 10$, where x represents the number, in hundreds, of belts produced per week. Because of sales and production constraints, Armstrong must produce between 300 and 400 belts per week.
   (a) Is there a local minima and/or maxima between 300 and 400?
   (b) Determine the optimum weekly production rate that is between 300 and 400 belts per week.
   (c) What is Armstrong's weekly profit at this production level?

7. The Humpty-Dumpty Egg Farm has determined that the cost of producing one gross of eggs is given by $C(x,y) = x^2 - 7x + 2y^2 - 20y + 100$, where x is the number of pounds of corn used and y is the number of pounds of wheat used.
   (a) Determine the number of pounds of corn and wheat that yield minimum cost.
   (b) Determine the minimum cost.

8. The Bi-Variate Function Factory has determined that its daily sales is given by $R(x,y) = 2x^2 + 5y^2 + 5xy + 20x + 1,000$. Their daily cost can be approximated by $C(x,y) = 5x^2 + 15y^2 + 10x - 12y$, where x is daily labor cost in thousands of dollars and y is inventory change in thousands of dollars.
   (a) Determine the profit function.
   (b) Determine the daily labor cost and inventory change that will yield maximum profit.
   (c) Determine the maximum profit.

9. The Humpty-Dumpty nutritional expert (from Problem 7) has determined that hens will produce healthier eggs if they use twice as much corn as they do wheat ($2x = y$).
   (a) What is the number of pounds of corn and wheat that will yield minimum cost?

(b) What is the minimum cost?

10. The Bi-Variate Factory (Problem 8) has decided that they want their daily inventory change to be equal to $5,000 less their daily labor cost ($y = 5 - x$).

(a) Determine the daily labor cost and inventory change that will yield maximum profit.

(b) Determine the maximum profit.

INTEGRAL CALCULUS: BASIC METHODOLOGY

13.1 INTRODUCTION

In the study of differential calculus, we are concerned with determining the instantaneous rate of change of a function; for example, we are interested in finding the marginal profit for a product when we are given a function that represents the total profit as a function of the number of units sold. In integral calculus, we are concerned with finding a function for which a given function represents the instantaneous rate of change, i.e., the derivative. For example, we may have a function that represents the marginal profit function for a product, and we may wish to determine the function that represents the total profit function for the product. In this chapter, first we consider integration as the inverse process of differentiation; next we consider some of the elementary rules of integration and apply these rules to marginal analysis; finally we consider the concept of a definite integral and some of its applications.

13.2 INDEFINITE INTEGRAL

Since the concept of integration involves the determination of a function whose derivative is given, we are seeking an *inverse* derivative, or, as it is more commonly called, an *antiderivative* of the given function. Thus by examination we can determine an antiderivative of most relatively simple functions; that is, we can use trial and error to determine an antiderivative. Let us begin by considering an example and establishing some basic terminology.

Example 13.1 Determine an antiderivative of $f(x) = 3x^2$.

Solution To determine an antiderivative of $f(x)$, we need to determine a function that has as its derivative $f(x) = 3x^2$. Upon inspection, we note that the derivative of x^3 is $3x^2$; thus an antiderivative of $f(x) = 3x^2$ is given by $F(x) = x^3$.

Sec. 13.2 / INDEFINITE INTEGRAL

From Example 13.1, we should note that $F(x) = x^3 + 1$, $F(x) = x^3 - 10$, and $F(x) = x^3 + 40$ are also antiderivatives of $f(x) = 3x^2$; that is, the function obtained by adding any constant to x^3 is also an antiderivative of $f(x)$. It may be proved that the most general form of an antiderivative is $F(x) + C$, where C is a constant.

The process by which a function is determined when its derivative is given is called *antidifferentiation*. If $f(x)$ is a given function and $F(x)$ is a function whose derivative is $f(x)$ such that $D_x F(x) = f(x)$, then $F(x)$ is *an antiderivative* of $f(x)$. However, since the most general form of an antiderivative is $F(x) + C$, we shall denote the process of determining the general antiderivative by the symbol $\int f(x)dx$, so that

$$\int f(x)\,dx = F(x) + C \qquad (13.1)$$

The function $f(x)$ is called the *integrand*; the symbol $\int$ is called the *integral sign*; the term dx is called the *differential*, which indicates the variable with respect to which the integration is performed; and the term C is called the *constant of integration*. The general form of the antiderivative of $f(x)$, that is, $F(x) + C$, is called the *indefinite integral* of $f(x)$. The symbol $\int f(x)dx$ is read "the integral of $f(x)$ with respect to x," or "the integral of $f(x)dx$."

In most problems involving integral calculus, after determining the function to be integrated, the next step is that of determining the indefinite integral of $f(x)$. In those situations involving relatively simple functions, we can determine the indefinite integral by obtaining an antiderivative and adding a general constant to it.

Example 13.2 Determine $\int 5x^4\,dx$.

Solution Since $D_x x^5 = 5x^4$, we have

$$\int 5x^4\,dx = x^5 + C$$

Example 13.3 Determine $\int e^x\,dx$.

Solution Since $D_x e^x = e^x$, we have

$$\int e^x\,dx = e^x + C$$

Example 13.4 Determine $\int 1/x\,dx$.

Solution Since $D_x(\ln x) = 1/x$, we have

$$\int \frac{1}{x}\,dx = \ln x + C$$

In order to determine a value for the constant of integration, we must have additional information. (We shall discuss this later.)

In Examples 13.2 to 13.4 the integrand was a *perfect* derivative of some function; i.e., the coefficient of the variable was in exact form. If the coefficient is not in exact form, we can adjust the coefficient of $F(x)$.

Example 13.5 Determine $\int 3x^5 dx$.

Solution Since $D_x x^6 = 6x^5$, we must divide $6x^5$ by 2 in order to obtain $f(x) = 3x^5$; thus

$$\int 3x^5 dx = \tfrac{1}{2}x^6 + C$$

Example 13.6 Determine $\int e^{4x} dx$.

Solution Since $D_x e^{4x} = 4e^{4x}$, we obtain

$$\int e^{4x} dx = \tfrac{1}{4}e^{4x} + C$$

To determine whether or not a function $F(x)$ is, in fact, the indefinite integral of $F(x)$, we need only to determine the derivative of $F(x)$ and verify that it is equal to $f(x)$. Thus we have a fairly easy check of our answer, and so we should never leave a problem without making certain whether or not our solution is correct.

EXERCISES

Determine the indefinite integral for Exercises 1 to 14 and check the answer by differentiation.

1. $\int 8x^7 dx$
2. $\int 4x^3 dx$
3. $\int -e^{-x} dx$
4. $\int 2e^{2x} dx$
5. $\int -1/x^2 dx$
6. $\int -2/x^3 dx$
7. $\int 2x^5 dx$
8. $\int 5x^9 dx$
9. $\int dx$
10. $\int x^2 dx$
11. $\int x^{1/2} dx$
12. $\int 2x^{3/2} dx$
13. $\int x^{-4} dx$
14. $\int x^n dx \, (n \neq -1)$

13.3 RULES OF INTEGRATION

Since differentiation and integration are intimately related, we would expect the rules of integration to be compatible and consistent with the rules of differentiation. This is true insofar as some of the related algebraic rules are concerned. However, there exist very few general rules of integration, and in most cases the integrals of relatively simple functions are determined by reversing the process of differentiation. Nevertheless, we can establish a few algebraic rules pertaining to integrals.

Sec. 13.3 / RULES OF INTEGRATION

The following rules are justified by the fact that the derivative on the right side of each equation is equal to the integrand on the left side of the equation.

Rule 13.1 The integral of a constant times a function is equal to the constant times the integral of the function:

$$\int kf(x)\,dx = k\int f(x)\,dx \qquad (13.2)$$

where k is a given constant.

This rule indicates that a constant may be placed either before or after the integral sign, but a variable cannot be moved in this way.

Rule 13.2 The integral of the sum of two functions is equal to the sum of their respective integrals:

$$\int [f(x)+g(x)]\,dx = \int f(x)\,dx + \int g(x)\,dx \qquad (13.3)$$

This rule can be extended to include the sum of three or more functions; furthermore, Rules 13.1 and 13.2 can be combined to prove that the integral of the difference of two functions is equal to the difference of the respective integrals.

Rule 13.3 If n is any constant that does not equal -1, the integral of x^n is obtained by increasing the exponent of x by 1 and then dividing by the new exponent:

$$\int x^n\,dx = \frac{x^{n+1}}{n+1} + C \qquad (13.4)$$

Rule 13.3 is known as the *power rule for integration* and it can be combined with the chain rule for differentiation to obtain the following rule.

Rule 13.4 If n is any constant that does not equal -1, the integral of $[f(x)]^n$ times the derivative of $f(x)$ with respect to x is obtained by increasing the exponent of $f(x)$ by 1 and dividing the result by the new exponent:

$$\int [f(x)]^n D_x f(x)\,dx = \frac{[f(x)]^{n+1}}{n+1} + C \qquad (13.5)$$

Example 13.7 Determine $\int 6x^2 dx$.

Solution Applying Rules 13.1 and 13.2, we have

$$\int 6x^2 dx = 6 \int x^2 dx = 6\frac{x^3}{3} + C = 2x^3 + C$$

Example 13.8 Determine $\int (3x+4)dx$.

Solution Applying Rules 13.1 to 13.3, we obtain

$$\int (3x+4)dx = 3\int x\,dx + 4\int dx$$
$$= 3\frac{x^2}{2} + 4x + C$$
$$= \tfrac{3}{2}x^2 + 4x + C$$

Example 13.9 Determine $\int (x^3 - 4x^2 - 2x + 3)dx$.

Solution Applying Rules 13.1 to 13.3, we obtain

$$\int (x^3 - 4x^2 - 2x + 3)dx = \int x^3 dx - 4\int x^2 dx - 2\int x\,dx + 3\int dx$$
$$= \frac{x^4}{4} - 4\frac{x^3}{3} - 2\frac{x^2}{2} + 3x + C$$
$$= \frac{x^4}{4} - \frac{4}{3}x^3 - x^2 + 3x + C$$

Example 13.10 Determine $\int 3(3x+4)^3 dx$.

Solution Since $D_x(3x+4) = 3$, we can apply Rule 13.4 to obtain

$$\int 3(3x+4)^3 dx = \frac{(3x+4)^4}{4} + C$$

Example 13.11 Determine $\int [3x^2 + 4x(x^2+4)^{-3}]dx$.

Solution Using all four rules, we have

$$\int \left[3x^2 + 4x(x^2+4)^{-3}\right]dx = 3\int x^2 dx + 2\int (2x)(x^2+4)^{-3} dx$$
$$= 3\frac{x^3}{3} + 2\frac{(x^2+4)^{-2}}{-2} + C$$
$$= x^3 - (x^2+4)^{-2} + C$$

EXERCISES

In each of Exercises 1 to 20, determine the integral and check the answer by differentiation.

1. $\int x^4 dx$

2. $\int x^9 dx$

Sec. 13.4 / MARGINAL ANALYSIS

3. $\int x^{-5} dx$
4. $\int y^{-7} dy$
5. $\int u^{3/2} du$
6. $\int v^{-2/3} dv$
7. $\int z^{3/4} dz$
8. $\int \dfrac{1}{\sqrt{x}} dx$
9. $\int \sqrt[3]{x}\, dx$
10. $\int (3x^3 + 2x - 4) dx$
11. $\int (2y^2 - 4y + 3) dy$
12. $\int \dfrac{1}{\sqrt[4]{y}} dy$
13. $\int [(3/y^4) - (4/y^5)] dy$
14. $\int (x^2 - 2/x^3) dx$
15. $\int (2\sqrt[3]{x} - 3\sqrt{x}) dx$
16. $\int \dfrac{(1-t^2)}{t} dt$
17. $\int \dfrac{(1+x)}{x^4} dx$
18. $\int \dfrac{(2+y)}{\sqrt{y}} dy$
19. $\int (1-y^3)^2 dy$
20. $\int t(2t^2 - 3)^2 dt$

13.4 MARGINAL ANALYSIS

In Sec. 12.3 we noted that the marginal cost function was obtained by determining the derivative of the total cost function; the same holds true for the marginal profit and marginal revenue functions. Now that we have the inverse process of differentiation, we can determine the total cost (or profit or revenue) function from the marginal cost (or profit or revenue) function. For example, if the marginal cost function is given by $c(x) = 2x^2 + 6x + 13$, the total cost function $C(x)$ is given by

$$C(x) = \int c(x) dx$$
$$= \int (2x^2 + 6x + 13) dx$$
$$= \tfrac{2}{3} x^3 + 3x^2 + 13x + C_0$$

where C_0 is the constant of integration. The question now arises as to how the constant of integration may be determined.

In the previous example, suppose that x represents the number of units produced; thus, if no units are produced, then $C(0) = C_0$. The cost associated with producing zero units is termed the *fixed cost* (*FC*) and is present in most cost function. Thus, if the cost of being able to produce a product for the above situation is $4,000 and the total cost is expressed in dollars, the total cost function is given by $C(x) = \tfrac{2}{3} x^3 + 3x^2 + 13x + 4{,}000$. Furthermore, the *variable cost* of producing x units $VC(x)$ is given by

$$VC(x) = \tfrac{2}{3} x^3 + 3x^2 + 13x$$

Example 13.12 The cost of manufacturing a product consists of $5,000 fixed cost and $3/unit variable cost for materials and labor. Determine the total cost function and the cost for manufacturing 6,000 units.

Solution Since $FC = \$5,000$, the total cost function is given by

$$C(x) = \int 3\,dx = 3x + 5,000$$

The cost of manufacturing 6,000 units is

$$C(6,000) = 3(6,000) + 5,000 = \$23,000$$

Example 13.13 Determine the total cost function if the marginal cost function for x tons of raw material is given by $c(x) = 3x^2 - 20x + 50$ and the fixed cost is $2,000. Also determine the total cost when 16 tons of raw material are used.

Solution Since $FC = \$2,000$, the total cost function is given by

$$C(x) = \int (3x^2 - 20x + 50)\,dx = x^3 - 10x^2 + 50x + 2,000$$

The cost of using 16 tons of raw material is

$$C(16) = (16)^3 - 10(16)^2 + 50(16) + 2,000 = \$4,336$$

Example 13.14 Given the marginal output function $MO(x) = 100x - x^2$ in units of finished product where x is the number of input units of raw material, determine the total output function and the total output when input is 40 units.

Solution Since there is no output unless there is some input, the total output function is

$$TO(x) = \int (100x - x^2)\,dx = 50x^2 - \frac{x^3}{3}$$

The total unit output when $x = 40$ units is

$$TO(40) = 50(40)^2 - \frac{(40)^3}{3} = 58,667$$

EXERCISES

1. The marginal cost in dollars for a particular operation is given by

$$c(x) = 3x^2 - 42x + 360$$

where x represents the number of units made. The fixed cost is $3,025. Determine the total cost function and the total cost for 10 units.

Sec. 13.4 / MARGINAL ANALYSIS

2. The marginal profit function in hundreds of dollars for a particular product is given by

$$p(x) = -3x^2 + 60x + 600$$

where x represents the dollars in thousands spent on advertising. A cost of $1,000 is incurred from miscellaneous expenses. Determine the total profit function and the total profit for spending $10,000 on advertising.

3. The marginal cost of making x units of a product is given by

$$c(x) = 0.0006x^2 - 0.6x + 50$$

The fixed cost is $100,000. Determine the total cost function and the total cost of producing 1,000 units.

4. If the marginal cost in dollars of making x units is $c(x) = 0.0027x^2 - 1.08x + 108$ and the fixed cost is $2,000, determine the total cost function and the cost of producing 200 units.

5. The marginal cost of manufacturing x units of a certain type of radio equipment for military aircraft is given by

$$c(x) = 3x^2 - 30x + 2{,}000$$

and the fixed cost is $4,000. Determine the total cost function and the cost of manufacturing 10 units.

6. The marginal profit in hundreds of dollars for a building contractor is given by

$$p(x) = 400 - 80x - 50e^{-x}$$

where x represents the number of four-unit apartment houses constructed. The contractor signed a contract that guarantees him a payment of $5,000 if he is not permitted to build any apartments. Determine the total profit function and the profit for building six apartment houses.

7. The Krebbie Vacuum Cleaner Company has determined that their marginal profit in hundreds of dollars is given by

$$p(x) = 90x^{1/2} - 200$$

where x represents the number of salespeople. When they have no salespeople, there is no profit (or loss). Determine the total profit function and the profit for 250 salespeople.

8. The manufacturer of Gungho Widgets has determined that the marginal cost in dollars of manufacturing x widgets is given by

$$c(x) = 4x - 10$$

and the marginal sales revenue in dollars is given by

$$r(x) = 2x + 40$$

A loss of $300 is incurred if no widgets are sold. Determine the total profit function and the profit for producing and selling 20 widgets.

9. An insurance salesman has determined that his marginal cost per month is given by

$$c(x) = 0.03x^2 + 0.2x + 0.05$$

where x represents the number of clients he sees per week. His fixed costs are $150. Determine the total cost function and the total cost if he sees 50 clients in a week.

10. A theater has determined that its marginal profit for a show is given by

$$p(x) = 0.5x^{-1/2} + 0.3x^{1/2} + 0.4$$

where x is the number of customers. The theater has fixed costs of $50 per show. Determine the total profit function and the total profit for 225 customers.

11. The Have-All Department Store has been told by a major television network that the marginal cost for tv advertising on their network is given by

$$c(x) = 10x^2 + 50x - 180$$

where x is the number of minutes of advertising per week. There is not a fixed charge for advertising on the network. Determine the total cost function and the total cost for 18 min of advertising per week.

12. A manufacturing company has determined that the marginal cost of manufacturing x bolts of cloth is given by

$$c(x) = 1.5x + 2$$

and the sales revenue in dollars is given by $r(x) = 2x$. Fixed costs are $5,000. Determine the total profit function and the total profit for producing and selling 500 bolts of cloth.

13.5 DEFINITE INTEGRAL

Mechanically speaking, there is little difference between determining the value for a *definite integral* and the function for an indefinite integral; however, the development of each of the concepts is quite different. We have established that the indefinite integral of $f(x)$ is the most general form of the antiderivative of $f(x)$ and is thus a function. The definite integral is defined differently. First we shall consider the definition of a definite integral and then relate the definition to the concept of an indefinite integral.

In order to define a definite integral, let $f(x)$ be a function that is continuous in the interval $a \leq x \leq b$, as shown in Fig. 13.1. Though it is not necessary for the function to be continuous, we consider only definite integrals of continuous functions. Next we divide the interval $a \leq x \leq b$ into n subintervals and let the $n-1$ points $x_1, x_2, \ldots, x_{n-1}$, where $a < x_1 < x_2 < \cdots < x_{n-1} < b$, represent the midpoints of the respective subintervals. If the subintervals are of equal width, then each of them will have width Δx_n, where

$$\Delta x_n = \frac{b-a}{n}$$

Sec. 13.5 / DEFINITE INTEGRAL

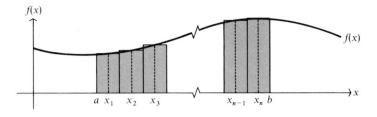

FIGURE 13.1

Thus we see that as n increases, Δx_n becomes smaller. Figure 13.1 also demonstrates this partitioning of the interval from a to b and the division of the area between the curve and the x axis into rectangles for which the base is Δx_n and the height is given by the value of the function at the midpoint x_i of the subinterval. That is, for the rectangle in the ith subinterval, the base is Δx_n and the height is $f(x_i)$.

For the partitioning in Fig. 13.1, we form the sum of the areas of the rectangles

$$S_n = f(x_1)\Delta x_n + f(x_2)\Delta x_n + \cdots f(x_n)\Delta x_n$$

which can be represented in summation notation as

$$S_n = \sum_{i=1}^{n} f(x_i)\Delta x_n$$

If we permit the division to become finer by letting n increase, it is likely that we shall obtain a different value of S_n. If we permit n to increase without bounds, that is, $n \to \infty$, we shall obtain the definition of a definite integral. Thus the limit is called the *definite integral* of $f(x)$ from a to b and is denoted by the symbol

$$\int_a^b f(x)\,dx$$

that is,

$$\int_a^b f(x)\,dx = \lim_{n \to \infty} \sum_{i=1}^{n} f(x_i)\Delta x_n \qquad (13.6)$$

In the definite integral $\int_a^b f(x)\,dx$, the function $f(x)$ is called the *integrand* and the constants a and b are called, respectively, the *lower limit* and the *upper limit* of the integral. The definite integral is read "the integral from a to b of $f(x)\,dx$." The evaluation of $\int_a^b f(x)\,dx$ from the definitional formula (13.6) is not a very functional method of obtaining the value of a definite integral. For example, consider the definite integral of $f(x) = x^2$ for the interval

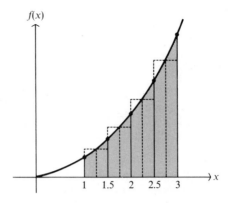

FIGURE 13.2

$1 \leqslant x \leqslant 3$. Partition the interval into four subintervals as given in Fig. 13.2. Thus the sum of the areas of the four rectangles is given by

$$S_4 = (1.25)^2(.5) + (1.75)^2(.5) + (2.25)^2(.5) + (2.75)^2(.5) = 8.625$$

The value $S_4 = 8.625$ compares favorably with the limit (i.e., the definite integral), which can later be verified to be $8\frac{2}{3}$. The best evaluation of a definite integral is determined through the use of the relationship of definite integrals and indefinite integrals. (This relationship is now given without proof because it is beyond the scope of this textbook and can be found in almost any textbook on advanced calculus.)

Fundamental Theorem of Integral Calculus If $f(x)$ is a continuous function in the interval $a \leqslant x \leqslant b$ and $F(x) = \int f(x)\,dx$ is any indefinite integral of $f(x)$, the value of the definite integral $\int_a^b f(x)\,dx$ is given by

$$\int_a^b f(x)\,dx = F(x)\Big|_a^b = F(b) - F(a) \qquad (13.7)$$

This result connects the two fundamental concepts of integral calculus: the definite and the indefinite integrals.

Denoting (13.7) by

$$\int_a^b f(x)\,dx = F(x)\Big|_a^b \qquad (13.8)$$

is usually most convenient because it is more compact.

Example 13.15 Determine the value of $\int_0^2 x\,dx$.

Sec. 13.5 / DEFINITE INTEGRAL

Solution First we determine the indefinite integral to be

$$F(x) = \int x\, dx = \frac{x^2}{2} + C$$

Thus

$$\int_0^2 x\, dx = \frac{x^2}{2} + C \Big|_0^2$$

$$= \left[\frac{(2)^2}{2} + C \right] - \left[\frac{(0)^2}{2} + C \right]$$

$$= 2$$

In Example 13.15, we see that the value of the constant of integration is irrelevant because it will always cancel itself. Thus, in (13.7) and (13.8), when we write the indefinite integral, we shall omit the constant of integration. Furthermore, since the value of a definite integral is determined through use of indefinite integrals, we note that any rule that is applicable for indefinite integrals is also applicable for definite integrals.

Example 13.16 Determine the value $\int_1^5 (3x^2 + 4x + 3)\, dx$.

Solution

$$\int_1^5 (3x^2 + 4x + 3)\, dx = x^3 + 2x^2 + 3x \Big|_1^5$$

$$= \left[(5)^3 + 2(5)^2 + 3(5) \right] - \left[(1)^3 + 2(1)^2 + 3(1) \right]$$

$$= 190 - 6$$

$$= 184$$

Example 13.17 Evaluate $\int_{-2}^1 (x^3 + 2x)\, dx$

Solution

$$\int_{-2}^1 (x^3 + 2x)\, dx = \frac{x^4}{4} + x^2 \Big|_{-2}^1$$

$$= \left[\frac{(1)^4}{4} + (1)^2 \right] - \left[\frac{(-2)^4}{4} + (-2)^2 \right]$$

$$= \tfrac{5}{4} - 8$$

$$= -\tfrac{27}{4}$$

Using the Definite Integral to Find Area

In referring to Fig. 13.2 and the definition of a definite integral, we may wish to conclude that the definite integral from a to b yields the area between the function $f(x)$ and the x axis bounded by $x = a$ and $x = b$. However, in general, this is *not* a correct interpretation because the function can possibly be negative, which, in turn, yields one or more of the function values as

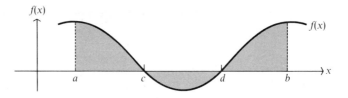

FIGURE 13.3

negative in the sum $\sum_{i=1}^{n} f(x_i)\Delta x_n$. Thus the definite integral would yield the difference between the area between $f(x)$ and the x axis when $f(x)$ is positive and the area between the x axis and $f(x)$ when $f(x)$ is negative. This is illustrated in Fig. 13.3 for a function that has both positive and negative values within the interval $a \leqslant x \leqslant b$.

Before considering the use of a definite integral to determine the area between a function and the x axis, let us consider some fundamental properties of definite integrals. Each of the following rules can be developed by resorting to Eq. (13.7) and the rules for indefinite integrals.

Rule 13.5 If c is any point such that $a \leqslant c \leqslant b$, we can express the definite integral of $f(x)$ from a to b as two separate definite integrals of $f(x)$, one from a to c and the second from c to b; that is

$$\int_a^b f(x)\,dx = \int_a^c f(x)\,dx + \int_c^b f(x)\,dx \qquad (13.9)$$

This rule can be verified be applying Eq. (13.7) to each term in (13.9). Since the integrand is $f(x)$ in each case, we have

$$F(b) - F(a) = F(c) - F(a) + F(b) - F(c)$$

The following rule is a direct result of Rule 13.1 and Eq. (13.7).

Rule 13.6 If k is any constant,

$$\int_a^b kf(x)\,dx = k\int_a^b f(x)\,dx \qquad (13.10)$$

Another rule that is quite useful at times and which results from Rule 13.2 and Eq. (13.7) is the following.

Rule 13.7 If $f(x)$ and $g(x)$ are continuous functions in the interval $a \leqslant x \leqslant b$,

$$\int_a^b [f(x) + g(x)]\,dx = \int_a^b f(x)\,dx + \int_a^b g(x)\,dx \qquad (13.11)$$

Sec. 13.5 / DEFINITE INTEGRAL

Another property of definite integrals that is worthy of mention is that, like derivatives, we may use any symbol for the independent variable because the definite integral depends only on the integrand and the limits a and b. Thus, for example, we have

$$\int_a^b f(x)\,dx = \int_a^b f(t)\,dt = \int_a^b f(u)\,du = \int_a^b f(z)\,dz$$

Let us now turn our attention to determining the area between the curve of a function and the x axis. As illustrated in Fig. 13.3 and in our preceding discussion, we cannot obtain the area between the curve and the x axis simply by determining the definite integral $\int_a^b f(x)\,dx$ unless $f(x)$ is nonnegative for all values in the interval $a \leq x \leq b$. Thus, if some of the heights (i.e., function values) in Fig. 13.2 were negative (as illustrated in Fig. 13.3), we could simply change their signs and determine the area (which will now be positive) between the curve and the x axis. For example, in Fig. 13.3, to determine the area between the curve and the x axis we would determine the area of the shaded region, which can be obtained by use of definite integrals as follows. Since $f(x) < 0$ for values in the interval $c \leq x \leq d$, we can multiply $f(x)$ by -1 within this interval and integrate $\int_c^d -f(x)\,dx$ to obtain the area below the x axis. The total area A is then obtained by

$$A = \int_a^c f(x)\,dx - \int_c^d f(x)\,dx + \int_d^b f(x)\,dx$$

Example 13.18 Determine the area between the curve of the function $f(x) = 2x - 6$ and the x axis bounded by $x = 2$ and $x = 4$.

Solution We determine that $f(x)$ is negative for $x < 3$ and positive for $x > 3$. Since $x = 3$ is within the interval of interest $2 \leq x \leq 4$, we sketch the graph of $f(x)$ in Fig. 13.4 and determine the area A of the shaded regions. Thus

$$A = -\int_2^3 (2x-6)\,dx + \int_3^4 (2x-6)\,dx$$

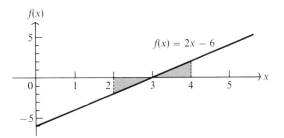

FIGURE 13.4

However,

$$F(x) = \int (2x-6)\,dx = x^2 - 6x$$

and hence the area is given by

$$A = -[F(3) - F(2)] + [F(4) - F(3)]$$
$$= F(2) - 2F(3) + F(4)$$

Determining the values of $F(x)$, we have

$$F(2) = (2)^2 - 6(2) = -8$$
$$F(3) = (3)^2 - 6(3) = -9$$
$$F(4) = (4)^2 - 6(4) = -8$$

Thus the area is determined to be

$$A = (-8) - 2(-9) + (-8) = 2$$

Example 13.19 Determine the area between the curve of the function $f(x) = x^2 - 3x + 2$ and the x axis bounded by $x = 0$ and $x = 3$.

Solution Examining $f(x) = x^2 - 3x + 2$, we determine that it has roots at $x = 1$ and $x = 2$. The graph of $f(x)$ is given in Fig. 13.5. We wish to determine the area of the shaded regions. We see that the area is determined by

$$A = \int_0^1 f(x)\,dx - \int_1^2 f(x)\,dx + \int_2^3 f(x)\,dx$$

However, excluding the constant of integration, we determine the indefinite integral to be

$$F(x) = \int (x^2 - 3x + 2)\,dx = \frac{x^3}{3} - \frac{3}{2}x^2 + 2x$$

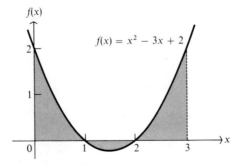

FIGURE 13.5

Thus, using Eq. (13.7) for each definite integral, we have

$$A = F(1) - F(0) - [F(2) - F(1)] + [F(3) - F(2)]$$
$$= F(3) - 2F(2) + 2F(1) - F(0)$$

Substituting into $F(x)$, we obtain

$$F(3) = \frac{(3)^3}{3} - \frac{3}{2}(3)^2 + 2(3) = \frac{3}{2}$$

$$F(2) = \frac{(2)^3}{3} - \frac{3}{2}(2)^2 + 2(2) = \frac{2}{3}$$

$$F(1) = \frac{(1)^3}{3} - \frac{3}{2}(1)^2 + 2(1) = \frac{5}{6}$$

$$F(0) = \frac{(0)^3}{3} - \frac{3}{2}(0)^2 + 2(0) = 0$$

Hence the shaded area in Fig. 13.5 is given by

$$A = \tfrac{3}{2} - 2(\tfrac{2}{3}) + 2(\tfrac{5}{6}) - 0 = \tfrac{11}{6}$$

The definite integral can be used to determine the area between two functions. For two functions $f(x)$ and $g(x)$, the area bounded by the two functions and $x = a$ and $x = b$ is given by

$$A = \int_a^b [f(x) - g(x)] \, dx$$

If $f(x)$ is greater than $g(x)$ between the limits of integration a and b, the area will represent the total area, as illustrated in Fig. 13.6; however, if there are values between a and b for which $f(x)$ is less than $g(x)$, the area A will represent the *net* area. Thus, if total area is desired, the integral must be evaluated as two or more integrals in order to determine the total for each area. Moreover, if the resulting integral in any case is negative, we must take its absolute value because negative area has no meaning.

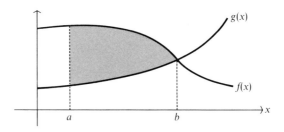

FIGURE 13.6

Example 13.20 Determine the net and total areas between the functions $f(x) = 11 - x$ and $g(x) = x^2 - 10x + 25$ bounded by $x = 0$ and $x = 10$.

Solution To determine the net area, we simply evaluate

$$\int_0^{10}(11 - x - x^2 + 10x - 25)\,dx = \int_0^{10}(-x^2 + 9x - 14)\,dx$$

$$= -\frac{x^3}{3} + \frac{9}{2}x^2 - 14x\Big|_0^{10}$$

$$= -\frac{1{,}000}{3} + 450 - 140$$

$$= -23\tfrac{1}{3}$$

Thus, taking the absolute value, we determine the net area to be $23\tfrac{1}{3}$. The negative sign indicates simply that there is more area between the two functions that is above $f(x) = 11 - x$ than there is below, which can be seen in Fig. 13.7 where the areas A_1 and A_3 are above $f(x) = 11 - x$ while A_2 is below.

To determine the total area between the two functions, we must determine the points of intersection for the curves of the two functions. To do this, we simply set the two functions as being equal and solve for x. Thus we obtain

$$x^2 - 10x + 25 = 11 - x$$

or

$$x^2 - 9x + 14 = 0$$

which factors into

$$(x - 2)(x - 7) = 0$$

Hence the points of intersection of the two functions occur at $x = 2$ and $x = 7$; specifically, the points are (2,9) and (7,4), as indicated in Fig. 13.7. The total area is found by determining the respective areas of A_1, A_2, and A_3 and adding.

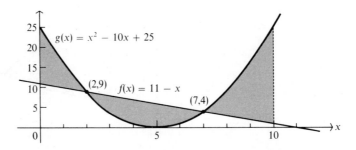

FIGURE 13.7

Sec. 13.5 / DEFINITE INTEGRAL

To obtain A_1, we determine

$$A_1 = \int_0^2 (x^2 - 10x + 25 - 11 + x)\,dx$$

$$= \int_0^2 (x^2 - 9x + 14)\,dx$$

$$= \left.\frac{x^3}{3} - \frac{9}{2}x^2 + 14x\right|_0^2$$

$$= \tfrac{8}{3} - 18 + 28$$

$$= 12\tfrac{2}{3}$$

For A_2, we have

$$A_2 = \int_2^7 (11 - x - x^2 + 10x - 25)\,dx$$

$$= \int_2^7 (-x^2 + 9x - 14)\,dx$$

$$= \left.-\frac{x^3}{3} + \frac{9}{2}x^2 - 14x\right|_2^7$$

$$= -\tfrac{343}{3} + \tfrac{441}{2} - 98 + \tfrac{8}{3} - 18 + 28$$

$$= 20\tfrac{5}{6}$$

For A_3, we have

$$A_3 = \int_7^{10} (x^2 - 10x + 25 - 11 + x)\,dx$$

$$= \int_7^{10} (x^2 - 9x + 14)\,dx$$

$$= \left.\frac{x^3}{3} - \frac{9}{2}x^2 + 14x\right|_7^{10}$$

$$= \frac{1{,}000}{3} - 450 + 140 - \frac{343}{3} + \frac{441}{2} - 98$$

$$= 31\tfrac{1}{2}$$

Thus the total area between the functions is given by

$$A_1 + A_2 + A_3 = 12\tfrac{2}{3} + 20\tfrac{5}{6} + 31\tfrac{1}{2} = 65$$

[We could have determined the net area between the functions by adding A_1 and A_3 and subtracting A_2 from this sum, simply separating the area between the curves above and below $f(x) = 11 - x$.]

The Definite Integral in Marginal Analysis

The definite integral can also be used in certain types of marginal analysis problems. For example, if we are given the marginal cost function $c(x)$ and wish to determine the total cost of increasing x by k units, we can determine the indefinite integral of the marginal cost function and then determine the difference between the cost for $x = a$ units and for $a + k$ units. However, this involves the same steps as determining the definite integral

$$\int_a^{a+k} c(x)\,dx = C(a+k) - C(a)$$

This concept can also be extended to total profit or revenue.

Example 13.21 The marginal cost in dollars of manufacturing x units by a particular operation is given by

$$c(x) = 3x^2 - 42x + 360$$

Determine the increase in cost of manufacturing 13 units as opposed to manufacturing 10 units.

Solution The increase in cost is given by

$$\int_{10}^{13} (3x^2 - 42x + 360)\,dx = x^3 - 21x^2 + 360x \Big|_{10}^{13}$$
$$= 3{,}328 - 2{,}500$$
$$= \$828$$

In Example 13.21 we should note that $3{,}328$ and $2{,}500$ are not total costs but total *variable* costs. In order to determine total costs, we would need to know the fixed costs also.

In the fundamental theorem of integral calculus, we assumed that the limits of integration a and b are constants. Consequently, if the interval $a \leq x \leq b$ is not bounded on both sides, the fundamental theorem of integral calculus is not applicable. Of course, the problem arises from the fact that at least one of the limits of integration is infinite. To evaluate a definite integral with an unbounded limit, such as $\int_a^\infty f(x)\,dx$, we simply supply an aribtrary upper limit of integration, say b; perform the integration; and take the limit as $b \to \infty$. Thus we have

$$\int_a^\infty f(x)\,dx = \lim_{b \to \infty} \int_a^b f(x)\,dx = \lim_{b \to \infty} F(b) - F(a)$$

This type of integral is called an *improper integral*, and, because it is a limit, it may or may not be defined. That is, the corresponding value of the integral may be finite or it may be infinite or undefined. A similar evaluation is employed for an integral that has an unbounded lower limit, except, of course, the arbitrary limit approaches $-\infty$.

Example 13.22 Evaluate, if possible, the following improper integrals:

(a) $\int_1^\infty \dfrac{1}{x^2}\,dx$ (b) $\int_{-\infty}^0 e^x\,dx$

Solution For (a) we let the upper limit be b and evaluate

$$\int_1^b \frac{1}{x^2}\,dx = -\frac{1}{x}\Big|_1^b = -\frac{1}{b}+1$$

Taking the limit of this result, we have

$$\lim_{b\to\infty}\left(-\frac{1}{b}+1\right)=1$$

Hence

$$\int_1^\infty \frac{1}{x^2}\,dx=1$$

For (b) we let the lower limit be a and evaluate

$$\int_a^0 e^x\,dx = e^x\Big|_a^0 = e^0-e^a = 1-e^a$$

Taking the limit, we get

$$\lim_{a\to-\infty}(1-e^a)=1-0=1$$

Therefore,

$$\int_{-\infty}^0 e^x\,dx=1$$

EXERCISES

In Exercises 1 to 20, evaluate the given definite integral.

1. $\int_1^4 3x^2\,dx$
2. $\int_0^3 5x^4\,dx$
3. $\int_0^3 (x^2-x)\,dx$
4. $\int_{-2}^0 (3y^2-2y)\,dy$
5. $\int_1^4 (z-1)^3\,dz$
6. $\int_0^2 t(t^2-4)^3\,dt$
7. $\int_1^{13} \dfrac{1}{\sqrt{2y-1}}\,dy$
8. $\int_{-2}^2 \sqrt{2-t}\,dt$
9. $\int_0^4 (2-\sqrt{x})^2\,dx$
10. $\int_{-1}^1 t(t^2+1)^3\,dt$
11. $\int_1^2 (2t^2+3)^2\,dt$
12. $\int_{-1}^1 (t^3-1)^2\,dt$
13. $\int_{-1}^1 y\sqrt{4-3y^2}\,dy$
14. $\int_0^{43} y\sqrt{y^2+9}\,dy$
15. $\int_0^3 \dfrac{z}{\sqrt{16+z^2}}\,dz$
16. $\int_1^4 \dfrac{3x+2}{\sqrt{x}}\,dx$

17. $\int_1^6 \dfrac{2}{\sqrt{x+3}}\,dx$

18. $\int_4^{16} \dfrac{2x^2+2x+1}{x^{3/2}}\,dx$

19. $\int_2^8 \dfrac{2}{x}\,dx$

20. $\int_1^3 2e^{3x}\,dx$

In Exercises 21 to 40, determine the area between the given function and the x axis bounded by the limits of the given interval.

21. $f(x)=2x+3 \quad (0\leqslant x \leqslant 4)$

22. $f(x)=x \quad (0\leqslant x \leqslant 6)$

23. $f(x)=x-1 \quad (1\leqslant x \leqslant 3)$

24. $f(x)=2x-4 \quad (0\leqslant x \leqslant 4)$

25. $f(x)=-4x+5 \quad (1\leqslant x \leqslant 3)$

26. $f(x)=-x \quad (-1\leqslant x \leqslant 1)$

27. $f(x)=x^2 \quad (0\leqslant x \leqslant 2)$

28. $f(x)=3x^2-4x \quad (2\leqslant x \leqslant 3)$

29. $f(x)=x^2-4x+5 \quad (0\leqslant x \leqslant 3)$

30. $f(x)=6x^2+2x+1 \quad (0\leqslant x \leqslant 2)$

31. $f(x)=-3x^2+4x-4 \quad (0\leqslant x \leqslant 3)$

32. $f(x)=6x^2-7x+2 \quad (0\leqslant x \leqslant 2)$

33. $f(x)=x^3 \quad (0\leqslant x \leqslant 1)$

34. $f(x)=x^3 \quad (-1\leqslant x \leqslant 1)$

35. $f(x)=x^3-2x^2-x \quad (0\leqslant x \leqslant 1)$

36. $f(x)=x^3-2x^2-x \quad (-1\leqslant x \leqslant 1)$

37. $f(x)=e^{2x} \quad (0\leqslant x \leqslant 3)$

38. $f(x)=2e^{-2x} \quad (0\leqslant x \leqslant 2)$

39. $f(x)=2/x \quad (1\leqslant x \leqslant 2)$

40. $f(x)=3/(x+4) \quad (0\leqslant x \leqslant 1)$

41. In Exercise 2, Sec. 13.4, determine the increase in profit obtained from increasing the advertising expenditure from $9,000 to $12,000.

42. In Exercise 3, Sec. 13.4, determine the increase in cost that results from increasing the production level from 1,000 to 1,100 units.

43. In Exercise 4, Sec. 13.4, determine the decrease in cost that results from decreasing the production level from 210 to 190 units.

44. In Exercise 5, Sec. 13.4, determine the increase in cost that results from increasing the production level of radios from 6 to 8.

45. In Exercise 6, Sec. 13.4, determine the increase in profit that results from increasing the construction of the number of apartment houses from 3 to 4.

46. In Exercise 7, Sec. 13.4, determine the decrease in profit that results from decreasing the number of sales people from 225 to 210.

47. In Exercise 8, Sec. 13.4, determine the increase in profit that results from increasing the number of widgets from 15 to 25.

In Exercises 48 to 52, determine the net area and the total area between f(x) and g(x) bounded by the given interval.

48. $f(x)=7-x;\ g(x)=7+x \quad (0\leqslant x \leqslant 7)$

49. $f(x)=2+3x;\ g(x)=5-2x \quad (0\leqslant x \leqslant 2)$

50. $f(x)=4;\ g(x)=6-2x \quad (0\leqslant x \leqslant 2)$

51. $f(x) = -x^2 + 2x - 15;\ g(x) = x \quad (0 \leq x \leq 2)$

52. $f(x) = x^2 - 3x + 2;\ g(x) = x^2 - 3x + 10 \quad (0 \leq x \leq 3)$

53. In Exercise 9, Sec. 13.4, determine the increase in cost that results from increasing the number of clients from 35 to 50 per week.

54. In Exercise 10, Sec. 13.4, determine the decrease in profit that results from the number of customers decreasing from 225 to 196 per show.

55. In Exercise 11, Sec. 13.4, determine the increase in cost that results from increasing advertising time on tv from 14 min to 21 min per week.

56. In Exercise 12, Sec. 13.4, determine the decrease in profit that results from decreasing production from 500 to 400 bolts of cloth.

57. The marginal time in hours required to manufacture x units of product A is given by the function

$$t(x) = 0.05x + 0.2$$

The set-up time is 2 hr. Determine the total time function and the additional time required to manufacture 50 units instead of 40 units.

13.6 PROBABILITY DENSITY FUNCTIONS

In Chap. 7, we discussed concepts of probability. In Sec. 7.10, we considered the properties that a function must possess in order for it to be a *probability density function* for a *continuous* random variable X. That is, the function $f(x)$ is a *probability density function* for the continuous random variable X if and only if the following properties are satisfied:

1. The function is nowhere negative; that is, $f(x) \geq 0$ for all possible values of X.
2. The probability $P(a \leq X \leq b)$, where a and b are any two possible values of X such that $a \leq b$, is determined by the area under the curve of $f(x)$ between the values a and b.
3. The total area under the curve of $f(x)$ is equal to unity (1).

Now that we have an understanding of integral calculus and its application to determining the area under the curve of a function, we may state properties 2 and 3 in terms of integrals. However, first we shall recall that probability density functions are usually stated for a specific domain, and in many cases this domain is a proper subset of the real numbers. However, we can express a probability density function such that its domain is always the set of real numbers, which is done by letting $f(x) = 0$ for those values of x that are not of interest. Let us now state the properties of a probability density function using integral calculus, and then we shall consider how they are used for specific probability density functions.

1. For all possible values of X,

$$f(x) \geq 0 \qquad (13.12)$$

2. For $a \leq b$,

$$P(a \leq X \leq b) = \int_a^b f(x)\,dx \qquad (13.13)$$

3.

$$\int_{-\infty}^{\infty} f(x)\,dx = 1 \qquad (13.14)$$

Thus, if $f(x)$ satisfies (13.12) and (13.14), we can use (13.13) to compute associated probabilities for X.

Example 13.23 The probability density function of the random variable X is given by

$$f(x) = \begin{cases} 1 & 0 \leq x \leq 1 \\ 0 & \text{otherwise} \end{cases}$$

Determine $P(\frac{1}{2} \leq X \leq \frac{2}{3})$.

Solution First we note that property 1 is satisfied because $f(x)$ can assume only two values (0 and 1), which are both nonnegative. Furthermore, the total area under $f(x)$ is given by

$$\int_{-\infty}^{\infty} f(x)\,dx = \int_{-\infty}^{0} f(x)\,dx + \int_{0}^{1} 1\,dx + \int_{1}^{\infty} f(x)\,dx$$

$$= \lim_{a \to -\infty} \int_a^0 0\,dx + x \Big|_0^1 + \lim_{b \to \infty} \int_1^b 0\,dx$$

$$= 0 + 1 + 0$$

$$= 1$$

Generally we do not write the function or the definite integrals for those values of x for which $f(x) = 0$; that is, for the previous function usually we would write $f(x) = 1$ for $0 \leq x \leq 1$ and

$$\int_{-\infty}^{\infty} f(x)\,dx = \int_0^1 1\,dx$$

To determine $P(\frac{1}{2} \leq X \leq \frac{2}{3})$, we use Eq. (13.13) and obtain

$$P(\tfrac{1}{2} \leq X \leq \tfrac{2}{3}) = \int_{1/2}^{2/3} 1\,dx = x \Big|_{1/2}^{2/3} = \tfrac{2}{3} - \tfrac{1}{2} = \tfrac{1}{6}$$

Example 13.24 For the probability density function

$$f(x) = \frac{x}{9} + \frac{1}{6} \qquad \text{for } 0 \leq x \leq 3$$

determine $P(0 \leq X \leq 2)$.

Sec. 13.6 / PROBABILITY DENSITY FUNCTIONS

Solution We note that $f(x) \geq 0$ for all values of x in the interval $0 \leq x \leq 3$ and that

$$\int_0^3 \left(\frac{x}{9} + \frac{1}{6}\right) dx = \frac{x^2}{18} + \frac{x}{6} \Big|_0^3 = \frac{9}{18} + \frac{3}{6} = 1$$

Thus

$$P(0 \leq X \leq 2) = \int_0^2 \left(\frac{x}{9} + \frac{1}{6}\right) dx = \frac{x^2}{18} + \frac{x}{6} \Big|_0^2 = \frac{5}{9}$$

Example 13.25 Show that $f(x) = x^2/8 + x/6 + \frac{1}{6}$ for $0 \leq x \leq 2$ satisfies (13.12) and (13.14) and determine $P(0 \leq X \leq 1)$.

Solution To show that (13.12) is satisfied, we determine the discriminant of the quadratic function to be

$$b^2 - 4ac = \left(\tfrac{1}{6}\right)^2 - 4\left(\tfrac{1}{8}\right)\left(\tfrac{1}{6}\right) = -\tfrac{1}{18}$$

Thus, recalling from Chap. 2 that a quadratic function with a negative discriminant has no real roots, $f(x)$ is either always positive or always negative; furthermore, $f(0) = \frac{1}{3}$, which is positive. Hence the function is always positive, thus satisfying (13.12).

To show that (13.14) is satisfied, we determine

$$\int_0^2 \left(\frac{x^2}{8} + \frac{x}{6} + \frac{1}{6}\right) dx = \frac{x^3}{24} + \frac{x^2}{12} + \frac{x}{6} \Big|_0^2 = 1$$

Furthermore,

$$P(0 \leq X \leq 1) = \int_0^1 \left(\frac{x^2}{8} + \frac{x}{6} + \frac{1}{6}\right) dx$$

$$= \frac{x^3}{24} + \frac{x^2}{12} + \frac{x}{6} \Big|_0^1$$

$$= \frac{7}{24}$$

Example 13.26 Determine the value of k such that $f(x) = k(x+2)$ for $0 \leq x \leq 3$ satisfies (13.12) and (13.14). Determine $P(1 \leq X \leq 2)$.

Solution By inspection, we see that $f(x) \geq 0$ for all x values in the interval $0 \leq x \leq 3$ provided $k \geq 0$; thus (13.12) is satisfied if $k \geq 0$. To determine k such that (13.14) is satisfied requires the following integration:

$$\int_0^3 k(x+2) dx = k \int_0^3 (x+2) dx$$

$$= k \left(\frac{x^2}{2} + 2x\right)\Big|_0^3$$

$$= k \frac{21}{2}$$

However, in order for the integral to equal unity, we must have

$$\tfrac{21}{2}k = 1$$

which is true if and only if $k = \tfrac{2}{21}$. Thus we obtain a probability density function if $k = \tfrac{2}{21}$, and it is given by

$$f(x) = \frac{2x}{21} + \frac{4}{21} \qquad \text{for } 0 \leqslant x \leqslant 3$$

We determine the probability to be

$$P(1 \leqslant X \leqslant 2) = \int_1^2 \left(\frac{2x}{21} + \frac{4}{21} \right) dx$$

$$= \frac{x^2}{21} + \frac{4x}{21} \bigg|_1^2$$

$$= \tfrac{12}{21} - \tfrac{5}{21}$$

$$= \tfrac{7}{21}$$

Expected Value of a Continuous Variable

In Chap. 7 we considered how to determine the *mean*, or *expected value*, of a probability distribution for a discrete random variable; the counterpart for a continuous random variable is given by

$$E(X) = \int_{-\infty}^{\infty} x f(x) \, dx \qquad (13.15)$$

Thus, if the domain of $f(x)$ is given by $a \leqslant x \leqslant b$, then the mean, or expected value, of the random variable X is determined by

$$E(X) = \int_a^b x f(x) \, dx \qquad (13.16)$$

The interpretation of $E(X)$ is the same as for a discrete random variable; that is, it is an average.

Example 13.27 Determine the expected value of the random variable X when the probability density function is given by $f(x) = 3x^2$ for $0 \leqslant x \leqslant 1$.

Solution   Using Eq. (13.16), we obtain

$$E(X) = \int_0^1 x(3x^2) \, dx$$

$$= \int_0^1 3x^3 \, dx$$

$$= \frac{3x^4}{4} \bigg|_0^1$$

$$= \tfrac{3}{4}$$

Sec. 13.6 / PROBABILITY DENSITY FUNCTIONS

Example 13.28 The Third National Bank has determined that the arrival pattern of its customers during the business day is given by $f(x) = kx(1-x)$ for $0 \leq x \leq 1$, where x represents the time of the customer's arrival on a *business-day* basis. Determine the value of k that makes $f(x)$ a probability density function, determine the average arrival time for the customers, and determine $P(0 \leq X \leq \frac{1}{4})$.

Solution To show that $f(x)$ is a probability density function, we note that $f(x) \geq 0$ for all values of x in the interval $0 \leq x \leq 1$ if $k \geq 0$. For the total area under the curve of $f(x)$ to equal unity we have

$$\int_0^1 kx(1-x)\,dx = k\int_0^1 (x - x^2)\,dx$$

$$= k\left(\frac{x^2}{2} - \frac{x^3}{3}\right)\bigg|_0^1$$

$$= k\tfrac{1}{6}$$

which must equal unity; thus $f(x)$ is a probability density function if and only if $k = 6$.

The average arrival time for the customers is given by

$$E(X) = \int_0^1 x \cdot 6x(1-x)\,dx$$

$$= 6\int_0^1 (x^2 - x^3)\,dx$$

$$= 6\left(\frac{x^3}{3} - \frac{x^4}{4}\right)\bigg|_0^1$$

$$= \tfrac{1}{2}$$

Thus the *average* arrival time for the customers would be in the middle of the working day.

To determine $P(0 \leq X \leq \frac{1}{4})$, we calculate

$$P(0 \leq X \leq \tfrac{1}{4}) = \int_0^{1/4} 6x(1-x)\,dx$$

$$= 6\left(\frac{x^2}{2} - \frac{x^3}{3}\right)\bigg|_0^{1/4}$$

$$= 6 \cdot \tfrac{5}{192}$$

$$= \tfrac{5}{32}$$

The probability density function of Example 13.28 is a special case of a more general family of probability density functions. That is, any probability density function of the form

$$f(x) = \frac{(\alpha + \beta + 1)!}{\alpha!\,\beta!} x^\alpha (1-x)^\beta \qquad \text{for } 0 \leq x \leq 1$$

where α and β are nonnegative integers, is called a *beta probability density function*; thus the random variable X is distributed according to the beta distribution. We state this merely to point out that general probability density functions exist for continuous random variables just as they do for discrete random variables, such as binomial and hypergeometric variables. For continuous random variables, often it is either difficult or impossible to evaluate the integral of a probability density function, in which case tables for selected values are determined. Nevertheless, even when using tables, it is most convenient to consider the probability for continuous variables as the area under the curve and remember that the total area must equal 1.

A probability distribution that is very widely used in statistical analysis is the *normal distribution*. The *normal probability density function* is given by

$$f(x) = \frac{1}{\sqrt{2\pi}\,\sigma} e^{-1/2[(x-\mu)/\sigma]^2} \qquad \text{for } -\infty < x < \infty$$

where μ and σ are parameters that represent the mean and standard deviation, respectively, of the distribution. The graph of this distribution has the familiar bell shape, given in Fig. 13.8. The probability of a possible value of the random variable X falling in the interval $a \leq x \leq b$ is given by

$$P(a \leq X \leq b) = \int_a^b \frac{1}{\sqrt{2\pi}\,\sigma} e^{-1/2[(x-\mu)/\sigma]^2}\, dx$$

and the shaded area in Fig. 13.8 represents this probability. The value of this integral must be determined by methods that are more advanced than those found in this book. However, the widespread use of the normal distribution requires the availability of tables from which we can obtain values of the definite integral for a wide variety of values of a and b. An in-depth familiarity with such tables will be gained by those who study statistics. (The study of the use of such tables and of other special probability density functions is beyond the scope of this book.)

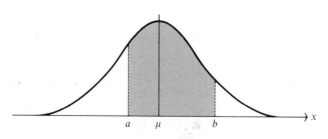

FIGURE 13.8

Sec. 13.6 / PROBABILITY DENSITY FUNCTIONS

EXERCISES

In each of the Exercises 1 to 12, determine whether or not the function is a probability density function. If it is not, state why.

1. $f(x) = \frac{1}{8}$ for $0 \leq x \leq 8$
2. $f(x) = 6$ for $0 \leq x \leq 6$
3. $f(x) = \frac{1}{10}$ for $0 \leq x \leq 5$
4. $f(x) = 2$ for $0 \leq x \leq \frac{1}{2}$
5. $f(x) = 1/n$ for $0 \leq x \leq n$ and $n > 0$
6. $f(x) = -x + 2$ for $0 \leq x \leq 2$
7. $f(x) = -2x/3 + \frac{1}{3}$ for $0 \leq x \leq 1$
8. $f(x) = 2x^2$ for $0 \leq x \leq 1$
9. $f(x) = 3x^2/16$ for $0 \leq x \leq 2$
10. $f(x) = 6x^2 - 12x + \frac{9}{2}$ for $0 \leq x \leq 2$
11. $f(x) = 12x^2 - 12x^3$ for $0 \leq x \leq 1$
12. $f(x) = 30(x^2 - 2x^4 + x^6)$ for $0 \leq x \leq 1$

In each of Exercises 13 to 22, determine the value of k such that f(x) is a probability density function. Also determine the expected value of X for each probability density function in Exercises 13 to 21.

13. $f(x) = k$ for $0 \leq x \leq 2$
14. $f(x) = 1/k$ for $0 \leq x \leq 5$
15. $f(x) = \frac{1}{3}$ for $0 \leq x \leq k$
16. $f(x) = k(-x + 3)$ for $0 \leq x \leq 2$
17. $f(x) = k(-x + 4)$ for $0 \leq x \leq 4$
18. $f(x) = kx^2$ for $0 \leq x \leq 2$
19. $f(x) = k(x^2 - x + 4)$ for $0 \leq x \leq 3$
20. $f(x) = k(x - x^2)$ for $2 \leq x \leq 5$
21. $f(x) = k(2x - x^2)$ for $0 \leq x \leq 2$
22. $f(x) = ke^{-2x}$ for $0 \leq x \leq \infty$

In each of Exercises 23 to 30, for the given probability density function, determine the respective probabilities.

23. $f(x) = \frac{1}{4}$ for $0 \leq x \leq 4$

(a) $P(0 \leq X \leq 1)$ (b) $P(1 \leq X \leq 2)$
(c) $P(2 \leq X \leq 4)$ (d) $P(\frac{1}{2} \leq X \leq \frac{5}{2})$

24. $f(x) = -x/4 + \frac{3}{4}$ for $0 \leq x \leq 2$
(a) $P(0 \leq X \leq 1)$ (b) $P(\frac{1}{2} \leq X \leq \frac{3}{2})$
(c) $P(X \geq \frac{3}{4})$ (d) $P(X \leq \frac{1}{2})$

25. $f(x) = -2x/16 + \frac{8}{16}$ for $0 \leq x \leq 4$
(a) $P(1 \leq X \leq 3)$ (b) $P(X \geq 2)$
(c) $P(X \leq 4)$ (d) $P(2 \leq X \leq 6)$

26. $f(x) = -2x/21 + \frac{8}{21}$ for $0 \leq x \leq 3$
(a) $P(1 \leq X \leq \frac{3}{2})$ (b) $P(X \geq 1)$
(c) $P(X \geq 4)$ (d) $P(2 \leq X \leq 3)$

27. $f(x) = \frac{3}{65} x^2$ for $1 \leq x \leq 4$
(a) $P(1 \leq X \leq 2)$ (b) $P(X \geq 2)$
(c) $P(X \leq 3)$ (d) $P(0 \leq X \leq 2)$

28. $f(x) = 20x^3 - 20x^4$ for $0 \leq x \leq 1$
(a) $P(0 \leq X \leq \frac{1}{2})$ (b) $P(X \geq \frac{1}{2})$
(c) $P(X \leq \frac{1}{3})$ (d) $P(X = \frac{1}{2})$

29. $f(x) = 12x - 24x^2 + 12x^3$ for $0 \leq x \leq 1$
(a) $P(X \leq \frac{1}{2})$ (b) $P(0 \leq X \leq 2)$
(c) $P(X \geq \frac{1}{4})$ (d) $P(X \leq 1)$

30. $f(x) = 3e^{-3x}$ for $0 \leq x \leq \infty$
(a) $P(X \leq 1)$ (b) $P(1 \leq X \leq 2)$
(c) $P(X \geq 3)$ (d) $P(0.2 \leq X \leq 0.4)$

31. The time in hours to failure of a battery in a transistor radio is described by the probability density function $f(t) = kt^3$ for $0 \leq t \leq 10$. Determine the value of k. Moreover, determine the probability that a randomly selected battery will fail within the first 4 hr of operation; also determine the probability that it will not fail before 8 hr of operation. Finally determine the average time to failure.

32. The time in minutes between arrivals of aircraft at Los Angeles International Airport is described by the probability density function $f(t) = ke^{-30t}$ for $0 \leq t \leq \infty$. Determine the value of k. Moreover, determine the probability that an aircraft will land within 2 min behind the aircraft preceding it. Finally determine the probability that an aircraft will land within 30 seconds behind the aircraft preceding it.

33. The time in minutes between customer arrivals at the checkout counter of the Super Saver Discount Store is described by the probability density function $f(t) = k(5t - t^2)$ for $0 \leq t \leq 5$. Determine the value of k. Also determine the probability that two customers arrive within 1 min of each other, and determine the probability that they arrive within 3 min of each other. Finally determine the average time between arrivals.

34. The distance in feet by which a shot from an antiaircraft gun misses its target is described by the probability density $f(x) = k$ for $0 \leq x \leq 50$. Determine the value of k.

Also determine the probability of hitting within 20 ft of the target, and determine the probability of hitting within 30 ft or less but not closer than 10 ft. Finally determine the average distance from the target.

35. The time in weeks to failure of a light bulb is described by the probability density function $f(x) = K(15 - x)$ for $0 \leq x \leq 15$. Determine the value of K. Also determine the probability that a randomly selected light bulb will fail within the first 2 weeks of operation; also determine the probability that it will fail in the fifth or sixth week of operation. Finally determine the average time to failure.

36. The time in hours between fire alarms in a small town is described by the probability density function $f(x) = K(10x - 0.5x^2)$ for $0 \leq x \leq 16$. Determine the value of K. Also determine the probability that there will be at least 14 hr between 2 alarms; also determine the probability that the time between 2 alarms is between 8 and 12 hr. Finally determine the average time between alarms.

37. The distance in yards between defects in weaving material is described by the probability density function $f(t) = K/t^2$ for $40 \leq t \leq 120$. Determine the value of K. Also determine the probability that 2 defects are within 50 yd of each other, and determine the probability that the distance between 2 defects is between 100 and 115 yd.

38. The Rubber Tire Company has determined that the distance in miles that their tires will last is described by the probability density function $f(x) = K(x^2 + x)$ where $1 \leq x \leq 6$ and x is in 10,000's. A trucking company will not purchase tires unless the probability of the tires lasting 40,000 miles is at least 0.75. Will the trucking company purchase tires from the Rubber Tire Company?

IMPORTANT TERMS AND CONCEPTS

Antiderivatives
Area under a curve
Beta probability density function
Constant of integration
Continuous random variable
Definite integral
Differential
Expected value
Fixed cost
Fundamental theorem of integral calculus

Improper integral
Integral calculus
Integrand
Integration
Lower limit
Normal probability density function
Probability density function
Upper limit
Variable cost

REVIEW PROBLEMS

1. The marginal cost for manufacturing x television sets is given by $c(x) = 6x^2 - 20x + 50$. There is also a fixed cost of $1,000.

(a) Determine the total cost function.

(b) What is the total cost of producing 10 television sets?

2. The Quick-Close Real Estate Company has determined that its marginal profit is given by $p(x) = 2(2x+4)^2$, where x represents the number of salespeople they employ. Quick-Close will incur a $2,000 fixed cost for rent regardless of the number they employ.

(a) Determine the total profit function.
(b) What is the total cost if they employ 15 salespeople?

3. The marginal cost for Texas Electronics, Inc., to manufacture a certain transistor is given by $c(x) = 0.15x^2 - 5x^{-2}$, where x is the number of transistors they manufacture. There is no fixed cost associated with the manufacturing operation.

(a) Determine the total cost function.
(b) What is the total cost to manufacture 5 transistors?

4. The manufacturing process for a certain type of cloth requires two separate operations. The marginal cost of the first operation is given by $c_1(x) = 6x - 20$ and the marginal cost of the second operation is given by $c_2(x) = 3x^2 - 4x + 5$. The x is the number of yards in thousands that pass through each operation. There is no fixed cost for the first operation but there is a $5,000 fixed cost associated with operation 2.

(a) Determine the total fixed cost function.
(b) What is the total fixed cost of producing 20,000 yd of cloth?

5. The marginal cost of operating an airplane is given by $c(x) = 20e^{2x}$, where x is the number of hours of flying time. There is also a $500 fixed cost. What is the total cost function?

6. Given the function $f(x) = x^2 - 6x + 8$, determine the area between the curve of the function and the x axis

(a) Bounded by $x = 5$ and $x = 7$
(b) Bounded by $x = 0$ and $x = 3$
(c) Bounded by $x = 1$ and $x = 5$

7. Given the two functions $f(x) = x^2 - 16x + 64$ and $g(x) = x + 4$,

(a) Determine the net area between the functions bounded by $x = 0$ and $x = 15$.
(b) Determine the total area between the functions bounded by $x = 0$ and $x = 15$.

8. Determine the increase in profit for the Quick-Close Real Estate Company (Problem 2) if they employ 20 salespersons instead of 15.

9. Determine the increase in cost of manufacturing, from Problem 4, for 30,000 yd of cloth instead of 20,000 yd.

10. The time in months to failure for your heavy duty tractor is described by the probability density function $f(t) = k(t^2 + 2)$ for $0 \leq t \leq 24$.

(a) Determine the value of k.
(b) Determine the average failure time.

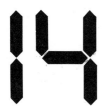

14 INTEGRAL CALCULUS: ADVANCED METHODOLOGY

14.1 INTRODUCTION

In Chap. 13, we considered a few rules of integration; however, the only method of integration that was discussed was that of inspection, i.e., determining the function whose derivative is the integrand. In this chapter we shall consider three methods of integration that give us increased flexibility in determining integrals: *change of variable*, *integration by parts*, and *integration through use of integral tables*.

The change-of-variable method is somewhat specialized in its use; however, it can be used to change a seemingly complex problem into a fairly simple one. Integration by parts can be used rather extensively for those functions that do not appear to be derivatives in their "perfect" form; it is developed through use of the property concerning the derivative of a product. Finally integration through use of integral tables is the simplest way to integrate difficult functions.

14.2 CHANGE OF VARIABLE

One of the easiest ways of solving any problem is to express it in terms of a simple problem with which we are familiar, and this is precisely the approach used when integrating a function by the change-of-variable method. For example, we can simply restate Rule 13.4 (Chap. 13) in a less complicated form; that is, in formula (13.5)

$$\int [f(x)]^n D_x f(x)\,dx = \frac{[f(x)]^{n+1}}{n+1} + C$$

we are integrating the derivative of a composite function. Thus, if we let $u = f(x)$ and $du = D_x f(x)\,dx$, the integrand becomes u^n and the integral is

$$\int u^n\,du = \frac{u^{n+1}}{n+1} + C$$

which is generally easier to recognize than when in the form of formula (13.5). It should be easy to see how the method received its name: We have simply changed the variable of integration from x to u.

The previous example demonstrates the concept associated with the method. However, for the general situation where we wish to determine the integral of a function that is of the chain-rule type, we can simplify the integration process by changing the variable of integration; that is, if we wish to determine the integral

$$\int f[\,g(x)\,]D_x g(x)\,dx \qquad (14.1)$$

we can let the variable u represent the function $g(x)$ and let the term du represent $D_x g(x)\,dx$. Thus the integral becomes

$$\int f(u)\,du \qquad (14.2)$$

which is less complicated to determine. After determining the antiderivative of $f(u)$, we simply substitute $g(x)$ for u in order to express the result as a function of x.

Example 14.1 Determine the indefinite integral

$$\int 2x(x^2+2)^3\,dx$$

Solution Let $u = x^2 + 2$; then $du = 2x\,dx$. Rearranging the terms and substituting, we have

$$\int (x^2+2)^3 2x\,dx = \int u^3\,du$$

Integrating with respect to u, we obtain

$$\int u^3\,du = \frac{u^4}{4} + C$$

Since we wish for the resulting integral to be a function of x, we substitute $x^2 + 2$ in place of u to obtain

$$\int 2x(x^2+3)^3\,dx = \frac{(x^2+2)^4}{4} + C$$

Example 14.2 Determine the indefinite integral

$$\int (1-x)^{3/2}\,dx$$

Sec. 14.2 / CHANGE OF VARIABLE

Solution Let $u = 1-x$; then $du = -dx$. Solving for dx, this becomes $dx = -du$. Substituting u for $1-x$ and $-du$ for dx, integrating with respect to u, and substituting $1-x$ in place of u in the resulting indefinite integral, we have

$$\int (1-x)^{3/2} dx = \int u^{3/2}(-du)$$
$$= -\int u^{3/2} du$$
$$= -\tfrac{2}{5} u^{5/2} + C$$
$$= -\tfrac{2}{5}(1-x)^{5/2} + C$$

Example 14.3 Determine the indefinite integral

$$\int (x^2 - 2x)^3 (x-1) dx$$

Solution Let $u = x^2 - 2x$; then $du = (2x-2)dx = 2(x-1)dx$, or $\tfrac{1}{2} du = (x-1)dx$. Substituting, integrating, and resubstituting, we have

$$\int (x^2-2x)^3 (x-1) dx = \int u^3 \tfrac{1}{2} du$$
$$= \frac{1}{2} \int u^3 du$$
$$= \frac{1}{2} \frac{u^4}{4} + C$$
$$= \frac{(x^2-2x)^4}{8} + C$$

Example 14.4 Determine the indefinite integral

$$\int \frac{x^2-1}{(x^3-3x)^2} dx$$

Solution Let $u = x^3 - 3x$; then $du = 3x^2 - 3 dx$, or $\tfrac{1}{3} du = (x^2-1)dx$. With this substitution, we obtain

$$\int \frac{x^2-1}{(x^3-3x)^2} dx = \int \frac{1}{u^2} \frac{1}{3} du$$
$$= \tfrac{1}{3} \int u^{-2} du$$
$$= -\tfrac{1}{3} u^{-1} + C$$
$$= -\frac{1}{3(x^3-3x)} + C$$

Definite Integrals In determining the previous indefinite integrals, we needed to express the resulting function as a function of the original variable of integration. In order to evaluate a definite integral from a to b, we can simply evaluate the indefinite integral in terms of the original variable of integration. For example, if we wish to evaluate the definite integral

$$\int_0^1 2x(x^2+2)^3\,dx$$

we can evaluate the result of Example 14.1 at a and b and take the appropriate difference; that is,

$$\int_1^2 2x(x^2+2)^3\,dx = \left.\frac{(x^2+2)^4}{4}\right|_1^2 = 324 - \frac{81}{4} = 303\frac{3}{4}$$

In many situations, the definite integral can more easily be evaluated for the new variable of integration than for the original variable. To employ the change-of-variable method to evaluate

$$\int_a^b f[\,g(x)\,]D_x g(x)\,dx$$

we can make the substitution $u = g(x)$. However, if we let u_1 and u_2 be values of u such that $u_1 = g(a)$ and $u_2 = g(b)$, then it can be proved that

$$\int_a^b f[\,g(x)\,]D_x g(x)\,dx = \int_{u_1}^{u_2} f(u)\,du \qquad (14.3)$$

provided that, as x changes continuously from a to b, $u(x)$ increases or decreases continuously from u_1 to u_2.

Example 14.5 Evaluate

$$\int_1^2 2x(x^2+2)^3\,dx$$

Solution Let $u = x^2 + 2$; then $du = 2x\,dx$. Furthermore, $x^2 + 2$ increases continuously as x increases from 1 to 2 such that $u(1) = 3$ and $u(2) = 6$. Thus we can use (14.3) to obtain

$$\int_1^2 2x(x^2+2)^3\,dx = \int_3^6 u^3\,du$$

$$= \left.\frac{u^4}{4}\right|_3^6$$

$$= 324 - \tfrac{81}{4}$$

$$= 303\tfrac{3}{4}$$

Sec. 14.2 / CHANGE OF VARIABLE

Example 14.6 Evaluate

$$\int_0^1 (1-x)^{3/2}\, dx$$

Solution Let $u = 1 - x$; then $du = -dx$, or $-du = dx$. Furthermore, $1 - x$ decreases continuously as x increases from 0 to 1 such that $u(0) = 1$ and $u(1) = 0$. Thus we can use (14.3) to obtain

$$\int_0^1 (1-x)^{3/2}\, dx = \int_1^0 -u^{3/2}\, du$$

$$= -\tfrac{2}{5} u^{5/2} \Big|_1^0$$

$$= -0 + \tfrac{2}{5}$$

$$= \tfrac{2}{5}$$

In Example 14.6 we should note that the limits of integration for the new variable u are reversed from the normal order. These limits must correspond with the appropriate limit values of x. [One way of determining whether $u(x)$ is increasing (or decreasing) continuously throughout the interval $a \leq x \leq b$ is to determine the first derivative of $u(x)$ and determine whether or not it has any roots between a and b.]

Integration by change of variable is helpful for many situations, particularly for integrands that involve composite functions. However, there are many situations for which this method is not particularly helpful, and in Sec. 14.3 we shall examine another method that may be more useful.

EXERCISES

Determine each of the indefinite integrals in Exercises 1 to 12.

1. $\int (3x+6)^4\, dx$

2. $\int \sqrt{x+4}\, dx$

3. $\int (x+2)^6\, dx$

4. $\int (x^2+3)^2 2x\, dx$

5. $\int 3x^2(x^3+2)^4\, dx$

6. $\int \dfrac{x+1}{\sqrt{2x^2+4x}}\, dx$

7. $\int \dfrac{5t}{\sqrt{2-3t^2}}\, dt$

8. $\int \dfrac{t}{(4-t^2)^3}\, dt$

9. $\int \dfrac{y+2}{(y^2+4y+5)}\, dy$

10. $\int 6y\sqrt{2y^2+2}\, dy$

11. $\int \dfrac{z^3}{\sqrt{z^4+2}}\, dz$

12. $\int \dfrac{(2+\sqrt{z})^5}{2\sqrt{z}}\, dz$

Evaluate each of the definite integrals in Exercises 13 to 20.

13. $\int_0^2 (x-2)^3 \, dx$

14. $\int_0^5 \sqrt{x+4} \, dx$

15. $\int_0^3 x(x^2+16)^{1/2} \, dx$

16. $\int_2^{10} \dfrac{t+1}{\sqrt{t-1}} \, dt$

17. $\int_0^5 \dfrac{t+2}{\sqrt{9-t}} \, dt$

18. $\int_1^3 \dfrac{t+2}{(t^2+4t)^2} \, dt$

19. $\int_1^6 \dfrac{y}{\sqrt{3+y}} \, dy$

20. $\int_0^1 y e^{y^2} \, dy$

21. The marginal cost function in dollars for manufacturing a particular model airplane is given by

$$c(x) = x(x^2+16)^{1/2}$$

where x is the number of units made; the fixed cost is $250. Determine the total cost function and the total cost for three units.

22. The marginal profit function in thousands of dollars for a brand of aftershave lotion is given by

$$p(x) = \dfrac{x+2}{\sqrt{x+4}}$$

where x represents the thousands of dollars spent on promotional advertising. Furthermore, historically, when $5,000 has been spent on advertising, the total profit has been about $40,000. Determine the total profit function and the total profit when $12,000 is spent on advertising. Also determine the increase in profit for increasing expenditures from $12,000 to $21,000.

23. For the probability density function $f(x) = kx(1-x^2)^3$ for $0 \leq x \leq 1$, determine the value of k. Also determine $P(0 \leq X \leq \tfrac{1}{2})$ and $P(X \geq 0.2)$.

24. For the probability density function $f(x) = kxe^{-x^2}$ for $0 \leq x < \infty$, determine the value of k. Also determine $P(0 \leq X \leq 2)$, $P(1 \leq X \leq 3)$, and $P(X \geq 3)$.

25. The telephone company has determined that the time in minutes between information calls is given by the probability density function

$$f(t) = \dfrac{kt}{\sqrt{t+2}} \quad \text{for } 0 \leq t \leq 8$$

Determine the value of k. Moreover, determine the probability that there will be no calls in a 2-min period; also determine the probability that there will be at least 6 min between calls.

26. The marginal profit for manufacturing and selling x thousand widgets is given by the function $f(x) = x^3(\sqrt{x^4 - 5{,}000})$. Fixed cost is $25,500. Determine the total

profit function and the total profit if 10,000 widgets are manufactured and sold. Also determine the increase in profit for increasing sales volume from 10,000 to 12,000 units.

27. A furniture company has determined that its marginal cost function for manufacturing chairs is given by

$$c(x) = \frac{x}{10}(x^2 - 175)$$

where x is the number of chairs manufactured in hundreds. Total fixed cost is $1,000. Determine the total cost function and the cost of manufacturing 1,500 chairs. Also determine the increase in cost for increasing production from 1,500 to 2,000 chairs.

28. The time in days between auto accidents at a busy corner is given by the probability density function $f(t) = k(t^{-2} + 4)^2 / t^3$ for $1 \leq t \leq 16$. Determine the value of k. Also determine the probability that two accidents will occur within 4 hr of each other, and determine the probability that two accidents occur within 4 to 9 hr of each other.

14.3 INTEGRATION BY PARTS

In many situations, we are confronted with the problem of integrating the product of two functions, neither of which appears to be the derivative of the other. When this happens, generally we can use the method of integration by parts, which is based on the inversion of the formula for the differentiation of a product.

From Rule 10.10 we have the equation for the derivative of a product; that is,

$$D_x[f(x)g(x)] = D_x f(x) g(x) + f(x) D_x g(x)$$

If we integrate both sides of this equation with respect to x, we obtain

$$\int D_x[f(x)g(x)] dx = \int g(x) D_x f(x) dx + \int f(x) D_x g(x) dx$$

We note that the integral to the left of the equality has an integrand for which the antiderivative is simply $f(x)g(x)$. In order to simplify the notation, let $u = f(x)$ and $v = g(x)$; then $du = D_x f(x) dx$ and $dv = D_x g(x) dx$. Substituting these changes into the equation, we have

$$uv = \int v \, du + \int u \, dv$$

Solving for one of the integrals, say, $\int u \, dv$, we obtain

$$\int u \, dv = uv - \int v \, du \qquad (14.4)$$

Thus, if we can choose u and dv such that we may integrate dv easily and such that the integral $\int v\,du$ is easier to determine than the original integral, we may use (14.4) to our advantage.

Example 14.7 Determine the indefinite integral

$$\int xe^x\,dx$$

Solution Let $u = x$ and $dv = e^x\,dx$; then $du = dx$ and $v = e^x$. Substituting into (14.4) and integrating the second term on the right, we obtain

$$\int xe^x\,dx = xe^x - \int e^x\,dx$$
$$= xe^x - e^x + C$$

The choice of u and dv is extremely important, as can be seen in Example 14.7. If we were to choose $u = e^x$ and $dv = x\,dx$, then $du = e^x\,dx$ and $v = x^2/2$, which, when substituted into (14.4), yields

$$\int xe^x\,dx = \frac{x^2}{2}e^x - \int x^2 e^x\,dx$$

In this case, the term $\int x^2 e^x\,dx$ is more difficult to integrate than the original term $\int xe^x\,dx$. However, if our first choice of u and dv does not work out, sometimes we can reverse the choice and obtain favorable results. Also, there are some functions that render themselves to a variety of choices. There is no single way of choosing u and dv that works every time. The ability to make good choices improves with practice.

Example 14.8 Determine the indefinite integral

$$\int \ln x\,dx$$

Solution Let $u = \ln x$ and $dv = dx$; then $du = 1/x\,dx$ and $v = x$. Substituting into (14.4), we obtain

$$\int \ln x\,dx = x \ln x - \int \frac{1}{x} x\,dx$$
$$= x \ln x - \int dx$$
$$= x \ln x - x + C$$

Example 14.9 Determine the indefinite integral

$$\int x(\ln x)\,dx$$

Sec. 14.3 / INTEGRATION BY PARTS

Solution Let $u = \ln x$ and $dv = x\,dx$; then $du = 1/x\,dx$ and $v = x^2/2$. Substituting, we obtain

$$\int x \ln x\,dx = \frac{x^2}{2} \ln x - \int \frac{x^2}{2} \frac{1}{x}\,dx$$

$$= \frac{x^2}{2} \ln x - \frac{x^2}{4} + C$$

Occasionally it is necessary to apply the process of integration by parts two or more times. When this is the case, we should be very careful that the new integral $\int v\,du$ is improving in a direction that will eventually lead to a function instead of an integral.

Example 14.10 Determine the indefinite integral

$$\int x^2 e^x\,dx$$

Solution Let $u = x^2$ and $dv = e^x\,dx$; then $du = 2x\,dx$ and $v = e^x$. Substituting, we obtain

$$\int x^2 e^x\,dx = x^2 e^x - \int e^x 2x\,dx$$

$$= x^2 e^x - 2 \int x e^x\,dx$$

Next we need to integrate $\int xe^x\,dx$. Since we did this in Example 14.7, we substitute the result and obtain

$$\int x^2 e^x\,dx = x^2 e^x - 2(xe^x - e^x + C)$$

$$= x^2 e^x - 2xe^x + 2e^x + C'$$

To determine the definite integral, we do nothing differently than we did when we first considered them because we are always integrating with respect to x and the limits are values of x; that is, the definite integral form of (14.4) is

$$\int_a^b u(x) D_x v(x)\,dx = u(x)v(x) \Big|_a^b - \int_a^b v(x) D_x u(x)\,dx \qquad (14.5)$$

Example 14.11 Determine the definite integral

$$\int_0^2 xe^{-2x}\,dx$$

Solution Let $u=x$ and $dv=e^{-2x}dx$; then $du=dx$ and $v=-\frac{1}{2}e^{-2x}$. Substituting, we obtain

$$\int_0^2 xe^{-2x}dx = \frac{1}{2}xe^{-2x}\Big|_0^2 - \int_0^2 -\frac{1}{2}e^{-2x}dx$$

$$= -e^{-4} + 0 - \frac{1}{4}e^{-2x}\Big|_0^2$$

$$= -e^{-4} - \frac{1}{4}e^{-4} + \frac{1}{4}e^0$$

$$= -\frac{5}{4}e^{-4} + \frac{1}{4}$$

$$= -\frac{5}{4}(0.01832) + 0.25000$$

$$= 0.22710$$

EXERCISES

In Exercises 1 to 12, determine the indefinite integral.

1. $\int x(x+2)^6 dx$

2. $\int (x+1)(x+3)^3 dx$

3. $\int xe^{3x} dx$

4. $\int x \ln x^2 dx$

5. $\int x^2(x+1)^4 dx$

6. $\int (\ln x)^2 dx$

7. $\int x^2 \ln x\, dx$

8. $\int x^3 \ln x\, dx$

9. $\int (x+2)^2(x+5)^8 dx$

10. $\int x^2 e^{3x} dx$

11. $\int x^3 e^{2x^2} dx$

12. $\int x^n \ln x\, dx$

In Exercises 13 to 16, determine the value of the definite integral.

13. $\int_0^1 x(x+1)^4 dx$

14. $\int_1^2 (x-1)^2(x-2)^4 dx$

15. $\int_0^2 3x^2 e^{2x} dx$

16. $\int_{-1}^1 (x+1)^2(x-1)^5 dx$

17. For the probability density function $f(x)=2e^{-2x}$ for $0 \leq x < \infty$, determine $E(X)$.

18. For the probability function $f(x)=56x(1-x)^6$ for $0 \leq x \leq 1$, determine $E(X)$.

19. For the probability density function $f(x)=105x^2(1-x)^4$ for $0 \leq x \leq 1$, determine $E(X)$.

20. For the probability density function $f(x)=2/\sqrt{2\pi}\, e^{-x^2/2}$ for $0 \leq x < \infty$, determine $E(X)$.

Sec. 14.4 / INTEGRAL TABLES

21. The number of people in the check-out lane at a large store is given by the probability density $f(n) = \frac{1}{64}(n-2)^3$ for $2 \leq n \leq 6$. Determine the number of people expected to be in the check-out lane.

22. The marginal profit function for growing and selling corn is given by

$$p(x) = (x-1)(x-5)^2$$

where x is the number of bushels grown and sold in tens. Fixed cost for the farmer is $25. Determine the total profit function and the total profit for selling 100 bu of corn. Determine the amount of increase in profits by increasing sales from 130 bu to 160 bu.

23. A sales department has a daily marginal cost function that is given by

$$c(x) = (x-3)^2(x+2)^2$$

where x is the number of sales people employed, and fixed cost is $75. Determine the total cost function and the cost of having 10 sales people employed. Determine the amount that could be saved by decreasing the sales force from 10 to 8 without affecting total sales.

14.4 INTEGRAL TABLES

The methods of integration discussed in Secs. 14.2 and 14.3 are far from exhaustive; however, we shall not consider the more specialized methods. In order to save time, *tables of integrals* have been constructed for the more common types of integral forms, and these tables present the integral formulas in a systematic order according to the type of function in the integrand. We should note that some of the formulas involve logarithms of the absolute value of an expression, because the expression must be positive for the logarithm to be defined. However, a table of integrals does not contain formulas for all the integrals that we may encounter. Thus we must supplement the use of an integral table with our knowledge of the different methods of integration.[1]

Table 14.1 is a short table of integrals. Use of any table of integrals involves simply making the proper substitution of the appropriate parameters and writing the answer.

Example 14.12 Determine the indefinite integral.

$$\int \frac{x}{2+3x} dx$$

---

[1] There are a number of books that contain extensive collections of integrals. Two of the better-known volumes are:

H. B. Dwight: *Tables of Integrals and Other Mathematical Data*, The Macmillan Company, New York, 1961.

B. O. Pierce: *A Short Table of Integrals*, Ginn and Company, Boston, 1957.

Table 14.1 Table of Integrals

Elementary Forms

1. $$\int kf(x)\,dx = k\int f(x)\,dx$$

2. $$\int [f(x)+g(x)]\,dx = \int f(x)\,dx + \int g(x)\,dx$$

3. $$\int [f(x)-g(x)]\,dx = \int f(x)\,dx - \int g(x)\,dx$$

4. $$\int x^n\,dx = \frac{x^{n+1}}{n+1} + C \quad \text{(for } n \neq 1\text{)}$$
$$= \ln|x| + C \quad \text{(for } n = -1\text{)}$$

5. $$\int [f(x)]^n D_x f(x)\,dx = \frac{[f(x)]^{n+1}}{n+1} + C \quad \text{(for } n \neq -1\text{)}$$
$$= \ln f(x) + C \quad \text{(for } n = -1\text{)}$$

Forms Containing $a + bx$

6. $$\int \frac{x}{a+bx}\,dx = \frac{1}{b^2}(a+bx - a\ln|a+bx|) + C$$

7. $$\int (a+bx)^n\,dx = \frac{1}{b(n+1)}(a+bx)^{n+1} + C \quad \text{(for } n \neq -1\text{)}$$
$$= \frac{1}{b}\ln|a+bx| + C \quad \text{(for } n = -1\text{)}$$

8. $$\int \frac{x}{(a+bx)^n}\,dx = \frac{1}{b^2}\left[\frac{-1}{(n-2)(a+bx)^{n-2}} + \frac{a}{(n-1)(a+bx)^{n-1}}\right] + C$$

$$\text{(for } n \neq 1 \text{ and } n \neq 2\text{)}$$

$$= \frac{1}{b^2}(a+bx - a\ln|a+bx|) + C \quad \text{(for } n=1\text{)}$$

$$= \frac{1}{b^2}\left(\ln|a+bx| + \frac{a}{a+bx}\right) + C \quad \text{(for } n=2\text{)}$$

9. $$\int x(a+bx)^n\,dx = \frac{1}{b^2(n+2)}(a+bx)^{n+2} - \frac{a}{b^2(n+1)}(a+bx)^{n+1} + C$$

$$\text{(for } n \neq -1, -2\text{)}$$

$$= \frac{1}{b^2}[a+bx - a\ln|a+bx|] + C \quad \text{(for } n=-1\text{)}$$

$$= \frac{1}{b^2}\left[\ln|a+bx| + \frac{a}{a+bx}\right] + C \quad \text{(for } n=-2\text{)}$$

Table 14.1 (continued)

Forms Containing $x^2 \pm a^2$, or $a^2 \pm x^2$

10. $\quad \int \dfrac{1}{x^2 - a^2} \, dx = \dfrac{1}{2a} \ln \left| \dfrac{x-a}{x+a} \right| + C$

11. $\quad \int \dfrac{1}{a^2 - x^2} \, dx = \dfrac{1}{2a} \ln \left| \dfrac{a+x}{a-x} \right| + C$

12. $\quad \int \sqrt{x^2 \pm a^2} \, dx = \tfrac{1}{2} \left[x\sqrt{x^2 \pm a^2} \pm a^2 \ln \left(x + \sqrt{x^2 \pm a^2} \right) \right] + C$

13. $\quad \int \dfrac{1}{\sqrt{x^2 \pm a^2}} \, dx = \ln \left(x + \sqrt{x^2 \pm a^2} \right) + C$

14. $\quad \int x\sqrt{x^2 \pm a^2} \, dx = \tfrac{1}{3} \sqrt{(x^2 \pm a^2)^3} + C$

15. $\quad \int \dfrac{x}{\sqrt{x^2 \pm a^2}} \, dx = \sqrt{x^2 \pm a^2} + C$

16. $\quad \int x\sqrt{a^2 - x^2} \, dx = -\tfrac{1}{3} \sqrt{(a^2 - x^2)^3} + C$

17. $\quad \int \dfrac{x}{\sqrt{a^2 - x^2}} \, dx = -\sqrt{a^2 - x^2} + C$

Exponential Forms

18. $\quad \int a^{bx} \, dx = \dfrac{a^{bx}}{b(\ln a)} + C$

19. $\quad \int e^{bx} \, dx = \dfrac{e^{bx}}{b} + C$

20. $\quad \int x e^{bx} \, dx = \dfrac{e^{bx}}{b^2} (bx - 1) + C$

Logarithmic Forms

21. $\quad \int x(\ln x) \, dx = \dfrac{x^2}{2} (\ln x) - \dfrac{x^2}{4} + C$

22. $\quad \int \dfrac{1}{x(\ln x)} \, dx = \ln (\ln x) + C$

23. $\quad \int (\ln x)^2 \, dx = x(\ln x)^2 - 2x(\ln x) + 2x + C$

24. $\quad \int \ln (ax) \, dx = x(\ln ax) - x + C$

Solution The integral is of the form of formula 6 in Table 14.1, where $a=2$ and $b=3$. Substituting, we have

$$\int \frac{x}{2+3x}dx = \frac{1}{9}(2+3x-2\ln|2+3x|)+C$$

Example 14.13 Determine the indefinite integral

$$\int (3+4x)^6 dx$$

Solution The integral is of the form of formula 7 in Table 14.1, where $a=3$, $b=4$, and $n=6$. Substituting, we obtain

$$\int (3+4x)^6 dx = \tfrac{1}{28}(3+4x)^7 + C$$

Example 14.14 Determine the indefinite integral

$$\int \frac{1}{x^2-25}dx$$

Solution The integral is of the form of formula 10 in Table 14.1, where $a=5$. Thus

$$\int \frac{1}{x^2-25}dx = \frac{1}{10}\ln\left|\frac{x-5}{x+5}\right| + C$$

Example 14.15 Determine the indefinite integral

$$\int 5^{3x} dx$$

Solution Formula 18 in Table 14.1 is appropriate, where $a=5$ and $b=3$. Substituting, we have

$$\int 5^{3x} dx = \frac{5^{3x}}{3\ln 5} + C$$

$$= \frac{5^{3x}}{4.8283} + C$$

EXERCISES

In each of Exercises 1 to 18 use the appropriate integral formula given in Table 14.1 to determine the integral.

1. $\int \frac{x}{5+x}dx$

2. $\int (10+2x)^3 dx$

3. $\int x(3+8x)^3 dx$

4. $\int x(2+5x)^{-4} dx$

5. $\displaystyle\int \frac{1}{\sqrt{x^2-16}}\,dx$

6. $\displaystyle\int \frac{x}{\sqrt{9-x^2}}\,dx$

7. $\displaystyle\int e^{-6x}\,dx$

8. $\displaystyle\int xe^{2x}\,dx$

9. $\displaystyle\int 3^{2x}\,dx$

10. $\displaystyle\int \ln(4x)\,dx$

11. $\displaystyle\int 3\ln(2x)\,dx$

12. $\displaystyle\int x(3-x^2)^{-1/2}\,dx$

13. $\displaystyle\int x(x^2+4)^{1/2}\,dx$

14. $\displaystyle\int \frac{x}{x+2}\,dx$

15. $\displaystyle\int_1^2 x\ln x\,dx$

16. $\displaystyle\int_4^6 \frac{1}{x(\ln x)}\,dx$

17. $\displaystyle\int (\ln 3x)^2\,dx$

18. $\displaystyle\int x\ln(4x)\,dx$

IMPORTANT TERMS AND CONCEPTS

Change of variable
Integral tables
Integration by parts

REVIEW PROBLEMS

1. The marginal profit for a new line of Elektronix CB radios is given by $p(x) = (x^2-1)(x^3-3x)$, where x is the number of radios sold in thousands. There is a fixed cost of \$30,000 that will be incurred, regardless of the sales volume.

   (a) Determine the total profit function.
   (b) Determine the total profit if 10,000 units are sold.

2. The East Publishing Company has determined that the marginal cost of publishing a textbook has been determined to be $c(x) = 4x(x^2+150)$, where x is the number of books produced in thousands. There is a fixed cost of \$10,000.

   (a) Determine the total cost function.
   (b) Determine the increase in cost if production were increased from 10,000 to 12,000 books.

3. The marginal cost to manufacture flowerpots at the Pottery Pot Maker Plant is given by $c(x) = x/\sqrt{x^2+10}$, where x is the number of flower pots produced. There is no associated fixed cost.

   (a) Determine the total cost function.
   (b) Determine the cost of manufacturing 20 flower pots.

4. The accounting department of Wood Products, Inc., has determined that the marginal cost of manufacturing pencils is given by $c(x) = 3x^2(x+1)$, where x is the number of pencils in thousands produced. There is a fixed cost of $500.

(a) Determine the total cost function.
(b) Determine the total cost of manufacturing 5,000 pencils.

5. The Sharpies State Bank has determined that its marginal profit is approximated by $p(x) = x(3x^2 + 1)$, where x is deposits in millions of dollars. There is a fixed cost of $5,000 associated with operating the bank.

(a) Determine the total profit function.
(b) Determine the increase in profits if deposits increase from $10,000,000 to $20,000,000.

6. The Green-Grow Plant Food Company has established that the marginal cost to produce fertilizer is given by $c(x) = x^2 e^x / 2$, where x is the number of pounds produced. Determine the total cost function.

7. The marginal cost to manufacture an automobile is given by $c(x) = (10 + 10x)^2$, where x is the number of automobiles produced. There is a $20,000 fixed cost associated with the production process.

(a) Determine the total cost function.
(b) Determine the cost of manufacturing 9 automobiles.

8. The marginal profit for a certain line of wool sweaters produced by Kayday Fabrics, Inc., is given by $p(x) = x/\sqrt{x^2 + 25}$, where x is the number of sweaters sold. There is no fixed cost.

(a) Determine the total profit function.
(b) Determine the profit if 15 sweaters are sold.

9. The marginal cost of manufacturing automobile windows at the Blackwell Glass Factory is given by $c(x) = 3^{2x}$, where x is the number of windows produced in thousands. There is a fixed cost of $2,000, regardless of the number of windows produced.

(a) Determine the total cost function.
(b) Determine the total cost if 4,000 windows are produced.

10. Fairfield Electronics Company has determined that the marginal cost of manufacturing transistors is given by $c(x) = e^{10x}/5$, where x is the number of transistors, in millions, produced. Determine the total cost function.

ANSWERS TO SELECTED EXERCISES

CHAPTER 1

Section 1.2 **1.** (a) T (c) F (e) T (g) T (i) T
(k) T (m) F (o) T (q) T (s) T
3. (a) (c)

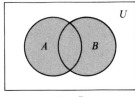

$A \cup B$

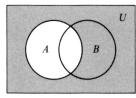

A'

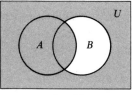

$A \cup B'$

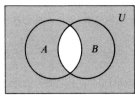

$A' \cup B'$

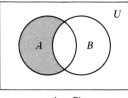

$A \cap B'$

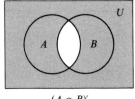

$(A \cap B)'$

476 ANSWERS TO SELECTED EXERCISES

9. 23
11. (a) 13   (b) 1   (c) 7   (d) 16
    (e) 33 (f) 21 (g) 27 (h) 34

Section 1.3
1. (a) 7   (c) 11   (e) $-\frac{15}{7}$
   (g) -6 (i) 6
3. $4.20   5. 300 miles
7. $48

Section 1.4
1. $X \times Y = \{(1,9),(1,10),(1,11),(2,9),(2,10),(2,11),(3,9),(3,10),(3,11)\}$
   $Y \times X = \{(9,1),(9,2),(9,3),(10,1),(10,2),(10,3),(11,1),(11,2),(11,3)\}$
3. $X \times Y = \{(1,10),(1,20),(2,10),(2,20),(3,10),(3,20),(4,10),(4,20)\}$
   $Y \times X = \{(10,1),(10,2),(10,3),(10,4),(20,1),(20,2),(20,3),(20,4)\}$
5. $X \times Y = \{(1,2),(1,3)\}$; $Y \times X = \{(2,1),(3,1)\}$
7. $R = \{-4, -5, -6\}$; $\{(16,-4),(25,-5),(36,-6)\}$
9. $R = \{8, 10, 12\}$; $\{(1,8),(2,10),(3,12)\}$
11. $R = \{1, \frac{1}{2}, \frac{1}{3}, \frac{1}{4}\}$; $\{(1,1),(2,\frac{1}{2}),(3,\frac{1}{3}),(4,\frac{1}{4})\}$
13. $\{(10,18),(16,27),(23,37.5)\}$

Section 1.5
1. (a) 13   (b) 16   (c) 25   (d) 7   (e) -5
3. (a) 217 (b) 65   (c) 2   (d) -7 (e) -63
5. $f(x) = 50x^2 - 100x + 200$
7. (a) -20 (b) 34 (c) 45 (d) 10 (e) 14
9. (a) $y = -3x + 9$ (b) $y = -2x + 5$ (c) $y = x + 5$
11. (a) $f(x) = 12x + 50{,}000$

CHAPTER 2

Section 2.2
1. $y = -2x + 6$; intercept $= 6$; slope $= -2$
3. $y = -\frac{2}{5}x + 2$; intercept $= 2$; slope $= -\frac{2}{5}$
5. $y = 4x - 2$; yes; no   7. $y = 3x - 2$; no; no
9. $y = -2x$; yes; no   11. $x = 3$
13. $x = 5$ 15. $x = \frac{9}{4}$
17. $x = -1$
19. $y = .5x - 200$
21. $y = 4x - 1200$

Section 2.3
1. $y = 8x^2 - 6x + 3$
3. $y = x^2 + 2x + 6$
5. (a) Upward   (b) $(\frac{5}{2}, -\frac{27}{4})$ (c) $(4,0)$ and $(1,0)$
7. (a) Downward   (b) $(\frac{11}{4}, \frac{25}{8})$ (c) $(\frac{3}{2},0)$ and $(4,0)$
9. (a) Upward   (b) $(0, -16)$ (c) $(-4,0)$ and $(4,0)$
11. (a) Upward   (b) Yes   (c) $(\frac{9}{2}, -\frac{9}{4})$ (d) $x = 3$, and $x = 6$
    (e)

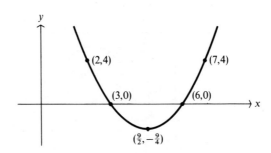

13. (a) Downward (b) No (c) (2, −1) (d) No real roots
(e)

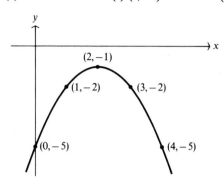

15. (a) Downward (b) Yes (c) (−2, 0) (d) $x = -2$
(e)

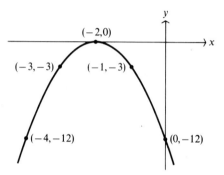

17. (a) $y = -50x^2 + 600x$

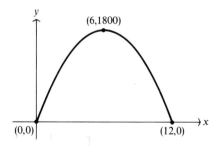

(b) $x = \$6$, and $y = \$1,800$ (c) 300 cakes

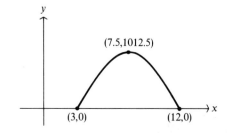

(e) $P=\$1,012.50$, and $x=\$7.50$ (f) 225 cakes (g) $112.50
19. $y = .06x^2 + 3x + 20$

CHAPTER 3

Section 3.2 1. (a) $\frac{8}{3}$ (b) 6 (c) $\frac{27}{2}$ (d) 8
3. (a) $\frac{3y^2}{xz}$ (b) $\frac{x^4}{y}$ (c) $(2x+1)(y+3)^6$ (d) $x^3 y^5 w^{10}$
5. (a) $y^{-2/3}$ (b) $x^{-3/2} y^{-1/2}$ (c) $x^{-1/2} y^{13/12}$ (d) $x^{3/4} y^{3/4}$

Section 3.3 1. 200 percent increase   3. 400 percent decrease

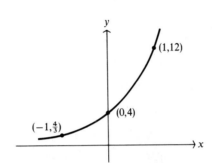

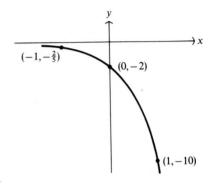

5. 582 percent increase   7. 1,100 percent increase

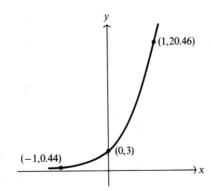

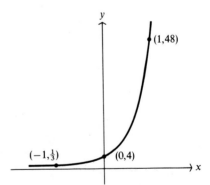

9. 36 percent decrease

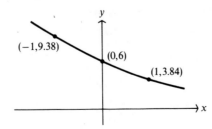

11. $f(x) = 2(0.5)^x$ 13. $f(x) = 1,500(2)^x$

ANSWERS TO SELECTED EXERCISES

Section 3.4   1. Yes   3. No   5. Yes   7. No   9. No
11. (a) 2.1004   (c) 2.8439   (e) 1.8591   (g) $0.3617-2$
12. (a) 72.3   (c) 126   (e) 5.61   (g) 698

Section 3.5   1. 1.82616   3. 0.28518   5. 1.29473   7. 4.49313   9. -0.65393
11. 5.03044   13. 13.13232   15. $e^{2.30259x}$ 17. $e^{1.60944x}$ 19. $e^{0.18232x}$

CHAPTER 4

Section 4.2   1. (a) $x=2, y=1$ (b) $x=3, y=-5$ (c) $w=1, z=1$
(d) $x=-4, y=3, z=5$ (e) $x_1=\frac{1}{2}, x_2=-\frac{1}{2}, x_3=\frac{5}{2}$
3. (a) Coincident   (c) (2,4)   (e) (2,1)
4. (a) $x=2, y=2, z=2$ (c) No solution   (e) No solution
5. Rubber: $1.50/lb; labor: $2.00/hr   7. 50 quarters; 65 dimes
9. 10 concerts; 15 dances
11. 30 hot dogs; 60 hamburgers
13. $1.90/short-sleeve shirt; $1.60/long-sleeve shirt
15. 25 regular; 5 quilted

Section 4.3   1. (a) 10   (c) 84   (e) 434   (g) 110
3. (a) $\sum_{i=1}^{4} 2i$ (c) $\sum_{i=2}^{5} ix_i$ (e) $\sum_{i=1}^{3}(x_i-y_i)$
5. (a) $\sum_{j=1}^{2} a_{ij}x_j = b_i \ (i=1,2,3)$ (c) $\sum_{j=1}^{3} c_{ij}x_j = b_i \ (i=1,2)$

Section 4.4   1. (a) $\begin{bmatrix} 4 & 7 & 4 \\ 7 & 1 & 9 \end{bmatrix}$ (c) $\begin{bmatrix} 5 & 15 & 10 \\ 30 & 0 & 20 \end{bmatrix}$
(e) $\begin{bmatrix} 5 & 1 & 1 \\ -4 & 1 & 4 \end{bmatrix}$ (g) $\begin{bmatrix} 15 & 21 & 13 \\ 22 & 3 & 30 \end{bmatrix}$
2. (a) **AB** is 1×1; **BA** is 2×2 (c) **AB** is 3×2; **BA** is undefined
(e) **AB** is 3×3; **BA** is 2×2 (g) **AB** is 3×3; **BA** is 3×3
3. (a) Undefined   (c) $\begin{bmatrix} 7 & 10 & 10 \\ 9 & 13 & 9 \end{bmatrix}$ (e) Undefined

5. (a) 3×2 (c) $\begin{bmatrix} 13 & 2 & 9 \\ -5 & 2 & -3 \\ 3 & 6 & 3 \end{bmatrix}$ (e) $\begin{bmatrix} 6 & 0 \\ 9 & 12 \end{bmatrix}$

6. (a) $[\frac{1}{4} \ \frac{1}{4} \ \frac{1}{4} \ \frac{1}{4}]$ (c) $\begin{bmatrix} 1 \\ 1 \\ 1 \\ 1 \end{bmatrix}$ (e) $\begin{bmatrix} 4 & 2 \\ 1 & 2 \\ 2 & 3 \\ 3 & 4 \end{bmatrix}$

7. (a) $[1 \ 1 \ 1]$ (c) $\begin{bmatrix} 3 \\ 2 \\ 4 \\ 10 \end{bmatrix}$ (e) $\begin{bmatrix} 3 & 4 & 9 \\ 2 & 6 & 3 \\ 4 & 7 & 8 \\ 10 & 3 & 1 \end{bmatrix}$

9. (a) $\begin{bmatrix} 3 & 4 \\ 2 & 1 \\ 4 & -8 \\ 6 & 4 \end{bmatrix} \begin{bmatrix} x \\ y \end{bmatrix} = \begin{bmatrix} 5 \\ 10 \\ 1 \\ 20 \end{bmatrix}$

480 ANSWERS TO SELECTED EXERCISES

11. (a) $\begin{bmatrix} 1 \\ 1 \\ 1 \end{bmatrix}$ (b) $\begin{bmatrix} 3 \\ 3 \\ 2 \\ 1 \\ 1 \end{bmatrix}$ (c) $\begin{bmatrix} 43 \\ 32 \\ 64 \end{bmatrix}$ (d) $\begin{bmatrix} 139 \\ 139 \\ 75 \\ 43 \\ 64 \end{bmatrix}$

Section 4.5

2. (a) $\begin{bmatrix} x \\ y \end{bmatrix} = \begin{bmatrix} 1 \\ 1 \end{bmatrix}$ (c) $\begin{bmatrix} x \\ y \end{bmatrix} = \begin{bmatrix} -\frac{1}{2} \\ -1 \end{bmatrix}$

3. (a) $\begin{bmatrix} x \\ y \\ z \end{bmatrix} = \begin{bmatrix} 5 \\ 0 \\ 0 \end{bmatrix}$ (c) $\begin{bmatrix} x \\ y \\ z \end{bmatrix} = \begin{bmatrix} 0 \\ 5 \\ 2 \end{bmatrix}$

4. $A^{-1} = \begin{bmatrix} 7 & -5 \\ -4 & 3 \end{bmatrix}$ (a) $\begin{bmatrix} x \\ y \end{bmatrix} = \begin{bmatrix} 1 \\ 1 \end{bmatrix}$ (c) $\begin{bmatrix} x \\ y \end{bmatrix} = \begin{bmatrix} 0 \\ 4 \end{bmatrix}$

5. $A^{-1} = \begin{bmatrix} \frac{11}{2} & -\frac{3}{2} & -1 \\ \frac{5}{4} & \frac{1}{4} & -\frac{1}{2} \\ -\frac{7}{4} & \frac{1}{4} & \frac{1}{2} \end{bmatrix}$ (a) $\begin{bmatrix} x_1 \\ x_2 \\ x_3 \end{bmatrix} = \begin{bmatrix} 1 \\ 1 \\ -1 \end{bmatrix}$ (c) $\begin{bmatrix} x_1 \\ x_2 \\ x_3 \end{bmatrix} = \begin{bmatrix} 5 \\ 3 \\ -2 \end{bmatrix}$

7. $\begin{bmatrix} x_1 \\ x_2 \\ x_3 \end{bmatrix} = \begin{bmatrix} 1 \\ 1 \\ 1 \end{bmatrix}$ 9. $A^{-1} = \begin{bmatrix} -3 & -3 & 10 \\ \frac{3}{2} & 1 & -4 \\ -\frac{1}{2} & 0 & 1 \end{bmatrix}; \begin{bmatrix} x_1 \\ x_2 \\ x_3 \end{bmatrix} = \begin{bmatrix} 1 \\ 1 \\ 1 \end{bmatrix}$

11. $\begin{bmatrix} x \\ y \\ z \end{bmatrix} = \begin{bmatrix} 3.0 \\ 3.5 \\ 2.5 \end{bmatrix}$ 13. $\begin{bmatrix} x \\ y \\ z \end{bmatrix} = \begin{bmatrix} .75 \\ .49 \\ .42 \end{bmatrix}$

CHAPTER 5

Section 5.1
1. (a) True (c) True (e) True (g) False
2. (a) $5x < 20$ (c) $x \leqslant 5$ (e) $x \geqslant -8$
3. (a) $x \leqslant 5$ (c) $x > 4$ (e) $x < 18$
5. $5 \leqslant x$
7. $x \geqslant 210$
9. $x \leqslant \$10.42$

Section 5.2 1. (b)

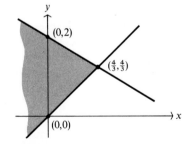

(c)

ANSWERS TO SELECTED EXERCISES

3. $2x + 5y \leqslant 50$
$20x + 20y \leqslant 260$
$10x + 4y \leqslant 100$
$x \geqslant 0$
$y \geqslant 0$

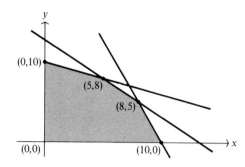

5. $3x + 5y \leqslant 98$
$x - 3y \geqslant 0$
$x \geqslant 0$
$y \geqslant 0$

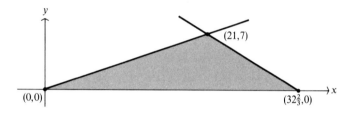

7. $x + 2y \leqslant 50$
$x + y \leqslant 40$
$x \geqslant 0$
$y \geqslant 0$

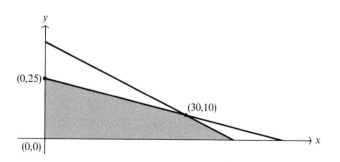

Section 5.3 **1.** Maximize $x_0 = 3x + 2y$ subject to

$3x + 4y \leqslant 30$ $x = 10$ paper clips
$x + 3y \leqslant 20$ $y = 0$ staples
$(x \geqslant 0; y \geqslant 0)$ $x_0 = \$30$

3. Minimize $x_0 = 1.4x + 1.2y$ subject to

$$12x + 20y \geq 240 \qquad x = 4\tfrac{1}{11} \text{ bushels of A}$$
$$25x + 5y \geq 150 \qquad y = 9\tfrac{6}{11} \text{ bushels of B}$$
$$(x \geq 0;\ y \geq 0) \qquad x_0 = \$17.18$$

5. Maximize $x_0 = .2x + .15y$

$$\tfrac{1}{3}x + \tfrac{2}{3}y \leq 20$$
$$\tfrac{1}{3}x + \tfrac{1}{2}y \leq 18$$
$$\tfrac{1}{3}x \qquad\quad \leq 4$$
$$x \geq 0, \qquad y \geq 0$$

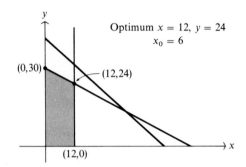

Optimum $x = 12$, $y = 24$
$x_0 = 6$

Section 5.4

7. $x_1 = 0;\ x_2 = \tfrac{8}{3};\ x_3 = \tfrac{9}{2};\ x_0 = 35\tfrac{5}{6}$
8. $x_1 = 0;\ x_2 = \tfrac{15}{16};\ x_3 = \tfrac{31}{8};\ x_0 = 22\tfrac{5}{16}$

CHAPTER 6

Section 6.1

3. $x_0 = \tfrac{10}{3};\ x_1 = \tfrac{8}{3};\ x_2 = 0;\ x_3 = \tfrac{1}{3};\ x_4 = \tfrac{7}{3};\ x_5 = 0;\ x_6 = 0$
5. $x_0 = 5;\ x_1 = 1;\ x_2 = \tfrac{1}{2};\ x_3 = \tfrac{1}{2};\ x_4 = 0;\ x_5 = 0$
7. $x_0 = 10;\ x_1 = 2;\ x_2 = 1;\ x_3 = 0;\ x_4 = 1;\ x_5 = 0$
9. (a) Maximize $x_0 = 10x_1 + 12x_2 + 4x_3$ subject to

$$3x_1 + 6x_2 + 12x_3 \leq 240$$
$$5x_1 + 15x_2 + 30x_3 \leq 1{,}800$$
$$(x_j \geq 0 \text{ for } j = 1, 2, 3)$$

(b) $x_0 = 800;\ x_1 = 80;\ x_2 = 0;\ x_3 = 0;\ x_4 = 0;\ x_5 = 1{,}400$

(c) $x_0 = \$800$ is weekly profit for the following mix:

$x_1 = 80$ units of product A to be assembled

$x_2 = 0$ units of product B to be assembled

$x_3 = 0$ units of product C to be assembled

$x_4 = 0$ hr of unused processing time in Department X

$x_5 = 1{,}400$ hr of unused processing time in Department Y

11. (a) Maximize: $x_0 = 50x_1 + 80x_2 + 100x_3$
Subject to:
$$10x_1 + 15x_2 \qquad\quad \leq 250$$
$$5x_1 + 2x_2 + 7x_3 \leq 150$$
$$4x_1 + 6x_2 + 10x_3 \leq 200$$
$$x_j \geq 0 \quad \text{for} \quad j = 1, 2, 3$$

(b) Optimal solution: $x_0 = \frac{7000}{3}$; $x_2 = \frac{50}{3}$; $x_3 = 10$; $x_5 = \frac{140}{3}$; $x_1 = x_4 = x_6 = 0$

(c) Max. profit is $2333.33 and can be obtained by producing no units of Model A, $16\frac{2}{3}$ units of Model B, and 10 units of Model C. $46\frac{2}{3}$ lb of padding will not be used. All available time and vinyl will be used.

13. (a) Minimize: $\quad -8x_1 - 10x_2 - 6x_3$
    Subject to: $\quad 5x_1 + 2x_2 + 4x_3 \leq 750$
    $\quad 2x_1 + 6x_2 + 3x_3 \leq 750$
    $\quad 3x_1 + 5x_2 + 7x_3 \leq 750$
    $\quad x_j \geq 0 \text{ for } j = 1, 2, 3$

(b) Optimal solution: $x_0 = -\frac{3375}{2}$; $x_1 = x_2 = x_4 = \frac{375}{4}$; $x_3 = x_5 = x_6 = 0$

(c) This pollution control plan will decrease pollution emission by 1687.5 units. Devices 1 and 2 will be used 93.75 hr per week. The cost on Product A machines will be $93.75 less than the maximum allowed.

Section 6.2

1. (a) Minimize $y_0 = 4y_1 + 15y_2$ subject to

$$y_1 + 3y_2 \geq 3$$
$$3y_1 + 2y_2 \geq 2$$
$$(y_1 \geq 0; y_2 \geq 0)$$

(c) Maximize $y_0 = 10y_1 + 6y_2 + 25y_3$ subject to

$$3y_1 + 3y_2 + y_3 \leq 4$$
$$y_1 - 4y_2 + 4y_3 \leq 5$$
$$(y_1 \geq 0; y_2 \geq 0; y_3 \geq 0)$$

3. Minimize $y_0 = 30y_1 + 20y_2$ subject to

$$3y_1 + y_2 \geq 3$$
$$4y_1 + 3y_2 \geq 2$$
$$(y_1 \geq 0; y_2 \geq 0)$$

Optimal solution: $y_0 = 30$; $y_1 = 1$; $y_2 = 0$; $y_3 = 0$; $y_4 = 2$

5. Minimize $y_0 = 6y_1 + 8y_2 + 9y_3$ subject to

$$y_1 + 2y_2 + 3y_3 \geq 1$$
$$6y_1 + 4y_2 - 3y_3 \geq -3$$
$$3y_1 + 8y_2 + 3y_3 \geq 2$$
$$(y_1 \geq 0; y_2 \geq 0; y_3 \geq 0)$$

Optimal solution: $y_0 = \frac{10}{3}$; $y_1 = 0$; $y_2 = \frac{1}{6}$; $y_3 = \frac{2}{9}$; $y_4 = 0$; $y_5 = 3$; $y_6 = 0$

7. Minimize $y_0 = 240y_1 + 1{,}800y_2$ subject to

$$3y_1 + 5y_2 \geq 10$$
$$6y_1 + 15y_2 \geq 12$$
$$12y_1 + 30y_2 \geq 4$$
$$(y_1 \geq 0; y_2 \geq 0)$$

Optimal solution: $y_0 = 800$; $y_1 = \frac{10}{3}$; $y_2 = 0$; $y_3 = 0$; $y_4 = 8$; $y_5 = 36$

9. Optimal primal values: $x_0 = \frac{189}{11}$; $x_1 = \frac{45}{11}$; $x_2 = \frac{105}{11}$; $x_3 = 0$; $x_4 = 0$

Optimal dual values: $y_0 = \frac{189}{11}$; $y_1 = \frac{23}{440}$; $y_2 = \frac{17}{550}$; $y_3 = 0$; $y_4 = 0$

11. Minimum: $y_0 = 250y_1 + 150y_2 + 200y_3$
    Subject to: $\quad 10y_1 + 5y_2 + 4y_3 \geq 50$
    $\quad 15y_1 + 2y_2 + 6y_3 \geq 80$
    $\quad 7y_2 + 10y_3 \geq 100$
    $\quad y_1 \geq 0, \quad y_2 \geq 0, \quad y_3 \geq 0$

Optimal solution: $y_0 = \frac{7000}{3}$; $y_1 = \frac{4}{3}$; $y_2 = 0$; $y_3 = 10$; $y_4 = \frac{10}{3}$; $y_5 = 0$; $y_6 = 0$

13. (a) Minimize: $x_0 = 250x_1 + 200x_2$
    Subject to: $80x_1 + 65x_2 \geq 500$
    $x_1 + x_2 \geq 20$
    $x_1 \geq 5$
    $x_2 \geq 0$

    (b) Maximize: $y_0 = 500y_1 + 20y_2 + 5y_3$
    Subject to: $80y_1 + y_2 + y_3 \leq 250$
    $65y_1 + y_2 \leq 200$
    $y_1 \geq 0$, $y_2 \geq 0$, $y_3 \geq 0$

    (c) Optimal solution: $y_0 = 4250$; $y_2 = 200$; $y_3 = 50$; $y_1 = y_4 = y_5 = 0$
    $x_1 = 5$; $x_2 = 15$; $x_3 = 875$; $x_4 = x_5 = 0$

CHAPTER 7

Section 7.1 6. (c) $P(\text{Elm}) = \frac{3}{10}$; $P(\text{Pine}) = \frac{5}{10}$; $P(\text{Pecan}) = \frac{2}{10}$

Section 7.2
3. (a) 495 (b) $12 \cdot 11 \cdot 10 \ldots 5 = 19{,}958{,}400$
5. $11 \cdot 10 \cdot 9 \cdot 7 \cdot 6 \cdot 5 \cdot 4 \cdot 3 = 2{,}494{,}800$
7. (a) $20!/5!5!5!5!$ (b) $20!/4!4!4!4!4!$ (c) $20!/10!10!$
9. 35
11. 1,680
13. (a) 84 (b) 45 (c) 24 (d) 28
15. 100

Section 7.4
3. (a) $P(E_1) = .20$; $P(E_2) = .16$; $P(E_3) = .17$; $P(E_4) = .14$; $P(E_5) = .18$; $P(E_6) = .15$
5. (a) .163 (b) .764 (c) .292 (d) .893 (e) .236
7. $\frac{1}{24}$
9. (a) .7 (b) .8 (c) .7 (d) .5

Section 7.5
3. (b) .59 (d) .03 (f) .50 (h) 90
5. (a) .544 (c) .098 (e) .231 (g) .435
7. (a) .1 (c) .9 (e) .667
9. (a) .4 (b) .9 (c) yes (d) .3003

Section 7.6
5. $\frac{5}{6}$
7. (a) $\frac{1}{5}$ (b) $\frac{4}{15}$ (c) $\frac{4}{15}$ (d) $\frac{4}{15}$
9. (a) $\frac{49}{225}$ (b) $\frac{56}{225}$ (c) $\frac{56}{225}$ (d) $\frac{64}{225}$
11. (a) $\frac{1}{4}$ (b) $\frac{1}{6}$
13. $\frac{1}{3}$
15. (a) .825 (b) .175
17. (a) $\frac{1}{225}$ (b) $\frac{16}{225}$ (c) $\frac{208}{225}$

Section 7.7
1. (a) $\{1,2,3,4,5,6\}$ (b) $\{6\}$ (c) $\{1,3,5,6,7\}$
   (d) $\{1,3\}$ (e) Empty (f) $\{1,2,3,4,5,6,7\}$
   (g) $\{1,3,6\}$ (h) $\{1,3,5,6\}$ (i) No (j) Yes
3. (a) Yes (b) Yes (c) Yes (d) No (e) No
5. (a) $\frac{1}{36}$ (b) $\frac{1}{18}$ (c) $\frac{1}{12}$ (d) $\frac{1}{9}$ (e) $\frac{5}{36}$ (f) $\frac{1}{6}$
   (g) $\frac{5}{36}$ (h) $\frac{1}{9}$ (i) $\frac{1}{12}$ (j) $\frac{1}{18}$ (k) $\frac{1}{36}$ (l) $\frac{1}{9}$
   (m) $\frac{2}{9}$ (n) $\frac{1}{6}$ (o) $\frac{2}{9}$ (p) $\frac{5}{18}$ (q) $\frac{1}{3}$ (r) $\frac{2}{3}$
7. (b) .311 (e) .639 (h) .053
9. (a) .05 (b) .258 (c) .343 (d) .57 (e) .05 (f) .161

ANSWERS TO SELECTED EXERCISES

11. (a) .267 (b) .2 (c) .467 (d) .572 (e) .428 (f) .533 (g) .4

Section 7.8

1. (a)

Outcome:	2	3	4	5	6	7	8	9	10	11	12
Probability:	$\frac{1}{36}$	$\frac{2}{36}$	$\frac{3}{36}$	$\frac{4}{36}$	$\frac{5}{36}$	$\frac{6}{36}$	$\frac{5}{36}$	$\frac{4}{36}$	$\frac{3}{36}$	$\frac{2}{36}$	$\frac{1}{36}$

(b) $\frac{1}{6}$ (c) $\frac{1}{2}$ (d) $\frac{1}{9}$ (e) $\frac{5}{9}$

3. (a)

Value (X):	2	3	4	5	6	7	8	9	10
Probability:	$\frac{1}{25}$	$\frac{2}{25}$	$\frac{3}{25}$	$\frac{4}{25}$	$\frac{5}{25}$	$\frac{4}{25}$	$\frac{3}{25}$	$\frac{2}{25}$	$\frac{1}{25}$

(b) $\frac{3}{5}$ (c) $\frac{12}{25}$ (d) $\frac{6}{25}$ (e) $\frac{21}{25}$

5. (a)

X	0	1	2	3
$P(X)$	$\frac{1}{8}$	$\frac{3}{8}$	$\frac{3}{8}$	$\frac{1}{8}$

(b)

X	0	1	2	3
$P(X)$	$\frac{2}{22}$	$\frac{9}{22}$	$\frac{9}{22}$	$\frac{2}{22}$

(c)

X	0	1	2	3
$P(X)$	$\frac{8}{27}$	$\frac{12}{27}$	$\frac{6}{27}$	$\frac{1}{27}$

(d)

X	0	1	2	3
$P(X)$	$\frac{14}{55}$	$\frac{28}{55}$	$\frac{12}{55}$	$\frac{1}{55}$

(e)

X	0	1	2	3
$P(X)$	$\frac{125}{216}$	$\frac{25}{72}$	$\frac{5}{72}$	$\frac{1}{216}$

(f)

X	0	1	2	3
$P(X)$	$\frac{6}{11}$	$\frac{9}{22}$	$\frac{1}{22}$	$\frac{0}{11}$

(g)

X	0	1	2	3
$P(X)$	$\frac{1}{8}$	$\frac{3}{8}$	$\frac{3}{8}$	$\frac{1}{8}$

(h)

X	0	1	2	3
$P(X)$	$\frac{1}{11}$	$\frac{9}{22}$	$\frac{9}{22}$	$\frac{1}{11}$

9.

X	3	4	5	6	7	8
$P(X \leq x)$	$\frac{1}{216}$	$\frac{4}{216}$	$\frac{10}{216}$	$\frac{20}{216}$	$\frac{35}{216}$	$\frac{56}{216}$

X	9	10	11	12	13
$P(X \leq x)$	$\frac{81}{216}$	$\frac{108}{216}$	$\frac{135}{216}$	$\frac{160}{216}$	$\frac{181}{216}$

X	14	15	16	17	18
$P(X \leq x)$	$\frac{196}{216}$	$\frac{206}{216}$	$\frac{212}{216}$	$\frac{215}{216}$	$\frac{216}{216}$

(a) $\frac{3}{8}$ (b) $\frac{5}{8}$

11. (a)

X	$P(X)$
0	.16807
1	.36015
2	.3087
3	.1323
4	.02835
5	.00243

(b)

X	$P(X)$
0	.16807
1	.52822
2	.83692
3	.96922
4	.99757
5	1.00000

(c) $P(X \geq 4) = 1 - P(X \leq 3) = 1 - .96922 = .03078$

(d) $P(X \leq 2) = .3087$

13. (a)

X	$P(X)$
0	.32
1	.24
2	.18
3	.13
4	.09
5 or more	.04

(b) $P(X \geq 3) = .13 + .09 + .04 = .26$
(c) $P(X \geq 4 | X \geq 2) = \dfrac{P(X \geq 4 \cap X \geq 2)}{P(X \geq 2)} = \dfrac{.13}{.44} = .295$
(d) $P(X \geq 1 \cap X < 4) = .24 + .18 + .13 = .55$

Section 7.9
3. 101,750
5. $-.07$
7. (a) 12.44   (b) 5.46
9. Expected value of expanding = $450
   Expected value of purchasing = $257.50
   Should expand factory capacity
11. 7

Section 7.10
3. (a) .25   (b) .4375   (c) .3125   (d) 0   (e) 0   (f) Yes
4. (a) .4   (c) .8   (e) 0

CHAPTER 8

Section 8.2
5. (a) .113   (b) .113

(c)

x	0	1	2	3	4	5
$P(X=x)$	.050	.206	.337	.276	.113	.018

7. .00856   9. (a) .0014   (c) .0020   (e) .1988   (g) .2847
10. (a) .0146   (c) 1.000   (e) .0000   (g) .9917
11. .05   12. (a) .6477   (c) .1643   (e) .3661   (g) .0013
13. (a) .0492   (c) 1.0   (e) .9362
14. (a) .0021   (c) .0163   (e) .9201   (g) .7206

Section 8.3
3. (a) .762   (b) .009   (c) .003   (d) .081
5. (a) .014   (b) .758   (c) .662
7. (a) .947   (b) .801   (c) .544   (d) .693   (e) .577
9. (a) .140   (b) .632   (c) .014   (d) .001   (e) .579   (f) .156
11. (a) .335   (b) .715   (c) .913   (d) .020   (e) .616   (f) .446

Section 8.4
3. (a) .2398   (b) .0000   (c) .0001   (d) .3074   (e) .0225
5. (a) .0468   (b) .9487   (c) .0513   (d) .8213
7. (a) .2048   (b) .0512   (c) .2627   (d) .9421
9. (a) .2569   (b) .0640   (c) .9360   (d) .0565
11. (a) .0640   (b) .2160   (c) .6480   (d) .7840   (e) .2880
13. (a) .3110   (b) .5443   (c) .0410   (d) .0467   (e) .5791

Section 8.5

9. $p^2 = \begin{bmatrix} \frac{5}{12} & \frac{7}{12} \\ \frac{7}{18} & \frac{11}{18} \end{bmatrix}$ $p^3 = \begin{bmatrix} \frac{29}{72} & \frac{43}{72} \\ \frac{43}{108} & \frac{65}{108} \end{bmatrix}$ 11. $p^2 = \begin{bmatrix} \frac{3}{4} & 0 & \frac{1}{4} \\ \frac{11}{18} & \frac{1}{9} & \frac{5}{18} \\ \frac{1}{2} & 0 & \frac{1}{2} \end{bmatrix}$ $p^3 = \begin{bmatrix} \frac{5}{8} & 0 & \frac{3}{8} \\ \frac{67}{108} & \frac{1}{27} & \frac{37}{108} \\ \frac{3}{4} & 0 & \frac{1}{4} \end{bmatrix}$

13. (a) $\begin{bmatrix} .80 & .15 & .05 \\ .50 & .35 & .15 \\ .20 & .60 & .20 \end{bmatrix}$ (b) .7250   (c) .6050   (d) .5000   (e) .64925

15. (a) $\begin{bmatrix} .55 & .14 & .31 \\ .3 & .3 & .4 \\ .3 & .14 & .56 \end{bmatrix}$ (b) .69   (c) .56

Section 8.6 1. No saddle point   3. player A strategy $(0,1)$; player B strategy $(0,1)$; $v = 4$

ANSWERS TO SELECTED EXERCISES

5. Player A strategy $(0,1)$; player B strategy $(0,1,0)$; $v=1$
7. No saddle point
9. Player A strategy $(1/7, 6/7)$; player B strategy $(5/7, 2/7)$; $v=23/7$
11. Player A strategy $(1/2, 0, 1/2)$; player B strategy $(1/2, 1/2)$; $v=2$
13. Stocks 30%; bonds 70%; banks 0%; expected return 69
15. Tom $(\frac{7}{12}, \frac{5}{12})$
    Sue $(\frac{7}{12}, \frac{5}{12})$

CHAPTER 9

Section 9.2
1. (a) $a_n = 2(n-1)$; 10, 12, 14 (c) $a_n = 3 + 4(n-1)$; 23, 27, 31
   (e) $a_n = 2^{n-5}$; 2, 4, 8 (g) $a_n = 3 \cdot 2^{n-1}$; 96, 192, 384
2. (a) $a_n = 2n + 4$; 6, 8, 10, 12, 14
   (c) $a_n = 1/n^2$; 1, $\frac{1}{4}$, $\frac{1}{9}$, $\frac{1}{16}$, $\frac{1}{25}$
   (e) $a_n = (5n+2)/(n+1)$; $\frac{7}{2}$, 4, $\frac{17}{4}$, $\frac{22}{5}$, $\frac{9}{2}$
3. (a) No (c) Yes; $d=4$ (e) Yes; $d=5b$
4. $a_8 = 37$; $s_8 = 156$ 6. $a_{15} = 150$; $s_{15} = 1{,}200$
7. (a) No (c) Yes, 5 (e) Yes, $(1.03)^3$
8. $a_5 = 1{,}250$; $s_5 = 1{,}562$ 10. $a_4 = 133.1$; $s_4 = 464.1$
12. $a_7 = 13{,}000$; $s_7 = 80{,}500$ 14. (a) $19 (b) $140 15. $15, $2,200
   (c) $23 (d) $192

Section 9.3
1. (a) $26.25 (c) $22.50 2. (a) $6,666.67 (c) $6,250
3. (a) 0.18 (c) 0.1184 4. (a) 1 year (c) 8 months
5. (a) $1,526.25; $6,180; $397.50; $1,875 (c) $4,360; $2,240; $4,100; $1,120
6. (a) $472.50 (c) $2,420 7. (a) $2,376 (c) 6
9. $1,054.78; $1,046.29; Airtight 11. 9 years, 14.3 years
13. 0.071 15. 27, 54, 108 17. $1,939.34

Section 9.4
1. (a) $S = \$13{,}181$; $I = \$3{,}181$ (c) $S = \$15{,}870.80$; $I = \$6{,}270.80$
   (e) $S = \$18{,}620.80$; $I = \$6{,}620.80$
2. (a) $1,057.50 (c) $6,329.50 (e) $2,950.20
3. $303.12 5. $699.01 7. $215.07 9. 23 Months
11. $2,552.76/yr; $2,594.72/yr; alternative 1

CHAPTER 10

Section 10.2
1. (a) 11 (c) 2 (e) $\frac{1}{3}$ (g) 2 (i) 2
3. (a) $\frac{5}{4}$ (c) 15 (e) 3 (g) 2
4. (a) 1 (c) 0 (e) 18 (g) $-\frac{1}{3}$
5. (a) -1 (c) 2, -2 (e) -1 (g) 0
6. (a) Continuous (c) Discontinuous (e) Continuous

Section 10.3
1. 0 3. 0 5. 0 7. $-1/x^2$ 9. $4x^3$
11. 0,0 13. 0,0 15. $-2x^{-3}$, 2 17. $x^{-1/2}/2$, $\frac{1}{4}$
19. $-4/x^5$ 21. $0.2/s^{0.8}$ 23. $0.05/w^{0.95}$
25. Does not exist 27. (a) $.85x$ (b) $.85$ (c) $.85, .85$
29. (a) $2\sqrt{x}$ (b) $1/\sqrt{x}$ (c) $\frac{1}{2}$, $\frac{1}{4}$

Section 10.4
1. $12x^3$ 3. $3 + 6x^2$ 5. $8x - 2$ 7. $12x - 6x^2 + 4x^3$
9. $8t + 5$ 11. $4x^3 + 9x^2 - 2x + \frac{1}{2}$ 13. 4 15. 4
17. $6x - 6$ 19. $8t^3 - 18t^2$ 21. $12z^2 - 4z$
23. $10x^4 - 24x^3 - 24x^2$ 25. 1 27. 1 29. $2t$

31. -1 **33.** $\frac{2}{27}$ **35.** 6 **37.** -1 **39.** $-\frac{1}{3}, -1$
41. $-5, 1$ **43.** (a) $10-2v$ (c) $-2, -4$
44. (a) $p = -0.05x^2 + 20x - 100$ (c) $5, 0, -5$
45. (a) $500 - 10x$ (b) $400, 250, 0$
47. (a) $85 + 1.5x - .01x^2$ (b) $1.5 - .02x$ (c) $1.5, 1, 0, -.5$ (d) 75 mi

Section 10.5 **1.** $2x^2, 4x^2$ **3.** x, x **5.** $(2x+7)^2 - 5; 2(x^2-5)+7$
7. $2(x+3)^{1/2} + x + 3; 2x^{1/2} + x + 3$
9. $\log_{10}(2x+3); 2(\log_{10}x)+3$ **11.** $3x+3$
13. $\log_{10}[(9x^2+6x)^{1/2}+4]$ **15.** $f(x)=x^3; g(x)=3x+1$
17. $f(x)=x^8; g(x)=2x+6$ **19.** $f(x)=x^4; g(x)=10x^2-3x+2$
21. $f(x)=2/x^4; g(x)=7x^2+3$
23. $f(x)=\log_4 x; g(x)=x^2+3x-1$
25. $f(x)=x^{-2}; g(x)=x^3; h(x)=x+2$
27. $f(x)=x^4; g(x)=x^{2/3}; h(x)=x^2-4$
29. $f(x)=\log_e x; g(x)=x^3; h(x)=x^2-4$
31. $32x+24$ **33.** $-15(5x+7)^{-4}$ **35.** $4(3x^2+4x-2)^3(6x+4)$
37. $-6(3x+4)^{-3}$ **39.** $\frac{1}{3}(2x^2+3x+1)^{-2/3}(4x+3)$
41. $9(x^2+4x)^2(2x+4)+4(x-5)^3$
43. $3(x^2+3x)^2(2x+3)(4x+1)^2+8(x^2+3x)^3(4x+1)$
45. $[6x(x^2-4)^2(2x+3)^2-4(x^2-4)^3(2x+3)]/(2x+3)^4$
47. $6x(x^2-4)^2(2x+3)^{-2}-4(x^2-4)^3(2x+3)^{-3}$
49. $\frac{3}{2}(10x^2+4x)^{1/2}(20x+4)$
51. (a) $20x-50$ (b) $\frac{1}{2}x$ (c) $10x-25$ (d) $\$975$
53. (a) $5x$ (b) $.9x$ (c) $4.5x$

Section 10.6 **1.** $f'(x)=2x+3; f'(-2)=-1$
$f''(x)=2; f''(-2)=2$
$f'''(x)=0; f'''(-2)=0$
3. $f'(x)=4x^3+15x^2-14x+3; f'(1)=8$
$f''(x)=12x^2+30x-14; f''(1)=28$
$f'''(x)=24x+30; f'''(1)=54$
$f^{iv}(x)=24; f^{iv}(1)=24$
$f^v(x)=0; f^v(1)=0$
5. $f'(x)=(x-1)^3+3x(x-1)^2; f'(1)=0$
$f''(x)=6(x-1)^2+6x^2-6x; f''(1)=0$
$f'''(x)=24x-18; f'''(1)=6$
$f^{iv}(x)=24; f^{iv}(1)=24$
$f^v(x)=0; f^v(0)=0$
7. $z'(t)=3t^2+6t+2; z'(-1)=-1$
$z''(t)=6t+6; z''(-1)=0$
$z'''(t)=6; z'''(-1)=6$
$z^{iv}(t)=0; z^{iv}(-1)=0$
9. $f'(x)=-\frac{1}{2}x^{-3/2}; f''(x)=\frac{3}{4}x^{-5/2}$ **11.** $f'(x)=(x^2-4)/x^2; f''(x)=8/x^3$
13. $f'(x)=\frac{3}{2}(3x-x^3)^{-1/2}(1-x^2); f''(x)=-\frac{9}{4}(3x-x^3)^{-3/2}(1-x^2)^2-3x(3x-x^3)^{-1/2}$
15. $f'(x)=2(x^2+3x+2)(2x+3); f''(x)=2(2x+3)^2+4(x^2+3x+2)$
17. (a) $20x^3+3x^2+6x$ (b) $29; 585; 2,605$ (c) $60x^2+6x+6$
19. (a) $4x(x+1)^3(x^2+5)+3(x^2+5)^2(x+1)^2$ (b) $624; 20,160; 226,800$
(c) $24x(x+1)^2(x^2+5)+8x^2(x+1)^3+4(x+1)^3(x^2+5)+6(x^2+5)^2(x+1)$
(d) $1,264; 29,024$ (e) $288; 12,544; 194,400$

CHAPTER 11

Section 11.2 **1.** $y=\frac{1}{3}x-\frac{5}{3}$ **3.** $x=4/(3y-2)$ **5.** $x=1/y$

ANSWERS TO SELECTED EXERCISES

7. $D_x y = 3$; $D_y x = \frac{1}{3}$
9. $D_x y = -4/3x^2$; $D_y x = -3x^2/4$
11. $D_x y = -1/x^2$; $D_y x = -x^2$
13. $D_x y = (8x+2)/(2y+3)$;
$D_y x = (2y+3)/(8x+2)$
15. $D_x y = (12x^2y^2 - 6y - 6xy^3 + 2)/(9x^2y^2 + 6x - 8x^3y)$;
$D_y x = (9x^2y^2 + 6x - 8x^3y)/(12x^2y^2 - 6y - 6xy^3 + 2)$
17. $D_x y = (10x - 3y^2 + 4)/(6xy + 12y)$; $D_y x = (6xy + 12y)/(10x - 3y^2 + 4)$
19. -1
21. $-\frac{4}{3}$
23. (a) $y = 20x - 50$ (b) $D_x y = 20$ (c) $D_y x = \frac{1}{20}$
25. (a) $.2xy + 4.6 = x^2 - y^2$ (b) 1.088 (c) $D_y x = (.2x + 2y)/(2x - .2y)$

Section 11.3
1. $2e^{2x}$
3. $2xe^{x^2}$
5. $(4x+3)e^{2x^2+3x-1}$
7. $4^x \ln 4$
9. $10^{3x} \cdot 3 \ln 10$
11. $8^{2x^2+x+5}(4x+1)\ln 8$
13. $1/x$
15. $3x/(3x^2+4)$
17. $\dfrac{1}{x \ln 5}$
19. $\dfrac{(6x+2)}{(3x^2+2x+7)\ln 3}$
21. $\dfrac{(4-12x-x^2)}{(x+6)(x^2+4)\ln 5}$
23. $\dfrac{(4x^2+4x+9)}{(x+4)(x^2+3)\ln 11}$
25. $\dfrac{1}{x(\ln 3x)\ln 10}$
27. $\dfrac{15x\left(\log_{10}\left[\ln(x^2+6)^5\right]^3\right)^{-1/2}}{(\ln 10)(x^2+6)[\ln(x^2+6)]}$
29. $4xe^{2x^2}$; $4e^{2x^2}(1+4x^2)$
31. $3x^2 e^{x^3}$; $3xe^{x^3}(2+3x^3)$
33. $\dfrac{4x}{(x^2+1)\ln 5}$; $\dfrac{4(1-x^2)}{(x^2+1)^2 \ln 5}$
35. $1 + \ln x + \dfrac{1}{x \ln 10}$; $-\dfrac{1}{x^2 \ln 10} + \dfrac{1}{x}$
37. $6x(x^2+4)^2 - x/(x^2+6)$; $6(x^2+4)(5x^2+4) - (6-x^2)/(x^2+6)^2$
39. (a) $(2x+4)e^{x^2+4x}$ (b) 4; $8e^{12}$, $14e^{45}$ (c) $(4x^2+16x+18)e^{x^2+4x}$
41. (a) $(.9x^2+2.4x)/(x^3+4x^2+5)$ (b) $.29, .18, .13$

Section 11.4
1. $x^{x+1}[\ln x + (x+1)/x]$
3. $2(3x-2)^{2x}[\ln(3x-2) + 3x/(3x-2)]$
5. $x^{x^2+4x+1}[(2x+4)\ln x + (x^2+4x+1)/x]$
7. $2(3x^2+6x+1)^{2x+3}[\ln(3x^2+6x+1) + (6x^2+15x+9)/(3x^2+6x+1)]$
9. $18x^2 e^{6x^3}$
11. $\ln(x^2+2x) + (2x^2+6x+4)/(x^2+2x)$
13. $b^x x^b[\ln b + b/x]$
15. $x^x[1+\ln x] = 4x^3$
17. $(x+2)^x/(x+3)[\ln(x+2) + (x^2+2x-2)/(x+2)(x+3)]$

Section 11.5
1. (a) $3x^2 + 3xy + x + 3y + 3y^2$
   (c) $2x^4 - 2x^2 + 2x^3 y + 6xy + 15xy^2 + x^2y^2 + 7xy^3 + 3y^3 + 2y^4$
2. (a) $e^{x^2+y^2}(e^{2xy}+1)$ (c) $e^{2(x^2+xy+y^2)}$
3. $\partial f/\partial x = 2$; $\partial f/\partial y = 0$
5. $\partial f/\partial x = 2x + 2y$; $\partial f/\partial y = 2x$
7. $\partial f/\partial x = 6y + 2x - 4$; $\partial f/\partial y = 2y + 6x + 3$
9. $\partial f/\partial x = 2xye^{x^2y}$; $\partial f/\partial y = x^2 e^{x^2 y}$
11. $\partial f/\partial x = (3x^2 + 2xy)^2 (18x + 6y)$; $\partial f/\partial y = 6x(3x^2 + 2xy)^2$
13. $\partial f/\partial x = 2x 4^{x^2-y^2} \ln 4$; $\partial f/\partial y = -2y 4^{x^2-y^2} \ln 4$
15. $\partial f/\partial x = y/z$; $\partial f/\partial y = x/z$; $\partial f/\partial z = -xy/z^2$
19. $\partial f/\partial x = 2x$; $\partial f/\partial y = 6y$; $\partial^2 f/\partial x^2 = 2$; $\partial^2 f/\partial y^2 = 6$; $\partial^2 f/\partial x \partial y = 0$
21. $\partial f/\partial x = y + 4x$; $\partial f/\partial y = x - 6y$; $\partial^2 f/\partial x^2 = 4$; $\partial^2 f/\partial y^2 = -6$; $\partial^2 f/\partial x \partial y = 1$
23. $\partial f/\partial x = 2 + 3/y$; $\partial f/\partial y = -4 - 3x/y^2$; $\partial^2 f/\partial x^2 = 0$; $\partial^2 f/\partial y^2 = 6x/y^3$;
$\partial^2 f/\partial x \partial y = -3/y^2$

25. $\partial f/\partial x = 4x + ye^x$; $\partial f/\partial y = -1 + e^x$; $\partial^2 f/\partial x^2 = 4 + ye^x$; $\partial^2 f/\partial y^2 = 0$; $\partial^2 f/\partial x\, \partial y = e^x$
27. $\partial f/\partial x = y/x + 4y^3$; $\partial f/\partial y = \ln x + 12xy^2$; $\partial^2 f/\partial x^2 = -y/x^2$; $\partial^2 f/\partial y^2 = 24xy$ $\partial^2 f/\partial x\, \partial y = 1/x + 12y^2$
29. (a) 8   (b) 12   31. (a) 17   (b) 0
33. (a) $x \ln y$ (b) $2x - 4y$ (c) $3y^2 - 4x$ (d) $-8, 51$ (e) $2, 6y, -4, -4$

CHAPTER 12

Section 12.2
1. -12.5 (minimum)
3. -1 (maximum); 5 (minimum)
5. -2 (minimum); 0 (maximum); 2 (minimum)
7. -2 (minimum)
9. -2 (minimum); 0 (maximum); 2 (minimum)
11. -2 (minimum)
13. 3 (minimum)
15. 0 (horizontal inflection point)
17. 0 (minimum)
19. -1 (minimum)
21. $-\frac{7}{9}$ (minimum); 1 (maximum)
23. 12,500; $312,500
25. (a) $P(x) = -.04x^2 + 8x - 80$ (b) 100   (c) $320
27. 26; $9,184
29. $-x^2 + 16x - 5$; 8 cases; $59
31. 5 boxes; $20
33. $1\frac{2}{3}$ in.; 74.074 sq in.
35. (a) 280   (b) $156,800   (c) $156,000
37. 44

Section 12.3
1. (a) $50   (b) $50,000
3. (a) $C(x) = 2,000 - 30x + 3x^2$
   (b) $2,000   (c) $69,500   (d) $1,925   (e) $59,750
5. (a) $157,500   (b) $1,150
7. 4,100; 4,033
9. 3 machines; $1,414
11. (a) $3x^2 - 90x + 750$ (b) 15   (c) $75   (d) $11,000

Section 12.4
1. $(0,2)$ minimum
3. $(0,0)$ saddle point
5. $(\frac{10}{3}, -\frac{2}{3})$ minimum
7. $(0,0)$ saddle point; $(1,1)$ minimum
9. $(0,0)$ saddle point
11. $(\frac{1}{9}, -\frac{1}{5})$ saddle point; $(\frac{1}{9}, 1)$ maximum; $(1, -\frac{1}{5})$ minimum; $(1,1)$ saddle point
13. (a) $(0,0)$ saddle point; $(32, 16)$ minimum   (b) 32 skilled; 16 unskilled
   (c) $69,520
15. (a) $P(x,y) = -9x^2 + 84x + 4xy + 8y - y^2$ (b) $(10, 24)$ maximum
   (c) $10,000 inventory value; 24,000 sq ft of floor space   (d) $516,000
17. (a) $(0,0)$ saddle point; $(.66, .55)$ minimum   (b) $.66 material; $.55 labor   (c) $2.34
19. (a) $(28.15, 20.17)$ maximum   (b) 20.17 boxes; 28.15 hr   (c) $292.02   (d) $.058

Section 12.5
1. $(4, \frac{17}{2})$ minimum; $\lambda = -\frac{47}{2}$
3. $(2, -20)$ minimum; $\lambda = -40$
5. $(3,5)$ minimum; $\lambda = -7$
7. Test fails
9. $(10, 10)$ minimum; $\lambda = -90$
11. 4,850 machines; 48.5 employees; $2,352.25 profit
13. $5 material; $15 price; 200 demand

ANSWERS TO SELECTED EXERCISES

CHAPTER 13

Section 13.2 1. $x^8 + C$ 3. $e^{-x} + C$ 5. $1/x + C$ 7. $x^6/3 + C$
9. $x + C$ 11. $2x^{3/2}/3 + C$ 13. $-x^{-3}/3 + C$

Section 13.3 1. $x^5/5 + C$ 3. $-x^{-4}/4 + C$ 5. $2u^{5/2}/5 + C$ 7. $4z^{7/4}/7 + C$
9. $3x^{4/3}/4 + C$ 11. $2y^3/3 - 2y^2 + 3y + C$ 13. $-y^{-3} + y^{-4} + C$
15. $3x^{4/3}/2 - 2x^{3/2} + C$ 17. $-\frac{1}{3}x^{-3} - \frac{1}{2}x^{-2} + C$ 19. $y - y^4/2 + y^7/7 + C$

Section 13.4 1. $C(x) = x^3 - 21x^2 + 360x + 3{,}025$; $C(10) = \$5{,}525$
3. $C(x) = .0002x^3 - .3x^2 + 50x + 100{,}000$; $C(1{,}000) = \$50{,}000$
5. $C(x) = x^3 - 15x^2 + 2{,}000x + 4{,}000$; $C(10) = \$23{,}500$
7. $P(x) = 60x^{3/2} - 200x$; $P(250) = \$18{,}715{,}000$
9. $c(x) = .01x^3 + .1x^2 + .05x + 150$; $\$1{,}652.50$
11. $c(x) = \frac{10}{3}x^3 + 25x^2 - 180x$; $\$24{,}300$

Section 13.5 1. 63 3. $4\frac{1}{2}$ 5. $20\frac{1}{4}$ 7. 4 9. $2\frac{2}{3}$
11. $61\frac{4}{5}$ 13. 0 15. 1 17. 4 19. 2.773
21. 28 23. 2 25. $6\frac{1}{4}$ 27. $2\frac{2}{3}$ 29. 6
31. 21 33. $\frac{1}{4}$ 35. $\frac{11}{12}$ 37. 201.215
39. 1.3863 41. \$2,691 43. \$1.80 45. \$11,843
47. \$100 49. 4 51. $30\frac{2}{3}$
53. \$949.50 55. \$26,588.33 57. 24.5 hr

Section 13.6 1. Yes 3. No (area not equal to 1) 5. Yes
7. No $[f(x) < 0]$ 9. No (area not equal to 1)
11. Yes 13. $k = \frac{1}{2}$; $E(X) = 1$ 15. $k = 3$; $E(X) = 1\frac{1}{2}$
17. $k = \frac{1}{8}$; $E(X) = 1\frac{1}{3}$ 19. $k = \frac{2}{33}$; $E(X) = 1\frac{51}{66}$
21. $k = \frac{3}{4}$; 23. (a) $\frac{1}{4}$ (b) $\frac{1}{4}$ (c) $\frac{1}{2}$ (d) $\frac{1}{2}$
$E(X) = 1$
25. (a) $\frac{1}{2}$ (b) $\frac{1}{4}$ (c) 1 (d) $\frac{1}{4}$
27. (a) $\frac{7}{65}$ (b) $\frac{56}{65}$ (c) $\frac{26}{65}$ (d) $\frac{7}{65}$
29. (a) $\frac{11}{16}$ (b) 1 (c) $\frac{189}{256}$ (d) 1
31. $k = 0.0004$; 0.0256; 0.5904; $E(T) = 8$
33. $k = \frac{6}{125}$; 0.104; 0.648; $E(T) = 2\frac{1}{2}$
35. $k = \frac{2}{225}, \frac{56}{225}, \frac{19}{225}$, 5 wk
37. $k = 60$; .3; $\frac{9}{115}$

CHAPTER 14

Section 14.2 1. $(3x + 6)^5/15 + C$ 3. $(x + 2)^7/7 + C$ 5. $(x^3 + 2)^5/5 + C$
7. $-5(2 - 3t^2)^{1/2}/3 + C$ 9. $\frac{1}{2}\ln(y^2 + 4y + 5) + C$
11. $(z^4 + 2)^{1/2}/2 + C$ 13. -4 15. $20\frac{1}{3}$ 17. $-9\frac{1}{3}$
19. $6\frac{2}{3}$ 21. $C(x) = (x^2 + 16)^{3/2}/3 + 250$; $C(3) = \$291.67$
23. $k = 8$; $\frac{175}{256}$; 0.8493 25. $k = .0819$; .9095; .3817
27. $c(x) = \frac{1}{40}(x^2 - 75)^2 + 1{,}000$; \$1,062.50; \$2,265.63

Section 14.3 1. $x(x + 2)^7/7 - (x + 2)^8/56 + C$ 3. $xe^{3x}/3 - e^{3x}/9 + C$
5. $x^2(x + 1)^5/5 - x(x + 1)^6/15 + (x + 1)^7/105 + C$

7. $x^3\ln x/3 - x^3/9 + C$
9. $(x+2)^2(x+5)^9/9 - (x+2)(x+5)^{10}/45 + (x+5)^{11}/495 + C$
11. $x^2e^{2x^2}/4 - e^{2x^2}/8 + C$ **13.** $4\frac{3}{10}$ **15.** 204 **17.** $\frac{1}{2}$
19. $\frac{3}{8}$ **21.** 5.2
23. $C(x) = [(x-3)^2(x+2)^3/3] - [(x-3)(x+2)^4/6] + [(x+2)^5/30] + 75$; $12,401.40; $3,993.07

Section 14.4
1. $5 + x - 5\ln|5+x| + C$
3. $(3+8x)^5/320 - 3(3+8x)^4/256 + C$
5. $\ln(x + \sqrt{x^2-16}) + C$ **7.** $-e^{-6x}/6 + C$ **9.** $3^{2x}/2\ln 3$
11. $3x\ln(2x) - 3x + C$ **13.** $(x^2+4)^{3/2}/3 + C$
15. 0.6363 **17.** $x(\ln 3x)^2 - 2x\ln(3x) + 2x + C$

APPENDIX

Table A.1 Common Logarithms of Numbers

Table A.2 Cumulative Binomial Distribution

Table A.3 Cumulative Poisson Distribution

Table A.4 Binomial Coefficients

Table A.5 Single-Payment Compound Amount Factor

Table A.6 Single-Payment Present Value Factor

Table A.7 Uniform-Payments Compound Amount Factor

Table A.8 Uniform-Payments Present Value Factor

Table A.9 Natural, or Naperian, Logarithms

Table A.10 Exponential Functions

Table A.1 Common Logarithms of Numbers

N	0	1	2	3	4	5	6	7	8	9	\multicolumn{5}{Proportional parts}
											1	2	3	4	5
10	0000	0043	0086	0128	0170	0212	0253	0294	0334	0374	4	8	12	17	21
11	0414	0453	0492	0531	0569	0607	0645	0682	0719	0755	4	8	11	15	19
12	0792	0828	0864	0899	0934	0969	1004	1038	1072	1106	3	7	10	14	17
13	1139	1173	1206	1239	1271	1303	1335	1367	1399	1430	3	6	10	13	16
14	1461	1492	1523	1553	1584	1614	1644	1673	1703	1732	3	6	9	12	15
15	1761	1790	1818	1847	1875	1903	1931	1959	1987	2014	3	6	8	11	14
16	2041	2068	2095	2122	2148	2175	2201	2227	2253	2279	3	5	8	11	13
17	2304	2330	2355	2380	2405	2430	2455	2480	2504	2529	2	5	7	10	12
18	2553	2577	2601	2625	2648	2672	2695	2718	2742	2765	2	5	7	9	12
19	2788	2810	2833	2856	2878	2900	2923	2945	2967	2989	2	4	7	9	11
20	3010	3032	3054	3075	3096	3118	3139	3160	3181	3201	2	4	6	8	11
21	3222	3243	3263	3284	3304	3324	3345	3365	3385	3404	2	4	6	8	10
22	3424	3444	3464	3483	3502	3522	3541	3560	3579	3598	2	4	6	8	10
23	3617	3636	3655	3674	3692	3711	3729	3747	3766	3784	2	4	6	7	9
24	3802	3820	3838	3856	3874	3892	3909	3927	3945	3962	2	4	5	7	9
25	3979	3997	4014	4031	4048	4065	4082	4099	4116	4133	2	4	5	7	9
26	4150	4166	4183	4200	4216	4232	4249	4265	4281	4298	2	3	5	7	8
27	4314	4330	4346	4362	4378	4393	4409	4425	4440	4456	2	3	5	6	8
28	4472	4487	4502	4518	4533	4548	4564	4579	4594	4609	2	3	5	6	8
29	4624	4639	4654	4669	4683	4698	4713	4728	4742	4757	1	3	4	6	7
30	4771	4786	4800	4814	4829	4843	4857	4871	4886	4900	1	3	4	6	7
31	4914	4928	4942	4955	4969	4983	4997	5011	5024	5038	1	3	4	5	7
32	5051	5065	5079	5092	5105	5119	5132	5145	5159	5172	1	3	4	5	7
33	5185	5198	5211	5224	5237	5250	5263	5276	5289	5302	1	3	4	5	7
34	5315	5328	5340	5353	5366	5378	5391	5403	5416	5428	1	2	4	5	6
35	5441	5453	5465	5478	5490	5502	5514	5527	5539	5551	1	2	4	5	6
36	5563	5575	5587	5599	5611	5623	5635	5647	5658	5670	1	2	4	5	6
37	5682	5694	5705	5717	5729	5740	5752	5763	5775	5786	1	2	4	5	6
38	5798	5809	5821	5832	5843	5855	5866	5877	5888	5899	1	2	3	5	6
39	5911	5922	5933	5944	5955	5966	5977	5988	5999	6010	1	2	3	4	5
40	6021	6031	6042	6053	6064	6075	6085	6096	6107	6117	1	2	3	4	5
41	6128	6138	6149	6160	6170	6180	6191	6201	6212	6222	1	2	3	4	5
42	6232	6243	6253	6263	6274	6284	6294	6304	6314	6325	1	2	3	4	5
43	6335	6345	6355	6365	6375	6385	6395	6405	6415	6425	1	2	3	4	5
44	6435	6444	6454	6464	6474	6484	6493	6503	6513	6522	1	2	3	4	5
45	6532	6542	6551	6561	6571	6580	6590	6599	6609	6618	1	2	3	4	5
46	6628	6637	6646	6656	6665	6675	6684	6693	6702	6712	1	2	3	4	5
47	6721	6730	6739	6749	6758	6767	6776	6785	6794	6803	1	2	3	4	5
48	6812	6821	6830	6839	6848	6857	6866	6875	6884	6893	1	2	3	4	5
49	6902	6911	6920	6928	6937	6946	6955	6964	6972	6981	1	2	3	4	4
50	6990	6998	7007	7016	7024	7033	7042	7050	7059	7067	1	2	3	3	4
51	7076	7084	7093	7101	7110	7118	7126	7135	7143	7152	1	2	3	3	4
52	7160	7168	7177	7185	7193	7202	7210	7218	7226	7235	1	2	3	3	4
53	7243	7251	7259	7267	7275	7284	7292	7300	7308	7316	1	2	2	3	4
54	7324	7332	7340	7348	7356	7364	7372	7380	7388	7396	1	2	2	3	4
N	0	1	2	3	4	5	6	7	8	9	1	2	3	4	5

Source: This table is reproduced from *Handbook of Probability and Statistics with Tables* by R. S. Burington and D. C. May, second edition. Copyright © 1970 by McGraw-Hill, Inc. Used with permission of McGraw-Hill Book Company.

Table A.1 Common Logarithms of Numbers (cont.)

N	0	1	2	3	4	5	6	7	8	9	Proportional parts				
											1	2	3	4	5
55	7404	7412	7419	7427	7435	7443	7451	7459	7466	7474	1	2	2	3	4
56	7482	7490	7497	7505	7513	7520	7528	7536	7543	7551	1	2	2	3	4
57	7559	7566	7574	7582	7589	7597	7604	7612	7619	7627	1	1	2	3	4
58	7634	7642	7649	7657	7664	7672	7679	7686	7694	7701	1	1	2	3	4
59	7709	7716	7723	7731	7738	7745	7752	7760	7767	7774	1	1	2	3	4
60	7782	7789	7796	7803	7810	7818	7825	7832	7839	7846	1	1	2	3	4
61	7853	7860	7868	7875	7882	7889	7896	7903	7910	7917	1	1	2	3	3
62	7924	7931	7938	7945	7952	7959	7966	7973	7980	7987	1	1	2	3	3
63	7993	8000	8007	8014	8021	8028	8035	8041	8048	8055	1	1	2	3	3
64	8062	8069	8075	8082	8089	8096	8102	8109	8116	8122	1	1	2	3	3
65	8129	8136	8142	8149	8156	8162	8169	8176	8182	8189	1	1	2	3	3
66	8195	8202	8209	8215	8222	8228	8235	8241	8248	8254	1	1	2	3	3
67	8261	8267	8274	8280	8287	8293	8299	8306	8312	8319	1	1	2	3	3
68	8325	8331	8338	8344	8351	8357	8363	8370	8376	8382	1	1	2	3	3
69	8388	8395	8401	8407	8414	8420	8426	8432	8439	8445	1	1	2	3	3
70	8451	8457	8463	8470	8476	8482	8488	8494	8500	8506	1	1	2	3	3
71	8513	8519	8525	8531	8537	8543	8549	8555	8561	8567	1	1	2	3	3
72	8573	8579	8585	8591	8597	8603	8609	8615	8621	8627	1	1	2	2	3
73	8633	8639	8645	8651	8657	8663	8669	8675	8681	8686	1	1	2	2	3
74	8692	8698	8704	8710	8716	8722	8727	8733	8739	8745	1	1	2	2	3
75	8751	8756	8762	8768	8774	8779	8785	8791	8797	8802	1	1	2	2	3
76	8808	8814	8820	8825	8831	8837	8842	8848	8854	8859	1	1	2	2	3
77	8865	8871	8876	8882	8887	8893	8899	8904	8910	8915	1	1	2	2	3
78	8921	8927	8932	8938	8943	8949	8954	8960	8965	8971	1	1	2	2	3
79	8976	8982	8987	8993	8998	9004	9009	9015	9020	9025	1	1	2	2	3
80	9031	9036	9042	9047	9053	9058	9063	9069	9074	9079	1	1	2	2	3
81	9085	9090	9096	9101	9106	9112	9117	9122	9128	9133	1	1	2	2	3
82	9138	9143	9149	9154	9159	9165	9170	9175	9180	9186	1	1	2	2	3
83	9191	9196	9201	9206	9212	9217	9222	9227	9232	9238	1	1	2	2	3
84	9243	9248	9253	9258	9263	9269	9274	9279	9284	9289	1	1	2	2	3
85	9294	9299	9304	9309	9315	9320	9325	9330	9335	9340	1	1	2	2	3
86	9345	9350	9355	9360	9365	9370	9375	9380	9385	9390	1	1	2	2	3
87	9395	9400	9405	9410	9415	9420	9425	9430	9435	9440	1	1	2	2	3
88	9445	9450	9455	9460	9465	9469	9474	9479	9484	9489	0	1	1	2	2
89	9494	9499	9504	9509	9513	9518	9523	9528	9533	9538	0	1	1	2	2
90	9542	9547	9552	9557	9562	9566	9571	9576	9581	9586	0	1	1	2	2
91	9590	9595	9600	9605	9609	9614	9619	9624	9628	9633	0	1	1	2	2
92	9638	9643	9647	9652	9657	9661	9666	9671	9675	9680	0	1	1	2	2
93	9685	9689	9694	9699	9703	9708	9713	9717	9722	9727	0	1	1	2	2
94	9731	9736	9741	9745	9750	9754	9759	9763	9768	9773	0	1	1	2	2
95	9777	9782	9786	9791	9795	9800	9805	9809	9814	9818	0	1	1	2	2
96	9823	9827	9832	9836	9841	9845	9850	9854	9859	9863	0	1	1	2	2
97	9868	9872	9877	9881	9886	9890	9894	9899	9903	9908	0	1	1	2	2
98	9912	9917	9921	9926	9930	9934	9939	9943	9948	9952	0	1	1	2	2
99	9956	9961	9965	9969	9974	9978	9983	9987	9991	9996	0	1	1	2	2
N	0	1	2	3	4	5	6	7	8	9	1	2	3	4	5

Table A.2 Cumulative Binomial Distribution

$$P_b(X \geq x \mid n, p)$$

$n = 1$

$x \backslash p$	01	02	03	04	05	06	07	08	09	10
1	0100	0200	0300	0400	0500	0600	0700	0800	0900	1000

$x \backslash p$	11	12	13	14	15	16	17	18	19	20
1	1100	1200	1300	1400	1500	1600	1700	1800	1900	2000

$x \backslash p$	21	22	23	24	25	26	27	28	29	30
1	2100	2200	2300	2400	2500	2600	2700	2800	2900	3000

$x \backslash p$	31	32	33	34	35	36	37	38	39	40
1	3100	3200	3300	3400	3500	3600	3700	3800	3900	4000

$x \backslash p$	41	42	43	44	45	46	47	48	49	50
1	4100	4200	4300	4400	4500	4600	4700	4800	4900	5000

$n = 2$

$x \backslash p$	01	02	03	04	05	06	07	08	09	10
1	0199	0396	0591	0784	0975	1164	1351	1536	1719	1900
2	0001	0004	0009	0016	0025	0036	0049	0064	0081	0100

$x \backslash p$	11	12	13	14	15	16	17	18	19	20
1	2079	2256	2431	2604	2775	2944	3111	3276	3439	3600
2	0121	0144	0169	0196	0225	0256	0289	0324	0361	0400

$x \backslash p$	21	22	23	24	25	26	27	28	29	30
1	3759	3916	4071	4224	4375	4524	4671	4816	4959	5100
2	0441	0484	0529	0576	0625	0676	0729	0784	0841	0900

$x \backslash p$	31	32	33	34	35	36	37	38	39	40
1	5239	5376	5511	5644	5775	5904	6031	6156	6279	6400
2	0961	1024	1089	1156	1225	1296	1369	1444	1521	1600

$x \backslash p$	41	42	43	44	45	46	47	48	49	50
1	6519	6636	6751	6864	6975	7084	7191	7296	7399	7500
2	1681	1764	1849	1936	2025	2116	2209	2304	2401	2500

$n = 3$

$x \backslash p$	01	02	03	04	05	06	07	08	09	10
1	0297	0588	0873	1153	1426	1694	1956	2213	2464	2710
2	0003	0012	0026	0047	0073	0104	0140	0182	0228	0280
3				0001	0001	0002	0003	0005	0007	0010

$x \backslash p$	11	12	13	14	15	16	17	18	19	20
1	2950	3185	3415	3639	3859	4073	4282	4486	4686	4880
2	0336	0397	0463	0533	0608	0686	0769	0855	0946	1040
3	0013	0017	0022	0027	0034	0041	0049	0058	0069	0080

$x \backslash p$	21	22	23	24	25	26	27	28	29	30
1	5070	5254	5435	5610	5781	5948	6110	6268	6421	6570
2	1138	1239	1344	1452	1563	1676	1793	1913	2035	2160
3	0093	0106	0122	0138	0156	0176	0197	0220	0244	0270

Source: This table is abridged from "Tables of the Cumulative Binomial Probability Distribution." *Annals of the Computational Laboratory of Harvard University*, Vol. 35 (1955). Used by permission of the publisher, Harvard University Press.

Table A.2 Cumulative Binomial Distribution (cont.)

p \ x	31	32	33	34	35	36	37	38	39	40
1	6715	6856	6992	7125	7254	7379	7500	7617	7730	7840
2	2287	2417	2548	2682	2818	2955	3094	3235	3377	3520
3	0298	0328	0359	0393	0429	0467	0507	0549	0593	0640

p \ x	41	42	43	44	45	46	47	48	49	50
1	7946	8049	8148	8244	8336	8425	8511	8594	8673	8750
2	3665	3810	3957	4104	4253	4401	4551	4700	4850	5000
3	0689	0741	0795	0852	0911	0973	1038	1106	1176	1250

$n = 4$

p \ x	01	02	03	04	05	06	07	08	09	10
1	0394	0776	1147	1507	1855	2193	2519	2836	3143	3439
2	0006	0023	0052	0091	0140	0199	0267	0344	0430	0523
3			0001	0002	0005	0008	0013	0019	0027	0037
4									0001	0001

p \ x	11	12	13	14	15	16	17	18	19	20
1	3726	4003	4271	4530	4780	5021	5254	5479	5695	5904
2	0624	0732	0847	0968	1095	1228	1366	1509	1656	1808
3	0049	0063	0079	0098	0120	0144	0171	0202	0235	0272
4	0001	0002	0003	0004	0005	0007	0008	0010	0013	0016

p \ x	21	22	23	24	25	26	27	28	29	30
1	6105	6298	6485	6664	6836	7001	7160	7313	7459	7599
2	1963	2122	2285	2450	2617	2787	2959	3132	3307	3483
3	0312	0356	0403	0453	0508	0566	0628	0694	0763	0837
4	0019	0023	0028	0033	0039	0046	0053	0061	0071	0081

p \ x	31	32	33	34	35	36	37	38	39	40
1	7733	7862	7985	8103	8215	8322	8425	8522	8615	8704
2	3660	3837	4015	4193	4370	4547	4724	4900	5075	5248
3	0915	0996	1082	1171	1265	1362	1464	1569	1679	1792
4	0092	0105	0119	0134	0150	0168	0187	0209	0231	0256

p \ x	41	42	43	44	45	46	47	48	49	50
1	8788	8868	8944	9017	9085	9150	9211	9269	9323	9375
2	5420	5590	5759	5926	6090	6252	6412	6569	6724	6875
3	1909	2030	2155	2283	2415	2550	2689	2831	2977	3125
4	0283	0311	0342	0375	0410	0448	0488	0531	0576	0625

$n = 5$

p \ x	01	02	03	04	05	06	07	08	09	10
1	0490	0961	1413	1846	2262	2661	3043	3409	3760	4095
2	0010	0038	0085	0148	0226	0319	0425	0544	0674	0815
3			0001	0005	0006	0012	0020	0031	0045	0063	0086
4							0001	0001	0002	0003	0005

p \ x	11	12	13	14	15	16	17	18	19	20
1	4416	4723	5016	5296	5563	5818	6061	6293	6513	6723
2	0965	1125	1292	1467	1648	1835	2027	2224	2424	2627
3	0112	0143	0179	0220	0266	0318	0375	0437	0505	0579
4	0007	0009	0013	0017	0022	0029	0036	0045	0055	0067
5				0001	0001	0001	0001	0002	0002	0003

p \ x	21	22	23	24	25	26	27	28	29	30
1	6923	7113	7293	7464	7627	7781	7927	8065	8196	8319
2	2833	3041	3251	3461	3672	3883	4093	4303	4511	4718
3	0659	0744	0836	0933	1035	1143	1257	1376	1501	1631
4	0081	0097	0114	0134	0156	0181	0208	0238	0272	0308
5	0004	0005	0006	0008	0010	0012	0014	0017	0021	0024

Table A.2 Cumulative Binomial Distribution (cont.)

p x	31	32	33	34	35	36	37	38	39	40
1	8436	8546	8650	8748	8840	8926	9008	9084	9155	9222
2	4923	5125	5325	5522	5716	5906	6093	6276	6455	6630
3	1766	1905	2050	2199	2352	2509	2670	2835	3003	3174
4	0347	0390	0436	0486	0540	0598	0660	0726	0796	0870
5	0029	0034	0039	0045	0053	0060	0069	0079	0090	0102

p x	41	42	43	44	45	46	47	48	49	50
1	9285	9344	9398	9449	9497	9541	9582	9620	9655	9688
2	6801	6967	7129	7286	7438	7585	7728	7865	7998	8125
3	3349	3525	3705	3886	4069	4253	4439	4625	4813	5000
4	0949	1033	1121	1214	1312	1415	1522	1635	1753	1875
5	0116	0131	0147	0165	0185	0206	0229	0255	0282	0313

$n = 6$

p x	01	02	03	04	05	06	07	08	09	10
1	0585	1142	1670	2172	2649	3101	3530	3936	4321	4686
2	0015	0057	0125	0216	0328	0459	0608	0773	0952	1143
3		0002	0005	0012	0022	0038	0058	0085	0118	0159
4					0001	0002	0003	0005	0008	0013
5										0001

p x	11	12	13	14	15	16	17	18	19	20
1	5030	5356	5664	5954	6229	6487	6731	6960	7176	7379
2	1345	1556	1776	2003	2235	2472	2713	2956	3201	3446
3	0206	0261	0324	0395	0473	0560	0655	0759	0870	0989
4	0018	0025	0034	0045	0059	0075	0094	0116	0141	0170
5	0001	0001	0002	0003	0004	0005	0007	0010	0013	0016
6										0001

p x	21	22	23	24	25	26	27	28	29	30
1	7569	7748	7916	8073	8220	8358	8487	8607	8719	8824
2	3692	3937	4180	4422	4661	4896	5128	5356	5580	5798
3	1115	1250	1391	1539	1694	1856	2023	2196	2374	2557
4	0202	0239	0280	0326	0376	0431	0492	0557	0628	0705
5	0020	0025	0031	0038	0046	0056	0067	0079	0093	0109
6	0001	0001	0001	0002	0002	0003	0004	0005	0006	0007

p x	31	32	33	34	35	36	37	38	39	40
1	8921	9011	9095	9173	9246	9313	9375	9432	9485	9533
2	6012	6220	6422	6619	6809	6994	7172	7343	7508	7667
3	2744	2936	3130	3328	3529	3732	3937	4143	4350	4557
4	0787	0875	0969	1069	1174	1286	1404	1527	1657	1792
5	0127	0148	0170	0195	0223	0254	0288	0325	0365	0410
6	0009	0011	0013	0015	0018	0022	0026	0030	0035	0041

p x	41	42	43	44	45	46	47	48	49	50
1	9578	9619	9657	9692	9723	9752	9778	9802	9824	9844
2	7819	7965	8105	8238	8364	8485	8599	8707	8810	8906
3	4764	4971	5177	5382	5585	5786	5985	6180	6373	6563
4	1933	2080	2232	2390	2553	2721	2893	3070	3252	3438
5	0458	0510	0566	0627	0692	0762	0837	0917	1003	1094
6	0048	0055	0063	0073	0083	0095	0108	0122	0138	0156

$n = 7$

p x	01	02	03	04	05	06	07	08	09	10
1	0679	1319	1920	2486	3017	3515	3983	4422	4832	5217
2	0020	0079	0171	0294	0444	0618	0813	1026	1255	1497
3		0003	0009	0020	0038	0063	0097	0140	0193	0257
4				0001	0002	0004	0007	0012	0018	0027
5								0001	0001	0002

Table A.2 Cumulative Binomial Distribution (cont.)

p\x	11	12	13	14	15	16	17	18	19	20
1	5577	5913	6227	6521	6794	7049	7286	7507	7712	7903
2	1750	2012	2281	2556	2834	3115	3396	3677	3956	4233
3	0331	0416	0513	0620	0738	0866	1005	1154	1313	1480
4	0039	0054	0072	0094	0121	0153	0189	0231	0279	0333
5	0003	0004	0006	0009	0012	0017	0022	0029	0037	0047
6					0001	0001	0001	0002	0003	0004

p\x	21	22	23	24	25	26	27	28	29	30
1	8080	8243	8395	8535	8665	8785	8895	8997	9090	9176
2	4506	4775	5040	5298	5551	5796	6035	6266	6490	6706
3	1657	1841	2033	2231	2436	2646	2861	3081	3304	3529
4	0394	0461	0536	0617	0706	0802	0905	1016	1134	1260
5	0058	0072	0088	0107	0129	0153	0181	0213	0248	0288
6	0005	0006	0008	0011	0013	0017	0021	0026	0031	0038
7					0001	0001	0001	0001	0002	0002

p\x	31	32	33	34	35	36	37	38	39	40
1	9255	9328	9394	9454	9510	9560	9606	9648	9686	9720
2	6914	7113	7304	7487	7662	7828	7987	8137	8279	8414
3	3757	3987	4217	4447	4677	4906	5134	5359	5581	5801
4	1394	1534	1682	1837	1998	2167	2341	2521	2707	2898
5	0332	0380	0434	0492	0556	0625	0701	0782	0869	0963
6	0046	0055	0065	0077	0090	0105	0123	0142	0164	0188
7	0003	0003	0004	0005	0006	0008	0009	0011	0014	0016

p\x	41	42	43	44	45	46	47	48	49	50
1	9751	9779	9805	9827	9848	9866	9883	9897	9910	9922
2	8541	8660	8772	8877	8976	9068	9153	9233	9307	9375
3	6017	6229	6436	6638	6836	7027	7213	7393	7567	7734
4	3094	3294	3498	3706	3917	4131	4346	4563	4781	5000
5	1063	1169	1282	1402	1529	1663	1803	1951	2105	2266
6	0216	0246	0279	0316	0357	0402	0451	0504	0562	0625
7	0019	0023	0027	0032	0037	0044	0051	0059	0068	0078

$n = 8$

p\x	01	02	03	04	05	06	07	08	09	10
1	0773	1492	2163	2786	3366	3904	4404	4868	5297	5695
2	0027	0103	0223	0381	0572	0792	1035	1298	1577	1869
3	0001	0004	0013	0031	0058	0096	0147	0211	0289	0381
4			0001	0002	0004	0007	0013	0022	0034	0050
5							0001	0001	0003	0004

p\x	11	12	13	14	15	16	17	18	19	20
1	6063	6404	6718	7008	7275	7521	7748	7956	8147	8322
2	2171	2480	2794	3111	3428	3744	4057	4366	4670	4967
3	0487	0608	0743	0891	1052	1226	1412	1608	1815	2031
4	0071	0097	0129	0168	0214	0267	0328	0397	0476	0563
5	0007	0010	0015	0021	0029	0038	0050	0065	0083	0104
6		0001	0001	0002	0002	0003	0005	0007	0009	0012
7									0001	0001

p\x	21	22	23	24	25	26	27	28	29	30
1	8483	8630	8764	8887	8999	9101	9194	9278	9354	9424
2	5257	5538	5811	6075	6329	6573	6807	7031	7244	7447
3	2255	2486	2724	2967	3215	3465	3718	3973	4228	4482
4	0659	0765	0880	1004	1138	1281	1433	1594	1763	1941
5	0129	0158	0191	0230	0273	0322	0377	0438	0505	0580
6	0016	0021	0027	0034	0042	0052	0064	0078	0094	0113
7	0001	0002	0002	0003	0004	0005	0006	0008	0010	0013
8									0001	0001

Table A.2 Cumulative Binomial Distribution (cont.)

p\x	31	32	33	34	35	36	37	38	39	40
1	9486	9543	9594	9640	9681	9719	9752	9782	9808	9832
2	7640	7822	7994	8156	8309	8452	8586	8711	8828	8936
3	4736	4987	5236	5481	5722	5958	6189	6415	6634	6846
4	2126	2319	2519	2724	2936	3153	3374	3599	3828	4059
5	0661	0750	0846	0949	1061	1180	1307	1443	1586	1737
6	0134	0159	0187	0218	0253	0293	0336	0385	0439	0498
7	0016	0020	0024	0030	0036	0043	0051	0061	0072	0085
8	0001	0001	0001	0002	0002	0003	0004	0004	0005	0007

p\x	41	42	43	44	45	46	47	48	49	50
1	9853	9872	9889	9903	9916	9928	9938	9947	9954	9961
2	9037	9130	9216	9295	9368	9435	9496	9552	9602	9648
3	7052	7250	7440	7624	7799	7966	8125	8276	8419	8555
4	4292	4527	4762	4996	5230	5463	5694	5922	6146	6367
5	1895	2062	2235	2416	2604	2798	2999	3205	3416	3633
6	0563	0634	0711	0794	0885	0982	1086	1198	1318	1445
7	0100	0117	0136	0157	0181	0208	0239	0272	0310	0352
8	0008	0010	0012	0014	0017	0020	0024	0028	0033	0039

$n = 9$

p\x	01	02	03	04	05	06	07	08	09	10
1	0865	1663	2398	3075	3698	4270	4796	5278	5721	6126
2	0034	0131	0282	0478	0712	0978	1271	1583	1912	2252
3	0001	0006	0020	0045	0084	0138	0209	0298	0405	0530
4			0001	0003	0006	0013	0023	0037	0057	0083
5						0001	0002	0003	0005	0009
6										0001

p\x	11	12	13	14	15	16	17	18	19	20
1	6496	6835	7145	7427	7684	7918	8131	8324	8499	8658
2	2599	2951	3304	3657	4005	4348	4685	5012	5330	5638
3	0672	0833	1009	1202	1409	1629	1861	2105	2357	2618
4	0117	0158	0209	0269	0339	0420	0512	0615	0730	0856
5	0014	0021	0030	0041	0056	0075	0098	0125	0158	0196
6	0001	0002	0003	0004	0006	0009	0013	0017	0023	0031
7						0001	0001	0002	0002	0003

p\x	21	22	23	24	25	26	27	28	29	30
1	8801	8931	9048	9154	9249	9335	9411	9480	9542	9596
2	5934	6218	6491	6750	6997	7230	7452	7660	7856	8040
3	2885	3158	3434	3713	3993	4273	4552	4829	5102	5372
4	0994	1144	1304	1475	1657	1849	2050	2260	2478	2703
5	0240	0291	0350	0416	0489	0571	0662	0762	0870	0988
6	0040	0051	0065	0081	0100	0122	0149	0179	0213	0253
7	0004	0006	0008	0010	0013	0017	0022	0028	0035	0043
8			0001	0001	0001	0001	0002	0003	0003	0004

p\x	31	32	33	34	35	36	37	38	39	40
1	9645	9689	9728	9762	9793	9820	9844	9865	9883	9899
2	8212	8372	8522	8661	8789	8908	9017	9118	9210	9295
3	5636	5894	6146	6390	6627	6856	7076	7287	7489	7682
4	2935	3173	3415	3662	3911	4163	4416	4669	4922	5174
5	1115	1252	1398	1553	1717	1890	2072	2262	2460	2666
6	0298	0348	0404	0467	0536	0612	0696	0787	0886	0994
7	0053	0064	0078	0094	0112	0133	0157	0184	0215	0250
8	0006	0007	0009	0011	0014	0017	0021	0026	0031	0038
9				0001	0001	0001	0001	0002	0002	0003

p\x	41	42	43	44	45	46	47	48	49	50
1	9913	9926	9936	9946	9954	9961	9967	9972	9977	9980
2	9372	9442	9505	9563	9615	9662	9704	9741	9775	9805
3	7866	8039	8204	8359	8505	8642	8769	8889	8999	9102
4	5424	5670	5913	6152	6386	6614	6836	7052	7260	7461
5	2878	3097	3322	3551	3786	4024	4265	4509	4754	5000
6	1109	1233	1366	1508	1658	1817	1985	2161	2346	2539
7	0290	0334	0383	0437	0498	0564	0637	0717	0804	0898
8	0046	0055	0065	0077	0091	0107	0125	0145	0169	0195
9	0003	0004	0005	0006	0008	0009	0011	0014	0016	0020

Table A.2 Cumulative Binomial Distribution (cont.)

$n = 10$

p\x	01	02	03	04	05	06	07	08	09	10
1	0956	1829	2626	3352	4013	4614	5160	5656	6106	6513
2	0043	0162	0345	0582	0861	1176	1517	1879	2254	2639
3	0001	0009	0028	0115	0188	0283	0401	0540	0702	
4			0001	0004	0010	0020	0036	0058	0088	0128
5					0001	0002	0003	0006	0010	0016
6									0001	0001

p\x	11	12	13	14	15	16	17	18	19	20
1	6882	7215	7516	7787	8031	8251	8448	8626	8784	8926
2	3028	3417	3804	4184	4557	4920	5270	5608	5932	6242
3	0884	1087	1308	1545	1798	2064	2341	2628	2922	3222
4	0178	0239	0313	0400	0500	0614	0741	0883	1039	1209
5	0025	0037	0053	0073	0099	0130	0168	0213	0266	0328
6	0003	0004	0006	0010	0014	0020	0027	0037	0049	0064
7			0001	0001	0001	0002	0003	0004	0006	0009
8									0001	0001

p\x	21	22	23	24	25	26	27	28	29	30
1	9053	9166	9267	9357	9437	9508	9570	9626	9674	9718
2	6536	6815	7079	7327	7560	7778	7981	8170	8345	8507
3	3526	3831	4137	4442	4744	5042	5335	5622	5901	6172
4	1391	1587	1794	2012	2241	2479	2726	2979	3239	3504
5	0399	0479	0569	0670	0781	0904	1037	1181	1337	1503
6	0082	0104	0130	0161	0197	0239	0287	0342	0404	0473
7	0012	0016	0021	0027	0035	0045	0056	0070	0087	0106
8	0001	0002	0002	0003	0004	0006	0007	0010	0012	0016
9							0001	0001	0001	0001

p\x	31	32	33	34	35	36	37	38	39	40
1	9755	9789	9818	9843	9865	9885	9902	9916	9929	9940
2	8656	8794	8920	9035	9140	9236	9323	9402	9473	9536
3	6434	6687	6930	7162	7384	7595	7794	7983	8160	8327
4	3772	4044	4316	4589	4862	5132	5400	5664	5923	6177
5	1679	1867	2064	2270	2485	2708	2939	3177	3420	3669
6	0551	0637	0732	0836	0949	1072	1205	1348	1500	1662
7	0129	0155	0185	0220	0260	0305	0356	0413	0477	0548
8	0020	0025	0032	0039	0048	0059	0071	0086	0103	0123
9	0002	0003	0003	0004	0005	0007	0009	0011	0014	0017
10								0001	0001	0001

p\x	41	42	43	44	45	46	47	48	49	50
1	9949	9957	9964	9970	9975	9979	9983	9986	9988	9990
2	9594	9645	9691	9731	9767	9799	9827	9852	9874	9893
3	8483	8628	8764	8889	9004	9111	9209	9298	9379	9453
4	6425	6665	6898	7123	7340	7547	7745	7933	8112	8281
5	3922	4178	4436	4696	4956	5216	5474	5730	5982	6230
6	1834	2016	2207	2407	2616	2832	3057	3288	3526	3770
7	0626	0712	0806	0908	1020	1141	1271	1410	1560	1719
8	0146	0172	0202	0236	0274	0317	0366	0420	0480	0547
9	0021	0025	0031	0037	0045	0054	0065	0077	0091	0107
10	0001	0002	0002	0003	0003	0004	0005	0006	0008	0010

$n = 11$

p\x	01	02	03	04	05	06	07	08	09	10
1	1047	1993	2847	3618	4312	4937	5499	6004	6456	6862
2	0052	0195	0413	0692	1019	1382	1772	2181	2601	3026
3	0002	0012	0037	0083	0152	0248	0370	0519	0695	0896
4			0002	0007	0016	0030	0053	0085	0129	0185
5					0001	0003	0005	0010	0017	0028
6								0001	0002	0003

Table A.2 Cumulative Binomial Distribution (cont.)

p / x	11	12	13	14	15	16	17	18	19	20
1	7225	7549	7839	8097	8327	8531	8712	8873	9015	9141
2	3452	3873	4286	4689	5078	5453	5811	6151	6474	6779
3	1120	1366	1632	1915	2212	2521	2839	3164	3494	3826
4	0256	0341	0442	0560	0694	0846	1013	1197	1397	1611
5	0042	0061	0087	0119	0159	0207	0266	0334	0413	0504
6	0005	0008	0012	0018	0027	0037	0051	0068	0090	0117
7		0001	0001	0002	0003	0005	0007	0010	0014	0020
8							0001	0001	0002	0002

p / x	21	22	23	24	25	26	27	28	29	30
1	9252	9350	9436	9511	9578	9636	9686	9730	9769	9802
2	7065	7333	7582	7814	8029	8227	8410	8577	8730	8870
3	4158	4488	4814	5134	5448	5753	6049	6335	6610	6873
4	1840	2081	2333	2596	2867	3146	3430	3719	4011	4304
5	0607	0723	0851	0992	1146	1313	1493	1685	1888	2103
6	0148	0186	0231	0283	0343	0412	0490	0577	0674	0782
7	0027	0035	0046	0059	0076	0095	0119	0146	0179	0216
8	0003	0005	0007	0009	0012	0016	0021	0027	0034	0043
9			0001	0001	0001	0002	0002	0003	0004	0006

p / x	31	32	33	34	35	36	37	38	39	40
1	9831	9856	9878	9896	9912	9926	9938	9948	9956	9964
2	8997	9112	9216	9310	9394	9470	9537	9597	9650	9698
3	7123	7361	7587	7799	7999	8186	8360	8522	8672	8811
4	4598	4890	5179	5464	5744	6019	6286	6545	6796	7037
5	2328	2563	2807	3059	3317	3581	3850	4122	4397	4672
6	0901	1031	1171	1324	1487	1661	1847	2043	2249	2465
7	0260	0309	0366	0430	0501	0581	0670	0768	0876	0994
8	0054	0067	0082	0101	0122	0148	0177	0210	0249	0293
9	0008	0010	0013	0016	0020	0026	0032	0039	0048	0059
10	0001	0001	0001	0002	0002	0003	0004	0005	0006	0007

p / x	41	42	43	44	45	46	47	48	49	50
1	9970	9975	9979	9983	9986	9989	9991	9992	9994	9995
2	9739	9776	9808	9836	9861	9882	9900	9916	9930	9941
3	8938	9055	9162	9260	9348	9428	9499	9562	9622	9673
4	7269	7490	7700	7900	8089	8266	8433	8588	8733	8867
5	4948	5223	5495	5764	6029	6288	6541	6787	7026	7256
6	2690	2924	3166	3414	3669	3929	4193	4460	4729	5000
7	1121	1260	1408	1568	1738	1919	2110	2312	2523	2744
8	0343	0399	0461	0532	0610	0696	0791	0895	1009	1133
9	0072	0087	0104	0125	0148	0175	0206	0241	0282	0327
10	0009	0012	0014	0018	0022	0027	0033	0040	0049	0059
11	0001	0001	0001	0001	0002	0002	0002	0003	0004	0005

$n = 12$

p / x	01	02	03	04	05	06	07	08	09	10
1	1136	2153	3062	3873	4596	5241	5814	6323	6775	7176
2	0062	0231	0486	0809	1184	1595	2033	2487	2948	3410
3	0002	0015	0048	0107	0196	0316	0468	0652	0866	1109
4		0001	0003	0010	0022	0043	0075	0120	0180	0256
5				0001	0002	0004	0009	0016	0027	0043
6							0001	0002	0003	0005
7										0001

p / x	11	12	13	14	15	16	17	18	19	20
1	7530	7843	8120	8363	8578	8766	8931	9076	9202	9313
2	3867	4314	4748	5166	5565	5945	6304	6641	6957	7251
3	1377	1667	1977	2303	2642	2990	3344	3702	4060	4417
4	0351	0464	0597	0750	0922	1114	1324	1552	1795	2054
5	0065	0095	0133	0181	0239	0310	0393	0489	0600	0726
6	0009	0014	0022	0033	0046	0065	0088	0116	0151	0194
7	0001	0002	0003	0004	0007	0010	0015	0021	0029	0039
8					0001	0001	0002	0003	0004	0006
9										0001

Table A.2 Cumulative Binomial Distribution (cont.)

p x	21	22	23	24	25	26	27	28	29	30
1	9409	9493	9566	9629	9683	9730	9771	9806	9836	9862
2	7524	7776	8009	8222	8416	8594	8755	8900	9032	9150
3	4768	5114	5450	5778	6093	6397	6687	6963	7225	7472
4	2326	2610	2904	3205	3512	3824	4137	4452	4765	5075
5	0866	1021	1192	1377	1576	1790	2016	2254	2504	2763
6	0245	0304	0374	0453	0544	0646	0760	0887	1026	1178
7	0052	0068	0089	0113	0143	0178	0219	0267	0322	0386
8	0008	0011	0016	0021	0028	0036	0047	0060	0076	0095
9	0001	0001	0002	0003	0004	0005	0007	0010	0013	0017
10						0001	0001	0001	0002	0002

p x	31	32	33	34	35	36	37	38	39	40
1	9884	9902	9918	9932	9943	9953	9961	9968	9973	9978
2	9256	9350	9435	9509	9576	9634	9685	9730	9770	9804
3	7704	7922	8124	8313	8487	8648	8795	8931	9054	9166
4	5381	5681	5973	6258	6533	6799	7053	7296	7528	7747
5	3032	3308	3590	3876	4167	4459	4751	5043	5332	5618
6	1343	1521	1711	1913	2127	2352	2588	2833	3087	3348
7	0458	0540	0632	0734	0846	0970	1106	1253	1411	1582
8	0118	0144	0176	0213	0255	0304	0359	0422	0493	0573
9	0022	0028	0036	0045	0056	0070	0086	0104	0127	0153
10	0003	0004	0005	0007	0008	0011	0014	0018	0022	0028
11				0001	0001	0001	0001	0002	0002	0003

p x	41	42	43	44	45	46	47	48	49	50
1	9982	9986	9988	9990	9992	9994	9995	9996	9997	9998
2	9834	9860	9882	9901	9917	9931	9943	9953	9961	9968
3	9267	9358	9440	9513	9579	9637	9688	9733	9773	9807
4	7953	8147	8329	8498	8655	8801	8934	9057	9168	9270
5	5899	6175	6443	6704	6956	7198	7430	7652	7862	8062
6	3616	3889	4167	4448	4731	5014	5297	5577	5855	6128
7	1765	1959	2164	2380	2607	2843	3089	3343	3604	3872
8	0662	0760	0869	0988	1117	1258	1411	1575	1751	1938
9	0183	0218	0258	0304	0356	0415	0481	0555	0638	0730
10	0035	0043	0053	0065	0079	0095	0114	0137	0163	0193
11	0004	0005	0007	0009	0011	0014	0017	0021	0026	0032
12				0001	0001	0001	0001	0001	0002	0002

$n = 13$

p x	01	02	03	04	05	06	07	08	09	10
1	1225	2310	3270	4118	4867	5526	6107	6617	7065	7458
2	0072	0270	0564	0932	1354	1814	2298	2794	3293	3787
3	0003	0020	0062	0135	0245	0392	0578	0799	1054	1339
4		0001	0005	0014	0031	0060	0103	0163	0242	0342
5				0001	0003	0007	0013	0024	0041	0065
6						0001	0001	0003	0005	0009
7									0001	0001

p x	11	12	13	14	15	16	17	18	19	20
1	7802	8102	8364	8592	8791	8963	9113	9242	9354	9450
2	4270	4738	5186	5614	6017	6396	6751	7080	7384	7664
3	1651	1985	2337	2704	3080	3463	3848	4231	4611	4983
4	0464	0609	0776	0967	1180	1414	1667	1939	2226	2527
5	0097	0139	0193	0260	0342	0438	0551	0681	0827	0991
6	0015	0024	0036	0053	0075	0104	0139	0183	0237	0300
7	0002	0003	0005	0008	0013	0019	0027	0038	0052	0070
8			0001	0001	0002	0003	0004	0006	0009	0012
9								0001	0001	0002

Table A.2 Cumulative Binomial Distribution (cont.)

p x	21	22	23	24	25	26	27	28	29	30
1	9533	9604	9666	9718	9762	9800	9833	9860	9883	9903
2	7920	8154	8367	8559	8733	8889	9029	9154	9265	9363
3	5347	5699	6039	6364	6674	6968	7245	7505	7749	7975
4	2839	3161	3489	3822	4157	4493	4826	5155	5478	5794
5	1173	1371	1585	1816	2060	2319	2589	2870	3160	3457
6	0375	0462	0562	0675	0802	0944	1099	1270	1455	1654
7	0093	0120	0154	0195	0243	0299	0365	0440	0527	0624
8	0017	0024	0032	0043	0056	0073	0093	0118	0147	0182
9	0002	0004	0005	0007	0010	0013	0018	0024	0031	0040
10			0001	0001	0001	0002	0003	0004	0005	0007
11									0001	0001

p x	31	32	33	34	35	36	37	38	39	40
1	9920	9934	9945	9955	9963	9970	9975	9980	9984	9987
2	9450	9527	9594	9653	9704	9749	9787	9821	9849	9874
3	8185	8379	8557	8720	8868	9003	9125	9235	9333	9421
4	6101	6398	6683	6957	7217	7464	7698	7917	8123	8314
5	3760	4067	4376	4686	4995	5301	5603	5899	6188	6470
6	1867	2093	2331	2581	2841	3111	3388	3673	3962	4256
7	0733	0854	0988	1135	1295	1468	1654	1853	2065	2288
8	0223	0271	0326	0390	0462	0544	0635	0738	0851	0977
9	0052	0065	0082	0102	0126	0154	0187	0225	0270	0321
10	0009	0012	0015	0020	0025	0032	0040	0051	0063	0078
11	0001	0001	0002	0003	0003	0005	0006	0008	0010	0013
12							0001	0001	0001	0001

p x	41	42	43	44	45	46	47	48	49	50
1	9990	9992	9993	9995	9996	9997	9997	9998	9998	9999
2	9895	9912	9928	9940	9951	9960	9967	9974	9979	9983
3	9499	9569	9630	9684	9731	9772	9808	9838	9865	9888
4	8492	8656	8807	8945	9071	9185	9288	9381	9464	9539
5	6742	7003	7254	7493	7721	7935	8137	8326	8502	8666
6	4552	4849	5146	5441	5732	6019	6299	6573	6838	7095
7	2524	2770	3025	3290	3563	3842	4127	4415	4707	5000
8	1114	1264	1426	1600	1788	1988	2200	2424	2659	2905
9	0379	0446	0520	0605	0698	0803	0918	1045	1183	1334
10	0096	0117	0141	0170	0203	0242	0287	0338	0396	0461
11	0017	0021	0027	0033	0041	0051	0063	0077	0093	0112
12	0002	0002	0003	0004	0005	0007	0009	0011	0014	0017
13							0001	0001	0001	0001

$n = 14$

p x	01	02	03	04	05	06	07	08	09	10
1	1313	2464	3472	4353	5123	5795	6380	6888	7330	7712
2	0084	0310	0645	1059	1530	2037	2564	3100	3632	4154
3	0003	0025	0077	0167	0301	0478	0698	0958	1255	1584
4		0001	0006	0019	0042	0080	0136	0214	0315	0441
5				0002	0004	0010	0020	0035	0059	0092
6						0001	0002	0004	0008	0015
7									0001	0002

p x	11	12	13	14	15	16	17	18	19	20
1	8044	8330	8577	8789	8972	9129	9264	9379	9477	9560
2	4658	5141	5599	6031	6433	6807	7152	7469	7758	8021
3	1939	2315	2708	3111	3521	3932	4341	4744	5138	5519
4	0594	0774	0979	1210	1465	1742	2038	2351	2679	3018
5	0137	0196	0269	0359	0467	0594	0741	0907	1093	1298
6	0024	0038	0057	0082	0115	0157	0209	0273	0349	0439
7	0003	0006	0009	0015	0022	0032	0046	0064	0087	0116
8		0001	0001	0002	0003	0005	0008	0012	0017	0024
9						0001	0001	0002	0003	0004

Table A.2 Cumulative Binomial Distribution (cont.)

x \ p	21	22	23	24	25	26	27	28	29	30
1	9631	9691	9742	9786	9822	9852	9878	9899	9917	9932
2	8259	8473	8665	8837	8990	9126	9246	9352	9444	9325
3	5887	6239	6574	6891	7189	7467	7727	7967	8188	8392
4	3366	3719	4076	4432	4787	5136	5479	5813	6137	6448
5	1523	1765	2023	2297	2585	2884	3193	3509	3832	4158
6	0543	0662	0797	0949	1117	1301	1502	1718	1949	2195
7	0152	0196	0248	0310	0383	0467	0563	0673	0796	0933
8	0033	0045	0060	0079	0103	0132	0167	0208	0257	0315
9	0006	0008	0011	0016	0022	0029	0038	0050	0065	0083
10	0001	0001	0002	0002	0003	0005	0007	0009	0012	0017
11						0001	0001	0001	0002	0002

x \ p	31	32	33	34	35	36	37	38	39	40
1	9945	9955	9963	9970	9976	9981	9984	9988	9990	9992
2	9596	9657	9710	9756	9795	9828	9857	9881	9902	9919
3	8577	8746	8899	9037	9161	9271	9370	9457	9534	9602
4	6747	7032	7301	7556	7795	8018	8226	8418	8595	8757
5	4486	4813	5138	5458	5773	6080	6378	6666	6943	7207
6	2454	2724	3006	3297	3595	3899	4208	4519	4831	5141
7	1084	1250	1431	1626	1836	2059	2296	2545	2805	3075
8	0381	0458	0545	0643	0753	0876	1012	1162	1325	1501
9	0105	0131	0163	0200	0243	0294	0353	0420	0497	0583
10	0022	0029	0037	0048	0060	0076	0095	0117	0144	0175
11	0003	0005	0006	0008	0011	0014	0019	0024	0031	0039
12		0001	0001	0001	0001	0002	0003	0003	0005	0006
13										0001

x \ p	41	42	43	44	45	46	47	48	49	50
1	9994	9995	9996	9997	9998	9998	9999	9999	9999	9999
2	9934	9946	9956	9964	9971	9977	9981	9985	9988	9991
3	9661	9713	9758	9797	9830	9858	9883	9903	9921	9935
4	8905	9039	9161	9270	9368	9455	9532	9601	9661	9713
5	7459	7697	7922	8132	8328	8510	8678	8833	8974	9102
6	5450	5754	6052	6344	6627	6900	7163	7415	7654	7880
7	3355	3643	3937	4236	4539	4843	5148	5451	5751	6047
8	1692	1896	2113	2344	2586	2840	3105	3380	3663	3953
9	0680	0789	0910	1043	1189	1348	1520	1707	1906	2120
10	0212	0255	0304	0361	0426	0500	0583	0677	0782	0898
11	0049	0061	0076	0093	0114	0139	0168	0202	0241	0287
12	0008	0010	0013	0017	0022	0027	0034	0042	0053	0065
13	0001	0001	0001	0002	0003	0003	0004	0006	0007	0009
14										0001

$n = 15$

x \ p	01	02	03	04	05	06	07	08	09	10
1	1399	2614	3667	4579	5367	6047	6633	7137	7570	7941
2	0096	0353	0730	1191	1710	2262	2832	3403	3965	4510
3	0004	0030	0094	0203	0362	0571	0829	1130	1469	1841
4		0002	0008	0024	0055	0104	0175	0273	0399	0556
5			0001	0002	0006	0014	0028	0050	0082	0127
6					0001	0001	0003	0007	0013	0022
7								0001	0002	0003

x \ p	11	12	13	14	15	16	17	18	19	20
1	8259	8530	8762	8959	9126	9269	9389	9490	9576	9648
2	5031	5524	5987	6417	6814	7179	7511	7813	8085	8329
3	2238	2654	3084	3520	3958	4392	4819	5234	5635	6020
4	0742	0959	1204	1476	1773	2092	2429	2782	3146	3518
5	0187	0265	0361	0478	0617	0778	0961	1167	1394	1642
6	0037	0057	0084	0121	0168	0227	0300	0387	0490	0611
7	0006	0010	0015	0024	0036	0052	0074	0102	0137	0181
8	0001	0001	0002	0004	0006	0010	0014	0021	0030	0042
9					0001	0001	0002	0003	0005	0008
10									0001	0001

Table A.2 Cumulative Binomial Distribution (cont.)

p x	21	22	23	24	25	26	27	28	29	30
1	9709	9759	9802	9837	9866	9891	9911	9928	9941	9953
2	8547	8741	8913	9065	9198	9315	9417	9505	9581	9647
3	6385	6731	7055	7358	7639	7899	8137	8355	8553	8732
4	3895	4274	4650	5022	5387	5742	6086	6416	6732	7031
5	1910	2195	2495	2810	3135	3469	3810	4154	4500	4845
6	0748	0905	1079	1272	1484	1713	1958	2220	2495	2784
7	0234	0298	0374	0463	0566	0684	0817	0965	1130	1311
8	0058	0078	0104	0135	0173	0219	0274	0338	0413	0500
9	0011	0016	0023	0031	0042	0056	0073	0094	0121	0152
10	0002	0003	0004	0006	0008	0011	0015	0021	0028	0037
11			0001	0001	0001	0002	0002	0003	0005	0007
12									0001	0001

p x	31	32	33	34	35	36	37	38	39	40
1	9962	9969	9975	9980	9984	9988	9990	9992	9994	9995
2	9704	9752	9794	9829	9858	9883	9904	9922	9936	9948
3	8893	9038	9167	9281	9383	9472	9550	9618	9678	9729
4	7314	7580	7829	8060	8273	8469	8649	8813	8961	9095
5	5187	5523	5852	6171	6481	6778	7062	7332	7587	7827
6	3084	3393	3709	4032	4357	4684	5011	5335	5654	5968
7	1509	1722	1951	2194	2452	2722	3003	3295	3595	3902
8	0599	0711	0837	0977	1132	1302	1487	1687	1902	2131
9	0190	0236	0289	0351	0422	0504	0597	0702	0820	0950
10	0048	0062	0079	0099	0124	0154	0190	0232	0281	0338
11	0009	0012	0016	0022	0028	0037	0047	0059	0075	0093
12	0001	0002	0003	0004	0005	0006	0009	0011	0015	0019
13					0001	0001	0001	0002	0002	0003

p x	41	42	43	44	45	46	47	48	49	50
1	9996	9997	9998	9998	9999	9999	9999	9999	10000	10000
2	9958	9966	9973	9979	9983	9987	9990	9992	9994	9995
3	9773	9811	9843	9870	9893	9913	9929	9943	9954	9963
4	9215	9322	9417	9502	9576	9641	9697	9746	9788	9824
5	8052	8261	8454	8633	8796	8945	9080	9201	9310	9408
6	6274	6570	6856	7131	7392	7641	7875	8095	8301	8491
7	4214	4530	4847	5164	5478	5789	6095	6394	6684	6964
8	2374	2630	2898	3176	3465	3762	4065	4374	4686	5000
9	1095	1254	1427	1615	1818	2034	2265	2510	2767	3036
10	0404	0479	0565	0661	0769	0890	1024	1171	1333	1509
11	0116	0143	0174	0211	0255	0305	0363	0430	0506	0592
12	0025	0032	0040	0051	0063	0079	0097	0119	0145	0176
13	0004	0005	0007	0009	0011	0014	0018	0023	0029	0037
14			0001	0001	0001	0002	0002	0003	0004	0005

$n = 16$

p x	01	02	03	04	05	06	07	08	09	10
1	1485	2762	3857	4796	5599	6284	6869	7366	7789	8147
2	0109	0399	0818	1327	1892	2489	3098	3701	4289	4853
3	0005	0037	0113	0242	0429	0673	0969	1311	1694	2108
4		0002	0011	0032	0070	0132	0221	0342	0496	0684
5			0001	0003	0009	0019	0038	0068	0111	0170
6					0001	0002	0005	0010	0019	0033
7							0001	0001	0003	0005
8										0001

p x	11	12	13	14	15	16	17	18	19	20
1	8450	8707	8923	9105	9257	9386	9493	9582	9657	9719
2	5386	5885	6347	6773	7161	7513	7830	8115	8368	8593
3	2545	2999	3461	3926	4386	4838	5277	5698	6101	6482
4	0907	1162	1448	1763	2101	2460	2836	3223	3619	4019
5	0248	0348	0471	0618	0791	0988	1211	1458	1727	2018
6	0053	0082	0120	0171	0235	0315	0412	0527	0662	0817
7	0009	0015	0024	0038	0056	0080	0112	0153	0204	0267
8	0001	0002	0004	0007	0011	0016	0024	0036	0051	0070
9				0001	0001	0002	0003	0004	0007	0010	0015
10							0001	0001	0002	0002

Table A.2 Cumulative Binomial Distribution (cont.)

x \ p	21	22	23	24	25	26	27	28	29	30
1	9770	9812	9847	9876	9900	9919	9935	9948	9958	9967
2	8791	8965	9117	9250	9365	9465	9550	9623	9686	9739
3	6839	7173	7483	7768	8029	8267	8482	8677	8851	9006
4	4418	4814	5203	5583	5950	6303	6640	6959	7260	7541
5	2327	2652	2991	3341	3698	4060	4425	4788	5147	5501
6	0992	1188	1405	1641	1897	2169	2458	2761	3077	3402
7	0342	0432	0536	0657	0796	0951	1125	1317	1526	1753
8	0095	0127	0166	0214	0271	0340	0420	0514	0621	0744
9	0021	0030	0041	0056	0075	0098	0127	0163	0206	0257
10	0004	0006	0008	0012	0016	0023	0031	0041	0055	0071
11	0001	0001	0001	0002	0003	0004	0006	0008	0011	0016
12						0001	0001	0001	0002	0003

x \ p	31	32	33	34	35	36	37	38	39	40
1	9974	9979	9984	9987	9990	9992	9994	9995	9996	9997
2	9784	9822	9854	9880	9902	9921	9936	9948	9959	9967
3	9144	9266	9374	9467	9549	9620	9681	9734	9778	9817
4	7804	8047	8270	8475	8661	8830	8982	9119	9241	9349
5	5846	6181	6504	6813	7108	7387	7649	7895	8123	8334
6	3736	4074	4416	4759	5100	5438	5770	6094	6408	6712
7	1997	2257	2531	2819	3119	3428	3746	4070	4398	4728
8	0881	1035	1205	1391	1594	1813	2048	2298	2562	2839
9	0317	0388	0470	0564	0671	0791	0926	1076	1242	1423
10	0092	0117	0148	0185	0229	0280	0341	0411	0491	0583
11	0021	0028	0037	0048	0062	0079	0100	0125	0155	0191
12	0004	0005	0007	0010	0013	0017	0023	0030	0038	0049
13		0001	0001	0001	0002	0003	0004	0005	0007	0009
14								0001	0001	0001

x \ p	41	42	43	44	45	46	47	48	49	50
1	9998	9998	9999	9999	9999	9999	9999	10000	10000	10000
2	9974	9979	9984	9987	9990	9992	9994	9995	9997	9997
3	9849	9876	9899	9918	9934	9947	9958	9966	9973	9979
4	9444	9527	9600	9664	9719	9766	9806	9840	9869	9894
5	8529	8707	8869	9015	9147	9265	9370	9463	9544	9616
6	7003	7280	7543	7792	8024	8241	8441	8626	8795	8949
7	5058	5387	5711	6029	6340	6641	6932	7210	7476	7728
8	3128	3428	3736	4051	4371	4694	5019	5343	5665	5982
9	1619	1832	2060	2302	2559	2829	3111	3405	3707	4018
10	0687	0805	0936	1081	1241	1416	1607	1814	2036	2272
11	0234	0284	0342	0409	0486	0574	0674	0786	0911	1051
12	0062	0078	0098	0121	0149	0183	0222	0268	0322	0384
13	0012	0016	0021	0027	0035	0044	0055	0069	0086	0106
14	0002	0002	0003	0004	0006	0007	0010	0013	0016	0021
15					0001	0001	0001	0001	0002	0003

$n = 17$

x \ p	01	02	03	04	05	06	07	08	09	10
1	1571	2907	4042	5004	5819	6507	7088	7577	7988	8332
2	0123	0446	0909	1465	2078	2717	3362	3995	4604	5182
3	0006	0044	0134	0286	0503	0782	1118	1503	1927	2382
4		0003	0014	0040	0088	0164	0273	0419	0603	0826
5			0001	0004	0012	0026	0051	0089	0145	0221
6					0001	0003	0007	0015	0027	0047
7							0001	0002	0004	0008
8										0001

x \ p	11	12	13	14	15	16	17	18	19	20
1	8621	8862	9063	9230	9369	9484	9579	9657	9722	9775
2	5723	6223	6682	7099	7475	7813	8113	8379	8613	8818
3	2858	3345	3836	4324	4802	5266	5711	6133	6532	6904
4	1087	1383	1710	2065	2444	2841	3251	3669	4091	4511
5	0321	0446	0598	0778	0987	1224	1487	1775	2087	2418
6	0075	0114	0166	0234	0319	0423	0548	0695	0864	1057
7	0014	0023	0037	0056	0083	0118	0163	0220	0291	0377
8	0002	0004	0007	0011	0017	0027	0039	0057	0080	0109
9		0001	0001	0002	0003	0005	0008	0012	0018	0026
10						0001	0001	0002	0003	0005
11										0001

Table A.2 Cumulative Binomial Distribution (cont.)

p\x	21	22	23	24	25	26	27	28	29	30
1	9818	9854	9882	9906	9925	9940	9953	9962	9970	9977
2	8996	9152	9285	9400	9499	9583	9654	9714	9765	9807
3	7249	7567	7859	8123	8363	8578	8771	8942	9093	9226
4	4927	5333	5728	6107	6470	6814	7137	7440	7721	7981
5	2766	3128	3500	3879	4261	4643	5023	5396	5760	6113
6	1273	1510	1770	2049	2347	2661	2989	3329	3677	4032
7	0479	0598	0736	0894	1071	1268	1485	1721	1976	2248
8	0147	0194	0251	0320	0402	0499	0611	0739	0884	1046
9	0037	0051	0070	0094	0124	0161	0206	0261	0326	0403
10	0007	0011	0016	0022	0031	0042	0057	0075	0098	0127
11	0001	0002	0003	0004	0006	0009	0013	0018	0024	0032
12				0001	0001	0002	0002	0003	0005	0007
13									0001	0001

p\x	31	32	33	34	35	36	37	38	39	40
1	9982	9986	9989	9991	9993	9995	9996	9997	9998	9998
2	9843	9872	9896	9917	9933	9946	9957	9966	9973	9979
3	9343	9444	9532	9608	9673	9728	9775	9815	9849	9877
4	8219	8437	8634	8812	8972	9115	9241	9353	9450	9536
5	6453	6778	7087	7378	7652	7906	8142	8360	8559	8740
6	4390	4749	5105	5458	5803	6139	6465	6778	7077	7361
7	2536	2838	3153	3479	3812	4152	4495	4839	5182	5522
8	1227	1426	1642	1877	2128	2395	2676	2971	3278	3595
9	0492	0595	0712	0845	0994	1159	1341	1541	1757	1989
10	0162	0204	0254	0314	0383	0464	0557	0664	0784	0919
11	0043	0057	0074	0095	0120	0151	0189	0234	0286	0348
12	0009	0013	0017	0023	0030	0040	0051	0066	0084	0106
13	0002	0002	0003	0004	0006	0008	0011	0015	0019	0025
14				0001	0001	0001	0002	0002	0003	0005
15										0001

p\x	41	42	43	44	45	46	47	48	49	50
1	9999	9999	9999	9999	10000	10000	10000	10000	10000	10000
2	9984	9987	9990	9992	9994	9996	9997	9998	9998	9999
3	9900	9920	9935	9948	9959	9968	9975	9980	9985	9988
4	9610	9674	9729	9776	9816	9849	9877	9901	9920	9936
5	8904	9051	9183	9301	9404	9495	9575	9644	9704	9755
6	7628	7879	8113	8330	8529	8712	8878	9028	9162	9283
7	5856	6182	6499	6805	7098	7377	7641	7890	8122	8338
8	3920	4250	4585	4921	5257	5590	5918	6239	6552	6855
9	2238	2502	2780	3072	3374	3687	4008	4335	4667	5000
10	1070	1236	1419	1618	1834	2066	2314	2577	2855	3145
11	0420	0503	0597	0705	0826	0962	1112	1279	1462	1662
12	0133	0165	0203	0248	0301	0363	0434	0517	0611	0717
13	0033	0042	0054	0069	0086	0108	0134	0165	0202	0245
14	0006	0008	0011	0014	0019	0024	0031	0040	0050	0064
15	0001	0001	0002	0002	0003	0004	0005	0007	0009	0012
16							0001	0001	0001	0001

$n = 18$

p\x	01	02	03	04	05	06	07	08	09	10
1	1655	3049	4220	5204	6028	6717	7292	7771	8169	8499
2	0138	0495	1003	1607	2265	2945	3622	4281	4909	5497
3	0007	0052	0157	0333	0581	0898	1275	1702	2168	2662
4		0004	0018	0050	0109	0201	0333	0506	0723	0982
5			0002	0006	0015	0034	0067	0116	0186	0282
6				0001	0002	0005	0010	0021	0038	0064
7						0001	0003	0006	0012
8									0001	0002

p\x	11	12	13	14	15	16	17	18	19	20
1	8773	8998	9185	9338	9464	9566	9651	9719	9775	9820
2	6042	6540	6992	7398	7759	8080	8362	8609	8824	9009
3	3173	3690	4206	4713	5203	5673	6119	6538	6927	7287
4	1282	1618	1986	2382	2798	3229	3669	4112	4554	4990
5	0405	0558	0743	0959	1206	1482	1787	2116	2467	2836

Table A.2 Cumulative Binomial Distribution (cont.)

p\x	11	12	13	14	15	16	17	18	19	20
6	0102	0154	0222	0310	0419	0551	0708	0889	1097	1329
7	0021	0034	0054	0081	0118	0167	0229	0306	0400	0513
8	0003	0006	0011	0017	0027	0041	0060	0086	0120	0163
9		0001	0002	0003	0005	0008	0013	0020	0029	0043
10					0001	0001	0002	0004	0006	0009
11								0001	0001	0002

p\x	21	22	23	24	25	26	27	28	29	30
1	9856	9886	9909	9928	9944	9956	9965	9973	9979	9984
2	9169	9306	9423	9522	9605	9676	9735	9784	9824	9858
3	7616	7916	8187	8430	8647	8839	9009	9158	9288	9400
4	5414	5825	6218	6591	6943	7272	7578	7860	8119	8354
5	3220	3613	4012	4414	4813	5208	5594	5968	6329	6673
6	1586	1866	2168	2488	2825	3176	3538	3907	4281	4656
7	0645	0799	0974	1171	1390	1630	1891	2171	2469	2783
8	0217	0283	0363	0458	0569	0699	0847	1014	1200	1407
9	0060	0083	0112	0148	0193	0249	0316	0395	0488	0596
10	0014	0020	0028	0039	0054	0073	0097	0127	0164	0210
11	0003	0004	0006	0009	0012	0018	0025	0034	0046	0061
12		0001	0001	0002	0002	0003	0005	0007	0010	0014
13						0001	0001	0001	0002	0003

p\x	31	32	33	34	35	36	37	38	39	40
1	9987	9990	9993	9994	9996	9997	9998	9998	9999	9999
2	9886	9908	9927	9942	9954	9964	9972	9978	9983	9987
3	9498	9581	9652	9713	9764	9807	9843	9873	9897	9918
4	8568	8759	8931	9083	9217	9335	9439	9528	9606	9672
5	7001	7309	7598	7866	8114	8341	8549	8737	8907	9058
6	5029	5398	5759	6111	6450	6776	7086	7379	7655	7912
7	3111	3450	3797	4151	4509	4867	5224	5576	5921	6257
8	1633	1878	2141	2421	2717	3027	3349	3681	4021	4366
9	0720	0861	1019	1196	1391	1604	1835	2084	2350	2632
10	0264	0329	0405	0494	0597	0714	0847	0997	1163	1347
11	0080	0104	0133	0169	0212	0264	0325	0397	0480	0576
12	0020	0027	0036	0047	0062	0080	0102	0130	0163	0203
13	0004	0005	0008	0011	0014	0019	0026	0034	0044	0058
14	0001	0001	0001	0002	0003	0004	0005	0007	0010	0013
15						0001	0001	0001	0002	0002

p\x	41	42	43	44	45	46	47	48	49	50
1	9999	9999	10000	10000	10000	10000	10000	10000	10000	10000
2	9990	9992	9994	9996	9997	9998	9998	9999	9999	9999
3	9934	9948	9959	9968	9975	9981	9985	9989	9991	9993
4	9729	9777	9818	9852	9880	9904	9923	9939	9952	9962
5	9193	9313	9418	9510	9589	9658	9717	9767	9810	9846
6	8151	8372	8573	8757	8923	9072	9205	9324	9428	9519
7	6582	6895	7193	7476	7742	7991	8222	8436	8632	8811
8	4713	5062	5408	5750	6085	6412	6728	7032	7322	7597
9	2928	3236	3556	3885	4222	4562	4906	5249	5591	5927
10	1549	1768	2004	2258	2527	2812	3110	3421	3742	4073
11	0686	0811	0951	1107	1280	1470	1677	1902	2144	2403
12	0250	0307	0372	0449	0537	0638	0753	0883	1028	1189
13	0074	0094	0118	0147	0183	0225	0275	0334	0402	0481
14	0017	0022	0029	0038	0049	0063	0079	0100	0125	0154
15	0003	0004	0006	0007	0010	0013	0017	0023	0029	0038
16		0001	0001	0001	0001	0002	0003	0004	0005	0007
17									0001	0001

$n = 19$

p\x	01	02	03	04	05	06	07	08	09	10
1	1738	3188	4394	5396	6226	6914	7481	7949	8334	8649
2	0153	0546	1100	1751	2453	3171	3879	4560	5202	5797
3	0009	0061	0183	0384	0665	1021	1439	1908	2415	2946
4		0005	0022	0061	0132	0243	0398	0602	0853	1150
5			0002	0007	0020	0044	0085	0147	0235	0352
6				0001	0002	0006	0014	0029	0051	0086
7						0001	0002	0004	0009	0017
8								0001	0001	0003

Table A.2 Cumulative Binomial Distribution (cont.)

p\x	11	12	13	14	15	16	17	18	19	20
1	8908	9119	9291	9431	9544	9636	9710	9770	9818	9856
2	6342	6835	7277	7669	8015	8318	8581	8809	9004	9171
3	3488	4032	4568	5089	5587	6059	6500	6910	7287	7631
4	1490	1867	2275	2708	3159	3620	4085	4549	5005	5449
5	0502	0685	0904	1158	1444	1762	2107	2476	2864	3267
6	0135	0202	0290	0401	0537	0700	0891	1110	1357	1631
7	0030	0048	0076	0113	0163	0228	0310	0411	0532	0676
8	0005	0009	0016	0026	0041	0061	0089	0126	0173	0233
9	0001	0002	0003	0005	0008	0014	0021	0032	0047	0067
10				0001	0001	0002	0004	0007	0010	0016
11							0001	0001	0002	0003

p\x	21	22	23	24	25	26	27	28	29	30
1	9887	9911	9930	9946	9958	9967	9975	9981	9985	9989
2	9313	9434	9535	9619	9690	9749	9797	9837	9869	9896
3	7942	8222	8471	8692	8887	9057	9205	9333	9443	9538
4	5877	6285	6671	7032	7369	7680	7965	8224	8458	8668
5	3681	4100	4520	4936	5346	5744	6129	6498	6848	7178
6	1929	2251	2592	2950	3322	3705	4093	4484	4875	5261
7	0843	1034	1248	1487	1749	2032	2336	2657	2995	3345
8	0307	0396	0503	0629	0775	0941	1129	1338	1568	1820
9	0093	0127	0169	0222	0287	0366	0459	0568	0694	0839
10	0023	0034	0047	0066	0089	0119	0156	0202	0258	0326
11	0005	0007	0011	0016	0023	0032	0044	0060	0080	0105
12	0001	0001	0002	0003	0005	0007	0010	0015	0021	0028
13				0001	0001	0001	0002	0003	0004	0006
14									0001	0001

p\x	31	32	33	34	35	36	37	38	39	40
1	9991	9993	9995	9996	9997	9998	9998	9999	9999	9999
2	9917	9935	9949	9960	9969	9976	9981	9986	9989	9992
3	9618	9686	9743	9791	9830	9863	9890	9913	9933	9945
4	8856	9022	9169	9297	9409	9505	9588	9659	9719	9770
5	7486	7773	8037	8280	8500	8699	8878	9038	9179	9304
6	5641	6010	6366	6707	7032	7339	7627	7895	8143	8371
7	3705	4073	4445	4818	5188	5554	5913	6261	6597	6919
8	2091	2381	2688	3010	3344	3690	4043	4401	4762	5122
9	1003	1186	1389	1612	1855	2116	2395	2691	3002	3325
10	0405	0499	0608	0733	0875	1035	1213	1410	1626	1861
11	0137	0176	0223	0280	0347	0426	0518	0625	0747	0885
12	0038	0051	0068	0089	0114	0146	0185	0231	0287	0352
13	0009	0012	0017	0023	0031	0041	0054	0070	0091	0116
14	0002	0002	0003	0005	0007	0009	0013	0017	0023	0031
15			0001	0001	0001	0002	0002	0003	0005	0006
16									0001	0001

p\x	41	42	43	44	45	46	47	48	49	50
1	10000	10000	10000	10000	10000	10000	10000	10000	10000	10000
2	9994	9995	9996	9997	9998	9999	9999	9999	9999	10000
3	9957	9967	9974	9980	9985	9988	9991	9993	9995	9996
4	9813	9849	9878	9903	9923	9939	9952	9963	9971	9978
5	9413	9508	9590	9660	9720	9771	9814	9850	9879	9904
6	8579	8767	8937	9088	9223	9342	9446	9537	9615	9682
7	7226	7515	7787	8039	8273	8488	8684	8862	9022	9165
8	5480	5832	6176	6509	6831	7138	7430	7706	7964	8204
9	3660	4003	4353	4706	5060	5413	5762	6105	6439	6762
10	2114	2385	2672	2974	3290	3617	3954	4299	4648	5000
11	1040	1213	1404	1613	1841	2087	2351	2631	2928	3238
12	0429	0518	0621	0738	0871	1021	1187	1372	1575	1796
13	0146	0183	0227	0280	0342	0415	0500	0597	0709	0835
14	0040	0052	0067	0086	0109	0137	0171	0212	0261	0318
15	0009	0012	0016	0021	0028	0036	0046	0060	0076	0096
16	0001	0002	0003	0004	0005	0007	0010	0013	0017	0022
17				0001	0001	0001	0001	0002	0003	0004

Table A.2 Cumulative Binomial Distribution (cont.)

$n = 20$

x \ p	01	02	03	04	05	06	07	08	09	10
1	1821	3324	4562	5580	6415	7099	7658	8113	8484	8784
2	0169	0599	1198	1897	2642	3395	4131	4831	5484	6083
3	0010	0071	0210	0439	0755	1150	1610	2121	2666	3231
4		0006	0027	0074	0159	0290	0471	0706	0993	1330
5			0003	0010	0026	0056	0107	0183	0290	0432
6				0001	0003	0009	0019	0038	0068	0113
7						0001	0003	0006	0013	0024
8								0001	0002	0004
9										0001

x \ p	11	12	13	14	15	16	17	18	19	20
1	9028	9224	9383	9510	9612	9694	9759	9811	9852	9885
2	6624	7109	7539	7916	8244	8529	8773	8982	9159	9308
3	3802	4369	4920	5450	5951	6420	6854	7252	7614	7939
4	1710	2127	2573	3041	3523	4010	4496	4974	5439	5886
5	0610	0827	1083	1375	1702	2059	2443	2849	3271	3704
6	0175	0260	0370	0507	0673	0870	1098	1356	1643	1958
7	0041	0067	0103	0153	0219	0304	0409	0537	0689	0867
8	0008	0014	0024	0038	0059	0088	0127	0177	0241	0321
9	0001	0002	0005	0008	0013	0021	0033	0049	0071	0100
10			0001	0001	0002	0004	0007	0011	0017	0026
11						0001	0001	0002	0004	0006
12									0001	0001

x \ p	21	22	23	24	25	26	27	28	29	30
1	9910	9931	9946	9959	9968	9976	9982	9986	9989	9992
2	9434	9539	9626	9698	9757	9805	9845	9877	9903	9924
3	8230	8488	8716	8915	9087	9237	9365	9474	9567	9645
4	6310	6711	7085	7431	7748	8038	8300	8534	8744	8929
5	4142	4580	5014	5439	5852	6248	6625	6981	7315	7625
6	2297	2657	3035	3427	3828	4235	4643	5048	5447	5836
7	1071	1301	1557	1838	2142	2467	2810	3169	3540	3920
8	0419	0536	0675	0835	1018	1225	1455	1707	1982	2277
9	0138	0186	0246	0320	0409	0515	0640	0784	0948	1133
10	0038	0054	0075	0103	0139	0183	0238	0305	0385	0480
11	0009	0013	0019	0028	0039	0055	0074	0100	0132	0171
12	0002	0003	0004	0006	0009	0014	0019	0027	0038	0051
13			0001	0001	0002	0003	0004	0006	0009	0013
14							0001	0001	0002	0003

x \ p	31	32	33	34	35	36	37	38	39	40
1	9994	9996	9997	9998	9998	9999	9999	9999	9999	10000
2	9940	9953	9964	9972	9979	9984	9988	9991	9993	9995
3	9711	9765	9811	9848	9879	9904	9924	9940	9953	9964
4	9092	9235	9358	9465	9556	9634	9700	9755	9802	9840
5	7911	8173	8411	8626	8818	8989	9141	9274	9390	9490
6	6213	6574	6917	7242	7546	7829	8090	8329	8547	8744
7	4305	4693	5079	5460	5834	6197	6547	6882	7200	7500
8	2591	2922	3268	3624	3990	4361	4735	5108	5478	5841
9	1340	1568	1818	2087	2376	2683	3005	3341	3688	4044
10	0591	0719	0866	1032	1218	1424	1650	1897	2163	2447
11	0220	0279	0350	0434	0532	0645	0775	0923	1090	1275
12	0069	0091	0119	0154	0196	0247	0308	0381	0466	0565
13	0018	0025	0034	0045	0060	0079	0102	0132	0167	0210
14	0004	0006	0008	0011	0015	0021	0028	0037	0049	0065
15	0001	0001	0001	0002	0003	0004	0006	0009	0012	0016
16						0001	0001	0002	0002	0003

x \ p	41	42	43	44	45	46	47	48	49	50
1	10000	10000	10000	10000	10000	10000	10000	10000	10000	10000
2	9996	9997	9998	9998	9999	9999	9999	10000	10000	10000
3	9972	9979	9984	9988	9991	9993	9995	9996	9997	9998
4	9872	9898	9920	9937	9951	9962	9971	9977	9983	9987
5	9577	9651	9714	9767	9811	9848	9879	9904	9924	9941

Table A.2 Cumulative Binomial Distribution (cont.)

x \ p	41	42	43	44	45	46	47	48	49	50
6	8921	9078	9217	9340	9447	9539	9619	9687	9745	9793
7	7780	8041	8281	8501	8701	8881	9042	9186	9312	9423
8	6196	6539	6868	7183	7480	7759	8020	8261	8482	8684
9	4406	4771	5136	5499	5857	6207	6546	6873	7186	7483
10	2748	3064	3394	3736	4086	4443	4804	5166	5525	5881
11	1480	1705	1949	2212	2493	2791	3104	3432	3771	4119
12	0679	0810	0958	1123	1308	1511	1734	1977	2238	2517
13	0262	0324	0397	0482	0580	0694	0823	0969	1133	1316
14	0084	0107	0136	0172	0214	0265	0326	0397	0480	0577
15	0022	0029	0038	0050	0064	0083	0105	0133	0166	0207
16	0004	0006	0008	0011	0015	0020	0027	0035	0046	0059
17	0001	0001	0001	0002	0003	0004	0005	0007	0010	0013
18						0001	0001	0001	0001	0002

$n = 50$

x \ p	01	02	03	04	05	06	07	08	09	10
1	3950	6358	7819	8701	9231	9547	9734	9845	9910	9948
2	0894	2642	4447	5995	7206	8100	8735	9173	9468	9662
3	0138	0784	1892	3233	4595	5838	6892	7740	8395	8883
4	0016	0178	0628	1391	2396	3527	4673	5747	6697	7497
5	0001	0032	0168	0490	1036	1794	2710	3710	4723	5688
6		0005	0037	0144	0378	0776	1350	2081	2928	3839
7		0001	0007	0036	0118	0289	0583	1019	1596	2298
8			0001	0008	0032	0094	0220	0438	0768	1221
9				0001	0008	0027	0073	0167	0328	0579
10					0002	0007	0022	0056	0125	0245
11						0002	0006	0017	0043	0094
12							0001	0005	0013	0032
13								0001	0004	0010
14									0001	0003
15										0001

x \ p	11	12	13	14	15	16	17	18	19	20
1	9971	9983	9991	9995	9997	9998	9999	10000	10000	10000
2	9788	9869	9920	9951	9971	9983	9990	9994	9997	9998
3	9237	9487	9661	9779	9858	9910	9944	9965	9979	9987
4	8146	8655	9042	9330	9540	9688	9792	9863	9912	9943
5	6562	7320	7956	8472	8879	9192	9428	9601	9726	9815
6	4760	5647	6463	7186	7806	8323	8741	9071	9327	9520
7	3091	3935	4789	5616	6387	7081	7686	8199	8624	8966
8	1793	2467	3217	4010	4812	5594	6328	6996	7587	8096
9	0932	1392	1955	2605	3319	4071	4832	5576	6280	6927
10	0435	0708	1074	1537	2089	2718	3403	4122	4849	5563
11	0183	0325	0535	0824	1199	1661	2203	2813	3473	4164
12	0069	0135	0242	0402	0628	0929	1309	1768	2300	2893
13	0024	0051	0100	0179	0301	0475	0714	1022	1405	1861
14	0008	0018	0037	0073	0132	0223	0357	0544	0791	1106
15	0002	0006	0013	0027	0053	0096	0164	0266	0411	0607
16	0001	0002	0004	0009	0019	0038	0070	0120	0197	0308
17			0001	0003	0007	0014	0027	0050	0087	0144
18				0001	0002	0005	0010	0019	0036	0063
19					0001	0001	0003	0007	0013	0025
20							0001	0002	0005	0009
21									0001	0003
22										0001

x \ p	21	22	23	24	25	26	27	28	29	30
1	10000	10000	10000	10000	10000	10000	10000	10000	10000	10000
2	9999	9999	10000	10000	10000	10000	10000	10000	10000	10000
3	9992	9995	9997	9998	9999	10000	10000	10000	10000	10000
4	9964	9978	9986	9992	9995	9997	9998	9999	9999	10000
5	9877	9919	9948	9967	9979	9987	9992	9995	9997	9998
6	9663	9767	9841	9893	9930	9954	9970	9981	9988	9993
7	9236	9445	9603	9720	9806	9868	9911	9941	9961	9975
8	8523	8874	9156	9377	9547	9676	9772	9842	9892	9927
9	7505	8009	8437	8794	9084	9316	9497	9635	9740	9817
10	6241	6870	7436	7934	8363	8724	9021	9260	9450	9598

Table A.2 / CUMULATIVE BINOMIAL DISTRIBUTION

Table A.2 Cumulative Binomial Distribution (cont.)

p \ x	21	22	23	24	25	26	27	28	29	30
11	4864	5552	6210	6822	7378	7871	8299	8663	8965	9211
12	3533	4201	4878	5544	6184	6782	7329	7817	8244	8610
13	2383	2963	3585	4233	4890	5539	6163	6749	7287	7771
14	1490	1942	2456	3023	3630	4261	4901	5534	6145	6721
15	0862	1181	1565	2013	2519	3075	3669	4286	4912	5532
16	0462	0665	0926	1247	1631	2075	2575	3121	3703	4308
17	0229	0347	0508	0718	0983	1306	1689	2130	2623	3161
18	0105	0168	0259	0384	0551	0766	1034	1359	1741	2178
19	0045	0075	0122	0191	0287	0418	0590	0809	1080	1406
20	0018	0031	0054	0088	0139	0212	0314	0449	0626	0848
21	0006	0012	0022	0038	0063	0100	0155	0232	0338	0478
22	0002	0004	0008	0015	0026	0044	0071	0112	0170	0251
23	0001	0001	0003	0006	0010	0018	0031	0050	0080	0123
24			0001	0002	0004	0007	0012	0021	0035	0056
25				0001	0001	0002	0004	0008	0014	0024
26						0001	0002	0003	0005	0009
27								0001	0002	0003
28									0001	0001

p \ x	31	32	33	34	35	36	37	38	39	40
1	10000	10000	10000	10000	10000	10000	10000	10000	10000	10000
2	10000	10000	10000	10000	10000	10000	10000	10000	10000	10000
3	10000	10000	10000	10000	10000	10000	10000	10000	10000	10000
4	10000	10000	10000	10000	10000	10000	10000	10000	10000	10000
5	9999	9999	10000	10000	10000	10000	10000	10000	10000	10000
6	9996	9997	9998	9999	9999	10000	10000	10000	10000	10000
7	9984	9990	9994	9996	9998	9999	9999	10000	10000	10000
8	9952	9969	9980	9987	9992	9995	9997	9998	9999	9999
9	9874	9914	9942	9962	9975	9984	9990	9994	9996	9998
10	9710	9794	9856	9901	9933	9955	9971	9981	9988	9992
11	9409	9563	9683	9773	9840	9889	9924	9949	9966	9978
12	8916	9168	9371	9533	9658	9753	9825	9878	9916	9943
13	8197	8564	8873	9130	9339	9505	9635	9736	9811	9867
14	7253	7732	8157	8524	8837	9097	9310	9481	9616	9720
15	6131	6698	7223	7699	8122	8491	8805	9069	9286	9460
16	4922	5530	6120	6679	7199	7672	8094	8462	8779	9045
17	3734	4328	4931	5530	6111	6664	7179	7649	8070	8439
18	2666	3197	3760	4346	4940	5531	6105	6653	7164	7631
19	1786	2220	2703	3227	3784	4362	4949	5533	6101	6644
20	1121	1447	1826	2257	2736	3255	3805	4376	4957	5535
21	0657	0882	1156	1482	1861	2289	2764	3278	3824	4390
22	0360	0503	0685	0912	1187	1513	1890	2317	2788	3299
23	0184	0267	0379	0525	0710	0938	1214	1540	1916	2340
24	0087	0133	0196	0282	0396	0544	0730	0960	1236	1562
25	0039	0061	0094	0141	0207	0295	0411	0560	0748	0978
26	0016	0026	0042	0066	0100	0149	0216	0305	0423	0573
27	0006	0011	0018	0029	0045	0070	0106	0155	0223	0314
28	0002	0004	0007	0012	0019	0031	0048	0074	0110	0160
29	0001	0001	0002	0004	0007	0012	0020	0032	0050	0076
30			0001	0002	0003	0005	0008	0013	0021	0034
31					0001	0002	0003	0005	0008	0014
32						0001	0001	0002	0003	0005
33								0001	0001	0002
34										0001

p \ x	41	42	43	44	45	46	47	48	49	50
1	10000	10000	10000	10000	10000	10000	10000	10000	10000	10000
2	10000	10000	10000	10000	10000	10000	10000	10000	10000	10000
3	10000	10000	10000	10000	10000	10000	10000	10000	10000	10000
4	10000	10000	10000	10000	10000	10000	10000	10000	10000	10000
5	10000	10000	10000	10000	10000	10000	10000	10000	10000	10000
6	10000	10000	10000	10000	10000	10000	10000	10000	10000	10000
7	10000	10000	10000	10000	10000	10000	10000	10000	10000	10000
8	10000	10000	10000	10000	10000	10000	10000	10000	10000	10000
9	9999	9999	10000	10000	10000	10000	10000	10000	10000	10000
10	9995	9997	9998	9999	9999	10000	10000	10000	10000	10000
11	9986	9991	9994	9997	9998	9999	9999	10000	10000	10000
12	9962	9975	9984	9990	9994	9996	9998	9999	9999	10000
13	9908	9938	9958	9973	9982	9989	9993	9996	9997	9998
14	9799	9858	9902	9933	9955	9970	9981	9988	9992	9995
15	9599	9707	9789	9851	9896	9929	9952	9968	9980	9987

Table A.2 Cumulative Binomial Distribution (cont.)

p x	41	42	43	44	45	46	47	48	49	50
16	9265	9443	9585	9696	9780	9844	9892	9926	9950	9967
17	8757	9025	9248	9429	9573	9687	9774	9839	9888	9923
18	8051	8421	8740	9010	9235	9418	9565	9680	9769	9836
19	7152	7617	8037	8406	8727	8998	9225	9410	9559	9675
20	6099	6638	7143	7608	8026	8396	8718	8991	9219	9405
21	4965	5539	6099	6635	7138	7602	8020	8391	8713	8987
22	3840	4402	4973	5543	6100	6634	7137	7599	8018	8389
23	2809	3316	3854	4412	4981	5548	6104	6636	7138	7601
24	1936	2359	2826	3331	3866	4422	4989	5554	6109	6641
25	1255	1580	1953	2375	2840	3343	3876	4431	4996	5561
26	0762	0992	1269	1593	1966	2386	2850	3352	3885	4439
27	0432	0584	0772	1003	1279	1603	1975	2395	2858	3359
28	0229	0320	0439	0591	0780	1010	1286	1609	1981	2399
29	0113	0164	0233	0325	0444	0595	0784	1013	1289	1611
30	0052	0078	0115	0166	0235	0327	0446	0596	0784	1013
31	0022	0034	0053	0079	0116	0167	0236	0327	0445	0595
32	0009	0014	0022	0035	0053	0079	0116	0166	0234	0325
33	0003	0005	0009	0014	0022	0035	0053	0078	0114	0164
34	0001	0002	0003	0005	0009	0014	0022	0034	0052	0077
35		0001	0001	0002	0003	0005	0008	0014	0021	0033
36				0001	0001	0002	0003	0005	0008	0013
37						0001	0001	0002	0003	0005
38								0001	0001	0002

$n = 100$

p x	01	02	03	04	05	06	07	08	09	10
1	6340	8674	9524	9831	9941	9979	9993	9998	9999	10000
2	2642	5967	8054	9128	9629	9848	9940	9977	9991	9997
3	0794	3233	5802	7679	8817	9434	9742	9887	9952	9981
4	0184	1410	3528	5705	7422	8570	9256	9633	9827	9922
5	0034	0508	1821	3711	5640	7232	8368	9097	9526	9763
6	0005	0155	0808	2116	3840	5593	7086	8201	8955	9424
7	0001	0041	0312	1064	2340	3936	5557	6968	8060	8828
8		0009	0106	0475	1280	2517	4012	5529	6872	7939
9		0002	0032	0190	0631	1463	2660	4074	5506	6791
10			0009	0068	0282	0775	1620	2780	4125	5487
11			0002	0022	0115	0376	0908	1757	2882	4168
12				0007	0043	0168	0469	1028	1876	2970
13				0002	0015	0069	0224	0559	1138	1982
14					0005	0026	0099	0282	0645	1239
15					0001	0009	0041	0133	0341	0726
16						0003	0016	0058	0169	0399
17						0001	0006	0024	0078	0206
18							0002	0009	0034	0100
19							0001	0003	0014	0046
20								0001	0005	0020
21									0002	0008
22									0001	0003
23										0001

p x	11	12	13	14	15	16	17	18	19	20
1	10000	10000	10000	10000	10000	10000	10000	10000	10000	10000
2	9999	10000	10000	10000	10000	10000	10000	10000	10000	10000
3	9992	9997	9999	10000	10000	10000	10000	10000	10000	10000
4	9966	9985	9994	9998	9999	10000	10000	10000	10000	10000
5	9886	9947	9977	9990	9996	9998	9999	10000	10000	10000
6	9698	9848	9926	9966	9984	9993	9997	9999	10000	10000
7	9328	9633	9808	9903	9953	9978	9990	9996	9998	9999
8	8715	9239	9569	9766	9878	9939	9970	9986	9994	9997
9	7835	8614	9155	9508	9725	9853	9924	9962	9982	9991
10	6722	7743	8523	9078	9449	9684	9826	9908	9953	9977
11	5471	6663	7663	8440	9006	9393	9644	9800	9891	9943
12	4206	5458	6611	7591	8365	8939	9340	9605	9773	9874
13	3046	4239	5446	6566	7527	8297	8876	9289	9567	9747
14	2076	3114	4268	5436	6526	7469	8234	8819	9241	9531
15	1330	2160	3173	4294	5428	6490	7417	8177	8765	9196
16	0802	1414	2236	3227	4317	5420	6458	7370	8125	8715
17	0456	0874	1492	2305	3275	4338	5414	6429	7327	8077
18	0244	0511	0942	1563	2367	3319	4357	5408	6403	7288
19	0123	0282	0564	1006	1628	2424	3359	4374	5403	6379
20	0059	0147	0319	0614	1065	1689	2477	3395	4391	5398

Table A.2 Cumulative Binomial Distribution (cont.)

p x	11	12	13	14	15	16	17	18	19	20
21	0026	0073	0172	0356	0663	1121	1745	2525	3429	4405
22	0011	0034	0088	0196	0393	0710	1174	1797	2570	3460
23	0005	0015	0042	0103	0221	0428	0754	1223	1846	2611
24	0002	0006	0020	0051	0119	0246	0462	0796	1270	1891
25	0001	0003	0009	0024	0061	0135	0271	0496	0837	1314
26		0001	0004	0011	0030	0071	0151	0295	0528	0875
27			0001	0005	0014	0035	0081	0168	0318	0558
28			0001	0002	0006	0017	0041	0091	0184	0342
29				0001	0003	0008	0020	0048	0102	0200
30					0001	0003	0009	0024	0054	0112
31						0001	0004	0011	0027	0061
32						0001	0002	0005	0013	0031
33							0001	0002	0006	0016
34								0001	0003	0007
35									0001	0003
36										0001
37										0001

p x	21	22	23	24	25	26	27	28	29	30
1	10000	10000	10000	10000	10000	10000	10000	10000	10000	10000
2	10000	10000	10000	10000	10000	10000	10000	10000	10000	10000
3	10000	10000	10000	10000	10000	10000	10000	10000	10000	10000
4	10000	10000	10000	10000	10000	10000	10000	10000	10000	10000
5	10000	10000	10000	10000	10000	10000	10000	10000	10000	10000
6	10000	10000	10000	10000	10000	10000	10000	10000	10000	10000
7	10000	10000	10000	10000	10000	10000	10000	10000	10000	10000
8	9999	10000	10000	10000	10000	10000	10000	10000	10000	10000
9	9996	9998	9999	10000	10000	10000	10000	10000	10000	10000
10	9989	9995	9998	9999	10000	10000	10000	10000	10000	10000
11	9971	9986	9993	9997	9999	9999	10000	10000	10000	10000
12	9933	9965	9983	9992	9996	9998	9999	10000	10000	10000
13	9857	9922	9959	9979	9990	9995	9998	9999	10000	10000
14	9721	9840	9911	9953	9975	9988	9994	9997	9999	9999
15	9496	9695	9823	9900	9946	9972	9986	9993	9997	9998
16	9153	9462	9671	9806	9889	9939	9967	9983	9992	9996
17	8668	9112	9430	9647	9789	9878	9932	9963	9981	9990
18	8032	8625	9074	9399	9624	9773	9867	9925	9959	9978
19	7252	7991	8585	9038	9370	9601	9757	9856	9918	9955
20	6358	7220	7953	8547	9005	9342	9580	9741	9846	9911
21	5394	6338	7189	7918	8512	8973	9316	9560	9726	9835
22	4419	5391	6320	7162	7886	8479	8943	9291	9540	9712
23	3488	4432	5388	6304	7136	7856	8448	8915	9267	9521
24	2649	3514	4444	5386	6289	7113	7828	8420	8889	9245
25	1933	2684	3539	4455	5383	6276	7091	7802	8393	8864
26	1355	1972	2717	3561	4465	5381	6263	7071	7778	8369
27	0911	1393	2009	2748	3583	4475	5380	6252	7053	7756
28	0588	0945	1429	2043	2776	3602	4484	5378	6242	7036
29	0364	0616	0978	1463	2075	2803	3621	4493	5377	6232
30	0216	0386	0643	1009	1495	2105	2828	3638	4501	5377
31	0123	0232	0406	0669	1038	1526	2134	2851	3654	4509
32	0067	0134	0247	0427	0693	1065	1554	2160	2873	3669
33	0035	0074	0144	0262	0446	0717	1091	1580	2184	2893
34	0018	0039	0081	0154	0276	0465	0739	1116	1605	2207
35	0009	0020	0044	0087	0164	0290	0482	0760	1139	1629
36	0004	0010	0023	0048	0094	0174	0303	0499	0780	1161
37	0002	0005	0011	0025	0052	0101	0183	0316	0515	0799
38	0001	0002	0005	0013	0027	0056	0107	0193	0328	0530
39		0001	0002	0006	0014	0030	0060	0113	0201	0340
40			0001	0003	0007	0015	0032	0064	0119	0210
41				0001	0003	0008	0017	0035	0068	0125
42				0001	0001	0004	0008	0018	0037	0072
43					0001	0002	0004	0009	0020	0040
44						0001	0002	0005	0010	0021
45							0001	0002	0005	0011
46								0001	0002	0005
47									0001	0003
48										0001
49										0001

Table A.2 Cumulative Binomial Distribution (cont.)

p\\x	31	32	33	34	35	36	37	38	39	40
1	10000	10000	10000	10000	10000	10000	10000	10000	10000	10000
2	10000	10000	10000	10000	10000	10000	10000	10000	10000	10000
3	10000	10000	10000	10000	10000	10000	10000	10000	10000	10000
4	10000	10000	10000	10000	10000	10000	10000	10000	10000	10000
5	10000	10000	10000	10000	10000	10000	10000	10000	10000	10000
6	10000	10000	10000	10000	10000	10000	10000	10000	10000	10000
7	10000	10000	10000	10000	10000	10000	10000	10000	10000	10000
8	10000	10000	10000	10000	10000	10000	10000	10000	10000	10000
9	10000	10000	10000	10000	10000	10000	10000	10000	10000	10000
10	10000	10000	10000	10000	10000	10000	10000	10000	10000	10000
11	10000	10000	10000	10000	10000	10000	10000	10000	10000	10000
12	10000	10000	10000	10000	10000	10000	10000	10000	10000	10000
13	10000	10000	10000	10000	10000	10000	10000	10000	10000	10000
14	10000	10000	10000	10000	10000	10000	10000	10000	10000	10000
15	9999	10000	10000	10000	10000	10000	10000	10000	10000	10000
16	9998	9999	10000	10000	10000	10000	10000	10000	10000	10000
17	9995	9998	9999	10000	10000	10000	10000	10000	10000	10000
18	9989	9995	9997	9999	9999	10000	10000	10000	10000	10000
19	9976	9988	9994	9997	9999	9999	10000	10000	10000	10000
20	9950	9973	9986	9993	9997	9998	9999	10000	10000	10000
21	9904	9946	9971	9985	9992	9996	9998	9999	10000	10000
22	9825	9898	9942	9968	9983	9991	9996	9998	9999	10000
23	9698	9816	9891	9938	9966	9982	9991	9995	9998	9999
24	9504	9685	9806	9885	9934	9963	9980	9990	9995	9997
25	9224	9487	9672	9797	9879	9930	9961	9979	9989	9994
26	8841	9204	9471	9660	9789	9873	9926	9958	9977	9988
27	8346	8820	9185	9456	9649	9780	9867	9922	9956	9976
28	7736	8325	8800	9168	9442	9638	9773	9862	9919	9954
29	7021	7717	8305	8781	9152	9429	9628	9765	9857	9916
30	6224	7007	7699	8287	8764	9137	9417	9618	9759	9852
31	5376	6216	6994	7684	8270	8748	9123	9405	9610	9752
32	4516	5376	6209	6982	7669	8254	8733	9110	9395	9602
33	3683	4523	5375	6203	6971	7656	8240	8720	9098	9385
34	2912	3696	4530	5375	6197	6961	7643	8227	8708	9087
35	2229	2929	3708	4536	5376	6192	6953	7632	8216	8697
36	1650	2249	2946	3720	4542	5376	6188	6945	7623	8205
37	1181	1671	2268	2961	3731	4547	5377	6184	6938	7614
38	0816	1200	1690	2285	2976	3741	4553	5377	6181	6932
39	0545	0833	1218	1708	2301	2989	3750	4558	5378	6178
40	0351	0558	0849	1235	1724	2316	3001	3759	4562	5379
41	0218	0361	0571	0863	1250	1739	2330	3012	3767	4567
42	0131	0226	0371	0583	0877	1265	1753	2343	3023	3775
43	0075	0136	0233	0380	0594	0889	1278	1766	2355	3033
44	0042	0079	0141	0240	0389	0605	0901	1290	1778	2365
45	0023	0044	0082	0146	0246	0397	0614	0911	1301	1789
46	0012	0024	0046	0085	0150	0252	0405	0623	0921	1311
47	0006	0012	0025	0048	0088	0154	0257	0411	0631	0930
48	0003	0006	0013	0026	0050	0091	0158	0262	0417	0638
49	0001	0003	0007	0014	0027	0052	0094	0162	0267	0423
50	0001	0001	0003	0007	0015	0029	0054	0096	0165	0271
51		0001	0002	0003	0007	0015	0030	0055	0098	0168
52			0001	0002	0004	0008	0016	0030	0056	0100
53				0001	0002	0004	0008	0016	0031	0058
54					0001	0002	0004	0008	0017	0032
55						0001	0002	0004	0009	0017
56							0001	0002	0004	0009
57								0001	0002	0004
58									0001	0002
59										0001

Table A.2 Cumulative Binomial Distribution (cont.)

x\p	41	42	43	44	45	46	47	48	49	50
1	10000	10000	10000	10000	10000	10000	10000	10000	10000	10000
2	10000	10000	10000	10000	10000	10000	10000	10000	10000	10000
3	10000	10000	10000	10000	10000	10000	10000	10000	10000	10000
4	10000	10000	10000	10000	10000	10000	10000	10000	10000	10000
5	10000	10000	10000	10000	10000	10000	10000	10000	10000	10000
6	10000	10000	10000	10000	10000	10000	10000	10000	10000	10000
7	10000	10000	10000	10000	10000	10000	10000	10000	10000	10000
8	10000	10000	10000	10000	10000	10000	10000	10000	10000	10000
9	10000	10000	10000	10000	10000	10000	10000	10000	10000	10000
10	10000	10000	10000	10000	10000	10000	10000	10000	10000	10000
11	10000	10000	10000	10000	10000	10000	10000	10000	10000	10000
12	10000	10000	10000	10000	10000	10000	10000	10000	10000	10000
13	10000	10000	10000	10000	10000	10000	10000	10000	10000	10000
14	10000	10000	10000	10000	10000	10000	10000	10000	10000	10000
15	10000	10000	10000	10000	10000	10000	10000	10000	10000	10000
16	10000	10000	10000	10000	10000	10000	10000	10000	10000	10000
17	10000	10000	10000	10000	10000	10000	10000	10000	10000	10000
18	10000	10000	10000	10000	10000	10000	10000	10000	10000	10000
19	10000	10000	10000	10000	10000	10000	10000	10000	10000	10000
20	10000	10000	10000	10000	10000	10000	10000	10000	10000	10000
21	10000	10000	10000	10000	10000	10000	10000	10000	10000	10000
22	10000	10000	10000	10000	10000	10000	10000	10000	10000	10000
23	10000	10000	10000	10000	10000	10000	10000	10000	10000	10000
24	9999	9999	10000	10000	10000	10000	10000	10000	10000	10000
25	9997	9999	9999	10000	10000	10000	10000	10000	10000	10000
26	9994	9997	9999	9999	10000	10000	10000	10000	10000	10000
27	9987	9994	9997	9998	9999	10000	10000	10000	10000	10000
28	9975	9987	9993	9997	9998	9999	10000	10000	10000	10000
29	9952	9974	9986	9993	9996	9998	9999	10000	10000	10000
30	9913	9950	9972	9985	9992	9996	9998	9999	10000	10000
31	9848	9910	9948	9971	9985	9992	9996	9998	9999	10000
32	9746	9844	9907	9947	9970	9984	9992	9996	9998	9999
33	9594	9741	9840	9905	9945	9969	9984	9991	9996	9998
34	9376	9587	9736	9837	9902	9944	9969	9983	9991	9996
35	9078	9368	9581	9732	9834	9900	9942	9968	9983	9991
36	8687	9069	9361	9576	9728	9831	9899	9941	9967	9982
37	8196	8678	9061	9355	9571	9724	9829	9897	9941	9967
38	7606	8188	8670	9054	9349	9567	9721	9827	9896	9940
39	6927	7599	8181	8663	9049	9345	9563	9719	9825	9895
40	6176	6922	7594	8174	8657	9044	9341	9561	9717	9824
41	5380	6174	6919	7589	8169	8653	9040	9338	9558	9716
42	4571	5382	6173	6916	7585	8165	8649	9037	9335	9557
43	3782	4576	5383	6173	6913	7582	8162	8646	9035	9334
44	3041	3788	4580	5385	6172	6912	7580	8160	8645	9033
45	2375	3049	3794	4583	5387	6173	6911	7579	8159	8644
46	1799	2384	3057	3799	4587	5389	6173	6911	7579	8159
47	1320	1807	2391	3063	3804	4590	5391	6174	6912	7579
48	0938	1328	1815	2398	3069	3809	4593	5393	6176	6914
49	0644	0944	1335	1822	2404	3074	3813	4596	5395	6178
50	0428	0650	0950	1341	1827	2409	3078	3816	4599	5398
51	0275	0432	0655	0955	1346	1832	2413	3082	3819	4602
52	0170	0278	0436	0659	0960	1350	1836	2417	3084	3822
53	0102	0172	0280	0439	0662	0963	1353	1838	2419	3086
54	0059	0103	0174	0282	0441	0664	0965	1355	1840	2421
55	0033	0059	0104	0175	0284	0443	0666	0967	1356	1841
56	0017	0033	0060	0105	0176	0285	0444	0667	0967	1356
57	0009	0018	0034	0061	0106	0177	0286	0444	0667	0967
58	0004	0009	0018	0034	0061	0106	0177	0286	0444	0666
59	0002	0005	0009	0018	0034	0061	0106	0177	0285	0443
60	0001	0002	0005	0009	0018	0034	0061	0106	0177	0284
61		0001	0002	0005	0009	0018	0034	0061	0106	0176
62			0001	0002	0005	0009	0018	0034	0061	0105
63				0001	0002	0005	0009	0018	0034	0060
64					0001	0002	0005	0009	0018	0033
65						0001	0002	0005	0009	0018
66							0001	0002	0004	0009
67								0001	0002	0004
68									0001	0002
69										0001

Table A.3 Cumulative Poisson Distribution

$$1000 \times P_p(X \leq k | \lambda)$$

1,000 × probability of k or less occurrences of event that has average number of occurrences equal to λ.

λ \ k	0	1	2	3	4	5	6	7	8	9
0.02	980	1,000								
0.04	961	999	1,000							
0.06	942	998	1,000							
0.08	923	997	1,000							
0.10	905	995	1,000							
0.15	861	990	999	1,000						
0.20	819	982	999	1,000						
0.25	779	974	998	1,000						
0.30	741	963	996	1,000						
0.35	705	951	994	1,000						
0.40	670	938	992	999	1,000					
0.45	638	925	989	999	1,000					
0.50	607	910	986	998	1,000					
0.55	577	894	982	998	1,000					
0.60	549	878	977	997	1,000					
0.65	522	861	972	996	999	1,000				
0.70	497	844	966	994	999	1,000				
0.75	472	827	959	993	999	1,000				
0.80	449	809	953	991	999	1,000				
0.85	427	791	945	989	998	1,000				
0.90	407	772	937	987	998	1,000				
0.95	387	754	929	984	997	1,000				
1.00	368	736	920	981	996	999	1,000			
1.1	333	699	900	974	995	999	1,000			
1.2	301	663	879	966	992	998	1,000			
1.3	273	627	857	957	989	998	1,000			
1.4	247	592	833	946	986	997	999	1,000		
1.5	223	558	809	934	981	996	999	1,000		
1.6	202	525	783	921	976	994	999	1,000		
1.7	183	493	757	907	970	992	998	1,000		
1.8	165	463	731	891	964	990	997	999	1,000	
1.9	150	434	704	875	956	987	997	999	1,000	
2.0	135	406	677	857	947	983	995	999	1,000	

Source: This table is reproduced from E. L. Grant, *Statistical Quality Control*, second edition, McGraw-Hill Book Company, New York, 1952. Used by permission.

Table A.3 Cumulative Poisson Distribution (cont.)

λ \ k	0	1	2	3	4	5	6	7	8	9
2.2	111	355	623	819	928	975	993	998	1,000	
2.4	091	308	570	779	904	964	988	997	999	1,000
2.6	074	267	518	736	877	951	983	995	999	1,000
2.8	061	231	469	692	848	935	976	992	998	999
3.0	050	199	423	647	815	916	966	988	996	999
3.2	041	171	380	603	781	895	955	983	994	998
3.4	033	147	340	558	744	871	942	977	992	997
3.6	027	126	303	515	706	844	927	969	988	996
3.8	022	107	269	473	668	816	909	960	984	994
4.0	018	092	238	433	629	785	889	949	979	992
4.2	015	078	210	395	590	753	867	936	972	989
4.4	012	066	185	359	551	720	844	921	964	985
4.6	010	056	163	326	513	686	818	905	955	980
4.8	008	048	143	294	476	651	791	887	944	975
5.0	007	040	125	265	440	616	762	867	932	968
5.2	006	034	109	238	406	581	732	845	918	960
5.4	005	029	095	213	373	546	702	822	903	951
5.6	004	024	082	191	342	512	670	797	886	941
5.8	003	021	072	170	313	478	638	771	867	929
6.0	002	017	062	151	285	446	606	744	847	916

λ \ k	10	11	12	13	14	15	16
2.8	1,000						
3.0	1,000						
3.2	1,000						
3.4	999	1,000					
3.6	999	1,000					
3.8	998	999	1,000				
4.0	997	999	1,000				
4.2	996	999	1,000				
4.4	994	998	999	1,000			
4.6	992	997	999	1,000			
4.8	990	996	999	1,000			
5.0	986	995	998	999	1,000		
5.2	982	993	997	999	1,000		
5.4	977	990	996	999	1,000		
5.6	972	988	995	998	999	1,000	
5.8	965	984	993	997	999	1,000	
6.0	957	980	991	996	999	999	1,000

Table A.3 Cumulative Poisson Distribution (cont.)

λ \ k	0	1	2	3	4	5	6	7	8	9
6.2	002	015	054	134	259	414	574	716	826	902
6.4	002	012	046	119	235	384	542	687	803	886
6.6	001	010	040	105	213	355	511	658	780	869
6.8	001	009	034	093	192	327	480	628	755	850
7.0	001	007	030	082	173	301	450	599	729	830
7.2	001	006	025	072	156	276	420	569	703	810
7.4	001	005	022	063	140	253	392	539	676	788
7.6	001	004	019	055	125	231	365	510	648	765
7.8	000	004	016	048	112	210	338	481	620	741
8.0	000	003	014	042	100	191	313	453	593	717
8.5	000	002	009	030	074	150	256	386	523	653
9.0	000	001	006	021	055	116	207	324	456	587
9.5	000	001	004	015	040	089	165	269	392	522
10.0	000	000	003	010	029	067	130	220	333	458

λ \ k	10	11	12	13	14	15	16	17	18	19
6.2	949	975	989	995	998	999	1,000			
6.4	939	969	986	994	997	999	1,000			
6.6	927	963	982	992	997	999	999	1,000		
6.8	915	955	978	990	996	998	999	1,000		
7.0	901	947	973	987	994	998	999	1,000		
7.2	887	937	967	984	993	997	999	999	1,000	
7.4	871	926	961	980	991	996	998	999	1,000	
7.6	854	915	954	976	989	995	998	999	1,000	
7.8	835	902	945	971	986	993	997	999	1,000	
8.0	816	888	936	966	983	992	996	998	999	1,000
8.5	763	849	909	949	973	986	993	997	999	999
9.0	706	803	876	926	959	978	989	995	998	999
9.5	645	752	836	898	940	967	982	991	996	998
10.0	583	697	792	864	917	951	973	986	993	997

λ \ k	20	21	22
8.5	1,000		
9.0	1,000		
9.5	999	1,000	
10.0	998	999	1,000

Table A.3 Cumulative Poisson Distribution (cont.)

λ \ k	0	1	2	3	4	5	6	7	8	9
10.5	000	000	002	007	021	050	102	179	279	397
11.0	000	000	001	005	015	038	079	143	232	341
11.5	000	000	001	003	011	028	060	114	191	289
12.0	000	000	001	002	008	020	046	090	155	242
12.5	000	000	000	002	005	015	035	070	125	201
13.0	000	000	000	001	004	011	026	054	100	166
13.5	000	000	000	001	003	008	019	041	079	135
14.0	000	000	000	000	002	006	014	032	062	109
14.5	000	000	000	000	001	004	010	024	048	088
15.0	000	000	000	000	001	003	008	018	037	070

λ \ k	10	11	12	13	14	15	16	17	18	19
10.5	521	639	742	825	888	932	960	978	988	994
11.0	460	579	689	781	854	907	944	968	982	991
11.5	402	520	633	733	815	878	924	954	974	986
12.0	347	462	576	682	772	844	899	937	963	979
12.5	297	406	519	628	725	806	869	916	948	969
13.0	252	353	463	573	675	764	835	890	930	957
13.5	211	304	409	518	623	718	798	861	908	942
14.0	176	260	358	464	570	669	756	827	883	923
14.5	145	220	311	413	518	619	711	790	853	901
15.0	118	185	268	363	466	568	664	749	819	875

λ \ k	20	21	22	23	24	25	26	27	28	29
10.5	997	999	999	1,000						
11.0	995	998	999	1,000						
11.5	992	996	998	999	1,000					
12.0	988	994	997	999	999	1,000				
12.5	983	991	995	998	999	999	1,000			
13.0	975	986	992	996	998	999	1,000			
13.5	965	980	989	994	997	998	999	1,000		
14.0	952	971	983	991	995	997	999	999	1,000	
14.5	936	960	976	986	992	996	998	999	999	1,000
15.0	917	947	967	981	989	994	997	998	999	1,000

Table A.3 Cumulative Poisson Distribution (cont.)

λ \ k	4	5	6	7	8	9	10	11	12	13
16	000	001	004	010	022	043	077	127	193	275
17	000	001	002	005	013	026	049	085	135	201
18	000	000	001	003	007	015	030	055	092	143
19	000	000	001	002	004	009	018	035	061	098
20	000	000	000	001	002	005	011	021	039	066
21	000	000	000	000	001	003	006	013	025	043
22	000	000	000	000	001	002	004	008	015	028
23	000	000	000	000	000	001	002	004	009	017
24	000	000	000	000	000	000	001	003	005	011
25	000	000	000	000	000	000	001	001	003	006

λ \ k	14	15	16	17	18	19	20	21	22	23
16	368	467	566	659	742	812	868	911	942	963
17	281	371	468	564	655	736	805	861	905	937
18	208	287	375	469	562	651	731	799	855	899
19	150	215	292	378	469	561	647	725	793	849
20	105	157	221	297	381	470	559	644	721	787
21	072	111	163	227	302	384	471	558	640	716
22	048	077	117	169	232	306	387	472	556	637
23	031	052	082	123	175	238	310	389	472	555
24	020	034	056	087	128	180	243	314	392	473
25	012	022	038	060	092	134	185	247	318	394

λ \ k	24	25	26	27	28	29	30	31	32	33
16	978	987	993	996	998	999	999	1,000		
17	959	975	985	991	995	997	999	999	1,000	
18	932	955	972	983	990	994	997	998	999	1,000
19	893	927	951	969	980	988	993	996	998	999
20	843	888	922	948	966	978	987	992	995	997
21	782	838	883	917	944	963	976	985	991	994
22	712	777	832	877	913	940	959	973	983	989
23	635	708	772	827	873	908	936	956	971	981
24	554	632	704	768	823	868	904	932	953	969
25	473	553	629	700	763	818	863	900	929	950

λ \ k	34	35	36	37	38	39	40	41	42	43
19	999	1,000								
20	999	999	1,000							
21	997	998	999	999	1,000					
22	994	996	998	999	999	1,000				
23	988	993	996	997	999	999	1,000			
24	979	987	992	995	997	998	999	999	1,000	
25	966	978	985	991	994	997	998	999	999	1,000

Table A.4 Binomial Coefficients

n	$_nC_0$	$_nC_1$	$_nC_2$	$_nC_3$	$_nC_4$	$_nC_5$	$_nC_6$	$_nC_7$	$_nC_8$	$_nC_9$	$_nC_{10}$
0	1										
1	1	1									
2	1	2	1								
3	1	3	3	1							
4	1	4	6	4	1						
5	1	5	10	10	5	1					
6	1	6	15	20	15	6	1				
7	1	7	21	35	35	21	7	1			
8	1	8	28	56	70	56	28	8	1		
9	1	9	36	84	126	126	84	36	9	1	
10	1	10	45	120	210	252	210	120	45	10	1
11	1	11	55	165	330	462	462	330	165	55	11
12	1	12	66	220	495	792	924	792	495	220	66
13	1	13	78	286	715	1287	1716	1716	1287	715	286
14	1	14	91	364	1001	2002	3003	3432	3003	2002	1001
15	1	15	105	455	1365	3003	5005	6435	6435	5005	3003
16	1	16	120	560	1820	4368	8008	11440	12870	11440	8008
17	1	17	136	680	2380	6188	12376	19448	24310	24310	19448
18	1	18	153	816	3060	8568	18564	31824	43758	48620	43758
19	1	19	171	969	3876	11628	27132	50388	75582	92378	92378
20	1	20	190	1140	4845	15504	38760	77520	125970	167960	184756

$$_nC_r = \binom{n}{r} = \frac{n!}{r!(n-r)!} = \binom{n}{n-r} = {_nC_{n-r}}, \qquad _nC_0 = 1$$

$$(q+p)^n = q^n + {_nC_1}pq^{n-1} + \cdots + {_nC_r}p^r q^{n-r} + \cdots + p^n$$

Source: This table is reproduced from *Handbook of Probability and Statistics with Tables* by R. S. Burington and D. C. May, second edition. Copyright © 1970 by McGraw-Hill, Inc. Used with permission of McGraw-Hill Book Company.

Table A.5 Single-Payment Compound Amount Factor

$$(1+r)^n$$

Conversion Periods	Rate r							
n	0.01	0.02	0.03	0.04	0.05	0.06	0.07	0.08
1	1.010	1.020	1.030	1.040	1.050	1.060	1.070	1.080
2	1.020	1.040	1.061	1.082	1.103	1.123	1.145	1.166
3	1.030	1.061	1.093	1.125	1.158	1.191	1.225	1.260
4	1.041	1.082	1.126	1.170	1.216	1.262	1.311	1.360
5	1.051	1.104	1.159	1.217	1.276	1.338	1.402	1.469
6	1.061	1.126	1.194	1.265	1.340	1.419	1.500	1.587
7	1.072	1.148	1.230	1.316	1.407	1.504	1.605	1.714
8	1.083	1.171	1.267	1.369	1.477	1.594	1.718	1.851
9	1.094	1.195	1.305	1.423	1.551	1.689	1.838	1.999
10	1.105	1.218	1.344	1.480	1.629	1.791	1.967	2.159
11	1.116	1.243	1.384	1.540	1.710	1.898	2.104	2.332
12	1.127	1.268	1.426	1.601	1.790	2.012	2.252	2.518
13	1.138	1.294	1.469	1.666	1.886	2.133	2.410	2.720
14	1.149	1.319	1.513	1.731	1.980	2.261	2.579	2.937
15	1.161	1.345	1.558	1.801	2.079	2.397	2.759	3.172
16	1.172	1.372	1.605	1.873	2.183	2.540	2.952	3.426
17	1.184	1.400	1.653	1.948	2.292	2.693	3.159	3.700
18	1.196	1.428	1.702	2.026	2.407	2.854	3.380	3.996
19	1.208	1.457	1.754	2.107	2.527	3.026	3.617	4.316
20	1.220	1.486	1.806	2.191	2.653	3.207	3.870	4.661
21	1.232	1.516	1.860	2.279	2.786	3.400	4.140	5.034
22	1.245	1.546	1.916	2.370	2.925	3.604	4.430	5.437
23	1.257	1.577	1.974	2.465	3.072	3.813	4.741	5.871
24	1.270	1.608	2.033	2.563	3.225	4.049	5.072	6.341

Source: This table is abridged from *Handbook of Financial Mathematics* by Justin H. Moore. Copyright © 1929 by Prentice-Hall, Inc.

Table A.5 Single-Payment Compound Amount Factor (cont.)
$$(1+r)^n$$

Conversion Periods	Rate r							
n	0.01	0.02	0.03	0.04	0.05	0.06	0.07	0.08
25	1.282	1.641	2.094	2.666	3.386	4.291	5.427	6.848
26	1.295	1.673	2.157	2.773	3.556	4.549	5.807	7.396
27	1.308	1.707	2.221	2.883	3.733	4.822	6.214	7.988
28	1.321	1.741	2.288	2.999	3.920	5.111	6.649	8.627
29	1.335	1.776	2.357	3.119	4.116	5.418	7.114	9.317
30	1.347	1.811	2.427	3.243	4.322	5.743	7.612	10.063
31	1.361	1.848	2.500	3.373	4.538	6.088	8.145	10.868
32	1.375	1.885	2.575	3.508	4.765	6.453	8.715	11.737
33	1.389	1.922	2.652	3.648	5.003	6.840	9.325	12.676
34	1.402	1.961	2.732	3.794	5.253	7.251	9.979	13.690
35	1.417	2.000	2.814	3.946	5.516	7.686	10.677	14.785
36	1.430	2.040	2.898	4.104	5.792	8.147	11.424	15.968
37	1.445	2.080	2.985	4.268	6.081	8.636	12.224	17.246
38	1.460	2.122	3.075	4.439	6.385	9.154	13.079	18.625
39	1.474	2.165	3.167	4.616	6.705	9.703	13.995	20.115
40	1.489	2.208	3.262	4.801	7.040	10.286	14.974	21.724
41	1.503	2.252	3.360	4.993	7.391	10.902	16.023	23.462
42	1.519	2.297	3.461	5.193	7.761	11.557	17.144	25.339
43	1.534	2.343	3.565	5.400	8.150	12.250	18.344	27.367
44	1.549	2.390	3.671	5.617	8.557	12.985	19.628	29.556
45	1.565	2.438	3.781	5.841	8.985	13.765	21.002	31.920
46	1.580	2.487	3.895	6.075	9.434	14.590	22.473	34.474
47	1.596	2.536	4.011	6.318	9.905	15.466	24.046	37.232
48	1.612	2.587	4.132	6.571	10.401	16.394	25.729	40.210
49	1.628	2.639	4.256	6.833	10.921	17.378	27.530	43.427
50	1.645	2.691	4.384	7.107	11.467	18.420	29.457	46.902

Table A.6 Single-Payment Present Value Factor

$$\frac{1}{(1+r)^n} = (1+r)^{-n}$$

Conversion Periods	Rate r							
n	0.01	0.02	0.03	0.04	0.05	0.06	0.07	0.08
1	0.991	0.980	0.970	0.962	0.952	0.943	0.934	0.926
2	0.980	0.961	0.942	0.925	0.907	0.890	0.873	0.857
3	0.971	0.942	0.915	0.889	0.864	0.840	0.817	0.794
4	0.961	0.924	0.888	0.956	0.823	0.792	0.763	0.735
5	0.951	0.906	0.863	0.822	0.784	0.747	0.713	0.681
6	0.942	0.888	0.837	0.790	0.746	0.705	0.666	0.630
7	0.933	0.871	0.813	0.760	0.710	0.665	0.623	0.583
8	0.923	0.853	0.789	0.730	0.677	0.627	0.582	0.540
9	0.914	0.837	0.766	0.702	0.645	0.592	0.544	0.500
10	0.905	0.820	0.744	0.676	0.614	0.558	0.508	0.463
11	0.896	0.804	0.722	0.649	0.585	0.527	0.475	0.429
12	0.887	0.788	0.701	0.624	0.557	0.497	0.444	0.397
13	0.879	0.773	0.681	0.600	0.530	0.469	0.415	0.368
14	0.870	0.758	0.661	0.577	0.505	0.442	0.388	0.340
15	0.861	0.743	0.642	0.555	0.481	0.417	0.362	0.315
16	0.853	0.728	0.623	0.534	0.458	0.394	0.339	0.292
17	0.844	0.714	0.605	0.513	0.436	0.371	0.317	0.270
18	0.836	0.700	0.587	0.494	0.416	0.350	0.296	0.250
19	0.828	0.686	0.570	0.475	0.396	0.331	0.277	0.232
20	0.820	0.673	0.554	0.456	0.377	0.312	0.258	0.215
21	0.811	0.660	0.538	0.439	0.359	0.294	0.242	0.198
22	0.803	0.647	0.522	0.421	0.342	0.278	0.226	0.184
23	0.795	0.634	0.507	0.406	0.326	0.262	0.211	0.170
24	0.788	0.622	0.492	0.390	0.310	0.247	0.197	0.158

Source: This table is abridged from *Handbook of Financial Mathematics* by Justin H. Moore, Copyright © 1929 by Prentice-Hall, Inc.

Table A.6 Single-Payment Present Value Factor (cont.)

$$\frac{1}{(1+r)^n} = (1+r)^{-n}$$

Conversion Periods n	0.01	0.02	0.03	0.04	Rate r 0.05	0.06	0.07	0.08
25	0.780	0.610	0.478	0.375	0.295	0.233	0.184	0.146
26	0.772	0.598	0.464	0.361	0.281	0.220	0.172	0.135
27	0.764	0.586	0.450	0.347	0.268	0.207	0.161	0.125
28	0.757	0.574	0.437	0.333	0.255	0.196	0.150	0.116
29	0.749	0.563	0.424	0.320	0.243	0.185	0.141	0.107
30	0.742	0.552	0.412	0.308	0.231	0.174	0.131	0.099
31	0.734	0.541	0.400	0.296	0.220	0.164	0.123	0.092
32	0.727	0.531	0.388	0.285	0.210	0.154	0.115	0.085
33	0.720	0.520	0.377	0.274	0.200	0.146	0.107	0.079
34	0.712	0.510	0.366	0.264	0.190	0.138	0.100	0.072
35	0.706	0.500	0.355	0.253	0.181	0.130	0.094	0.068
36	0.699	0.490	0.345	0.244	0.172	0.123	0.088	0.063
37	0.692	0.481	0.335	0.234	0.164	0.116	0.082	0.058
38	0.685	0.471	0.325	0.225	0.157	0.109	0.076	0.054
39	0.678	0.462	0.316	0.217	0.149	0.103	0.071	0.050
40	0.671	0.453	0.307	0.208	0.142	0.097	0.067	0.046
41	0.665	0.444	0.298	0.200	0.135	0.092	0.062	0.043
42	0.658	0.435	0.289	0.192	0.129	0.087	0.058	0.039
43	0.651	0.427	0.281	0.185	0.123	0.082	0.055	0.037
44	0.645	0.418	0.272	0.178	0.117	0.077	0.051	0.034
45	0.639	0.410	0.264	0.171	0.111	0.073	0.048	0.031
46	0.633	0.402	0.257	0.164	0.106	0.069	0.044	0.029
47	0.626	0.394	0.249	0.158	0.101	0.065	0.041	0.027
48	0.620	0.387	0.242	0.152	0.096	0.061	0.039	0.025
49	0.614	0.378	0.235	0.146	0.091	0.058	0.036	0.023
50	0.608	0.372	0.228	0.141	0.087	0.054	0.034	0.021

Table A.7 Uniform-Payments Compound Amount Factor

$$[\frac{(1+r)^n - 1}{r}]$$

Conversion Periods	Rate r							
n	0.01	0.02	0.03	0.04	0.05	0.06	0.07	0.08
1	1.000	1.000	1.000	1.000	1.000	1.000	1.000	1.000
2	2.010	2.020	2.030	2.040	2.050	2.060	2.070	2.080
3	3.030	3.060	3.091	3.122	3.153	3.184	3.215	3.246
4	4.060	4.122	4.184	4.246	4.310	4.375	4.440	4.506
5	5.101	5.204	5.309	5.416	5.526	5.637	5.751	5.867
6	6.152	6.308	6.468	6.633	6.802	6.975	7.153	7.336
7	7.214	7.434	7.662	7.898	8.142	8.394	8.654	8.923
8	8.286	8.583	8.892	9.214	9.549	9.897	10.260	10.637
9	9.367	9.755	10.159	10.583	11.027	11.491	11.978	12.488
10	10.462	10.950	11.464	12.006	12.578	13.181	13.816	14.487
11	11.567	12.169	12.808	13.486	14.207	14.972	15.784	16.645
12	12.682	13.412	14.192	15.026	15.917	16.870	17.888	18.977
13	13.809	14.680	15.618	16.627	17.713	18.882	20.141	21.495
14	14.947	15.974	17.086	18.292	19.599	21.015	22.550	24.215
15	16.097	17.293	18.599	20.024	21.579	23.276	25.129	27.152
16	17.258	18.639	20.157	21.825	23.657	25.673	27.888	30.324
17	18.430	20.012	21.762	23.698	25.840	28.213	30.840	33.750
18	19.615	21.412	23.414	25.645	28.132	30.906	33.999	37.450
19	20.810	22.840	25.117	27.671	30.539	33.760	37.379	41.446
20	22.019	24.297	26.870	29.778	33.066	36.786	40.995	45.762
21	23.239	25.783	28.676	31.969	35.719	39.993	44.865	50.423
22	24.471	27.299	30.537	34.248	38.505	43.392	49.006	55.457
23	25.716	28.845	32.453	36.618	41.430	46.996	53.436	60.893
24	26.973	30.422	34.426	39.083	44.502	50.816	58.177	66.765

Source: This table is abridged from *Handbook of Financial Mathematics* by Justin H. Moore. Copyright © 1929 by Prentice-Hall, Inc.

Table A.7 Uniform-Payments Compound Amount Factor (cont.)

$$\left[\frac{(1+r)^n - 1}{r}\right]$$

Conversion Periods	Rate r							
n	0.01	0.02	0.03	0.04	0.05	0.06	0.07	0.08
25	28.243	32.030	36.459	41.646	47.727	54.865	63.249	73.106
26	29.526	33.671	38.553	44.312	51.113	59.156	68.676	79.954
27	30.821	35.344	40.710	47.084	54.669	63.706	74.484	87.351
28	32.129	37.051	42.931	49.968	58.403	68.528	80.698	95.339
29	33.450	38.792	45.219	52.966	62.323	73.640	87.347	103.966
30	34.785	40.568	47.575	56.085	66.439	79.058	94.461	113.283
31	36.133	42.379	50.003	59.328	70.761	84.802	102.073	123.346
32	37.494	44.227	52.503	62.701	75.299	90.890	110.218	134.213
33	38.869	46.111	55.073	66.210	80.064	97.343	118.933	145.951
34	40.258	48.034	57.730	69.858	85.067	104.184	128.259	158.627
35	41.660	49.994	60.462	73.652	90.320	111.435	138.237	172.317
36	43.077	51.994	63.276	77.598	95.836	119.121	148.913	187.102
37	44.508	54.034	66.174	81.702	101.628	127.268	160.337	203.070
38	45.953	56.115	69.159	85.970	107.710	135.904	172.561	220.316
39	47.412	58.237	72.234	90.409	114.095	145.058	185.640	238.941
40	48.886	60.402	75.401	95.026	120.800	154.762	199.635	259.057
41	50.375	62.610	78.663	99.827	127.840	165.048	214.610	280.781
42	51.879	64.862	82.023	104.820	135.232	175.951	230.632	304.244
43	53.398	67.159	85.484	110.012	142.993	187.508	247.776	329.583
44	54.932	69.503	89.048	115.412	151.143	199.758	266.121	356.950
45	56.481	71.893	92.720	121.029	159.700	212.743	285.749	386.506
46	58.046	74.330	96.501	126.871	168.685	226.508	306.752	418.426
47	59.626	76.817	100.397	132.945	178.119	241.099	329.224	452.900
48	61.223	79.354	104.408	139.263	188.025	256.564	353.270	490.132
49	62.835	81.940	108.541	145.834	198.427	272.958	378.999	530.343
50	64.463	84.580	112.797	152.667	209.348	290.336	406.529	573.770

Table A.8 Uniform-Payments Present Value Factor

$$\left[\frac{(1+r)^n - 1}{r(1+r)^n}\right] = \left[\frac{1-(1+r)^{-n}}{r}\right]$$

Conversion Periods	Rate r							
n	0.01	0.02	0.03	0.04	0.05	0.06	0.07	0.08
1	0.990	0.980	0.971	0.962	0.952	0.943	0.935	0.926
2	1.970	1.942	1.913	1.886	1.859	1.833	1.808	1.783
3	2.941	2.884	2.829	2.775	2.723	2.673	2.624	2.577
4	3.902	3.808	3.717	3.630	3.546	3.465	3.387	3.312
5	4.853	4.713	4.580	4.452	4.329	4.212	4.100	3.993
6	5.795	5.601	5.417	5.242	5.076	4.917	4.767	4.623
7	6.728	6.472	6.230	6.002	5.786	5.582	5.389	5.206
8	7.652	7.325	7.020	6.733	6.463	6.210	5.971	5.747
9	8.566	8.162	7.786	7.435	7.108	6.802	6.515	6.247
10	9.471	8.983	8.530	8.111	7.722	7.360	7.024	6.710
11	10.368	9.787	9.253	8.760	8.306	7.887	7.449	7.139
12	11.255	10.575	9.954	9.385	8.863	7.384	7.943	7.536
13	12.134	11.348	10.635	9.986	9.394	8.853	8.358	7.904
14	13.004	12.106	11.296	10.563	9.899	9.295	8.745	8.244
15	13.865	12.849	11.938	11.118	10.380	9.712	9.108	8.559
16	14.718	13.578	12.561	11.652	10.838	10.106	9.447	8.851
17	15.562	14.292	13.166	12.166	11.274	10.477	9.763	9.122
18	16.398	14.992	13.754	12.659	11.690	10.828	10.059	9.372
19	17.226	15.678	14.324	13.134	12.085	11.158	10.336	9.603
20	18.046	16.351	14.877	13.590	12.462	11.470	10.594	9.818
21	18.857	17.011	15.415	14.029	12.821	11.764	10.836	10.017
22	19.660	17.658	15.937	14.451	13.163	12.041	11.061	10.201
23	20.456	18.292	16.444	14.857	13.489	12.303	11.272	10.371
24	21.243	18.914	16.936	15.247	13.799	12.550	11.469	10.529

Source: This table is abridged from *Handbook of Financial Mathematics* by Justin H. Moore. Copyright © 1929 by Prentice-Hall, Inc.

Table A.8 Uniform-Payments Present Value Factor (cont.)

$$\left[\frac{(1+r)^n - 1}{r(1+r)^n}\right] = \left[\frac{1-(1+r)^{-n}}{r}\right]$$

Conversion Periods	Rate r							
n	0.01	0.02	0.03	0.04	0.05	0.06	0.07	0.08
25	22.023	19.523	17.413	15.622	14.094	12.783	11.654	10.675
26	22.795	20.121	17.877	15.983	14.375	13.003	11.826	10.810
27	23.560	20.707	18.327	16.330	14.643	13.210	11.987	10.935
28	24.316	21.281	18.764	16.663	14.898	13.406	12.137	11.051
29	25.066	21.844	19.188	16.984	15.141	13.591	12.278	11.158
30	25.808	22.396	19.600	17.292	15.372	13.765	12.410	11.258
31	26.542	22.938	20.000	17.588	15.593	13.929	12.532	11.350
32	27.270	23.468	20.389	17.874	15.803	14.084	12.647	11.435
33	27.990	23.989	20.766	18.148	16.003	14.230	12.754	11.514
34	28.703	24.499	21.132	18.411	16.193	14.368	12.854	11.587
35	29.409	24.999	21.487	18.665	16.374	14.498	12.948	11.655
36	30.108	25.489	21.832	18.908	16.547	14.621	13.035	11.717
37	30.800	25.969	22.167	19.143	16.711	14.737	13.117	11.775
38	31.485	26.441	22.492	19.368	16.868	14.846	13.193	11.829
39	32.163	26.903	22.808	19.584	17.017	14.949	13.265	11.879
40	32.835	27.355	23.115	19.793	17.159	15.046	13.332	11.925
41	33.500	27.799	23.412	19.993	17.294	15.138	13.394	11.967
42	34.158	28.235	23.701	20.186	17.423	15.225	13.452	12.007
43	34.810	28.662	23.982	20.371	17.546	15.306	13.507	12.043
44	35.455	29.080	24.254	20.549	17.663	15.383	13.558	12.077
45	36.095	29.490	24.519	20.720	17.774	15.456	13.606	12.109
46	36.727	29.892	24.775	20.885	17.880	15.524	13.650	12.137
47	37.354	30.287	25.025	21.043	17.981	15.589	13.692	12.164
48	37.974	30.673	25.267	21.195	18.077	15.650	13.730	12.189
49	38.588	31.052	25.502	21.341	18.169	15.708	13.767	12.212
50	39.196	31.424	25.730	21.482	18.256	15.762	13.801	12.233

Table A.9 Natural, or Naperian, Logarithms

To find the natural logarithm of a number which is 1/10, 1/100, 1/1000, etc. of a number whose logarithm is given, subtract from the given logarithm $\log_e 10$, $2 \log_e 10$, $3 \log_e 10$, etc.

To find the natural logarithm of a number which is 10, 100, 1000, etc. times a number whose logarithm is given, add to the given logarithm $\log_e 10$, $2 \log_e 10$, $3 \log_e 10$, etc.

$$\log_e 10 = 2.30259 \qquad 6 \log_e 10 = 13.81551$$
$$2 \log_e 10 = 4.60517 \qquad 7 \log_e 10 = 16.11810$$
$$3 \log_e 10 = 6.90776 \qquad 8 \log_e 10 = 18.42068$$
$$4 \log_e 10 = 9.21034 \qquad 9 \log_e 10 = 20.72327$$
$$5 \log_e 10 = 11.51293 \qquad 10 \log_e 10 = 23.02585$$

N	0	1	2	3	4	5	6	7	8	9
1.0	0.00000	.00995	.01980	.02956	.03922	.04879	.05827	.06766	.07696	.08618
.1	.09531	.10436	.11333	.12222	.13103	.13976	.14842	.15700	.16551	.17395
.2	.18232	.19062	.19885	.20701	.21511	.22314	.23111	.23902	.24686	.25464
.3	.26236	.27003	.27763	.28518	.29267	.30010	.30748	.31481	.32208	.32930
.4	.33647	.34359	.35066	.35767	.36464	.37156	.37844	.38526	.39204	.39878
.5	.40547	.41211	.41871	.42527	.43178	.43825	.44469	.45108	.45742	.46373
.6	.47000	.47623	.48243	.48858	.49470	.50078	.50682	.51282	.51879	.52473
.7	.53063	.53649	.54232	.54812	.55389	.55962	.56531	.57098	.57661	.58222
.8	.58779	.59333	.59884	.60432	.60977	.61519	.62058	.62594	.63127	.63658
.9	.64185	.64710	.65233	.65752	.66269	.66783	.67294	.67803	.68310	.68813
2.0	0.69315	.69813	.70310	.70804	.71295	.71784	.72271	.72755	.73237	.73716
.1	.74194	.74669	.75142	.75612	.76081	.76547	.77011	.77473	.77932	.78390
.2	.78846	.79299	.79751	.80200	.80648	.81093	.81536	.81978	.82418	.82855
.3	.83291	.83725	.84157	.84587	.85015	.85442	.85866	.86289	.86710	.87129
.4	.87547	.87963	.88377	.88789	.89200	.89609	.90016	.90422	.90826	.91228
.5	.91629	.92028	.92426	.92822	.93216	.93609	.94001	.94391	.94779	.95166
.6	.95551	.95935	.96317	.96698	.97078	.97456	.97833	.98208	.98582	.98954
.7	.99325	.99695	*.00063	*.00430	*.00796	*.01160	*.01523	*.01885	*.02245	*.02604
.8	1.02962	.03318	.03674	.04028	.04380	.04732	.05082	.05431	.05779	.06126
.9	.06471	.06815	.07158	.07500	.07841	.08181	.08519	.08856	.09192	.09527
3.0	1.09861	.10194	.10526	.10856	.11186	.11514	.11841	.12168	.12493	.12817
.1	.13140	.13462	.13783	.14103	.14422	.14740	.15057	.15373	.15688	.16002
.2	.16315	.16627	.16938	.17248	.17557	.17865	.18173	.18479	.18784	.19089
.3	.19392	.19695	.19996	.20297	.20597	.20896	.21194	.21491	.21788	.22083
.4	.22378	.22671	.22964	.23256	.23547	.23837	.24127	.24415	.24703	.24990
.5	.25276	.25562	.25846	.26130	.26413	.26695	.26976	.27257	.27536	.27815
.6	.28093	.28371	.28647	.28923	.29198	.29473	.29746	.30019	.30291	.30563

Source: This table is reproduced from S. Selby, *Standard Mathematical Tables*, 20th edition (Cleveland, Ohio: The Chemical Rubber Company, 1972). Used by permission.

Table A.9 Natural, or Naperian, Logarithms (cont.)

N	0	1	2	3	4	5	6	7	8	9
.7	.30833	.31103	.31372	.31641	.31909	.32176	.32442	.32708	.32972	.33237
.8	.33500	.33763	.34025	.34286	.34547	.34807	.35067	.35325	.35584	.35841
.9	.36098	.36354	.36609	.36864	.37118	.37372	.37624	.37877	.38128	.38379
4.0	1.38629	.38879	.39128	.39377	.39624	.39872	.40118	.40364	.40610	.40854
.1	.41099	.41342	.41585	.41828	.42070	.42311	.42552	.42792	.43031	.43270
.2	.43508	.43746	.43984	.44220	.44456	.44692	.44927	.45161	.45395	.45629
.3	.45862	.46094	.46326	.46557	.46787	.47018	.47247	.47476	.47705	.47933
.4	.48160	.48387	.48614	.48840	.49065	.49290	.49515	.49739	.49962	.50185
.5	.50408	.50630	.50851	.51072	.51293	.51513	.51732	.51951	.52170	.52388
.6	.52606	.52823	.53039	.53256	.53471	.53687	.53902	.54116	.54330	.54543
.7	.54756	.54969	.55181	.55393	.55604	.55814	.56025	.56235	.56444	.56653
.8	.56862	.57070	.57277	.57485	.57691	.57898	.58104	.58309	.58515	.58719
.9	.58924	.59127	.59331	.59534	.59737	.59939	.60141	.60342	.60543	.60744
5.0	1.60944	.61144	.61343	.61542	.61741	.61939	.62137	.62334	.62531	.62728
.1	.62924	.63120	.63315	.63511	.63705	.63900	.64094	.64287	.64481	.64673
.2	.64866	.65058	.65250	.65441	.65632	.65823	.66013	.66203	.66393	.66582
.3	.66771	.66959	.67147	.67335	.67523	.67710	.67896	.68083	.68269	.68455
.4	.68640	.68825	.69010	.69194	.69378	.69562	.69745	.69928	.70111	.70293
.5	.70475	.70656	.70838	.71019	.71199	.71380	.71560	.71740	.71919	.72098
.6	.72277	.72455	.72633	.72811	.72988	.73166	.73342	.73519	.73695	.73871
.7	.74047	.74222	.74397	.74572	.74746	.74920	.75094	.75267	.75440	.75613
.8	.75786	.75958	.76130	.76302	.76473	.76644	.76815	.76985	.77156	.77326
.9	.77495	.77665	.77834	.78002	.78171	.78339	.78507	.78675	.78842	.79009
6.0	1.79176	.79342	.79509	.79675	.79840	.80006	.80171	.80336	.80500	.80665
.1	.80829	.80993	.81156	.81319	.81482	.81645	.81808	.81970	.82132	.82294
.2	.82455	.82616	.82777	.82938	.83098	.83258	.83418	.83578	.83737	.83896
.3	.84055	.84214	.84372	.84530	.84688	.84845	.85003	.85160	.85317	.85473
.4	.85630	.85786	.85942	.86097	.86253	.86408	.86563	.86718	.86872	.87026
.5	.87180	.87334	.87487	.87641	.87794	.87947	.88099	.88251	.88403	.88555
.6	.88707	.88858	.89010	.89160	.89311	.89462	.89612	.89762	.89912	.90061
.7	.90211	.90360	.90509	.90658	.90806	.90954	.91102	.91250	.91398	.91545
.8	.91692	.91839	.91986	.92132	.92279	.92425	.92571	.92716	.92862	.93007
.9	.93152	.93297	.93442	.93586	.93730	.93874	.94018	.94162	.94305	.94448
7.0	1.94591	.94734	.94876	.95019	.95161	.95303	.95445	.95586	.95727	.95869
.1	.96009	.96150	.96291	.96431	.96571	.96711	.96851	.96991	.97130	.97269
.2	.97408	.97547	.97685	.97824	.97962	.98100	.98238	.98376	.98513	.98650
.3	.98787	.98924	.99061	.99198	.99334	.99470	.99606	.99742	.99877	*.00013
.4	2.00148	.00283	.00418	.00553	.00687	.00821	.00956	.01089	.01223	.01357
.5	.01490	.01624	.01757	.01890	.02022	.02155	.02287	.02419	.02551	.02683
.6	.02815	.02946	.03078	.03209	.03340	.03471	.03601	.03732	.03862	.03992
.7	.04122	.04252	.04381	.04511	.04640	.04769	.04898	.05027	.05156	.05284
.8	.05412	.05540	.05668	.05796	.05924	.06051	.06179	.06306	.06433	.06560

Table A.9 Natural, or Naperian, Logarithms (cont.)

N	0	1	2	3	4	5	6	7	8	9
.9	.06686	.06813	.06939	.07065	.07191	.07317	.07443	.07568	.07694	.07819
8.0	2.07944	.08069	.08194	.08318	.08443	.08567	.08691	.08815	.08939	.09063
.1	.09186	.09310	.09433	.09556	.09679	.09802	.09924	.10047	.10169	.10291
.2	.10413	.10535	.10657	.10779	.10900	.11021	.11142	.11263	.11384	.11505
.3	.11626	.11746	.11866	.11986	.12106	.12226	.12346	.12465	.12585	.12704
.4	.12823	.12942	.13061	.13180	.13298	.13417	.13535	.13653	.13771	.13889
.5	.14007	.14124	.14242	.14359	.14476	.14593	.14710	.14827	.14943	.15060
.6	.15176	.15292	.15409	.15524	.15640	.15756	.15871	.15987	.16102	.16217
.7	.16332	.16447	.16562	.16677	.16791	.16905	.17020	.17134	.17248	.17361
.8	.17475	.17589	.17702	.17816	.17929	.18042	.18155	.18267	.18380	.18493
.9	.18605	.18717	.18830	.18942	.19054	.19165	.19277	.19389	.19500	.19611
9.0	2.19722	.19834	.19944	.20055	.20166	.20276	.20387	.20497	.20607	.20717
.1	.20827	.20937	.21047	.21157	.21266	.21375	.21485	.21594	.21703	.21812
.2	.21920	.22029	.22138	.22246	.22354	.22462	.22570	.22678	.22786	.22894
.3	.23001	.23109	.23216	.23324	.23431	.23538	.23645	.23751	.23858	.23965
.4	.24071	.24177	.24284	.24390	.24496	.24601	.24707	.24813	.24918	.25024
.5	.25129	.25234	.25339	.25444	.25549	.25654	.25759	.25863	.25968	.26072
.6	.26176	.26280	.26384	.26488	.26592	.26696	.26799	.26903	.27006	.27109
.7	.27213	.27316	.27419	.27521	.27624	.27727	.27829	.27932	.28034	.28136
.8	.28238	.28340	.28442	.28544	.28646	.28747	.28849	.28950	.29051	.29152
.9	.29253	.29354	.29455	.29556	.29657	.29757	.29858	.29958	.30058	.30158

Table A.10 Exponential Functions

x	e^x	e^{-x}	x	e^x	e^{-x}	x	e^x	e^{-x}
0.00	1.0000	1.000000	0.40	1.4918	0.670320	0.80	2.2255	0.449329
0.01	1.0101	0.990050	0.41	1.5068	.663650	0.81	2.2479	.444858
0.02	1.0202	.980199	0.42	1.5220	.657047	0.82	2.2705	.440432
0.03	1.0305	.970446	0.43	1.5373	.650509	0.83	2.2933	.436049
0.04	1.0408	.960789	0.44	1.5527	.644036	0.84	2.3164	.431711
0.05	1.0513	0.951229	0.45	1.5683	0.637628	0.85	2.3396	0.427415
0.06	1.0618	.941765	0.46	1.5841	.631284	0.86	2.3632	.423162
0.07	1.0725	.932394	0.47	1.6000	.625002	0.87	2.3869	.418952
0.08	1.0833	.923116	0.48	1.6161	.618783	0.88	2.4109	.414783
0.09	1.0942	.913931	0.49	1.6323	.612626	0.89	2.4351	.410656
0.10	1.1052	0.904837	0.50	1.6487	0.606531	0.90	2.4596	0.406570
0.11	1.1163	.895834	0.51	1.6653	.600496	0.91	2.4843	.402524
0.12	1.1275	.886920	0.52	1.6820	.594521	0.92	2.5093	.398519
0.13	1.1388	.878095	0.53	1.6989	.588605	0.93	2.5345	.394554
0.14	1.1503	.869358	0.54	1.7160	.582748	0.94	2.5600	.390628
0.15	1.1618	0.860708	0.55	1.7333	0.576950	0.95	2.5857	0.386741
0.16	1.1735	.852144	0.56	1.7507	.571209	0.96	2.6117	.382893
0.17	1.1853	.843665	0.57	1.7683	.565525	0.97	2.6379	.379083
0.18	1.1972	.835270	0.58	1.7860	.559898	0.98	2.6645	.375311
0.19	1.2092	.826959	0.59	1.8040	.554327	0.99	2.6912	.371577
0.20	1.2214	0.818731	0.60	1.8221	0.548812	1.00	2.7183	0.367879
0.21	1.2337	.810584	0.61	1.8404	.543351	1.01	2.7456	.364219
0.22	1.2461	.802519	0.62	1.8589	.537944	1.02	2.7732	.360595
0.23	1.2586	.794534	0.63	1.8776	.532592	1.03	2.8011	.357007
0.24	1.2712	.786628	0.64	1.8965	.527292	1.04	2.8292	.353455
0.25	1.2840	0.778801	0.65	1.9155	0.522046	1.05	2.8577	0.349938
0.26	1.2969	.771052	0.66	1.9348	.516851	1.06	2.8864	.346456
0.27	1.3100	.763379	0.67	1.9542	.511709	1.07	2.9154	.343009
0.28	1.3231	.755784	0.68	1.9739	.506617	1.08	2.9447	.339596
0.29	1.3364	.748264	0.69	1.9937	.501576	1.09	2.9743	.336216
0.30	1.3499	0.740818	0.70	2.0138	0.496585	1.10	3.0042	0.332871
0.31	1.3634	.733447	0.71	2.0340	.491644	1.11	3.0344	.329559
0.32	1.3771	.726149	0.72	2.0544	.486752	1.12	3.0649	.326280
0.33	1.3910	.718924	0.73	2.0751	.481909	1.13	3.0957	.323033
0.34	1.4049	.711770	0.74	2.0959	.477114	1.14	3.1268	.319819
0.35	1.4191	0.704688	0.75	2.1170	0.472367	1.15	3.1582	0.316637
0.36	1.4333	.697676	0.76	2.1383	.467666	1.16	3.1899	.313486
0.37	1.4477	.690734	0.77	2.1598	.463013	1.17	3.2220	.310367
0.38	1.4623	.683861	0.78	2.1815	.458406	1.18	3.2544	.307279
0.39	1.4770	.677057	0.79	2.2034	.453845	1.19	3.2871	.304221

Table A.10 Exponential Functions (cont.)

x	e^x	e^{-x}	x	e^x	e^{-x}	x	e^x	e^{-x}
1.20	3.3201	0.301194	1.60	4.9530	0.201897	2.00	7.3891	0.135335
1.21	3.3535	.298197	1.61	5.0028	.199888	2.01	7.4633	.133989
1.22	3.3872	.295230	1.62	5.0531	.197899	2.02	7.5383	.132655
1.23	3.4212	.292293	1.63	5.1039	.195930	2.03	7.6141	.131336
1.24	3.4556	.289384	1.64	5.1552	.193980	2.04	7.6906	.130029
1.25	3.4903	0.286505	1.65	5.2070	0.192050	2.05	7.7679	0.128735
1.26	3.5254	.283654	1.66	5.2593	.190139	2.06	7.8460	.127454
1.27	3.5609	.280832	1.67	5.3122	.188247	2.07	7.9248	.126186
1.28	3.5966	.278037	1.68	5.3656	.186374	2.08	8.0045	.124930
1.29	3.6328	.275271	1.69	5.4195	.184520	2.09	8.0849	.123687
1.30	3.6693	0.272532	1.70	5.4739	0.182684	2.10	8.1662	0.122456
1.31	3.7062	.269820	1.71	5.5290	.180866	2.11	8.2482	.121238
1.32	3.7434	.267135	1.72	5.5845	.179066	2.12	8.3311	.120032
1.33	3.7810	.264477	1.73	5.6407	.177284	2.13	8.4149	.118837
1.34	3.8190	.261846	1.74	5.6973	.175520	2.14	8.4994	.117655
1.35	3.8574	0.259240	1.75	5.7546	0.173774	2.15	8.5849	0.116484
1.36	3.8962	.256661	1.76	5.8124	.172045	2.16	8.6711	.115325
1.37	3.9354	.254107	1.77	5.8709	.170333	2.17	8.7583	.114178
1.38	3.9749	.251579	1.78	5.9299	.168638	2.18	8.8463	.113042
1.39	4.0149	.249075	1.79	5.9895	.166960	2.19	8.9352	.111917
1.40	4.0552	0.246597	1.80	6.0496	0.165299	2.20	9.0250	0.110803
1.41	4.0960	.244143	1.81	6.1104	.163654	2.21	9.1157	.109701
1.42	4.1371	.241714	1.82	6.1719	.162026	2.22	9.2073	.108609
1.43	4.1787	.239309	1.83	6.2339	.160414	2.23	9.2999	.107528
1.44	4.2207	.236928	1.84	6.2965	.158817	2.24	9.3933	.106459
1.45	4.2631	0.234570	1.85	6.3598	0.157237	2.25	9.4877	0.105399
1.46	4.3060	.232236	1.86	6.4237	.155673	2.26	9.5831	.104350
1.47	4.3492	.229925	1.87	6.4883	.154124	2.27	9.6794	.103312
1.48	4.3929	.227638	1.88	6.5535	.152590	2.28	9.7767	.102284
1.49	4.4371	.225373	1.89	6.6194	.151072	2.29	9.8749	.101266
1.50	4.4817	0.223130	1.90	6.6859	0.149569	2.30	9.9742	0.100259
1.51	4.5267	.220910	1.91	6.7531	.148080	2.31	10.074	.099261
1.52	4.5722	.218712	1.92	6.8210	.146607	2.32	10.176	.098274
1.53	4.6182	.216536	1.93	6.8895	.145148	2.33	10.278	.097296
1.54	4.6646	.214381	1.94	6.9588	.143704	2.34	10.381	.096328
1.55	4.7115	0.212248	1.95	7.0287	0.142274	2.35	10.486	0.095369
1.56	4.7588	.210136	1.96	7.0993	.140858	2.36	10.591	.094420
1.57	4.8066	.208045	1.97	7.1707	.139457	2.37	10.697	.093481
1.58	4.8550	.205975	1.98	7.2427	.138069	2.38	10.805	.092551
1.59	4.9037	.203926	1.99	7.3155	.136695	2.39	10.913	.091630

Table A.10 Exponential Functions (cont.)

x	e^x	e^{-x}	x	e^x	e^{-x}	x	e^x	e^{-x}
2.40	11.023	0.090718	2.80	16.445	0.060810	3.20	24.533	0.040762
2.41	11.134	.089815	2.81	16.610	.060205	3.21	24.779	.040357
2.42	11.246	.088922	2.82	16.777	.059606	3.22	25.028	.039955
2.43	11.359	.088037	2.83	16.945	.059013	3.23	25.280	.039557
2.44	11.473	.087161	2.84	17.116	.058426	3.24	25.534	.039164
2.45	11.588	0.086294	2.85	17.288	0.057844	3.25	25.790	0.038774
2.46	11.705	.085435	2.86	17.462	.057269	3.26	26.050	.038388
2.47	11.822	.084585	2.87	17.637	.056699	3.27	26.311	.038006
2.48	11.941	.083743	2.88	17.814	.056135	3.28	26.576	.037628
2.49	12.061	.082910	2.89	17.993	.055576	3.29	26.843	.037254
2.50	12.182	0.082085	2.90	18.174	0.055023	3.30	27.113	0.036883
2.51	12.305	.081268	2.91	18.357	.054476	3.31	27.385	.036516
2.52	12.429	.080460	2.92	18.541	.053934	3.32	27.660	.036153
2.53	12.554	.079659	2.93	18.728	.053397	3.33	27.938	.035793
2.54	12.680	.078866	2.94	18.916	.052866	3.34	28.219	.035437
2.55	12.807	0.078082	2.95	19.106	0.052340	3.35	28.503	0.035084
2.56	12.936	.077305	2.96	19.298	.051819	3.36	28.789	.034735
2.57	13.066	.076536	2.97	19.492	.051303	3.37	29.079	.034390
2.58	13.197	.075774	2.98	19.688	.050793	3.38	29.371	.034047
2.59	13.330	.075020	2.99	19.886	.050287	3.39	29.666	.033709
2.60	13.464	0.074274	3.00	20.086	0.049787	3.40	29.964	0.033373
2.61	13.599	.073535	3.01	20.287	.049292	3.41	30.265	.033041
2.62	13.736	.072803	3.02	20.491	.048801	3.42	30.569	.032712
2.63	13.874	.072078	3.03	20.697	.048316	3.43	30.877	.032387
2.64	14.013	.071361	3.04	20.905	.047835	3.44	31.187	.032065
2.65	14.154	0.070651	3.05	21.115	0.047359	3.45	31.500	0.031746
2.66	14.296	.069948	3.06	21.328	.046888	3.46	31.817	.031430
2.67	14.440	.069252	3.07	21.542	.046421	3.47	32.137	.031117
2.68	14.585	.068563	3.08	21.758	.045959	3.48	32.460	.030807
2.69	14.732	.067881	3.09	21.977	.045502	3.49	32.786	.030501
2.70	14.880	0.067206	3.10	22.198	0.045049	3.50	33.115	0.030197
2.71	15.029	.066537	3.11	22.421	.044601	3.51	33.448	.029897
2.72	15.180	.065875	3.12	22.646	.044157	3.52	33.784	.029599
2.73	15.333	.065219	3.13	22.874	.043718	3.53	34.124	.029305
2.74	15.487	.064570	3.14	23.104	.043283	3.54	34.467	.029013
2.75	15.643	0.063928	3.15	23.336	0.042852	3.55	34.813	0.028725
2.76	15.800	.063292	3.16	23.571	.042426	3.56	35.163	.028439
2.77	15.959	.062662	3.17	23.807	.042004	3.57	35.517	.028156
2.78	16.119	.062039	3.18	24.047	.041586	3.58	35.874	.027876
2.79	16.281	.061421	3.19	24.288	.041172	3.59	36.234	.027598

Table A.10 Exponential Functions (cont.)

x	e^x	e^{-x}	x	e^x	e^{-x}	x	e^x	e^{-x}
3.60	36.598	0.027324	4.00	54.598	0.018316	4.40	81.451	0.012277
3.61	36.966	.027052	4.01	55.147	.018133	4.41	82.269	.012155
3.62	37.338	.026783	4.02	55.701	.017953	4.42	83.096	.012034
3.63	37.713	.026516	4.03	56.261	.017774	4.43	83.931	.011914
3.64	38.092	.026252	4.04	56.826	.017597	4.44	84.775	.011796
3.65	38.475	0.025991	4.05	57.397	0.017422	4.45	85.627	0.011679
3.66	38.861	.025733	4.06	57.974	.017249	4.46	86.488	.011562
3.67	39.252	.025476	4.07	58.557	.017077	4.47	87.357	.011447
3.68	39.646	.025223	4.08	59.145	.016907	4.48	88.235	.011333
3.69	40.045	.024972	4.09	59.740	.016739	4.49	89.121	.011221
3.70	40.447	0.024724	4.10	60.340	0.016573	4.50	90.017	0.011109
3.71	40.854	.024478	4.11	60.947	.016408	4.51	90.922	.010998
3.72	41.264	.024234	4.12	61.559	.016245	4.52	91.836	.010889
3.73	41.679	.023993	4.13	62.178	.016083	4.53	92.759	.010781
3.74	42.098	.023754	4.14	62.803	.015923	4.54	93.691	.010673
3.75	42.521	0.023518	4.15	63.434	0.015764	4.55	94.632	0.010567
3.76	42.948	.023284	4.16	64.072	.015608	4.56	95.583	.010462
3.77	43.380	.023052	4.17	64.715	.015452	4.57	96.544	.010358
3.78	43.816	.022823	4.18	65.366	.015299	4.58	97.514	.010255
3.79	44.256	.022596	4.19	66.023	.015146	4.59	98.494	.010153
3.80	44.701	0.022371	4.20	66.686	0.014996	4.60	99.484	0.010052
3.81	45.150	.022148	4.21	67.357	.014846	4.61	100.48	.009952
3.82	45.604	.021928	4.22	68.033	.014699	4.62	101.49	.009853
3.83	46.063	.021710	4.23	68.717	.014552	4.63	102.51	.009755
3.84	46.525	.021494	4.24	69.408	.014408	4.64	103.54	.009658
3.85	46.993	0.021280	4.25	70.105	0.014264	4.65	104.58	0.009562
3.86	47.465	.021068	4.26	70.810	.014122	4.66	105.64	.009466
3.87	47.942	.020858	4.27	71.522	.013982	4.67	106.70	.009372
3.88	48.424	.020651	4.28	72.240	.013843	4.68	107.77	.009279
3.89	48.911	.020445	4.29	72.966	.013705	4.69	108.85	.009187
3.90	49.402	0.020242	4.30	73.700	0.013569	4.70	109.95	0.009095
3.91	49.899	.020041	4.31	74.440	.013434	4.71	111.05	.009005
3.92	50.400	.019841	4.32	75.189	.013300	4.72	112.17	.008915
3.93	50.907	.019644	4.33	75.944	.013168	4.73	113.30	.008826
3.94	51.419	.019448	4.34	76.708	.013037	4.74	114.43	.008739
3.95	51.935	0.019255	4.35	77.478	0.012907	4.75	115.58	0.008652
3.96	52.457	.019063	4.36	78.257	.012778	4.76	116.75	.008566
3.97	52.985	.018873	4.37	79.044	.012651	4.77	117.92	.008480
3.98	53.517	.018686	4.38	79.838	.012525	4.78	119.10	.008396
3.99	54.055	.018500	4.39	80.640	.012401	4.79	120.30	.008312

Table A.10 Exponential Functions (cont.)

x	e^x	e^{-x}
4.80	121.51	0.008230
4.81	122.73	.008148
4.82	123.97	.008067
4.83	125.21	.007987
4.84	126.47	.007907
4.85	127.74	0.007828
4.86	129.02	.007750
4.87	130.32	.007673
4.88	131.63	.007597
4.89	132.95	.007521
4.90	134.29	0.007447
4.91	135.64	.007372
4.92	137.00	.007299
4.93	138.38	.007227
4.94	139.77	.007155
4.95	141.17	0.007083
4.96	142.59	.007013
4.97	144.03	.006943
4.98	145.47	.006874
4.99	146.94	.006806
5.00	148.41	0.006738

Values beyond the range of this table may be obtained by

$$e^{x+y} = e^x \cdot e^y$$

GLOSSARY OF TERMS AND FORMULAS

Abscissa The value of the x coordinate in the point (x,y).

Addition rule for probabilities For any two events A and B, $P(A \cup B) = P(A) + P(B) - P(A \cap B)$.

Algebraic function A function that can be expressed as a result of sums, differences, products, quotients, and roots of polynomial functions.

Amortization The retirement of a debt by equal periodic payments at a compound interest rate over a specified length of time.

Annuity An amount payable at regular intervals.

Antiderivative An antiderivative of $f(x)$ is any function having $f(x)$ as its derivative.

Antilog The antilog of x is y if $\log y = x$.

Arithmetic progression A sequence obtained by adding a value d, called the *common difference*, to each succeeding term.

Associative law for addition $A + (B + C) = (A + B) + C$

Associative law for multiplication $A(BC) = (AB) \cdot C$.

Asymptote A line which is the limiting position that a curve approaches but never touches.

Axiom A statement that is a self-evident truth.

Base b is called the *base of the expression* b^n.

Basic solution A solution for a set of m simultaneous linear equations in n variables ($n > m$) for which no more than m variables have values different from zero.

Basic variable One of the m variables in a basic solution that may have a value different from zero.

Basis The set of m basic variables in a basic solution.

Bernoulli probability function $f(x) = p^x(1-p)^{1-x}$ $(x = 0, 1)$.

Bernoulli process A random experiment that generates two mutually exclusive outcomes.

Bernoulli trial A trial generated by a Bernoulli process.

Beta probability density function $f(x) = (\alpha + \beta + 1)!/(\alpha!\beta!)x^\alpha(1-x)^\beta$ for $0 \leq x \leq 1$.

Binomial probability function $f(x) = {}_nC_x p^x (1-p)^{n-x}$ $(x = 0, 1, \ldots, n)$.

Calculus Any method of calculating or investigating by algebraic symbols.

Cartesian coordinates The coordinates of points obtained by taking the cartesian product of the set of real numbers with itself.

GLOSSARY OF TERMS AND FORMULAS

Cartesian product The cartesian product of two sets X and Y, denoted by $X \times Y$, is the set of all ordered pairs (x,y) such that the first element is from the set X and the second element is from the set Y.

Characteristic The exponent of 10 when a number is expressed in scientific form.

Coincident lines Two or more lines which coincide.

Commutative law for addition $A + B = B + A$.

Commutative law for multiplication $AB = BA$.

Complement The complement of a set A relative to a universe I is the set of all elements of I that are not elements of A. The complement of A is denoted by A'.

Composite function A function of a function.

Compound amount The total amount payable at the end of the period for which the interest has been left on deposit to earn further interest.

Compound event An event that can be represented by a collection of two or more simple events.

Compound interest The total interest payable if the interest is left on deposit to earn further interest.

Compound maturity value Same as compound amount.

Conditional probability The probability of the occurrence of event A when sampling from a population for which each event is an occurrence of the event B. [Denoted $P(A|B)$.]

Continuous function A function $f(x)$ is continuous at $x = a$ if $f(a)$ exists, $\lim_{x \to a} f(x)$ exists and $\lim_{x \to a} f(x) = f(a)$.

Coordinate axes Two perpendicular lines along which the values of the respective variables are measured.

Critical values Values of x satisfying $f'(x) = 0$.

Cubic function Any function for which the rule is given by $ax^3 + bx^2 + cx + d$ and $a \neq 0$.

Definite integral The definite integral of $f(x)$ from a to b, denoted $\int_a^b f(x)dx$, is $\int_a^b f(x)dx = \lim_{n \to \infty} \sum_{i=1}^{n} f(x_i)\Delta x_n$ provided the limit exists.

Dependent variable A variable, the value of which depends on the value of another variable.

Derivative The derivative with respect to x of a function $f(x)$ is the limit of the ratio $\Delta f(x)/\Delta x$ as Δx approaches zero and is given by $D_x f(x) = \lim_{\Delta x \to 0} [f(x + \Delta x) - f(x)]/\Delta x$.

Discontinuous function A function that is not continuous throughout the domain.

Distributive law of multiplication over addition $A(B + C) = AB + AC$.

Domain The set of elements that represent the values of the variable used to define a function.

Empty set A set that contains no elements.

Equal sets Two sets A and B are equal if and only if every element of A is an element of B and every element of B is an element of A.

Exhaustive events A set of events is exhaustive for a population if, when the events are viewed collectively, no simple event of the population is omitted from the set.

Expected value of X $E(X) = \sum_{i=1}^{n} x_i P(X = x_i)$ if X is a discrete variable. $E(X) = \int_{-\infty}^{\infty} x f(x) dx$ if X is a continuous variable.

Explicit function A function that explicitly states one variable as a function of another variable.

Exponential function Any function of the form $f(x) = ab^x$, where a and b are nonzero real numbers and b is positive and not equal to 1.

Feasible solution A solution that satisfies all constraints.

Function A function is a rule that, for each element of a set D, called the *domain*, assigns one and only one element to a set R, called the *range*.

Functional form of a linear equation $y = mx + b$, where $m \neq 0$.

Functional form of a quadratic equation $y = ax^2 + bx + c$, where $a \neq 0$.

Fundamental theorem of integral calculus If $f(x)$ is a continuous function in the interval $a \leqslant x \leqslant b$ and $F(x) = \int f(x)dx$ is any indefinite integral of $f(x)$, the value of the definite integral $\int_a^b f(x)dx$ is given by $\int_a^b f(x)dx = F(b) - F(a)$.

Geometric progression A sequence obtained by multiplying each succeeding term by a value r, called the *common ratio*.

Hypergeometric probability function $h(x;n,a,b) = (_aC_x)(_bC_{n-x})/_{a+b}C_n$ $(x = 0, 1, \ldots, a)$.

Identity matrix A square matrix having ones on the diagonal and zeros elsewhere such that $AI = A$ and $IA = A$.

Implicit function A function in which it is not apparent which variable is the dependent variable.

Inconsistent equations A set of equations for which there is no simultaneous solution.

Independent events Events A and B are independent if $P(A \cap B) = P(A)P(B)$.

Independent variable A variable, the value of which does not depend on the value of any other variable.

Inflection point A point at which the slope of the function becomes zero and then continues its preceding trend.

Intercept The constant b in the equation $y = mx + b$.

Interest The amount paid for the use of a quantity of money over a period of time.

Intersection The intersection of sets A and B, denoted $A \cap B$, is the set of elements that belong to both A and B.

Inverse function For a function $f(x)$, the rule $g(y)$, which is obtained by solving the equation $y = f(x)$ for y in terms of x to obtain $x = g(y)$, is the inverse function of $f(x)$ provided $g(y)$ is indeed a function.

Inverse of a matrix If A and B are square matrices such that $AB = I$, where I is the identity matrix of the same order as A, then B is called the *inverse of A* and is denoted by A^{-1}.

Joint events Events A and B are joint events if they both occur on a given trial.

Joint probability The probability that two events both occur on a given trial.

Laws of exponents $b^m b^n = b^{m+n}$ and $(b^m)^n = b^{mn}$.

Laws of logarithms $\log_b MN = \log_b M + \log_b N$; $\log_b M^N = N \log_b M$.

Linear equation An equation of the form $y = mx + b$.

Linear function A function for which the rule is $mx + b$.

Logarithm If b is a positive number such that $y = b^x$, the logarithm of y to the base b is x and is written $\log_b y = x$.

Logarithmic function A function of the form $f(x) = \log_b x$.

Mantissa The decimal part of common logarithms.

Marginal probability If the event A can occur jointly with all or any subset of k mutually exclusive and exhaustive events $B_1, B_2, \ldots, B_k$, then the marginal probability of the event A is $P(A) = P(A \cap B_1) + P(A \cap B_2) + \cdots + P(A \cap B_k)$.

Mathematical expectation Same as expected value.

Matrix A rectangular array of elements.

Maturity value Total amount of principal and interest to be paid on due date.

Multiplication rule for probabilities For any events A and B: $P(A \cap B) = P(A)P(B|A)$, and $P(A \cap B) = P(B)P(A|B)$.

Multivariate function A function of two or more variables.

Mutually exclusive events Two events are mutually exclusive if they cannot occur jointly on one trial of the experiment of interest.

Natural logarithm A logarithm for which the base is the natural number e.

Nonbasic variable One of the $n-m$ variables that is not a basic variable in a basic solution.

Normal probability density function $f(x) = 1/\sqrt{2\pi}\sigma \; e^{-1/2[(x-\mu)/\sigma]^2}$ for $-\infty < x < \infty$.

Null matrix A matrix of all zero elements, denoted by $\mathbf{0}$.

Null set A set containing no elements (also called an *empty set*).

Ordered pair An ordered pair is an element (x,y), where x is called the *first element* and y is called the *second element*. The ordered pairs (x,y) and (s,t) are equal if and only if $x=s$ and $y=t$.

Ordinate The value of the y coordinate in the point (x,y).

Overconstrained system A system of linear equations for which the number of equations exceeds the number of unknowns and there exists no simultaneous solution.

Parabola The graph of a quadratic equation.

Parameter A quantity to which the operator may assign arbitrary values, as distinguished from a *variable*, which can assume only those values that the form of the function makes possible.

Partial derivative The derivative of a multivariate function obtained by considering all variables constant except the variable of differentiation.

Poisson probability function $p(x;\lambda) = e^{-\lambda}\lambda^x/x!$ $(x=0,1,\ldots)$.

Polynomial function The function $f(x) = a_n x^n + a_{n-1} x^{n-1} + \cdots + a_1 x + a_0$, where n is a nonnegative integer $(a_n \neq 0)$ and $a_n, a_{n-1}, \ldots, a_1, a_0$ are all real constants, is a polynomial function of degree n.

Population The collection of all possible simple events that can occur as the result of a given random experiment.

Postulate A statement that is a self-evident truth, also referred to as an *axiom*.

Postulates of probability
Postulate 1: $P(A)$ is a real number such that $P(A) \geq 0$ for any event in S.
Postulate 2: If A_i and A_j are mutually exclusive events in S, then $P(A_i \cup A_j) = P(A_i) + P(A_j)$.
Postulate 3: If $A_1, A_2, \ldots, A_n$ are mutually exclusive and exhaustive events in S, then
$$P(S) = P(A_1 \cup A_2 \cup \ldots \cup A_n)$$
$$= P(A_1) + P(A_2) + \cdots + P(A_n)$$
$$= 1$$

Principal The amount invested or borrowed.

Probability density function A function $f(x)$ of the continuous random variable X such that
1. For all possible values of X, $f(x) \geq 0$
2. For $a \leq b$, $P(a \leq X \leq b) = \int_a^b f(x)dx$.
3. $\int_{-\infty}^{\infty} f(x)dx = 1$.

Probability Function A function of a discrete random variable X such that
1. $f(x) \geq 0$ for any possible value of X.
2. $f(x)$ represents the probability that X assumes the possible value x.
3. $\sum f(x) = 1$, where the summation is over all possible values of X.

Proper subset The set B is a proper subset of the set A if B is a subset of A and A is not a subset of B.

Quadrant One of the four regions resulting from the intersection of the two coordinate axes.

Quadratic equation An equation of the form $y = ax^2 + bx + c$ $(a \neq 0)$.

Quadratic function A function for which the rule is $ax^2 + bx + c$ $(a \neq 0)$.

Random event An event generated by a random experiment.

Random experiment A process, or operation, resulting in one of a number of possible outcomes such that the outcome is determined only by chance and that it is impossible to predict the exact outcome.

Random variable A variable, the values of which are determined by the outcome of a random event and can be ordered according to numerical value.

Range The set of all elements obtained from evaluating a function for each of the elements in its domain.

Rectangular coordinates Such as cartesian coordinates.

Root Any value of x for which $f(x) = 0$.

Sample A set of simple events (or observations) that is taken from a population of events.

Sample space Same as a population of events.

Scientific form of a number Any number expressed in the form $N\,10^k$, where k is an integer and N is a decimal number having one digit that must be nonzero to the left of the decimal.

Sequence A set of elements arranged in a specific order according to some definite pattern or law of formation.

Set A collection of objects called *elements*.

Simple event An event that cannot be represented as a collection of two or more events.

Simple interest The amount paid only for the use of an original quantity and does not include interest left on deposit to accrue further interest.

Simple random sample A random sample such that any simple event remaining in the population at any given instant has the same chance of appearing in the sample as any other remaining simple event.

Simplex algorithm The algebraic procedure utilizing the gaussian elimination method in solving linear programming problems.

Slack variable A nonnegative variable that is added to an inequality to convert it into an equation.

Slope The ratio of the change in y to the change in x.

Standard form of a linear equation $ax + by = c$.

Standard form of a quadratic equation Same as functional form.

Subset A set A is a subset of the set B if every element of A is an element of B.

Surplus variable A nonnegative variable subtracted from an inequality to convert it to an equation.

Transcendental function Any nonalgebraic function.

Underconstrained system A system of linear equations for which solutions exist that are not unique.

Union The union of sets A and B, denoted $A \cup B$, is the set of all elements belonging either to A or B or to both A and B.

Univariate function A function of one variable.

Universal set The set containing all elements of concern in a given situation.

Venn diagram A graphic representation of sets.

Vertex of a parabola The point $[-b/2a, -(b^2 - 4ac)/4a]$.

Zero matrix Same as null matrix.

INDEX

A

abscissa 25
amortization 318
and 6
annuities 313
antiderivative 428
antilog 74
area 439
   net 443
   total 443
associative law 14
asymptote 65

B

base 58
basic solution 154
basis 155
Bernoulli
   probability function 253
   process 253
   trial 253
binomial
   coefficients 258
   distribution 253
   probability function 258

C

calculus 322
   differential 322
   integral 429

cartesian
   coordinates 25
   product 19
chain rule 353
change of variables 459
characteristic 73
coefficient 34
coincident lines 84
column vector 105
combinations 187
common
   difference 298
   logarithm 73
   ratio 301
commutative law 14
complement 4
compound amount 306
constant 34
constrained optima 418
constraints 140
conversion period 306
coordinate axes 25
critical values 393

D

degree 53
derivative 331
   higher-order 358
differential 429
discriminant 49
distributive law 15

disjoint sets 9
domain 20
double root 49
dual
    problem 169
    theorem 171

E

empty set 2
equal sets 3
equation 24
event
    complement 201
    compound 196
    independent 218
    joint 205
    random 195
    simple 195
    exhaustive 197
expected value 237, 452
exponents 58
extreme
    point 136
    test 414, 421
    value 412

F

feasible solution 141
first derivative test 397
function 19, 20, 23
    algebraic 365
    bivariate 412
    composite 351
    continous 328
    cubic 53
    cumulative probability
      distribution 234
    discontinuous 329
    explicit 368
    exponential 62, 372
    implicit 368
    inverse 365
    linear 33
    logarithmic 79, 374
    objective 140
    polynomial 52
    probability 232
    probability density 244, 449
    quadratic 33
    rational 330
    transcendental 365
fundamental theorem of integral
    calculus 438
future value 313

G

game theory 284
gaussian elimination method 89, 158
graphs 23
greater than 131

H

hypergeometric
    distribution 273
    probability function 275
    random variable 275

I

implicit differentiation 368
inclusive or 5
inconsistent equations 95
inequalities 132
    systems of 135
inflection points 393
initial basic solution 154
integral 428
    definite 462
    indefinite 429
    tables 469
integrand 429
integration by parts 465
intercept 36
interest 305
intersection 6

L

lagrangian multiplier 420

INDEX

less than 131
limit 323
linear equation 24, 33
    functional form 24
    standard form 33
    systems 83
linear programming 140
    graphic solution 140
    simplex method 152
logarithmic
    computations 73
    differentiation 380
    functions 79
logarithms 68
    common 69
    natural 76
long-term regularity 181

M

mantissa 72
marginal analysis 405, 433
    optimality condition 408
Markov chains 280
mathematical expectation 237
mathematics of finance 296
matrix 106
    addition 107
    algebra 105
    augmented 119
    identity 117
    inverse 117
    multiplication 108
    null 108
    zero 108
maximize 140
maximum 392
    local 393
method of
    elimination 88
    gaussian elimination 89, 158
    inspecting extreme points 145
    substitution 86
minimize 142
minimum 44, 392
    local 393

mutually exclusive 6, 197

N

natural constant 76
necessary condition 420
null
    matrix 108
    set 2

O

objective function 140
or 5
ordered pair 19
origin 24
overconstrained sytem 86

P

parabola 46
parameters 82
partial derivative 382
    cross 388
permutations 187
point-slope formula 37
population 195
postulates 14
    of probability 198
power 58
predictability 181
present value 309
primal problem 169
principal 305
probability 181
    a posteriori 183
    a priori 182
    addition rule 223
    beta 453
    binomial 253
    classical 182
    conditional 207
    density function 232
    distribution 230
    frequency 183
    hypergeometric 273
    joint 205